Cell Biology

Charlotte J. Avers
RUTGERS UNIVERSITY

D. Van Nostrand Company

NEW YORK   CINCINNATI   TORONTO   LONDON   MELBOURNE

To my friend and colleague David Axelrod

D. Van Nostrand Company Regional Offices:
New York   Cincinnati

D. Van Nostrand Company International Offices:
London   Toronto   Melbourne

Copyright © 1976 by Litton Educational Publishing, Inc.

Library of Congress Catalog Card Number: 75-16522
ISBN: 0-442-20382-9

Published by D. Van Nostrand Company
450 West 33rd Street, New York, N.Y. 10001

10 9 8 7 6 5 4

Preface

Cell Biology has been designed to provide a perspective on structure and function within the cell, a perspective from which the cell can be viewed as a dynamic, intricately attuned system.

The text is divided into three parts. Part One (Foundations for Cell Study) provides fundamental information on cell organization, important organic molecules and cellular energetics, enzymes, membranes, and the genetic system. The five chapters in Part One may be presented as a core of introductory information or selected individually to serve as introductions to various series of topics in Part Two and Three.

Part Two (Cytoplasmic Compartments of the Cell) includes chapters on ribosomes and protein synthesis; mitochondria and energy flow; chloroplasts and photosynthesis; organelle genetic systems; Golgi apparatus and other single-membrane structures, such as lysosomes and microbodies; and cell movements based on the unifying structural theme of microtubules and microfilaments. In each of the six chapters the structural components are presented and then followed with discussions of the functions carried out by such components. The chapter on organelle genetics (Chapter 9) is a follow-through on the preceding three chapters concerning ribosomes, mitochondria, and chloroplasts; otherwise, the chapters of Part Two may be covered in any sequence.

Part Three (The Nuclear Compartment) focuses on the nucleus, particularly on chromosomes. Chapter 12 (Chromosomes and Other Nuclear Compartments) discusses the rapidly expanding field of studies in which the geneticist, biochemist, and cell biologist interact to piece together the structure, organization, and activities of chromosomes. Chapter 13 (Mitosis and Meiosis) concerns cell reproduction and the mechanisms by which cell diversity is achieved, particularly crossing over. Chapter 14 (Cytogenetics: Classical and Human) pulls together some current approaches to medical cytogenetics and the classical foundation upon which modern studies depend. Part Three may be presented before Part Two. Some information has been repeated throughout the text to insure choice of sequence presentation.

Several chapters of the text are unique in presentation and emphasis. The topics of organelle genetics, cell movements, and cytogenetics are included as individual chapters because they are in the forefront of current research and because they enhance the appreciation of the diversity of cellular themes with which cell biologists are concerned. Throughout the text, more emphasis than is usual has been given to the genetic aspects of cells, from the viewpoint of genetic relationships to expressions of cell structure, function, and reproduction. Whenever possible, emphasis has been given to the importance of observation, experiment, and interpretation of data.

Students are expected to have completed courses in Introductory Biology and Introductory Chemistry. Chapter 2 (Biological Molecules and Bioenergetics) provides sufficient background for the student who has not taken Organic Chemistry and may serve as a review for the student who has completed the course. The student is not expected to have taken courses in Biochemistry or Genetics; these subjects are discussed when necessary to the understanding of the topics presented. The focus has been on the sophomore–junior courses for Biology majors, but under some circumstances the text could serve for seniors.

It is my pleasure to acknowledge the assistance of many friends and colleagues. I am grateful to David Axelrod, Bill Davis, Renee Miller, Carol Sauers, Faye Schwelitz, Leonard Sciorra, and John Sinclair, who read all or parts of the manuscript at various stages of its development and made useful and pertinent comments and suggestions. Photographs were provided by many people who went to great trouble to find and send materials, often on rather short notice. In particular, I want to thank Ruth Dippell, Micheline Federman, Joseph Gall, Ursula Goodenough, Charles Hux, Myron Ledbetter, Peter Moens, Hilton Mollenhauer, Eldon Newcomb, Keith Porter, Marcus Rhoades, Hans Ris, Norman Rothwell, Thomas Tegenkamp, and David Wolstenholme for providing as much as, and more than, I had requested of a varied assortment of photographs. To everyone else who sent materials I am also deeply indebted and very grateful. I alone, however, am responsible for any errors or omissions in the text and illustrations.

I am also glad to acknowledge the patience and help of the publisher's staff, whose hard work and attention to detail were essential in assembling the final product. I enjoyed the project and learned a great deal in the process. I hope this is implicit in the writing and that some of my enthusiasm will be communicated to the reader.

Charlotte J. Avers

Contents

Part
One

Foundations
for Cell Study

Chapter 1

An Overview of the Cell

The accumulation of knowledge and understanding does not proceed in a straight line. Each significant step forward usually becomes possible when different lines of information are brought together to create a new and deeper perspective. If this perspective leads to new investigations then we acknowledge its success. If the new conception only permits untestable predictions, then the ideas may be shelved or discarded. Many discoveries in modern cell biology are the result of new methods or new applications of old methods. New methodological tools provide means for seeking answers to questions based on hypotheses, or working ideas. Questions that lead only to philosophical debate may be interesting, but they are not especially productive. Hypotheses that generate new research are most valuable in making progress toward problem-solving and toward a fuller understanding of the world around us.

Cell biologists focus on the microscopic features of the living world. In this discipline, molecules are important subjects for study along with cellular structures that are, themselves, made up of highly organized assemblies of molecules. The study of the cell by microscopy, known as **cytology,** has been joined with biochemical, physiological, and genetic studies to produce the discipline of cell biology. This merger of ideas and methods from different spheres of biological study has led us to examine cells as dynamic living units with astonishingly complex and interacting elements of structure, function, and regulation. The following section describes some historical landmarks which produced the modern approach to studying the biology of the cell as an integrated and coordinated living system.

BRIEF HISTORICAL BACKGROUND

Microscopy and the Cell Concept

If cells and their components are to be studied, artificial aids are needed to magnify and bring them into sharp focus because the human eye cannot *resolve* two points or objects as separate entities if they are closer together than about 200 micrometers (μm). Two such points would appear as a single point no matter how greatly we might magnify the material. Magnification without **resolution** (distinguishing separated points that exist) produces only fuzzy enlarge-

ments and we still would not be able to identify the enlarged object. The beginnings of cell study must therefore be traced to the invention of lenses and microscopes that allowed early scientists to see cells ranging in average size between 10 and 100 micrometers (Table 1.1).

The invention of the first useful **compound microscope,** which had two lenses to increase total magnification and reduce optical aberrations, is credited to Z. and H. Janssen, an uncle and nephew. Their microscope, which became available in 1590, could magnify an object 30 times its actual size, without loss of resolution. We credit the first significant information gained by using microscopy to Robert Hooke who published his observations in *Micrographia* in 1665. Hooke noted the occurrence of "cells" or "pores" in various plant tissues, such as cork from the bark of trees (Fig. 1.1). Hooke also observed that cells of other kinds of tissues were filled with "juices," but the high visibility of the thick cell wall drew his attention most and continued to be the focus for cell studies by biologists for 150–200 years afterward.

Nehemiah Grew added his observations in publications beginning in 1672. His two volumes of information and illustrations on microscopic plant

Table 1.1 Units of measurement used in biology and biochemistry

A. LENGTH				
METER (m)	MILLIMETER (mm)	MICROMETER (MICRON) (μm)	NANOMETER (MILLIMICRON) (nm)	ÅNGSTROM (Å)
1	1,000 (1×10^3)	1,000,000 (1×10^6)	1,000,000,000 (1×10^9)	1×10^{10}
0.001	1	1,000	1,000,000	1×10^7
0.000001	0.001	1	1,000	1×10^4
1×10^{-9}	1×10^{-6}	0.001	1	10
1×10^{-10}	1×10^{-7}	1×10^{-4}	0.1	1

B. WEIGHT				
GRAM (g)	MILLIGRAM (mg)	MICROGRAM (μg)	NANOGRAM (ng)	PICOGRAM (pg)
1	1,000	1,000,000	1×10^9	1×10^{12}
0.001	1	1,000	1×10^6	1×10^9
1×10^{-6}	0.001	1	1×10^3	1×10^6
1×10^{-9}	1×10^{-6}	0.001	1	1×10^3
1×10^{-12}	1×10^{-9}	1×10^{-6}	0.001	1

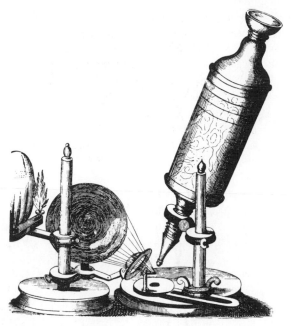

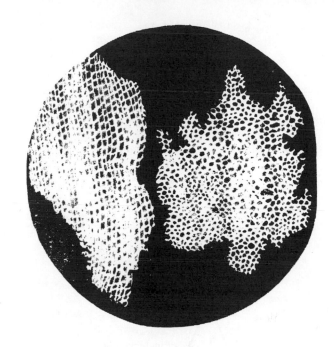

Figure 1.1
The crude microscope used by Robert Hooke to view materials such as cork tissue, in which he observed box-like compartments he called "cells."

anatomy, published in 1682, laid the foundations for the cell concept, that is, the idea that the cell is the underlying unit of structure in organisms. Another microscopist, Antonie van Leeuwenhoek, regularly sent letters to the Royal Society in London between 1673 and the early 1700s describing the many objects and moving forms he saw using a simple microscope (one magnifying lens) of his own manufacture. The single lens must have been of unusually high quality to permit van Leeuwenhoek to record the wealth of detail contained in his letters to London.

Little new information about cellular structure appeared until the 1830s when improved microscope lenses with enhanced resolution and greater magnification became available. At this time it was possible to see objects in sharp focus even when they were separated by only about 1 micrometer (μm). Further improvements in the later 1800s produced lenses that

corrected halos and blurs which had been particular problems at high magnifications. It was now possible to achieve resolution of 0.2 μm with ordinary white light as the source of illumination (Fig. 1.2). The microtome, which was invented in 1870, allowed controlled sectioning of tissues and thin slices became readily available for microscopical study. In addition to new and improved instruments, newly manufactured stains and dyes made it possible to obtain high contrast among cells and cell structures. Industrial chemists in Germany were especially helpful in producing new aniline dyes that could be used by biologists.

These technical advances led to a burst of cell and tissue studies that ultimately led to the development of an essentially modern concept of the cell. These major developments took place between 1830 and the early 1900s in a continuing progression. The first

Light microscope

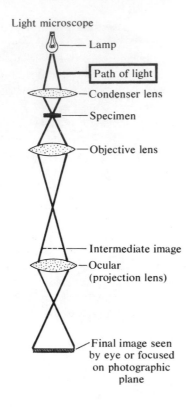

Lamp

Path of light

Condenser lens

Specimen

Objective lens

Intermediate image

Ocular
(projection lens)

Final image seen
by eye or focused
on photographic
plane

Electron microscope

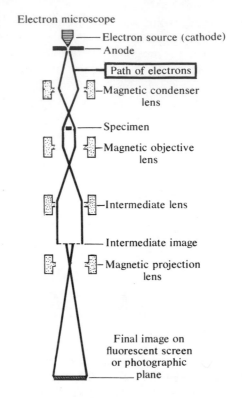

Electron source (cathode)

Anode

Path of electrons

Magnetic condenser
lens

Specimen

Magnetic objective
lens

Intermediate lens

Intermediate image

Magnetic projection
lens

Final image on
fluorescent screen
or photographic
plane

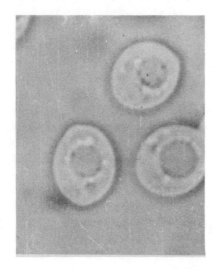

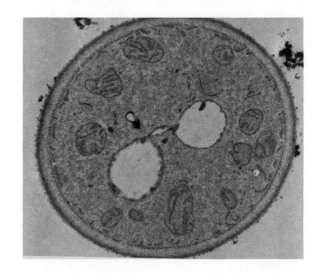

major step forward was the recognition of the **nucleus** as a significant part of the cell. Improved microscope lenses made it possible to see cell contents more clearly and a report by Robert Brown in 1833 was the first important publication to emphasize the cell nucleus. Some attention was then directed to the previously neglected nucleus, but because the later biologists were greatly influenced by the seventeenth-century studies of Hooke and Grew, emphasis continued to be placed on the cell wall and not on cell contents.

Animal cell boundaries are far less conspicuous than plant cell walls, and this structural feature delayed formulation of a unifying cell concept for all life forms. Cells of plants and animals appeared profoundly different when viewed through the microscope. The postulate that the cell is an underlying unit of structure of all organisms, or the **cell concept,** was formalized in the cell theory first proposed by T. Schwann and M. Schleiden in 1838–39. Schwann was the first to describe animal cells as similar to plant cells. He studied cartilage tissues in which substantial amounts of fibrous cartilage around the periphery of the cells delineate these cell boundaries as clearly as cell walls do in plants. Schleiden's generalizations about cell constructs of plant tissues in 1838 and Schwann's information about animal cells in 1839 led to the simple and unified cell concept for both plant and animal life. Much of the impact of the cell theory can be attributed to Schwann's perceptive studies and interpretations which united the separate lines of plant and animal cell investigations.

In 1840 J. Purkinje coined the term **protoplasm** to refer to the fluid contents of cells. Even though Brown had pointed out the nucleus as a regular cell feature,

Figure 1.2
Schematic drawings showing the basic optical systems of the light microscope and the electron microscope. Resolution of structure is much lower in the yeast cells photographed through the light microscope (*left,* × 3,000) than in a comparable cell that has been sectioned and photographed using the electron microscope (*right,* × 25,000).

its relation to cell activities was not properly understood. One reason for the delay in understanding the role of the nucleus at this time was the influence of Schleiden's ideas on many biologists. Schleiden had proposed that daughter cells formed within the mother cell by aggregations of protoplasmic granules in "free-cell formation." Because this idea was accepted by others, attention was diverted away from the nucleus and its contributions to new cell generations.

Cell divisions, however, had been observed as early as 1835 by Hugo von Mohl, who was especially critical of Schleiden's ideas on cell formation. Further observations during the 1840s led Karl Nägeli to propose the alternative of **cell division** instead of free-cell formation as proposed by Schleiden. Since the cell wall was still considered to be the most important part of the cell, Schleiden's theory of protoplasmic granule aggregation was modified to state that only the nucleus was put together anew in each division cycle, while the rest of the cell was produced by a cell division process. By 1855 there had been enough observation of both plant and animal cells for R. Virchow to postulate that all cells arise from pre-existing cells (*"omnis cellula e cellula"*). This directed attention to the cell as an important factor in the transmission of inherited traits from one generation to the next.

The importance of the nucleus to cell continuity was firmly established in the 1870s. By this time evidence had accumulated against the prevailing notion that nuclei dissolved and reformed anew in each cell division cycle. E. Strasburger, who worked with plants and W. Flemming, who worked with animal materials, both presented microscopical evidence showing that nuclei maintained physical continuity from one generation to the next. Flemming coined the term **mitosis** to describe the process in which nucleoprotein threads, or **chromatin,** elongated and then "split" lengthwise to be distributed to the two daughter cells at the end of cell division. Chromosome numbers were shown to be constant for a species, further supporting the view that nuclei were preserved from generation to generation. By 1885 Rabl was able to demonstrate that **chromosomes** remained physically intact between generations, even though they seemed to enter an invisible stage between cell divisions. Since chromosomes held their same positions in a cell at the end of one division and the beginning of the next, it seemed improbable that such consistency could occur if chromosomes formed anew in each generation. Even more significantly, Oskar Hertwig had shown in 1875–1876 that two nuclei must fuse if an egg is to develop into an embryo. He further showed that one nucleus was present within the egg cell while the second was introduced by the sperm at fertilization. Strasburger confirmed these observations in plants in 1877.

In order to complete the modern version of the cell concept, chromosome behavior had to be related to the physical link between generations created by the formation and subsequent fusion of sex cells, or **gametes.** E. van Beneden reported in 1883 that gametes of the roundworm *Ascaris* had one chromosome each, whereas the zygote formed when egg and sperm fused possessed two chromosomes. (*Ascaris* germ lineages are unusual since all the separate chromosomes join end-to-end to form one huge chromosome structure in the gametes. This feature makes chromosome analysis a relatively simple matter in this organism.) August Weismann had predicted that a compensating reduction division must precede gamete formation because doubling of chromosomes could not take place at fertilizations and still yield species with a constant chromosome number generation after generation unless chromosome number was reduced in gametes. It was known from Hertwig's studies that chromosome number doubled at fertilization. Hertwig also showed that reduction in chromosome number took place just before gamete formation in animals. T. Boveri reported the same observations in the late 1880s. Strasburger shortly afterward confirmed that these events also took place in plants, although there was some difference in timing. It was not until 1905, however, that J. B. Farmer and J. E.

Moore coined the term *maiosis* (**meiosis**) and more fully described the division events leading to halving of the chromosome number in sexually reproducing species. Continuity between generations thus resided in the nucleus. It underwent reduction in chromosome number at meiosis and was restored in chromosome number at fertilization in a sexual cycle.

Genetic and Biochemical Aspects of the Cell

Wilhelm Roux suggested in 1883 that chromosomes were the bearers of the hereditary units. In 1884 Hertwig and Strasburger presented a more formal postulate stating that the nucleus contained factors which controlled heredity. Weismann lent the weight of his prestige to these views in 1885 with *The Germ Plasma, A Theory of Heredity*. In this book, he asserted that chromatin was the most important component of the nucleus and that the cell apparatus was disposed to ensure nuclear division and the distribution of the hereditary material. He later expanded his proposals to state that there was a "continuity of the germ plasm." Germ plasm was designated as the only carrier of heredity and was quite distinct from body cells, whose variations were not transmitted to the next generation. One line of evidence cited in support of this theory was an experiment in which mice had their tails amputated over a period of fifty generations. Each new litter born to these mice, however, was made up of mice with perfectly normal and intact tails.

The twentieth century began with the momentous "rediscovery" of Gregor Mendel's 1865 studies of inheritance in garden peas. Mendel had conducted well designed breeding experiments, and he had concluded from his quantitative analyses of the inheritance patterns that unit factors (genes) governed the development of the seven characteristics he studied in the pea plants. At this time, biologists were ill-prepared to appreciate an abstract approach to heredity because the nucleus was not believed to be a vital cellular component. By 1900, however, when Hugo de Vries, Carl Correns, and Ernst von Tschermak independently cited Mendel's experiments in reports of their own inheritance studies, the state of biology was sufficiently advanced to accept this concept and the new discipline of genetics. By 1902 W. A. Cannon, E. B. Wilson, and W. S. Sutton had provided microscopical evidence in support of gene location within chromosomes. Sutton formally proposed the Chromosome Theory of Heredity in 1902. It was not until 1916, however, that Calvin Bridges was able to demonstrate the precise correspondence between the distribution pattern for the white-eye gene and the X chromosome in the fruit fly, *Drosophila melanogaster.*

The rules of Mendelian inheritance were rapidly extended to many other plant and animal species during the first years of our century. Combined studies of gene behavior and chromosome behavior led to the new discipline of **cytogenetics,** which blended information from cytology and genetics into a coherent and cross-supporting body of evidence. Cytogenetic studies were essential during the 1930s and 1940s to the developing concepts of the chromosome theory of heredity, but there was a temporary decline in these activities until the advent of **molecular cytogenetics** in the 1960s. This discipline forms a substantial portion of modern cell biology and will be discussed in many of the later chapters.

Although there was very little exchange between chemists and biologists before the 1940s and 1950s, significant information about cellular chemistry was obtained during the eighteenth and nineteenth centuries. The principal difficulty in relating chemistry to biology was overcome in 1828 by Friedrich Wöhler who synthesized the organic compound urea from inorganic ammonium cyanate. Before this event it had been believed that the substances and processes of the inorganic world were entirely distinct from those of living organisms. Urea was known to be excreted by certain animals and to be involved in living processes, so that its synthesis from an inorganic substance showed the continuity between the two worlds. Numerous studies afterward pointed out

that the laws of physics and chemistry applied to living organisms, as well as to the nonliving world, and by the end of the nineteenth century there had been many syntheses of biologically important compounds.

Emil Fisher and other German chemists essentially established the foundations for chemical analysis and description of principal groups of organic compounds by the early 1900s. The relationships between chemical activities involving organic compounds and living organisms had been clearly demonstrated earlier by Louis Pasteur who showed that fermentation of sugar to alcohol would proceed only if certain microorganisms were also present. H. and E. Buchner accidentally discovered that sugars could be fermented in cell-free extracts of yeast and this led to systematic studies and the description of **enzymes,** which are unique catalysts in living cells. The first enzyme to be crystallized was urease, an event reported in 1926 by J. B. Sumner.

Studies of the chemical basis of heredity began over 100 years ago. In 1871 Friedrich Miescher announced the discovery of "nuclein," which he had extracted from preparations of pure nuclei isolated from white blood cells. Nuclein was an unusual compound because it had a high content of phosphorus and was highly acidic. In 1874 Miescher described *protamine,* a nuclein-protein compound he had isolated from salmon sperm. These reports were considered significant since the nucleus was a central focus of interest about this time. In 1889 R. Altmann was able to purify and analyze nuclein, which he called **nucleic acid,** and to identify the specific sugars and nitrogenous bases that made up the larger molecules. There was a general belief at this time, which persisted well into the 1930s and 1940s despite evidence to the contrary, that one kind of nucleic acid was typical of animal cells (deoxyribonucleic acid or **DNA**), while the other was characteristic of yeasts and plants (ribonucleic acid or **RNA**). "Yeast nucleic acid" was still referred to in the literature long after DNA and RNA were shown to be present in all cells.

One line of evidence supporting the concept that DNA occurred specifically in chromosomes was derived from observations of specially stained microscopic preparations of cells and tissues. In this test, first developed in 1914 by R. Feulgen, cells were stained with basic fuchsin after they had been hydrolyzed in dilute hydrochloric acid at an elevated temperature. DNA takes on an unmistakable magenta color with this stain and this color was confined to the chromosomes in the nucleus according to many studies reported in the 1920s and 1930s. Since DNA was present only in the gene-carrying chromosomes, it provided support for the proposition that DNA was associated with heredity. In spite of many avenues of investigation linking DNA with heredity, it was not until the early 1950s that DNA was accepted as the genetic material. Many scientists tenaciously clung to the idea that genes were most likely proteins since protein diversity matched gene diversity. DNA, which appeared to be a molecule with little or no inherent variety, was an unlikely candidate. Chromosomes also contained proteins as well as DNA so that either substance could serve very nicely as the chemical constituent of the gene.

By the early 1950s the scientific community leaned toward the acceptance of DNA as the genetic material because of evidence provided by chemical studies by E. Chargaff and others who showed that DNA was a highly variable molecule type. In 1953 the molecular model of DNA proposed by J. D. Watson and F. H. C. Crick was widely accepted, and molecular genetics and modern notions of the gene were firmly established. The profusion of studies sparked by the Watson-Crick DNA model attests to the fact that the scientific community was geared up and ready to go at that moment. Less than 20 years later, purified genes had been isolated.

Advances in elucidating other structured components of the nucleated cell, in addition to the nucleus itself, continued steadily in these same 100 years. In 1886 R. Altmann described granular components in the cytoplasm that he believed to be involved in cellular oxidation, or energy-yielding burning of foods. These granules, seen with light microscopy, were called **mitochondria** by C. Benda in 1898. During the 1920s O. Warburg, D. Keilin, and other biochemists contributed substantially to our understanding of the enzymes involved in metabolism of foodstuffs in respiratory activities carried out in mitochondria and other parts of the cell. Plants had long been known to release oxygen and manufacture foods during photosynthesis activities localized in **chloroplasts,** organelles that were readily visible in the light microscope. The **spindle** of dividing cells was described by H. Fol in 1873 but was not widely accepted as real until the development of electron microscopy and other instrumentation in the 1950s. T. Boveri described the **centriole** in 1888, but the complexity of its structure was not revealed until the development of the electron microscope since it appears as a tiny granule in most cells, seen under the light microscope, even at the highest magnification. In 1898 Camillo Golgi described a network of granules and filaments in nerve cells that could be visualized with a highly temperamental silver-staining technique. This structure, now called the **Golgi apparatus,** also required electron microscopy to verify its reality.

The light microscope provided an essential entryway into studies of cellular structure and function, but detailed investigation absolutely requires the high magnifications and resolution of the electron microscope. Each microscope is useful in certain kinds and levels of study, so that they complement rather than compete with each other in the repertory of biological instrumentation. The fact that we still rely on light microscopy for various kinds of cell study is adequate testimony to its continued usefulness more than 40 years after the invention of the electron microscope.

The essential features that distinguish cell biology from related disciplines are not clear-cut. The union of cytology, genetics, and biochemistry also underlies recent advances in cell physiology, molecular genetics, molecular biology, and other areas of study. This blurring of boundaries is a positive indication of advances in understanding. Just as the study of biology allows us to see fundamental similarities among plants, animals, and microorganisms which we would miss if we studied botany, zoology, or microbiology separately, the union of biological disciplines has allowed us to seek the fundamental qualities and similarities of all living systems, rather than merely elucidating their superficial differences. We will discuss some fruits of this union of disciplines in the next section.

CELL STRUCTURAL ORGANIZATIONS

In traditional systems of classification, all organisms are assigned either to the plant or the animal kingdom. While many species can easily be accommodated within these two major categories, others pose difficulties because they have some features that are plantlike and others that are animallike. One example of such an organism is the unicell **Euglena** which has chloroplasts like plants but lacks a rigid cell wall and moves by means of flagella like some animals. Bacteria and fungi lack chloroplasts, but these organisms were classified as plants in traditional systems because they possess a rigid cell wall structure. The difficulty of sorting organisms in evolutionarily related groups has prompted various attempts to modify kingdom classifications so that evolutionary relationships are emphasized rather than selected structural traits (Table 1.2).

Some recent revisions in kingdom classifications are based upon the radically different cell plan found

Table 1.2 Systems of kingdom classifications*

"TRADITIONAL"	DODSON, 1971	STANIER ET AL., 1970	COPELAND, 1956	WHITTAKER, 1969
Plantae	**Monera**	**Protista**	**Monera**	**Monera**
Bacteria	Bacteria	Bacteria	Bacteria	Bacteria
Blue-green	Blue-green	Blue-green	Blue-green	Blue-green
algae	algae	algae	algae	algae
Chrysophytes		Protozoa		
Green algae	**Plantae**	Chrysophytes	**Protoctista**	**Protista**
Red algae	Chrysophytes	Green algae	Protozoa	Protozoa
Brown algae	Green algae	Red algae	Chrysophytes	Chrysophytes
Slime molds	Red algae	Brown algae	Green algae	
True fungi	Brown algae	Slime molds	Red algae	**Fungi**
Bryophytes	Slime molds	True fungi	Brown algae	Slime molds
Tracheophytes	True fungi		Slime molds	True fungi
	Bryophytes	**Plantae**	True fungi	
	Tracheophytes	Bryophytes		**Plantae**
Animalia		Tracheophytes	**Plantae**	Green algae
Protozoa	**Animalia**		Bryophytes	Red algae
Metazoa	Protozoa	**Animalia**	Tracheophytes	Brown algae
	Metazoa	Metazoa		Bryophytes
			Animalia	Tracheophytes
			Metazoa	
				Animalia
				Metazoa

* References:
Copeland, H. F. 1956. *Classification of the Lower Organisms.* Palo Alto: Pacific Books.
Dodson, E. O. 1971. The kingdoms of organisms. *Systematic Zoology* 20:265–281.
Stanier. R., Adelberg, E., and Doudoroff, M. 1970. *The Microbial World.* Englewood Cliffs, New Jersey: Prentice-Hall.
Whittaker, R. H. 1969. New concepts of the kingdoms of organisms. *Science* 163:150–160.

in **prokaryotes** and **eukaryotes.** Prokaryotes lack a membrane-bound cell nucleus while eukaryotic cells have membranes which separate the chromosomes and other constituents of the nucleus from the cytoplasm in the remainder of the cell. In addition to the difference in nuclear organization, which is a fundamental distinction, other features also identify an organism as prokaryotic or eukaryotic (Table 1.3). A discussion of the new prokaryote—eukaryote dichotomy versus the older plant—animal distinction is outside the province of this book, but the fundamental evolutionary divergence of prokaryotes and eukaryotes is of paramount significance in cell studies. The terms prokaryote (*pro:* before; *karyon:* nucleus) and eukaryote (*eu:* true) were suggested by Hans Ris in the early 1960s and have since been widely accepted.

Prokaryotic Cell Organization

Blue-green algae and the many kinds of bacteria are the major prokaryotic groups. These organisms have been placed in the kingdom Monera by several authorities. There is widespread acceptance of this classification since monerans have a very distinctive cellular plan of organization. The contents of the prokaryotic cell are dispersed into two recognizable regions: **cytoplasm** and **nucleoid.** These regions are bounded by a cell membrane, or **plasmalemma,** which in turn is enclosed by a rigid, or semi-rigid, cell wall that provides support and shape (Fig. 1.3).

The average size of the prokaryotic cell is 1–10 μm, but bacteria of the mycoplasma, rickettsia, and psittacosis groups are as small as 0.2–0.3 μm and cells of the blue-green alga *Oscillatoria princeps* are 60 μm in diameter. In some cases, the products of cell division remain associated in chains, filaments, or other groups. Each cell in this type of group is capable of producing new cells like itself, usually by binary fission. Some species produce spores and a few reproduce by budding.

One of the smallest prokaryote types, the mycoplasmas, exists as a protoplast lacking a wall. Even mycoplasmas, however, have a plasmalemma or plasma membrane as the cell boundary. No cell, prokaryote or eukaryote, can survive loss or damage of its plasma membrane. The plasmalemma of all cells is about 100 angstroms (Å) thick and displays a characteristic dark-light-dark staining pattern when seen with the electron microscope (Figure 1.4). In certain prokaryote groups the plasmalemma infolds at some

Table 1.3 Some major differences between prokaryotes and eukaryotes

CHARACTERISTIC	PROKARYOTES	EUKARYOTES
Cell size	Mostly small (1–10 μm)	Mostly larger (10–100 μm)
Genetic system	DNA not associated with proteins in chromosomes	DNA complexed with proteins in chromosomes
	Nucleoid not membrane-bounded	Membrane-bounded nucleus
	One linkage group	Two or more linkage groups
	Little or no repetitious DNA	Repetitious DNA
Internal membranes (organelles)	Transient, if present	Numerous types and differentiations, e.g., mitochondrion, chloroplast, lysosome, Golgi, etc.
Tissue formation	Absent	Present in many groups
Cell division	Binary fission, budding, or other means; no mitosis	Various means, associated with mitosis
Sexual system	Unidirectional transfer of genes from donor to recipient, if present	Complete nuclear fusion of equal gamete genomes; associated with meiosis
Motility organelle	Simple flagella in bacteria, if present	Complex cilia or flagella, if present
Nutrition	Principally absorption, some photosynthesizers	Absorption, ingestion, photosynthesis

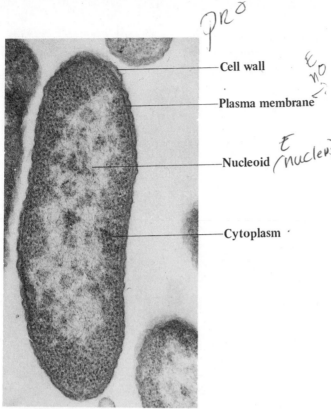

pro

Cell wall

E no ?

Plasma membrane

Nucleoid *E (nucleus)*

Cytoplasm

Figure 1.3
Electron micrograph of a thin section of the rod-shaped bacterium *Pseudomonas aeruginosa*. The central nucleoid region is surrounded by a denser cytoplasmic area. A relatively thin cell wall encloses the living protoplast. × 60,000. (Photograph by H.-P. Hoffmann)

locations, but because the entire membrane system and its infoldings are continuous, these folded portions are not considered to be separate internal compartments of the cell. Photosynthetic prokaryotes such as the blue-green algae and purple bacteria have plasmalemma infoldings that contain tightly bound pigments and enzymes of the light-capturing reactions (Fig. 1.5). Other sorts of plasmalemma infolds, called **mesosomes,** appear to function in various cellular activities such as respiration or cell division.

Except for the plasmalemma and its occasional infoldings, there is no other permanent membrane system in the prokaryotic cell. The prokaryotic cellular plan is noncompartmentalized. Localized concentrations of activities or chemicals are not set off by membranes from other regions of the cell and this is a major distinction between prokaryotes and eukaryotes. The cytoplasmic region of the prokaryotic cell contains dense concentrations of **ribosome** particles that measure 150–200 Å in diameter. Ribosomes are made up of three different-sized molecules of RNA and about 50 different proteins. Their morphological simplicity is deceptive in view of the chemical and functional complexity involved in ribosomal coordination and assistance in protein synthesis. All living cells have ribosomes, but they are smaller in prokaryotic than in eukaryotic cytoplasm. Prokaryotic and eukaryotic ribosomes also differ in the specific RNA and protein molecules from which they are assembled.

Within this ribosome-rich cytoplasm, there are one or more less dense regions of irregular shape in which tangles of thin fibrils are visible in electron micrographs. These irregular **nucleoids** or "bacterial nuclei" contain fibrils of DNA that exist as naked double helixes lacking specific associated proteins. Where it can be seen by electron microscopy, DNA in prokaryotic cells apparently exists as a circular molecule about 25 Å in diameter and hundreds of angstroms in length. Since there is no membrane around the nucleoid, the prokaryotic cell can always be distinguished from eukaryotic counterparts, whose genetic materials are separated from the cytoplasm by a complex nuclear membrane; this is another basic distinction between the two groups of organisms.

Only three cellular components occur in all prokaryotic cells: plasmalemma, ribosomes, and nucleoid. All other features are found in some but not all prokaryotes. We already mentioned the infolded plasmalemma found in some. In addition, a cell wall is absent in mycoplasmas, and many prokaryotes lack encapsulating or sheath materials around the cell wall. Some prokaryotes are spore-forming but most are not; some move by means of flagella but many do

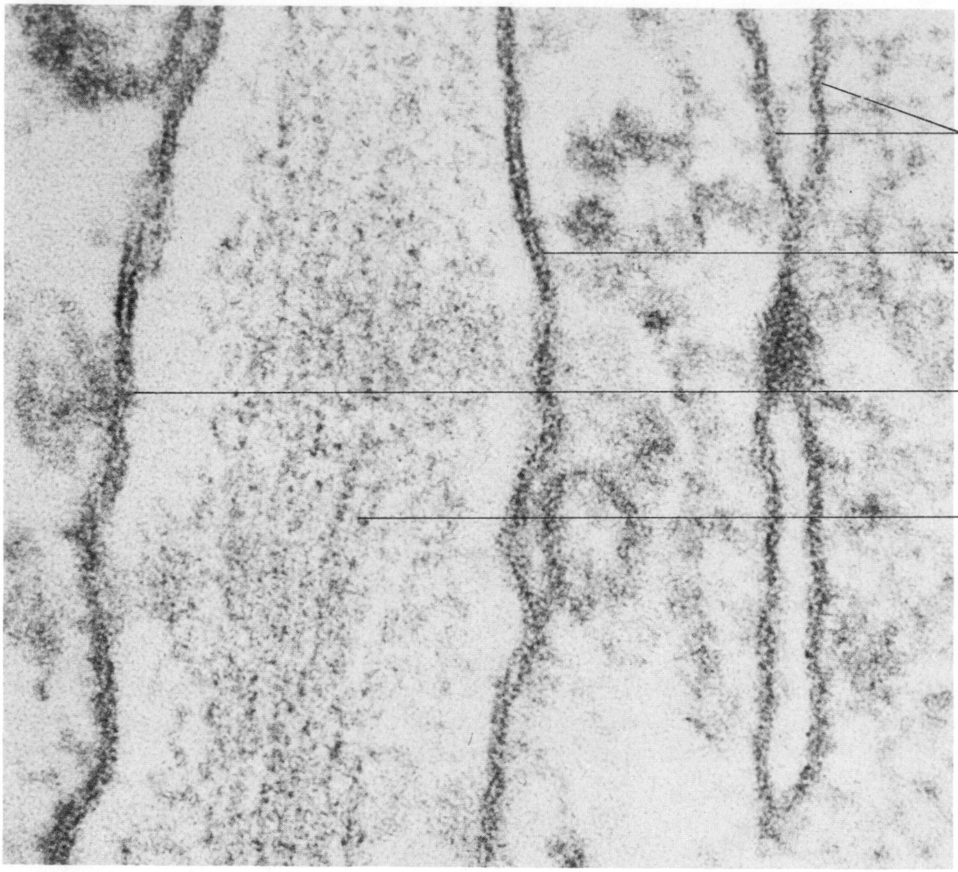

Nuclear membranes

Plasmalemma

Plasmalemma

Cell wall

Figure 1.4
Electron micrograph of a thin section through parts of two adjacent cells of *Arabidopsis thaliana* root. The trilaminate staining pattern is evident in the plasmalemma of each cell and in the nearby membranes of the nuclear envelope at the right. × 274,000. (Courtesy of M. C. Ledbetter)

not; and some form gas vacuoles or other inclusion bodies not found in all species.

Except for the relatively few photosynthesizing species that can synthesize cellular materials using light energy and simple raw materials, other prokaryotes *absorb* organic nutrients to obtain building blocks and energy for growth. They therefore differ *in general* from animals that ingest their food, and they differ from algae and plants that are primarily photosynthetic in their nutrition (Fig. 1.6). Although certain species in various eukaryote groups may absorb nutrients for growth, only the fungi, as an entire group of eukaryotes, routinely obtain carbon and energy from nutrients absorbed from their surroundings and are thus nutritionally like most prokaryotes.

Prokaryotes are generally accepted as the ancestors of eukaryote groups. Prokaryotes are not only simpler in structure, organization, and functions than eukaryotes, but they alone are found in the most

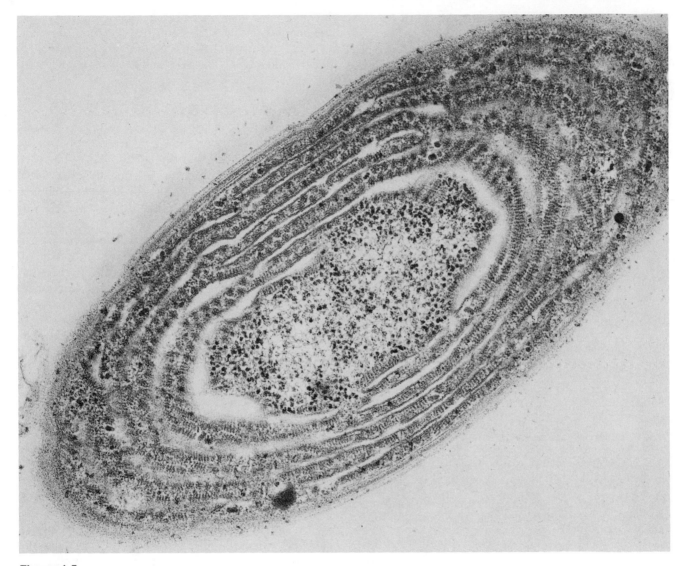

Figure 1.5
Electron micrograph of a thin section through the blue-green alga *Synechococcus lividus*. The concentrically folded photosynthetic membranes are infolded differentiations of the plasma membrane. × 62,400. (Courtesy of E. Gantt, from Edwards, M. E., and E. Gantt, 1971, *J. Cell Biol.* **50**:896–900, Fig. 1.)

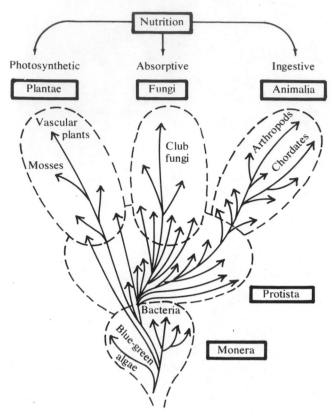

Figure 1.6
The classification scheme suggested by R. H. Whittaker. This scheme includes all cellular life in five separate kingdoms. All prokaryotes are assigned to the Monera, while different types of eukaryotes are members of the remaining four kingdoms. (Adapted with permission from Whittaker, R. H., *Science* **163**:150–160, Fig. 3, 10 January 1969. Copyright © 1969 by the American Association for the Advancement of Science.)

ancient fossil deposits. Eukaryotes occur in deposits no older than about 1 billion years whereas prokaryote remains may be 3 billion years old. Each type, of course, is also found in younger deposits. There is considerable homology between molecules and functions in prokaryotes and eukaryotes, a condition which implies common ancestry. In addition, all life employs the same genetic code, nucleic acid genetic materials,

ribosomal machinery for protein synthesis, and other fundamental traits that point to common descent through evolution. The particular evolutionary lineages, however, are still subjects of lively controversies.

Eukaryotic Cell Organization

The major trademark of the eukaryotic cell is its membrane-bounded compartments that are physically separate from the plasmalemma (Fig. 1.7). The most conspicuous compartment is the nucleus in which DNA is organized into complex nucleoprotein bodies called chromosomes. Unlike the single naked DNA molecule in the prokaryotic nucleoid, eukaryotes always have more than one chromosome within which the genes are located. The minimum number of different chromosomes in a cell is two, but chromosome numbers range into the hundreds in some species.

The nuclear membrane, or **nuclear envelope,** is a double layer system pockmarked with nuclear openings called **pores.** These nuclear pores are filled with a dense material of unknown composition. The outer face of the membrane facing the cytoplasm is studded with ribosomes and shows occasional continuities with cytoplasmic membranes (Fig. 1.8). In addition to chromosomes, one or more **nucleoli** always occur within the nucleus. These globular structures are produced at specific regions of certain chromosomes and are essential for cell existence. Ribosome precursors are produced within the nucleolus and protein synthesis cannot continue to take place unless new ribosomes become available during the life of the cell. The remainder of the nucleus, apart from nucleoli and chromosomes, is chemically complex but amorphous **nucleoplasm.** Of all the parts of the nucleus, only the chromosomes and their inherent capacities are transmitted from one cell generation to the next.

There is a plasmalemma boundary to the eukaryotic cell. Between the plasmalemma and the nuclear envelope is the cytoplasm with its profusion

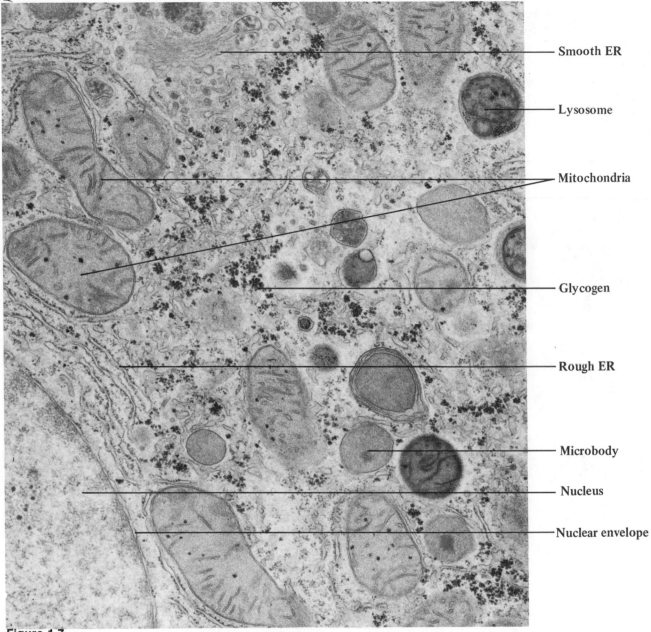

Smooth ER

Lysosome

Mitochondria

Glycogen

Rough ER

Microbody

Nucleus

Nuclear envelope

Figure 1.7
Thin section of rat liver cell. A small portion of the nucleus is visible along with various membranous and fibrous cytoplasmic structures. The opaque granules of glycogen are stored carbohydrate food material in the cytoplasm of these cells. × 24,300. (Courtesy of K. R. Porter)

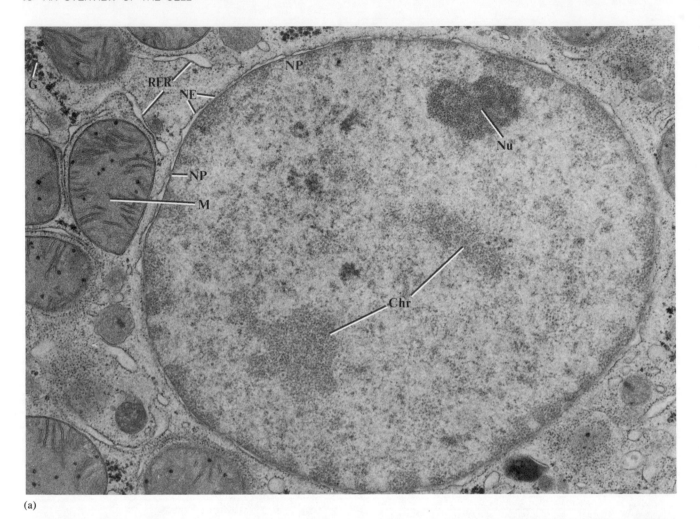

(a)

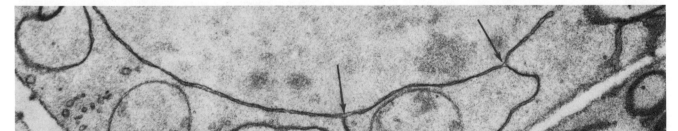

(b)

of membranes and particles. Occasional continuities probably occur among all membranous compartments of the cytoplasm, but none of these is continuous with the plasmalemma. Anastomosing throughout the cytoplasm is a system of membranous channels called the **endoplasmic reticulum, or ER.** The *ER* exists as an extensive network of flattened sacs, or **cisternae,** which are formed by the encircling membrane sheet. In cross section it seems that there are two membranes to a cisterna, but this is a deceptive appearance caused by sectioning through an enclosed system and seeing only the cut ends of the enclosure (Fig. 1.9). Regions of the *ER* may possess ribosomes that are attached to the outer face of the cisternae. These regions are called **rough ER.** Where there are no attached ribosomes, the cisternae form part of the **smooth ER.** The membranes of the endoplasmic reticulum together with the ribosomes form the structural basis of a system that synthesizes and distributes proteins.

One particular region of smooth *ER* is the **Golgi apparatus.** This membranous compartment acts as a processing and packaging station for many proteins destined for export from the cell. In addition to secretion droplets of various kinds, the Golgi apparatus

Figure 1.8
Electron micrographs of the eukaryotic nucleus: (a) cell from bat liver, fixed in osmium tetroxide, which preserves fibrous components. The nucleolus (Nu) and two chromosome (Chr) areas are within the nucleoplasm of the nucleus. Ribosomes are attached only to the outer membrane of the nuclear envelope (NE). Note the resemblance between the rough endoplasmic reticulum (RER) elements and the nuclear envelope. Two nuclear pores (NP) are present in the plane of this section through the nuclear envelope. Glycogen (G) and mitochondrial profiles (M) are in the cytoplasm surrounding the nucleus. × 23,660. (Courtesy of K. R. Porter) (b) Favorable thin sections reveal continuities between the endoplasmic reticulum and the nuclear envelope (at arrows), as in this thin section from anther of African violet (*Saintpaulia ionantha*). The material was fixed in permanganate, which does not preserve fibrous components. × 36,000. (Courtesy of M. C. Ledbetter)

also packages proteins in membranous vesicles that remain and function in the same cell which produced them. Chief among these vesicle-enclosed materials are **lysosomes** which contain digestive enzymes. Lysosomes are intracellular "garbage disposals." They also have been implicated in various symptoms of disease processes. Lysosomal enzymes can digest every kind of organic material in the cell, but digestion is controlled because these powerful enzymes are separated from the cell by the organelle membrane. If the lysosome releases its contents into the cell, death of the cell usually follows. Ordinarily, digestion proceeds within the lysosome itself after it has fused with some material, native or foreign to the cell.

Except for some unusual anaerobic protozoa, all other eukaryotic cells possess the double-membrane **mitochondrion** as a conspicuous cytoplasmic organelle. The outer membrane is smooth in outline while the inner membrane is infolded into numerous tubular projections called **cristae;** these are identifying features of all mitochondria (Fig. 1.10). The innermost part of the mitochondrion is the unstructured **matrix** region. Enzymes located in the membranes and matrix of the mitochondrion act jointly to carry out the energy-yielding steps by which aerobic life is sustained. The production of **adenosine triphosphate (ATP),** the most significant energy-storing molecule of all cells, takes place within the mitochondrion during aerobic respiration reactions in eukaryotes. Mitochondria contain their own DNA and ribosomal protein-synthesizing machinery.

In addition to a variety of membranous compartments, eukaryotic cells contain several varieties of nonmembranous elements, such as the ribosomes, **centrioles** (also known as **basal bodies**), and systems of **microtubules** and **microfilaments** (Fig. 1.11). Microtubules and microfilaments contribute to many cellular movement phenomena, including chromosome movement to the poles of the dividing cell, streaming of the cytoplasm, and cell propulsion through liquid surroundings by means of **cilia** and **flagella.** The centriole, usually associated with cell division

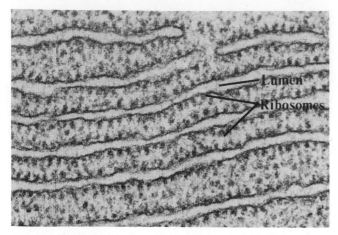

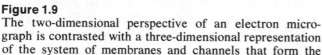

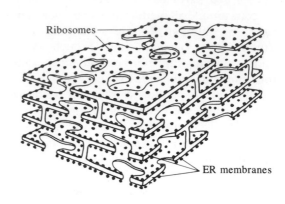

Figure 1.9
The two-dimensional perspective of an electron micrograph is contrasted with a three-dimensional representation of the system of membranes and channels that form the rough endoplasmic reticulum of eukaryotic cells. × 85,000. (Courtesy of K. R. Porter)

events, and the basal body, usually associated with cilia and flagella, were revealed to be identical in ultrastructure by electron microscopical studies. In fact, there are numerous organisms whose centrioles may act at the poles of the cell during nuclear division and then migrate to the cell periphery, there assuming a role in generating a new cilium or flagellum. Some eukaryotes lack cilia or flagella in all cells. Among these species are a few protozoa, certain kinds of algae, and seed-producing plants of the more highly evolved groups. Microtubules and microfilaments, on the other hand, are a ubiquitous component of all eukaryotic cells.

Certain features are found uniquely in some eukaryotic groups and not in the others. All eukaryotic algae, green **protists,** and tissue-forming plants possess one or more kinds of **plastid.** The familiar **chloroplast** in photosynthesizing green cells is separated from its cytoplasmic surroundings by two encircling membranes that are essentially smooth in outline and closely appressed. All aerobic life on this planet depends on photosynthetic species for continual replenishment of oxygen in the atmosphere, this gas being a byproduct of their light-capturing activities. Photosynthetic species also provide an endless supply of food and fuel by converting trapped light energy to usable chemical energy. Photosynthesis takes place within the chloroplast. (Fig. 1.12).

The algae, true fungi, and land plants have a prominent **cell wall** that surrounds the living **protoplast** or remains as a structural framework after a cell has died. Another feature common to algae, fungi, and land plants is the occurrence of large **vacuoles** containing various kinds of molecules in dilute solution or suspension (Fig. 1.13). Cell enlargement is accompanied by increasing size of vacuoles, and virtually the entire volume of a mature cell may be occupied by a vacuole that is surrounded by a narrow band of protoplasm into which the nucleus and other organelles are squeezed. Animal cells often contain small vacuoles, but more typically the cytoplasm is densely packed with particles and organelles suspended in the unstructured cytosol material. The appearance of most animal cells is thus quite different from plant cells. Furthermore since they lack a rigid and confining cell wall, animal cells undergo changes in shape

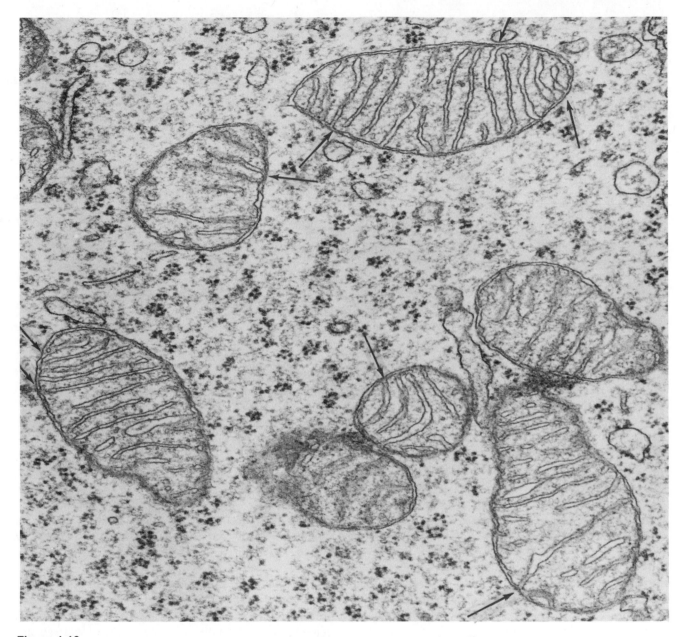

Figure 1.10
High magnification electron micrograph showing mitochondrial profiles in a spermatogonial cell of rat testis. The
cristae can be seen to be infolded regions of the inner
membrane in favorable planes of the section (at arrows).
× 65,000. (Courtesy of H. H. Mollenhauer)

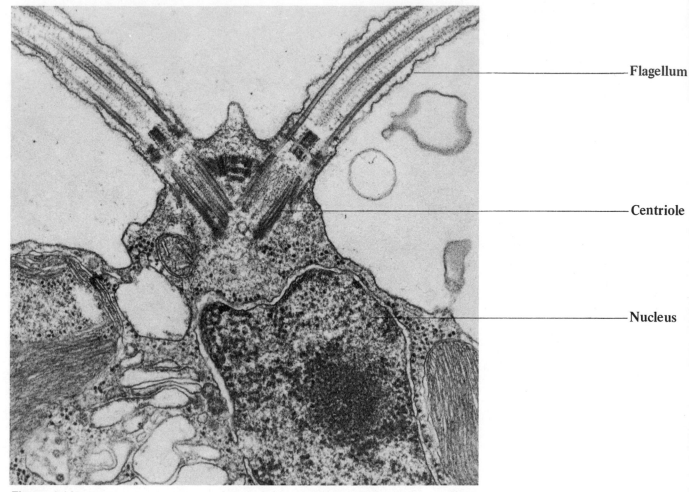

Flagellum

Centriole

Nucleus

Figure 1.11
Thin section through a portion of the green, unicellular flagellate *Chlamydomonas reinhardi.* The pair of flagella extrude from the anterior end of the cell and a centriole underlies each flagellum. Microtubules are a prominent component of centrioles and flagella. × 66,600. (Courtesy of R. L. Weiss and U. W. Goodenough)

rather readily, if suitable conditions are provided. Migration of cells is an underlying feature of tissue development in animals, a process which does not occur in plant development.

The cell surface often shows unique features in animal tissues. The plasmalemma may be variously folded, sometimes into fingerlike projections called **microvilli,** which substantially increase the surface area of this membrane. The adjacent plasmalemmas of cells in an animal tissue often show localized differentiations called **junctions,** which are important structural features in intercellular communication

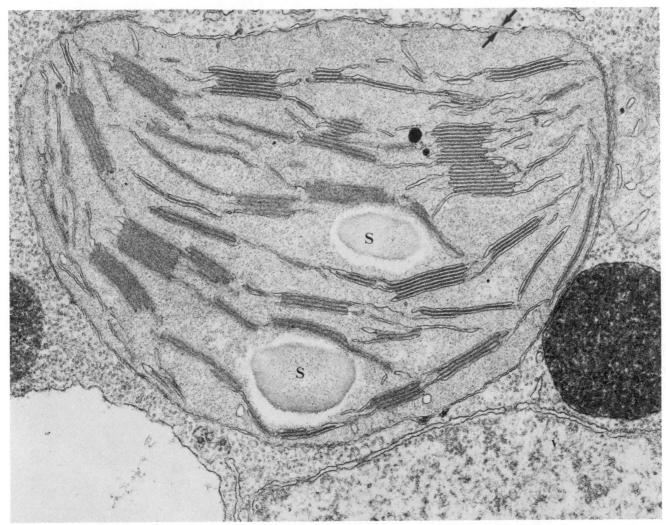

Figure 1.12
Thin section through a chloroplast from tobacco leaf mesophyll cell. The two closely appressed enveloping membranes (at arrows) enclose a third system of stacked internal chloroplast membranes. Photosynthetic reactions take place within these stacked membranes and in the amorphous matrix in which they are suspended. Two sites of starch (S) deposit can also be seen within the chloroplast matrix. × 46,000. (Courtesy of E. H. Newcomb, from Frederick, S. E., and E. H. Newcomb, 1969, *J. Cell Biol.* **43**:343.)

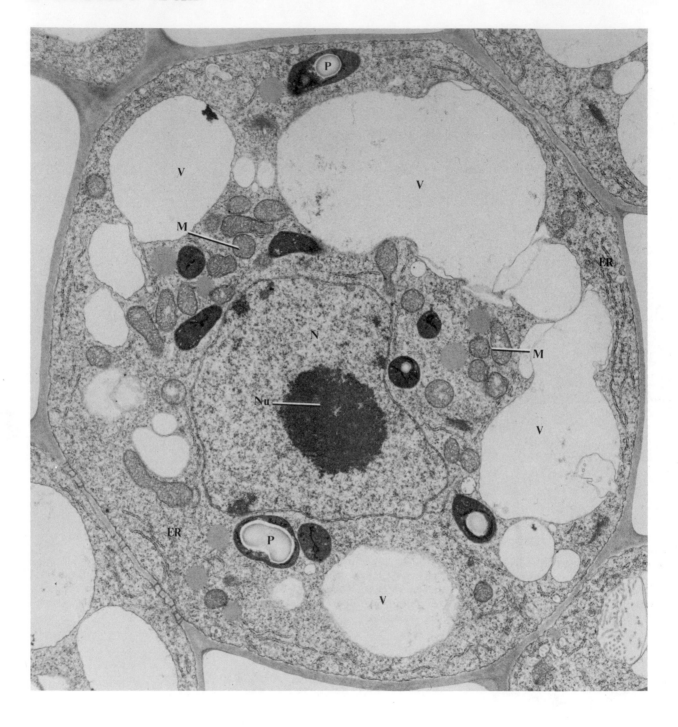

events. Another surface feature is the **desmosome,** which is believed to be a region of intercellular binding (Figure 1.14).

THE GENETIC SYSTEM

All of cellular life and most viruses store their genetic information in linear sequences of DNA. It is known that DNA occurs as a double helix in which each of the two intertwining strands consists of an invariant sequence of alternating 5-carbon sugars and phosphate groups, with any one of four kinds of nitrogenous bases bonded to each sugar in the chain. The nitrogenous bases are paired by bonds across the lengths of the two strands, producing a vertical sequence of "steps" (Fig. 1.15). The four kinds of bases, **adenine, thymine, guanine,** and **cytosine,** form specific pairs (adenine with thymine and guanine with cytosine) across the width of the molecule but they can occur in any sequence in the vertical plane. Theoretically, therefore, many different kinds of DNA sequences can form simply depending on the number of base-pairs in the vertical sequence. If there were 500 base-pairs in a particular molecule made of four different pair possibilities (A-T, T-A, G-C, and C-G), and if these could be arranged in all possible sequences vertically, then 4^{500} different molecules could occur. This remarkable potential for diversity underlies the gene variety in life forms.

Expression of the stored information in genes can occur only through the mediation of various kinds of

Figure 1.13
Large vacuoles (V) are a distinctive feature of mature plant cells, as this one from *Potamogeton natans* root. The cytoplasm also contains profiles of mitochondria (M), starch-filled plastids (P), endoplasmic reticulum (ER), and ribosomes. There is a promin   olus (Nu) in the centrally situated nucleus (N). × 13,0   Courtesy of M. C. Ledbetter)

RNA, certain enzymes and other proteins, and the ribosomal apparatus (Fig. 1.16). DNA remains in place in the prokaryotic nucleoid or the eukaryotic chromosomes, while the protein products of active genes are formed in the adjacent cytoplasm. Clearly, an intermediary is required to carry informational sequences from DNA to the ribosomes where synthesis proceeds under appropriate conditions. One of the kinds of RNA, known as **messenger RNA** (mRNA) is **transcribed** as a faithful complementary copy of the gene, and it is the molecule upon which the **translation** of genetic information into a specific protein is based. The details of protein synthesis are discussed in Chapter 6.

The genetic information is encoded as triplets of bases, each triplet or **codon** spelling one **amino acid** which is to go into a protein. The sequence of codons in DNA determines the sequence of amino acids in the protein, that is, DNA and protein are *co-linear* (Fig. 1.17). The great variety of genes coincides with the great variety of proteins known in the living world. The only enzyme proteins which can be synthesized are those whose composition is specified in the stored information of the DNA, so that organisms with different genes will produce different gene products, that is, RNA and proteins. Diversity among life forms ultimately is the outcome of genetic diversity which is itself the outcome of mutations and other genetic changes that have accumulated during more than 3 billion years of cellular evolution on this planet. A remarkable fact is that the same genetic code serves for viruses and all cellular life, and the same transcriptional and translational processes are responsible for gene expression. The *universality of the genetic code* implies common ancestry for all life on Earth. The continuity of life from its ancient beginnings to the present day is an outcome of the precision with which DNA is replicated in each cell generation. Identical genes are transmitted from parent to progeny, except for occasional mutational modifications and other genetic alterations. DNA replicational fidelity depends in large part on the restrictions in

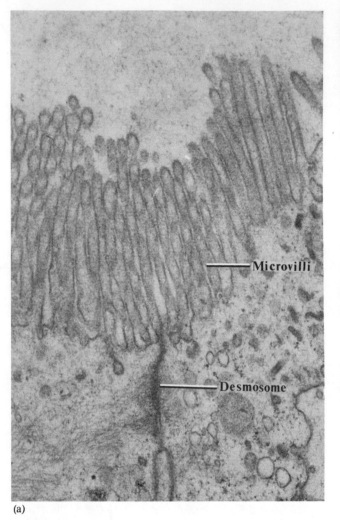

(a)

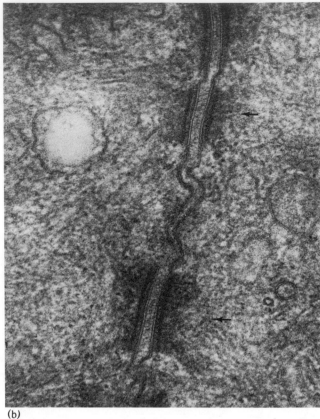

(b)

Figure 1.14

Cell surface differentiations: (a) Microvilli provide increased absorptive surface area in these cells from proximal tubule of frog kidney. A desmosome is visible below the microvilli, probably serving as an adhesion region between the two contiguous cell membranes. × 30,000. (b) Higher magnification view showing two desmosomes (at arrows) in rat intestinal epithelium. The desmosomes occur as symmetrical plaques between two adjacent cells. × 156,000. (Courtesy of N. B. Gilula, from Gilula, N. B., 1974, *Cell Communications* [ed. R. P. Cox], John Wiley & Sons, pp. 1–29, Fig. 12.)

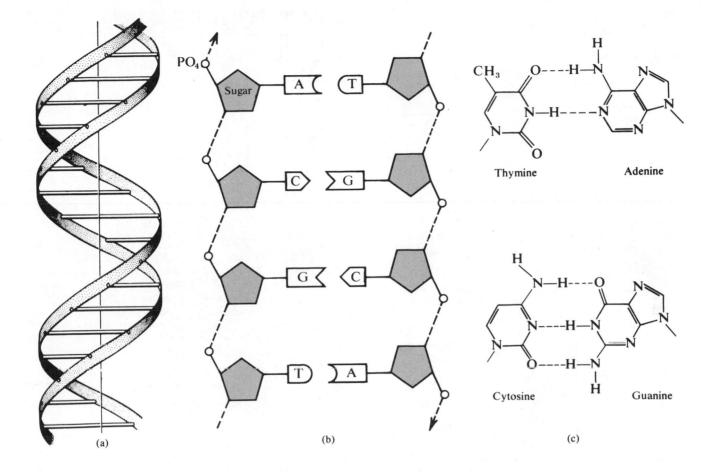

Figure 1.15
The molecular structure of DNA: (a) The intertwining strands of the double helix have "steplike" connections at regular intervals along the length of the molecule; (b) complementary pairs of nitrogen-containing bases, each base attached to the pentose sugar of the twisting sugar— phosphate "backbone," keep the two strands of the double helix at a constant distance (diameter) for the entire length of the molecule; and (c) the pairing is specific between each kind of purine and pyrimidine, but there is no restriction on the vertical sequence of base-pairs that can be incorporated into a DNA molecule.

pairing of the nitrogenous bases (Fig. 1.18). Since adenine pairs with thymine and cytosine pairs with guanine, the new molecule is an accurate replica of the original duplex DNA strands. Great strides have been made recently in understanding the molecular features of DNA replication, a topic discussed in Chapters 5 and 13.

It is immediately obvious that cells in the same organism may be as different as cells belonging to different forms of life, even though the same genes

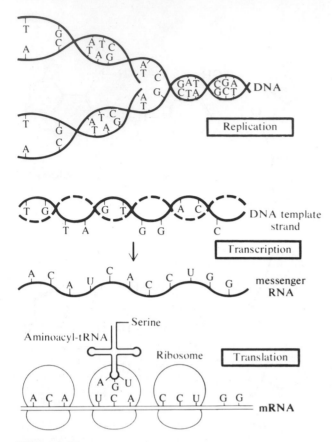

Figure 1.16
DNA may replicate and may also transcribe its genetic information into single strands of messenger RNA. The messenger RNA transcript carries the genetic blueprint which is translated into protein at the ribosomes. Base-pair specificity underlies all interactions between nucleic acid molecules.

are present in all the cells of an individual organism. These differences in structure and function result from complex mechanisms that regulate the synthesis of enzymes and other proteins in different cells and in the same cells at different stages in development or activity. In its broadest outlines we may consider regulation of cell structure and function as the outcome of control mechanisms that influence three variables of enzymes. These mechanisms affect the *type*

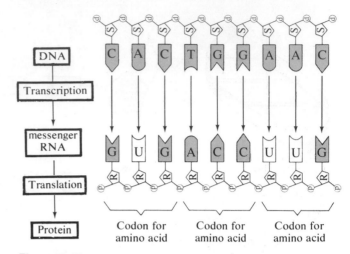

Figure 1.17
The messenger RNA copy of genetic information contained in DNA is translated into a co-linear protein. Each 3-letter codeword in the nucleic acid is translated into one of the 20 kinds of amino acids in proteins. The precise kinds and sequence of amino acids in the protein are specified by the particular codons in a particular sequence in the genetic material.

of enzyme synthesized, the *amount* of a particular enzyme synthesized, and the *activity* of an enzyme in the cell.

The type of enzyme manufactured in the cell depends on the specific coded information in the genes that are present. The metabolic potential depends upon the genetic potential in each species since a particular functional enzyme protein can be synthesized only when the appropriate gene exists in the organism's DNA. Even when the genetic potential exists it may not be expressed in a particular cell. For example, hemoglobin proteins are made only in certain cells even though all the cells of the organism also possess the same set of genes. There must be some mechanism that turns gene expression on and off and in some situations, turns off this expression indefinitely. The amount of enzyme manufactured must be modulated in such cases over a range from zero to some maximum number of molecules. Control

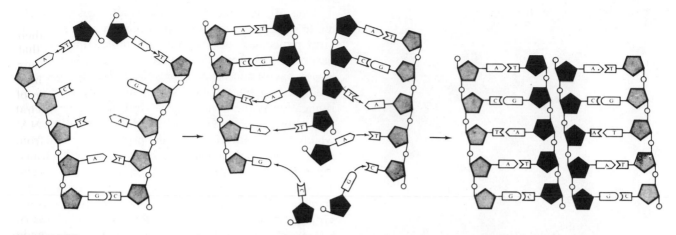

Figure 1.18

The duplex DNA molecule replicates to produce a new complementary strand from each of the original parental strands. At the end of the process, two duplex molecules, identical to each other and to the original parental molecule, have been formed. Faithful replication ensures genetic continuity from generation to generation.

over enzyme amount is the subject of vigorous study, and the available evidence indicates that regulation occurs at the level of gene transcription and translation. This topic will be discussed in Chapter 5.

Mechanisms that control enzyme activity also modulate the metabolic activities of cells. These mechanisms make possible rapid responses and are believed to serve as a fine control superimposed on coarser controls over type and amount of enzyme proteins. The chaos that would arise if all genes acted at once and all molecules were formed, regardless of usefulness or efficiency, is never realized. Cell structure and function is coordinated to produce ordered reactions and interactions by control mechanisms that monitor cellular regulation of activity.

VIRUSES

Viruses are not cellular and thus are not prokaryotic or eukaryotic. It is still debated, in fact, whether they are living or nonliving systems. In the extracellular or infectious state, viruses are metabolically inert. They may even be crystallized much like molecules, although some kinds of viruses can only be purified but not crystallized. When viruses gain entry to a host cell they may initiate the intracellular state of their existence and proceed to direct the replication of new virus particles. To do this viruses rely in part on their own programmed genetic potential and in part on exploitation of the biosynthetic machinery, raw materials, and catalysts of the host cell. Since viral genes direct the replication of unique viral molecules, they resemble living systems. Since they lack the metabolic equipment to express their genetic potential, however, and must utilize host capabilities, some would consider them to be nonliving entities.

Viruses are an extremely heterogeneous group, varying in size from 300 Å to 3000 Å, or from about the size of a ribosome almost to the limit of resolution of the light microscope. The simpler viruses consist only of one kind of nucleic acid, either DNA or RNA but never both, enclosed within a protein coat. There may be one or more kinds of proteins that contribute to the geometrical forms assumed by mature virus

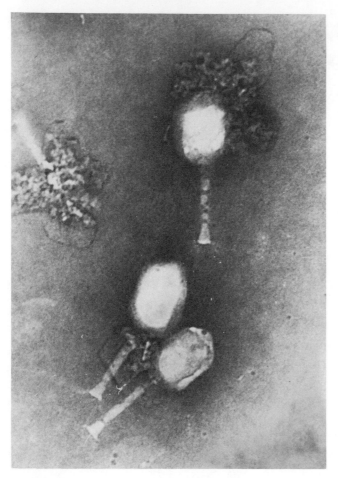

Figure 1.19
Negatively stained T4 bacteriophages, partially disrupted by the method of preparation. × 139,000.

particles. Viruses also vary in shape; they may be like rods, spheres with modified triangular faces, with tails, or without such appendages (Fig. 1.19). Viruses are named in a somewhat random fashion according to the disease they cause (poliomyelitis virus), the tissue affected (adenovirus infects adenoid tissue), the host organism (bacterial virus, also known as **bacteriophage** or simply **phage**), or some coded system (T1, T4, P1 phages of *Escherichia coli*).

One of the unique features of all viruses is their mode of replication which proceeds in stages so that viral molecules are formed, assembled into mature particles, and then released from the host cell. These stages have been especially well studied for some of the bacteriophages. The significant feature of viral replication is the synthesis of viral messenger RNA. The virus genes provide the encoded information from which mRNA for virus-specific proteins will be translated on host ribosomes using host amino acids, transfer RNAs, and other constituents (see Fig. 1.16). These first messenger RNAs are translated into virus-specified enzymes that later catalyze the formation of virus protein from messenger RNAs formed later in infection. The subversion of host biosynthetic machinery to manufacture progeny viruses is a characteristic of all viruses, which basically supply only the encoded genetic information for their continuity. It is this striking distinction in its mode of replication that clearly identifies a virus and distinguishes it from all cellular species.

Genetic mechanisms are very similar in viruses and prokaryotic microorganisms. Viruses undergo mutations, their genes recombine by an exchange mechanism that probably occurs in all organisms, and they clearly are capable of evolutionary modifications. With newer analytical methods, viruses have been shown to contain either double-stranded or sometimes single-stranded DNA, or single-stranded but sometimes double-stranded RNA genetic systems. The nature of viral nucleic acid is now considered a primary characteristic for discerning relationships among viruses, along with other traits.

SUGGESTED READING

Books, Monographs, and Symposia

DuPraw, E. J. 1972. *The Biosciences: Cell and Molecular Biology.* Stanford: Cell and Molecular Biology Council.

Fawcett, D. W. 1966. *The Cell. An Atlas of Fine Structure.* Philadelphia: Saunders.

Ledbetter, M. C., and Porter, K. R. 1970. *Introduction to*

the *Fine Structure of Plant Cells*. New York: Springer-Verlag.

Lenhoff, E. S. 1966. *Tools of Biology*. New York: Macmillan.

Lima-de-Faria, A., ed. 1969. *Handbook of Molecular Cytology*. Amsterdam: North-Holland.

Porter, K. R., and Bonneville, M. A. 1968. *Fine Structure of Cells and Tissues*. Philadelphia: Lea & Febiger.

Roller, A. 1974. *Discovering the Basis of Life: An Introduction to Molecular Biology*. New York: McGraw-Hill.

Articles and Reviews

Baserga, R., and Kisieleski, W. E. 1963. Autobiographies of cells. *Scientific American* 209(2):103–110.

Brachet, J. 1961. The living cell. *Scientific American* 205(3):50–61.

Changeux, J. 1965. The control of biochemical reactions. *Scientific American* 212(4):36–45.

Fernández-Morán, H. 1970. Cell fine structure and function—past and present. *Experimental Cell Research* 62:90–101.

Mirsky, A. E. 1968. The discovery of DNA. *Scientific American* 218(6):78–88.

Porter, K. R., and Novikoff, A. B. 1974. The 1974 Nobel prize for physiology or medicine. *Science* 186:516–520.

Chapter 2

Biological Molecules and Bioenergetics

Modern organic chemistry is the chemistry of carbon. In earlier times it was believed that the chemistry of living systems was quite different from that of the inorganic world, so that organic chemistry in those days centered around biological systems. With the synthesis of urea by Wöhler and other syntheses of biological molecules from inorganic materials by chemists shortly afterward, it became clear that the uniqueness of organic compounds resided in special properties of the carbon atom and not in the source of the chemical. The term "organic" chemistry has been retained, but modern biochemistry refers in general to the chemical dynamics of living systems. This framework has been increasingly broadened to include physics and chemistry of cell structure and function. In this way, the goals of biochemistry overlap with cell physiology, molecular biology, cell biology, and genetics. All these approaches converge on central problems of relationships between cell structure and function and the analysis of mechanisms of control over gene expression, metabolism, and development.

THE CHEMISTRY OF CARBON ATOMS

Carbon atoms are the principal components of biologically important molecules, and they possess unique properties that contribute to special qualities of organic compounds. Organic compounds occur in virtually unlimited numbers that are widely varied in their properties, and they are relatively sluggish in reacting with each other and with water or molecular oxygen. The versatility and stability of organic compounds are due to particular features of carbon atom interactions with other carbons and with the relatively few other elements from which cellular organic compounds are constructed.

Carbon can form **covalent bonds** with other carbon atoms or with hydrogen, oxygen, nitrogen, phosphorus, sulfur, and the few other elements generally found in cellular organic molecules. The fundamental carbon skeleton that is the framework of organic compounds can exist in chains, rings, networks, or combinations of these forms. This is due to the fact that a carbon atom is *tetravalent* and each carbon atom can combine with one to four other carbon atoms in covalent linkages (Fig. 2.1). Structural diversity of

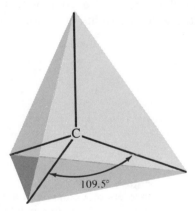

Figure 2.1
The tetrahedral nature of carbon is a result of the four valencies of the atom which are disposed in space in particular angles and lengths.

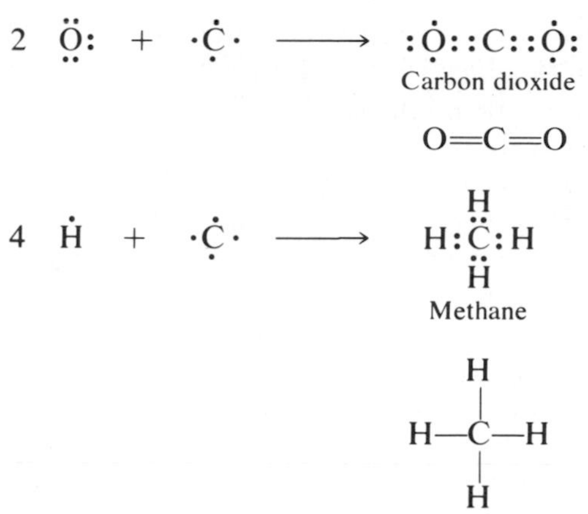

$$2 \quad \ddot{\mathrm{O}}\colon \; + \; \cdot\dot{\mathrm{C}}\cdot \quad \longrightarrow \quad \colon\ddot{\mathrm{O}}\colon\colon\mathrm{C}\colon\colon\dot{\mathrm{O}}\colon$$

Carbon dioxide

$$O{=}C{=}O$$

$$4 \quad \dot{\mathrm{H}} \; + \; \cdot\dot{\mathrm{C}}\cdot \quad \longrightarrow \quad \mathrm{H}\colon\ddot{\mathrm{C}}\colon\mathrm{H}$$

Methane

Figure 2.2
Covalent bonds form when electrons of different atoms share a common electron shell. Carbon compounds are very stable when the outer shell of the carbon atom is completely filled with eight electrons. Covalent interactions are very strong when compared with hydrogen bonds or other atomic interactions over greater atomic distances.

organic compounds is further enhanced since carbon atoms can react with electronegative elements such as nitrogen, oxygen, phosphorus, and sulfur, as well as with electropositive hydrogen atoms. Since carbon can form single, double, or triple bonds with other carbon atoms and single or double bonds with oxygen and nitrogen atoms, versatility of organic molecules is also increased by this feature.

The stability of carbon compounds results from the tetravalency of the carbon atom itself and the strength of the C—C, C—O, C—H, and C—N bonds of the carbon skeleton, among other characteristics. Since the carbon atom is tetravalent, its outer shell becomes completely filled with the stable configuration of eight electrons when four covalent bonds are formed (Fig. 2.2). The valences are in a tetrahedral arrangement, which confers a basic geometrical symmetry to the carbon atom and greater stability of bonding. The covalent interaction between carbon and atoms of oxygen, nitrogen, and hydrogen is strong as shown by bond energy values of about 100 kilo-calories (kcal) per mole (Table 2.1). Higher amounts of energy are required to break the covalent bonds between carbons double-bonded to oxygen or to other carbon atoms. Organic molecules with double-bonded structures generally are even more stable than carbon chains consisting of atoms linked by single bonds.

The tetrahedral structure of carbon can lead to asymmetry in many organic molecules. When a carbon atom is bonded to four different atoms, two spatial arrangements of the molecule can be constructed, with one alternative being the mirror image of the other (Fig. 2.3). The structure of such asymmetric molecules is conventionally drawn so that two sides of the tetrahedron face the viewer and the other two sides lie away from the viewer. The alternative forms

of the same compound are called **isomers.** If the two members of a mirror image pair are exposed in solution to polarized light, each may be able to rotate the plane of polarized light equally but in opposite directions. In this case the mirror image alternatives of a compound are *optical isomers.* Isomers of many asymmetric molecules are optically inactive, but if the isomers can rotate the plane of polarized light, it is possible to distinguish between the isomers by this simple test. The significance of the potential asymmetry of each carbon atom in an organic molecule is that the same compound may occur in a number of isomeric forms, which increases the versatility of carbon compounds. Most molecules assume only a fraction of the number of possible isomer configurations, however, these being the more energetically favorable alternatives.

The four major classes of biologically significant organic compounds are carbohydrates, proteins, lipids, and nucleic acids. Each of these is a high-molecular weight **polymer** made up of **monomer** subunits, or building blocks. There may be only one kind of monomer in some carbohydrates or as many as twenty different kinds of monomer in proteins.

The molecular weight of a compound is equal to the sum of the atomic weights of all the atoms in the

Table 2.1 Values for some covalent bond energies*

SINGLE BONDS		DOUBLE BONDS		TRIPLE BONDS	
O—H	110				
H—H	104				
C—H	99				
C—O	84	C=O	170		
C—C	83	C=C	146	C≡C	195
C—N	70	C=N	147	C≡N	212

* Energy (kcal/mole) required to break the bond.

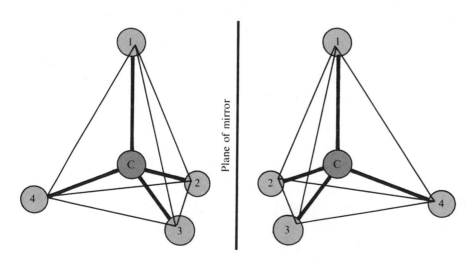

Plane of mirror

Figure 2.3
Isomers of a compound are found when a carbon atom is bonded at each of its four valencies to a different atom or group of atoms. The alternative forms of the compound are mirror images of one another.

molecule. Molecular weights of these compounds range from thousands up to hundreds of millions of **daltons,** the unit of measurement for atoms, molecules, or particles. One dalton is approximately equal to the weight of one hydrogen atom. Another useful unit is grams molecular weight, or **moles.** A mole is equal to the weight in grams of the compound which is numerically equal to the molecular weight. One mole of H_2O has a weight of 18 grams, whereas one molecule of H_2O has a weight of 18 daltons ($H = 1$, $O = 16$, $H_2O = 18$). One mole of a substance contains 6.02×10^{23} molecules (Avogadro's number). Simple arithmetic provides information about molecular weights of small molecules with known atomic composition, but physical methods for measurement are required to establish molecular weights of large molecules or unknowns. Average values for monomer molecular weights often are used to estimate the molecular weight of polymers, or conversely, polymer molecular weight can be used to estimate the number of monomer subunits in the compound. For example, proteins are constructed from amino acids that have an average molecular weight of 120 daltons. If a protein contains 150 amino acids, then the protein would have an estimated molecular weight of 18,000 daltons (150 amino acids × 120 = 18,000). If a protein is known to have a molecular weight of 18,000

daltons, then it is obvious that we would estimate it to have about 150 amino acids in its construction.

CARBOHYDRATES

Carbohydrates are compounds of the general formula $(CH_2O)_n$. Some biologically important carbohydrate derivatives also contain nitrogen and sulfur atoms. Among all carbohydrates in living systems, the most widespread molecule is the six-carbon sugar D-glucose. This compound and other simple sugars that are single units are called **monosaccharides.**

Monosaccharides

Monosaccharides are categorized according to the number of carbon atoms in the molecule; a *hexose,* like glucose, has 6 carbons, a *pentose* sugar has 5 carbon atoms, a *triose* has 3, and so on. Because of asymmetric carbon atoms, monosaccharides can occur in a number of isomeric forms. Although 16 different isomers of glucose can be formed theoretically on the basis of four asymmetric carbons, only three of these isomers are found in nature (Fig. 2.4). Monosaccharides are identified as D- or L-forms according to a convention based on the configuration of the triose

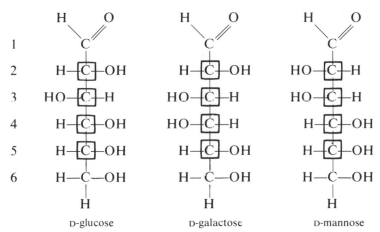

Figure 2.4
Three naturally occurring isomers of the hexose sugar D-glucose. There are asymmetric carbon atoms at positions 2, 3, 4, and 5 in the molecule.

D-glyceraldehyde. A monosaccharide is a D-variety if the hydroxyl group on the bottom-most asymmetric carbon atom is on the right (when the carbonyl group, $\diagdown C = O$, is at the top of the formula) as in D-glyceraldehyde (Fig. 2.5). If this hydroxyl group is on the left, then the molecule is of the L-variety. Glucose in its naturally-occurring form is a D-hexose according to this convention.

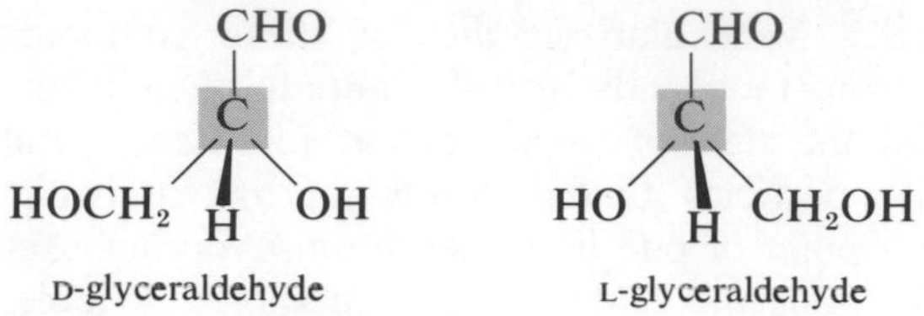

Figure 2.5
The D- and L-isomers of glyceraldehyde are shown in relation to the asymmetric carbon atom of the molecule.

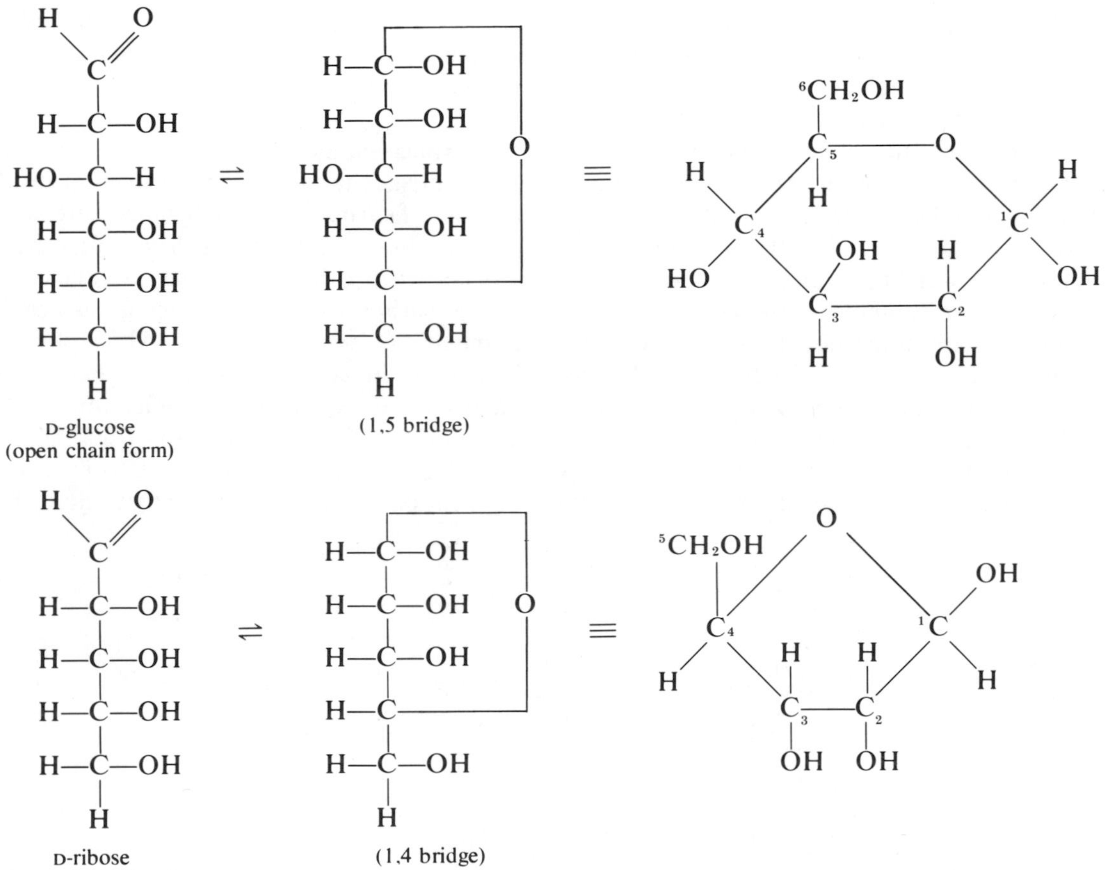

Figure 2.6
The linear and predominant ring forms of pentose and hexose sugars exist in solution in an equilibrium mixture. The conventional hexagon and pentagon formulations refer to the ring form of the molecules.

Pentose and hexose sugars in solution exist largely in a ring form in equilibrium with a small amount of the linear form of the molecule (Fig. 2.6). When the oxygen bridge is located between carbons 1 and 5 (pyranose form), it is conventional to depict the molecule as a hexagon. When the oxygen bridge is a 1,4 bond (furanose form), a pentagonal notation is used to represent the molecule. The notations show a heavier line across the bottom of the pentagon or hexagon figure to indicate that this is the part of the molecule nearest the viewer, and the plane of the molecule is perpendicular to the plane of the paper. The hydrogens and hydroxyls are oriented up or down and are indicated by vertical lines at each of the carbons (Fig. 2.7). These conventional representations, called Haworth formulae, make it very simple to specify and recognize a particular isomer at a glance. The same notation applies equally to pentoses and hexoses because all

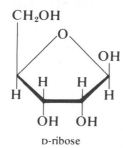

$^6CH_2OH$ α-D-glucose

$^6CH_2OH$ β-D-glucose

Figure 2.8
α- and β-isomers of D-glucose are distinguished on the basis of the position of the hydroxyl group at carbon-1 relative to the hydroxyl group at carbon-2 of the molecule.

share the carbon atoms whose bond angles provoke the chain to bend into a ring rather than to remain in a linear configuration. Because of spatial restrictions, the monosaccharide ring can only be 5- or 6-membered; a 7-membered carbon chain would not form a ring because excessive strain would result.

With the formation of the oxygen bridge, the cyclic molecule gains asymmetry at carbon atom 1. The hydroxyl of carbon-1 is either adjacent to the hydroxyl of carbon-2 or is rotated 180° relative to it (Fig. 2.8). These isomers are designated as the α- and β-forms of the cyclic monosaccharides. The importance of α- and β-isomers is related to the nature of the bonds between monosaccharide units that make up larger compound molecules.

Polysaccharides

Polysaccharides, which have the general formula $(C_6H_{10}O_5)_n$, form on condensation of smaller units,

CH$_2$OH

D-glucose

CH$_2$OH

D-ribose

Figure 2.7
Haworth formulae depicting 5- and 6-membered ring forms of sugars.

with elimination of H_2O for each **glycosidic bond** produced (Fig. 2.9). Compounds containing 2–6 monosaccharide units are generally called *oligosaccharides* (Gr. *oligos:* few). The most important of these are the two-sugar compounds, or *disaccharides,* such as sucrose (table sugar), lactose (milk sugar), and maltose (degradation product of starch) (Fig. 2.10).

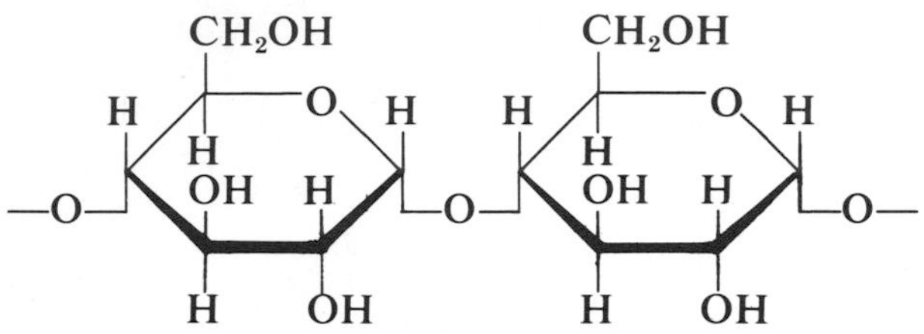

Figure 2.9
Portion of a polysaccharide showing the glycosidic bond joining the monomer units.

Two kinds of bonds can link saccharide monomers, depending on the α- or β-position of the hydroxyl group on carbon atom 1. These are generally referred to as α- and β-glycosidic linkages, but in the specific cases of glucose monomers they are specified as α- and β-glucosidic bonds. Monomers can be joined more directly by α-linkages whereas formation of a β-glycosidic bond requires a rotation in space of 180° for the hydroxyl of one unit to come into an appropriate spatial relationship to the hydroxyl of its neighbor monomer (Fig. 2.11). These linkages are physiologically important for at least three reasons:

1. They provide a means for joining two or more subunits in construction of a variety of larger molecules with different functions and specificities.
2. Different enzymes attack α- and β-glycosidic links allowing cells additional ways of discriminating among compounds used in structure and function.
3. The molecular potential in the cell is different for each type, with α-glycosides being readily

mobilized for metabolism and β-glycosides contributing to formation of many structural molecules.

Polysaccharides usually contain hundreds or thousands of monosaccharide residues even though as few as ten of these subunits are enough to define a molecule as the polymer form. The major functions served by polysaccharides in living systems are concerned with food storage and structure. Two principal food reserve polysaccharides in eukaryotic organisms are *starch* and *glycogen,* both of which are hydrolyzed to their constituent glucose monomer units by specific

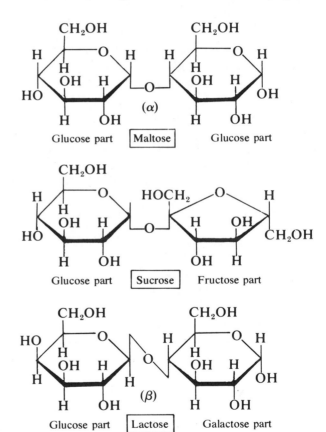

Figure 2.10
Three common disaccharide sugars.

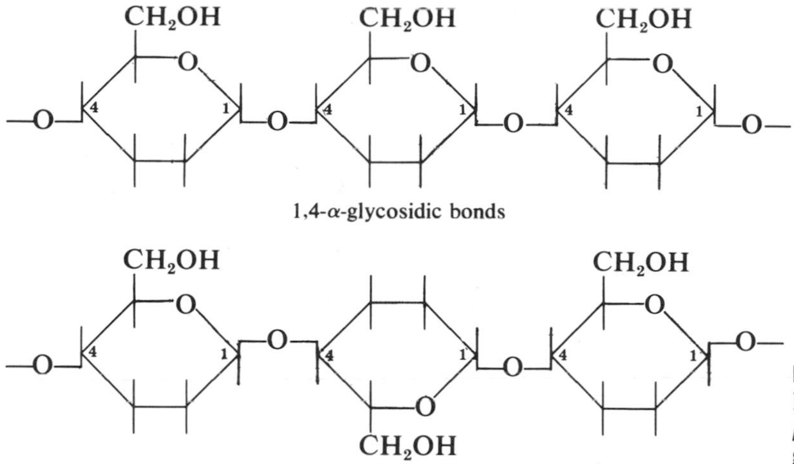

Figure 2.11
Polysaccharide fragments showing α- and β-glycosidic bonds; in these examples glycosidic bonds occur between C_1 and C_4 of adjacent monomer units.

enzyme actions. Starch is deposited in large granules in the chloroplasts of some green cells or in the colorless leucoplasts of root, stem, and other plant tissues. These starch granules can be seen with the light microscope and identified in several ways, including a simple staining test using an iodine solution. There are two types of molecules in the starch granule, both of which are constructed from α-glucosidic bonds. One kind of molecule is an unbranched chain containing 250–300 glucose units. The other type of molecule in starch has occasional branches along the 1000-residue length of the polymer. While starch is a major food storage form in algae and land plants, glycogen is the principal food reserve in animals and fungi. Glycogen is deposited in the cytoplasm rather than within an organelle as is starch. Glycogen occurs in particularly large amounts in the liver of animals. The liver glycogen polymer is a long branched chain containing about 30,000 glucose units connected by α-glucosidic bonds between carbon atoms 1 and 4 in the backbone of the molecule and 1,6-links at the frequent points of branching (Fig. 2.12). The entire structure of the glycogen granule resembles a flattened ellipsoid, but the molecules themselves are feathery or bushy in form as a result of numerous branches.

Structural polysaccharides in eukaryotes include *cellulose* and *chitin*, which have β-glycosidic links between monomer subunits. Cellulose occurs alone or with other materials in the cell wall of most algae and all higher plants, as well as certain fungi and protists. The 1,4-β-glycoside contains about 8,000 glucose units arranged in a long unbranched chain. These long-chain polymers fold in such a way as to form long fibrils aggregated into bundles easily seen with the electron microscope (Fig. 2.13). The polymer consists of 1,3-β-links in some of the fungi and 1,4-β-glycosides in other fungal groups (Fig. 2.14). Chitin is a nitrogen-containing polysaccharide that is constructed of N-acetylglucosamine residues joined by 1,4-β-glycosidic bonds (Fig. 2.15). The chitin found in the cell walls of many fungi and in the rigid exoskeleton of insects, crustacea, and certain other invertebrate animals is chemically identical. The sugar residues in the long, unbranched chitin chain are glucose derivatives.

The polysaccharide chains in bacterial cell walls are composed of disaccharide units, joined together into long unbranched molecules by 1,4-β-glycosidic bonds. The disaccharide unit itself is invariably composed of two sugar derivatives, N-acetylglu-

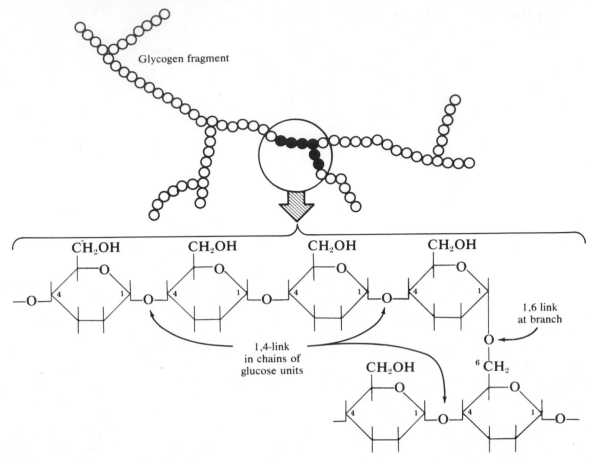

Figure 2.12
1,4- and 1,6-glycosidic links in glycogen, a polymer made up of about 30,000 glucose units.

cosamine and *N*-acetylmuramic acid (Fig. 2.16). The polysaccharide chains are connected by cross-links of small peptides made of 4–5 amino acid units, so that the basic structure is a *peptidoglycan* sheet. The particular cross-linkages vary considerably among bacterial species, whereas the polysaccharide (glycan) portion is essentially uniform throughout the prokaryotic group. The amount of cell wall that is composed of peptidoglycan varies from about 90 percent in Gram-positive bacteria to as low as 5 percent in some Gram-negative species. Other polysaccharides as well as lipids and proteins usually are present in bacterial cell walls, contributing to the complexity of organization and function of these structures in prokaryotes.

LIPIDS

Lipids are a diverse group of substances that have the common property of solubility in nonpolar, organic solvents. Included in this group of organic compounds are fats, fatty acids, waxes, steroids, phosphoglycerides, glycolipids, and sphingolipids, as examples of classes of biologically important substances.

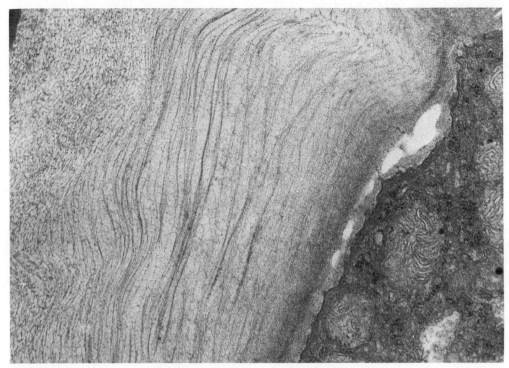

Figure 2.13
Electron micrograph showing parallel bundles of cellulose fibers in the cell wall of wheat (*Triticum aestivum*). × 37,000. (Courtesy of M. C. Ledbetter)

Fatty Acids

Naturally-occurring **fatty acids** are unbranched hydrocarbon chains with a carboxyl group at one end (Fig. 2.17). Since fatty acids are synthesized from two-carbon acetyl units, they usually have an even number of carbon atoms, the commonest numbers being 16 and 18. When all carbon atoms of a fatty acid chain are joined by single bonds the compound is "saturated" (with hydrogens at both valences not involved in the —C—C— chain). An "unsaturated"

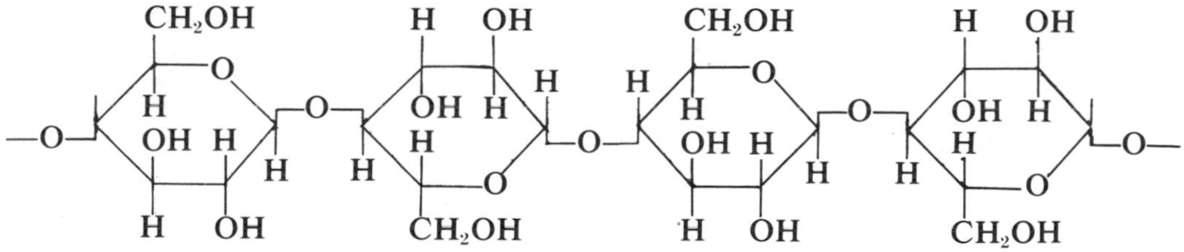

Figure 2.14
1,4-β-glycosidic links join the glucose monomers in the cellulose found in some species, while 1,3-β-glycosidic links occur in other species.

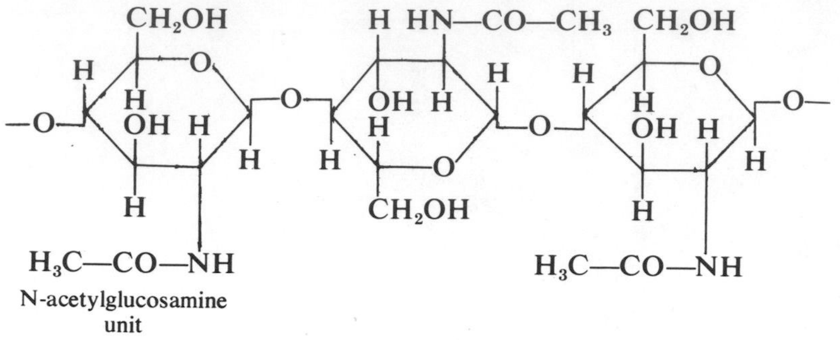

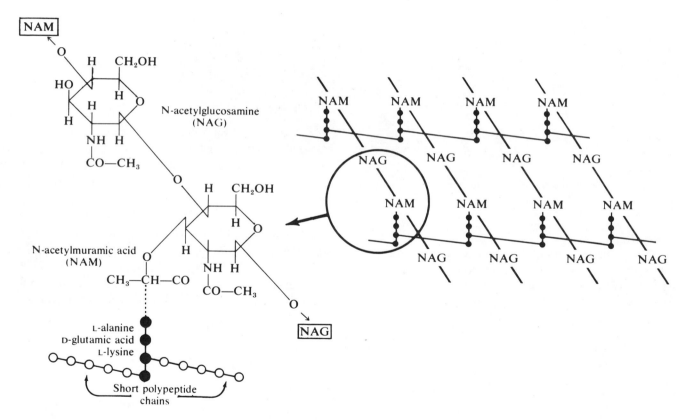

N-acetylglucosamine unit

Figure 2.15
Portion of a chitin molecule showing 1,4-β-glycosidic links between *N*-acetylglucosamine monomer units that make up the polymer.

Figure 2.16
The peptidoglycan sheet structure of bacterial cell walls is made from carbohydrate chains connected by short peptide cross-links that usually contain one or more unusual D-amino acids.

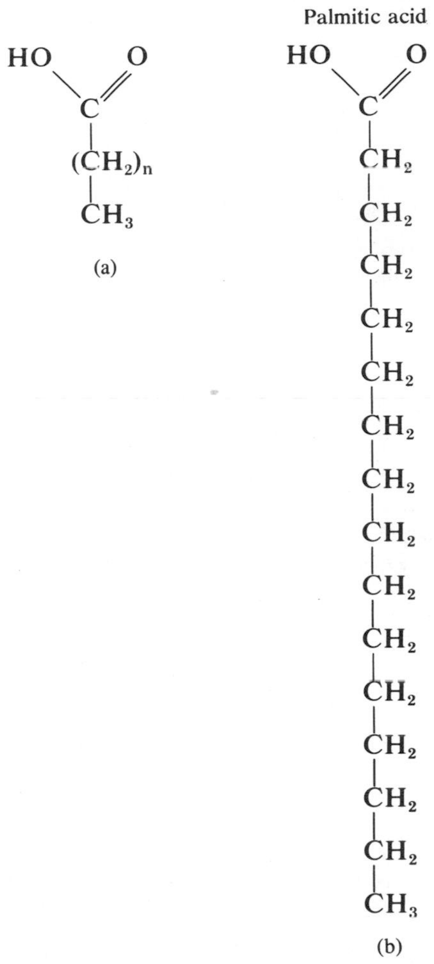

Palmitic acid

(a)

(b)

Figure 2.17
The saturated fatty acid: (a) general nature of the molecule with a carboxyl group at one end and a long chain of carbon atoms fully saturated with hydrogens; and (b) the molecule of palmitic acid, whose formula is $C_{16}H_{32}O_2$, or $COOH—(CH_2)_{14}—CH_3$.

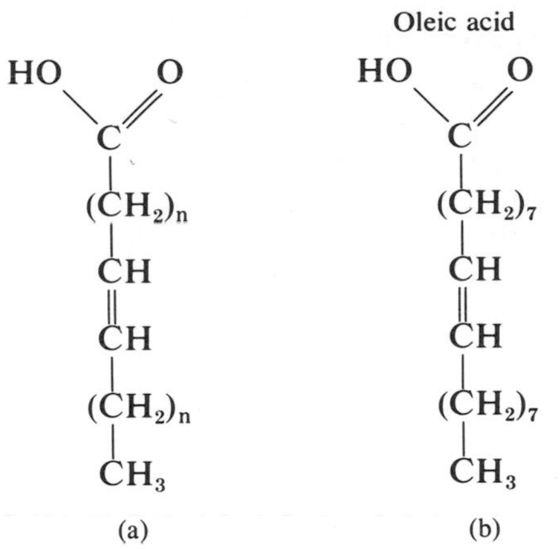

Oleic acid

(a)

(b)

Figure 2.18
The unsaturated fatty acid: (a) general type of molecule having only one double bond in the hydrocarbon chain; and (b) oleic acid, whose formula is $C_{18}H_{34}O_2$, or $COOH—(CH_2)_7—CH=CH—(CH_2)_7—CH_3$.

fatty acids interact with water, the soluble carboxyl end is contained within the water as a layer while the hydrocarbon tails of these molecules remain outside the water surface (Fig. 2.19). Although fatty acids occur in trace amounts in cells and tissues they are important as building blocks of several classes of lipids.

fatty acid has one or more double bonds between carbons in the backbone of the chain (Fig. 2.18). The carboxyl end of the fatty acid molecule is water-soluble and highly polar while the hydrocarbon portion of the chain is water-insoluble and highly nonpolar. When

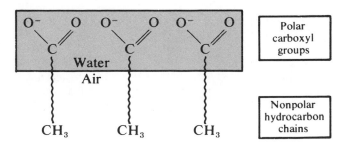

Figure 2.19
When fatty acids interact with water, the polar carboxyl end of the fatty acid undergoes hydrophilic interaction with water molecules while the nonpolar hydrocarbon chain is sequestered away from water because of its hydrophobic nature.

Neutral Fats (Glycerides)

Fatty acid esters of the alcohol *glycerol* are **neutral fats,** or **glycerides** (Fig. 2.20). One, two, or three different kinds of fatty acids can combine with the hydroxyl groups of glycerol to form a molecule of fat. These glycerides are the major storage form of fats in both plants and animals, and they may be formed from excess carbohydrate, protein, or lipid in cells and tissues. When fats are oxidized they provide more than twice the energy (in calories per gram) of carbohydrates or proteins, because the overall state of oxidation of the long hydrocarbon chains of the fatty acids is very low. Fats, therefore, have much further to go before they are completely processed to carbon dioxide and water.

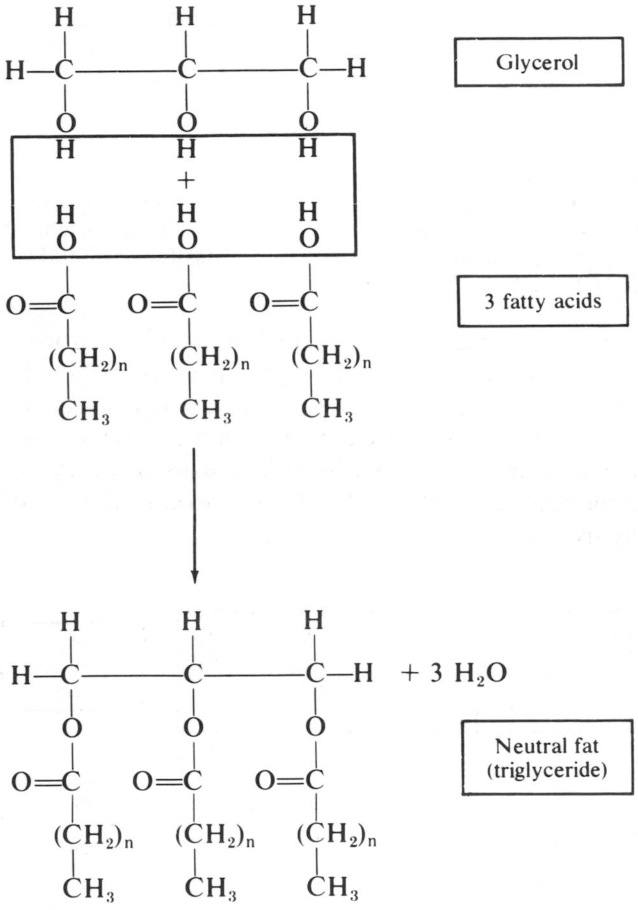

When fatty acids are esterified by reaction with glycerol, the water-soluble carboxyl end of the molecules become internal and inaccessible components of the newly formed glyceride. Glycerides are nonpolar molecules and do not dissolve in water. Neutral fats are hydrolyzed in cells by the action of enzymes called *lipases,* or when boiled with acids or bases. Hydrolysis with alkali, called *saponification,* results in a mixture of fatty acid soaps plus glycerol. Animal fats can be converted to soap in this way, a practice still continued in some societies. Oils are a form of fat that liquefy at room temperature. The degree of saturation of the constituent fatty acids determines the melting point of the fat, and the higher the degree of saturation the higher the melting point. Vegetable oils are saturated to convert them to the "hard" fat form of margarine.

Phospholipids (Phosphoglycerides)

Phospholipids are almost entirely found in membranes and may occur in only small amounts in some storage fats. Fatty acids are attached by ester bridges to two of the hydroxyl groups of glycerol while the third hydroxyl group is esterified to phosphoric acid instead of a fatty acid (Fig. 2.21). The phospholipids, also called phosphoglycerides, share one important property: they have a hydrophobic "tail" region consisting of the two fatty acid chains, and a hydrophilic "head" consisting of the negatively charged phosphoric acid residue and a positively charged residue that is bonded to this phosphoric acid group (Fig. 2.22). Phospholipids are therefore **amphipathic** lipids, since they have both hydrophobic and hydrophilic regions in the molecule. They are the most polar of all groups of lipids, which makes phospholipids important

Figure 2.20
Neutral fats form by a dehydration reaction between the alcohol glycerol and the carboxyl groups of fatty acid chains, producing a fatty acid ester of glycerol. This molecule is a triglyceride.

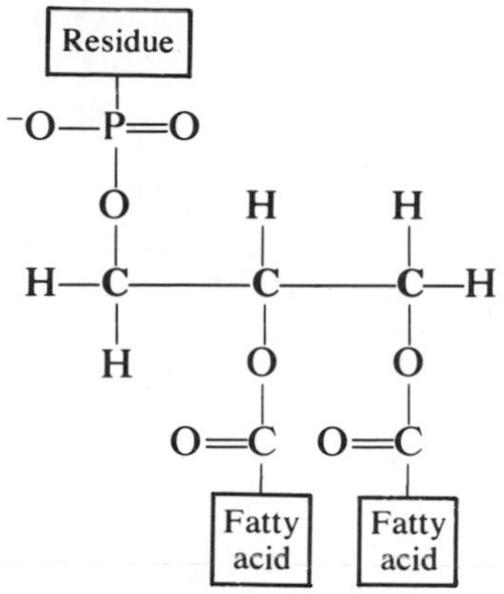

Figure 2.21
General formulation of a phospholipid molecule showing the uncharged fatty acid chains at two carbons and a phosphate residue at the third carbon atom. A positively charged residue is bonded through the negatively charged phosphate to produce an amphipathic molecule with spatially separated hydrophilic "head" and hydrophobic "tails."

mediators between membranes and the hydrophobic and hydrophilic substances on both sides of the membrane. Phospholipids serve as vital structural links between the aqueous and nonaqueous phases outside and inside the cell, and they also play a functional role in certain enzyme activities.

Among the major phospholipid components of most membranes in animal cells are *ethanolamine phosphoglyceride* (cephalin) and *choline phosphoglyceride* (lecithin). Another important compound is *cardiolipin,* which is found in membranes of bacterial cells, mitochondria, and chloroplasts (Table 2.2). When choline phosphoglycerides interact with water they spontaneously aggregate to form bimolecular "leaflets," with the hydrophilic ends of the molecules in water and the hydrophobic tails in the air phase (Fig. 2.23). These aggregates form convoluted "mye-

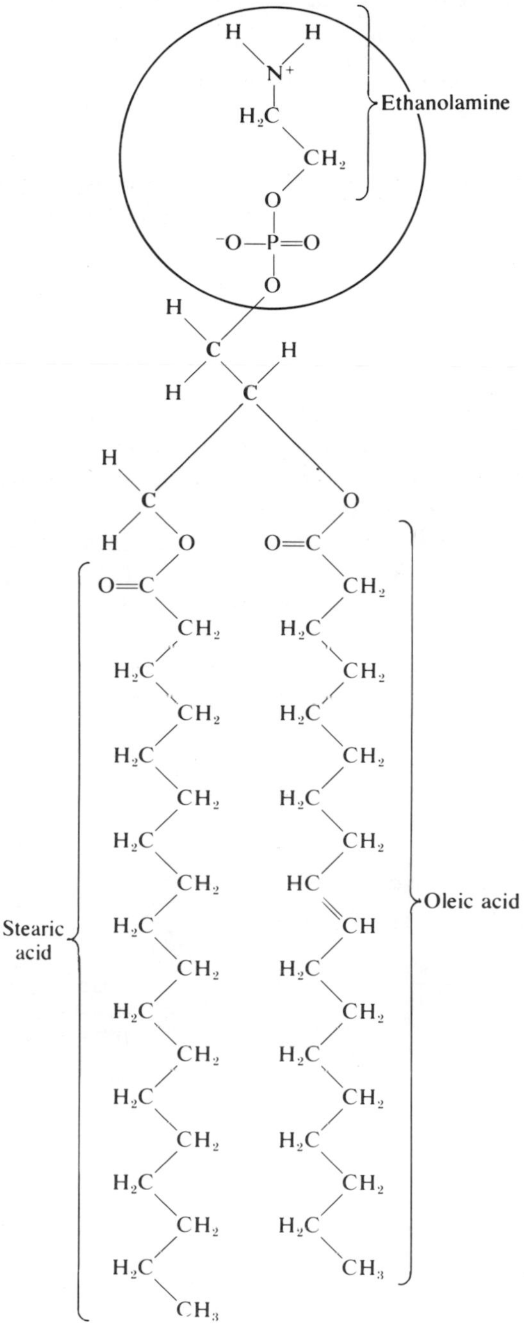

Figure 2.22
The phospholipid compound ethanolamine phosphoglyceride, or phosphatidyl ethanolamine, a major component of cell membranes. The three carbon atoms of the glycerol region are shown in boldface.

Table 2.2 Major types of lipids in cells

LIPID GROUP	SOME MAJOR TYPES	IMPORTANT CELLULAR LOCATION
Fatty acids	Oleic acid, palmitic acid, stearic acid	Cytosol, mitochondria, glyoxysomes of fatty seeds
Glycerides (neutral fats)	Coconut oil, beef tallow	Fat storage depots
Phosphoglycerides (phospholipids)	Ethanolamine phosphoglyceride, choline phosphoglyceride, cardiolipin	Membranes
Sphingolipids	Sphingomyelin	Membranes
Glycolipids	Cerebrosides, gangliosides	Membranes
Steroids	Cholesterol	Membranes
Terpenes	Essential oils, carotenoids	Plant cytosol, chloroplasts

lin" figures that resemble cell membranes in some ways. They are even more like cellular membranes if proteins are added to the phospholipid and then combined with water. The amphipathic properties of phospholipids are thus important in membrane conformation as well as membrane functions in interaction with water, proteins, and lipids.

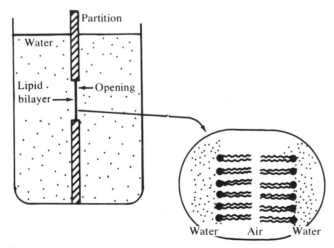

Figure 2.23
When phospholipid molecules interact with water, a spontaneous aggregation into a bimolecular layer occurs. The hydrophilic heads of the amphipathic phospholipid interact with water, and the hydrophobic tails of the molecules are sequestered away from water and within the air phase of the system.

Sphingolipids and Glycolipids

Sphingolipids and glycolipids resemble phosphoglycerides in having an amphipathic construction consisting of two hydrophobic residues in a "tail" and one hydrophilic residue in a "head" region of the molecule. These two classes of lipids are constituents of membranes, but are somewhat more restricted in their distribution among different kinds of cells than are the phospholipids. All sphingolipids lack a glycerol component, and they contain a *sphingosine* residue in place of one of the two fatty acid chains. The fatty acid tail and the sphingosine tail serve as the hydrophobic region, while some kind of hydrophilic residue attached to sphingosine supplies a polar property to the lipid (Fig. 2.24). Sphingolipids are found in plant and animal membranes but are especially prominent components of cell membranes in brain and nerve tissues; *sphingomyelin* is the most abundant compound in this group. Glycolipids contain polar hydrophilic carbohydrate head groups, usually D-glucose or D-galactose.

Other kinds of glycolipids have glycerol or sphingosine components. Two particular classes of these compounds are *cerebrosides* and *gangliosides,* which are either glycolipids because they contain a carbohydrate component or sphingolipids because sphingosine is present. They are often referred to as glycosphingolipids because of their component combination. Cerebrosides have both a sugar and a sphingo-

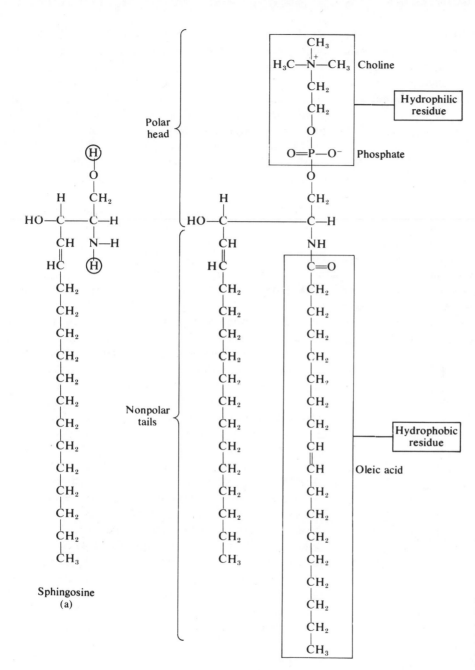

Sphingosine
(a)

Sphingomyelin
(b)

Figure 2.24
Sphingolipids: (a) sphingosine, a component of all sphingolipids; (b) the sphingolipid sphingomyelin that contains choline in the hydrophilic head portion and oleic acid as the fatty acid tail, along with the sphingosine tail, in the hydrophobic region of the amphipathic molecule.

sine residue and are found particularly in the myelin sheath of nervous tissue. Gangliosides are abundant in the outer surface of the cell membrane. They are distinguished by having a polar oligosaccharide head group instead of a monosaccharide as in other glycolipids.

Steroids and Terpenes

Lipids of the nonsaponifiable groups include **steroids** and **terpenes.** Both types are derived from common five-carbon building blocks and are therefore related groups of compounds. The most familiar steroid compounds are male and female sex hormones in vertebrates, bile acids, and adrenocortical hormones. Most steroids occur in trace amounts, but *sterols* are a relatively abundant class of these compounds. Sterols are steroids that occur as free alcohols or as long-chain fatty acid esters. The most abundant of these substances in animal tissues is **cholesterol,** which is found particularly in the plasma membrane (Fig. 2.25). Plants and fungi contain other kinds of sterols; sterols have not been reported to occur in bacteria.

Terpenes are especially evident constituents of certain plant species, and are responsible for characteristic odors and flavors. They are a major component of "essential oils" derived from such plants; for ex-

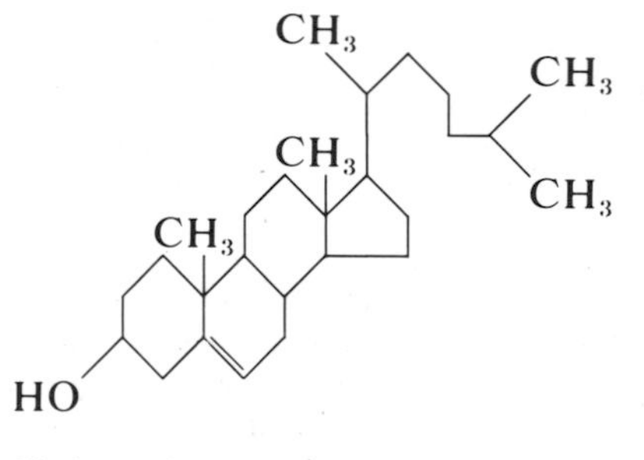

Cholesterol

Figure 2.25
Cholesterol, a sterol member of the steroid group of lipids and an important constituent of animal cell plasma membrane.

ample, the terpenes camphor, limonene, and menthol occur in oil of camphor, lemon, and mint, respectively. The linear component of Vitamin A and of the pigment chlorophyll is a terpenoid alcohol called *phytol* (Fig. 2.26). Some other terpenoid compounds in plants are natural rubber and *carotenoid* pigments that absorb light energy in photosynthesis and also contribute to yellow and orange colors of carrots, autumn foliage, and other materials. The fat-soluble vitamins A, D, E, and K are synthesized from precursors of the same five-carbon **isoprene** unit as are the terpene compounds (Fig. 2.27). Carotenes from plants are precursors of Vitamin A, which is found only in animals.

PROTEINS

The enormous variety of proteins serve as major building materials and as regulatory molecules that control the diverse activities of living systems. Major types of fibrous proteins that act as structural elements include *actin* and *myosin* of muscle and other contractile systems, *collagens* that form connective ligaments within the body, *keratins* that form protective coverings such as skin, hair, claws, horns, feathers, and other structures of land vertebrates, and a number of other compounds. Proteins that regulate the numerous processes and activities of the organism include *enzymes* that modulate chemical reactions of metabolism, *antibodies* that provide immunity against infection, *hormones,* and various other substances that make each life form respond appropriately to the constantly changing environment. Regulatory proteins are globular in shape, in contrast with the fibrous structural proteins. These conformations have an important bearing on the character of the molecule and its activity or function, as we shall see shortly.

Amino Acids

There is a common plan of construction for the thousands of kinds of proteins in living systems. The

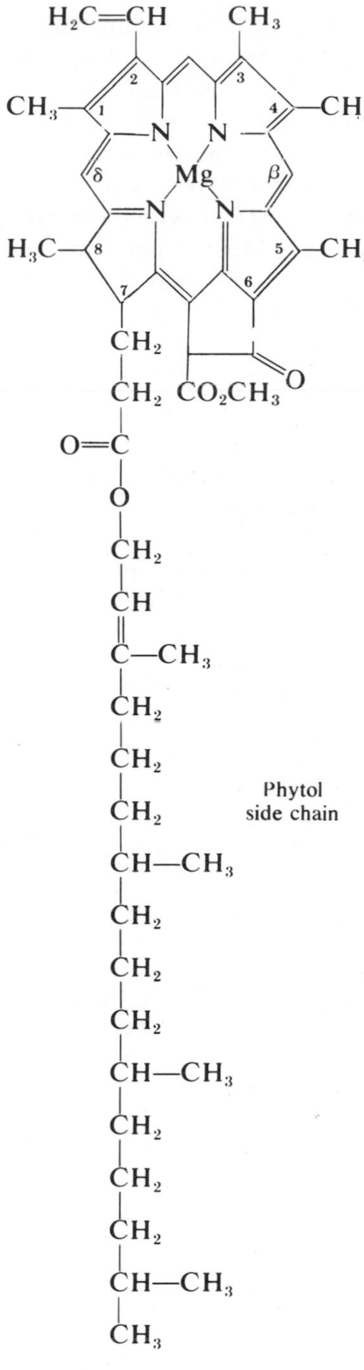

H$_2$C=CH   CH$_3$

CH$_3$—1   2   3   4—CH$_2$CH$_3$

N   N

δ   Mg   β

N   N

H$_3$C—8   7   6   5—CH$_3$

CH$_2$

CH$_2$ CO$_2$CH$_3$ O

O=C

O

CH$_2$

CH

C—CH$_3$

CH$_2$

CH$_2$

CH$_2$ Phytol side chain

CH—CH$_3$

CH$_2$

CH$_2$

CH$_2$

CH—CH$_3$

CH$_2$

CH$_2$

CH$_2$

CH—CH$_3$

CH$_3$

20 common types of **amino acid** monomers are strung together in the unbranched, linear polymer chain of naturally-occurring proteins. These are the 20 amino acids specified in the genetic code that is universal in viruses, prokaryotes, and eukaryotes. Some other kinds of amino acids are also known, but they are either degradation products or residues that have been modified from one of the 20 commonly occurring amino acids after the latter has been inserted into the polymer chain. Hydroxyproline is a major amino acid constituent of collagen, but proline residues are initially included in the protein and become hydroxylated after polymerization. Hydroxyproline is not one of the encoded amino acids, but proline is. Many proteins contain fewer than 20 kinds of amino acids. The relative proportions and the absolute numbers of the amino acid repertory vary from one protein to another, as a reflection of the specific information in the gene that is the blueprint for the protein construction.

All 20 amino acids have the same basic structure. (Fig. 2.28). There is a carboxyl group (—COOH) and an amino group (—NH$_2$) joined to the first, or α-carbon atom, and a hydrogen atom at the third valence of this carbon in all amino acids. Except for glycine, which has a second hydrogen joined to the α-carbon, the other 19 amino acids have a fourth group that differs from those at the other three valences of the first carbon atom. Except for glycine, therefore, the amino acids have an asymmetric α-carbon atom and are optically active molecules. The L-isomer of these 19 amino acids occurs in almost all natural proteins, although some D-amino acids have been found in certain molecules from plants and bacteria. For example, D-alanine and D-glutamic acid are constituents of the peptidoglycan in bacterial cell walls (see Fig. 2.16).

The side-chain differences are responsible for the

Figure 2.26
A molecule of chlorophyll *a* showing the terpenoid alcohol *phytol* chain of the molecule.

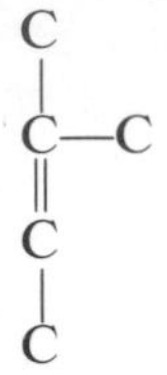

Figure 2.27
The 5-carbon *isoprene* skeleton underlies construction of terpene compounds of various biological functions.

varying properties of different amino acids. These side-chains also confer particular properties on the proteins in which they occur. When individual amino acids are in solution at neutral pH, added alkali or added acid is neutralized by the amino acid (Fig. 2.29). The amino acid is a "zwitterion" with simultaneous negatively- and positively-charged groups. These effects are cancelled when amino acids condense to form a *peptide* unit because of the nature of the bond that joins the linear assemblage of amino acid monomers. The formation of a **peptide bond** or **amide** involves the linking of an amino group of one amino acid with the carboxyl group of the adjoining amino acid (Fig. 2.30). Since this linkage involves one charged group of each amino acid, the "zwitterion" property no longer exists.

An amino acid polymer may still display acidic or basic properties because of the presence of acidic or basic side-chains in its constituent amino acid units. At pH 7.0, aspartic acid and glutamic acid residues confer acidic properties on a polymer region, while the positively charged polar groups of histidine, arginine, and lysine contribute basic properties to a protein (Table 2.3).

When acid is added to protein in solution, a net positive charge develops because —COO^- changes to —COOH. Addition of base causes —NH_3^+ to change to —NH_2, leaving the protein with a net negative charge. At a particular intermediate pH called the **isoelectric point** there are equal numbers of positive

and negative charges in the protein, and the net charge is zero (Fig. 2.31). Positively and negatively charged proteins migrate in an electrical field, but proteins at their isoelectric point do not move toward either cathode or anode. Proteins are most easily precipitated by appropriate solvents when in solution at their isoelectric pH, the pH at which the protein is at its minimum solubility. Most proteins have an isoelectric point on the acid side and therefore carry a net negative charge at physiological pH. Basic proteins, such as *histones* found in chromosomes, carry a net positive charge under normal cellular conditions. Large amounts of lysine and arginine contribute to the

Table 2.3 The 20 common amino acids grouped by nature of side-chain at pH 6 to 7

A. NONPOLAR, HYDROPHOBIC	
Alanine	Proline
Valine	Phenylalanine
Leucine	Tryptophan
Isoleucine	Methionine

B. POLAR, UNCHARGED (NEUTRAL)	
Glycine*	Tyrosine
Serine	Asparagine
Threonine	Glutamine
Cysteine	

C. POLAR, NEGATIVELY CHARGED (ACIDIC)	
Aspartic acid	Glutamic acid

D. POLAR, POSITIVELY CHARGED (BASIC)	
Lysine	Histidine
Arginine	

* Sometimes considered nonpolar.

Figure 2.28
The 20 amino acids specified by the genetic code, arranged to show the α-carbon atom with its variable R groups and common residues at the other three valency positions.

Nonpolar R group			Uncharged polar R group			Positively charged[*] polar R group		

Alanine

$$CH_3 - \overset{\overset{\text{H}}{|}}{\underset{\underset{+}{NH_3}}{C}} - COO^-$$

Glycine

$$H - \overset{\overset{\text{H}}{|}}{\underset{\underset{+}{NH_3}}{C}} - COO^-$$

Lysine

$$H_3N^+ - CH_2 - CH_2 - CH_2 - CH_2 - \overset{\overset{\text{H}}{|}}{\underset{\underset{+}{NH_3}}{C}} - COO^-$$

Valine

$$\overset{CH_3}{\underset{CH_3}{\diagdown}} CH - \overset{\overset{\text{H}}{|}}{\underset{\underset{+}{NH_3}}{C}} - COO^-$$

Serine

$$HO - CH_2 - \overset{\overset{\text{H}}{|}}{\underset{\underset{+}{NH_3}}{C}} - COO^-$$

Arginine

$$H_2N - \underset{\underset{+}{NH_3}}{\overset{\|}{C}} - NH - CH_2 - CH_2 - CH_2 - \overset{\overset{\text{H}}{|}}{\underset{\underset{+}{NH_3}}{C}} - COO^-$$

Leucine

$$\overset{CH_3}{\underset{CH_3}{\diagdown}} CH - CH_2 - \overset{\overset{\text{H}}{|}}{\underset{\underset{+}{NH_3}}{C}} - COO^-$$

Threonine

$$CH_3 - \underset{\underset{OH}{|}}{CH} - \overset{\overset{\text{H}}{|}}{\underset{\underset{+}{NH_3}}{C}} - COO^-$$

Histidine

$$HC = C - CH_2 - \overset{\overset{\text{H}}{|}}{\underset{\underset{+}{NH_3}}{C}} - COO^-$$

Isoleucine

$$CH_3 - CH_2 - \underset{\underset{CH_3}{|}}{CH} - \overset{\overset{\text{H}}{|}}{\underset{\underset{+}{NH_3}}{C}} - COO^-$$

Cysteine

$$HS - CH_2 - \overset{\overset{\text{H}}{|}}{\underset{\underset{+}{NH_3}}{C}} - COO^-$$

						Negatively charged[*] polar R group		

Proline

$$C - COO^-$$

Tyrosine

$$HO - \diagup\!\diagdown - CH_2 - \overset{\overset{\text{H}}{|}}{\underset{\underset{+}{NH_3}}{C}} - COO^-$$

Aspartic acid

$$\overset{O^-}{\underset{O}{\diagdown}} C - CH_2 - \overset{\overset{\text{H}}{|}}{\underset{\underset{+}{NH_3}}{C}} - COO^-$$

Phenylalanine

$$\diagup\!\diagdown - CH_2 - \overset{\overset{\text{H}}{|}}{\underset{\underset{+}{NH_3}}{C}} - COO^-$$

Asparagine

$$\overset{NH_2}{\underset{O}{\diagdown}} C - CH_2 - \overset{\overset{\text{H}}{|}}{\underset{\underset{+}{NH_3}}{C}} - COO^-$$

Glutamic acid

$$\overset{O^-}{\underset{O}{\diagdown}} C - CH_2 - CH_2 - \overset{\overset{\text{H}}{|}}{\underset{\underset{+}{NH_3}}{C}} - COO^-$$

Tryptophan

$$C - CH_2 - \overset{\overset{\text{H}}{|}}{\underset{\underset{+}{NH_3}}{C}} - COO^-$$

Glutamine

$$\overset{NH_2}{\underset{O}{\diagdown}} C - CH_2 - CH_2 - \overset{\overset{\text{H}}{|}}{\underset{\underset{+}{NH_3}}{C}} - COO^-$$

[*] at pH 6.0–7.0.

Methionine

$$CH_3 - S - CH_2 - CH_2 - \overset{\overset{\text{H}}{|}}{\underset{\underset{+}{NH_3}}{C}} - COO^-$$

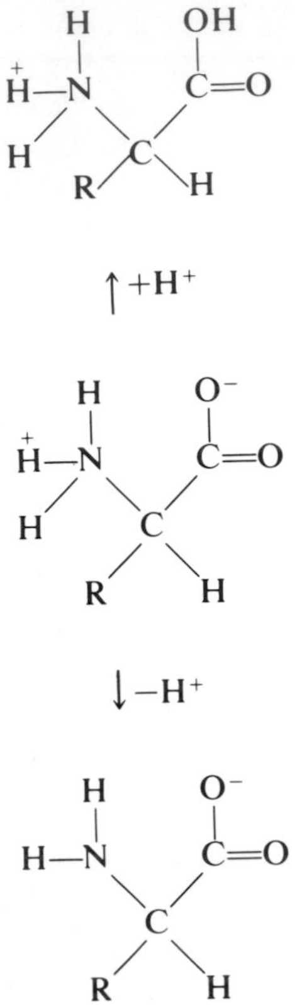

histone positive charge and contribute to chromosome structure since positively charged histones readily combine with negatively charged DNA to form nucleoprotein complexes of a highly stable nature.

Polypeptides

The formation of a peptide bond is part of the extremely complex sequence of events that take place during protein synthesis at the ribosomes. These processes are discussed at some length in Chapter 6. A *dipeptide* is formed when two amino acids are joined by amide linkage, a *tripeptide* involves three amino acids, and a **polypeptide** contains a variable number of amino acid monomers, perhaps as many as 1000. The same peptide linkage that joins two amino acids is responsible for the addition of each adjoining amino acid in the linear chain (Fig. 2.32). Most polypeptides are long-chain molecules, but some biologically important peptides may have only 8–10 amino acid units in the molecule. The neurohormone *oxytocin* consists of only 8 amino acids. This molecule influences muscle contractions during the labor stages of birth and in milk flow during suckling in mammals.

Proteins may be composed of one or more polypeptide chains held together by various forces in the functional protein molecule. In some cases the individual polypeptides are the same, as in the enzyme *phosphorylase* that acts in storage and degradation of

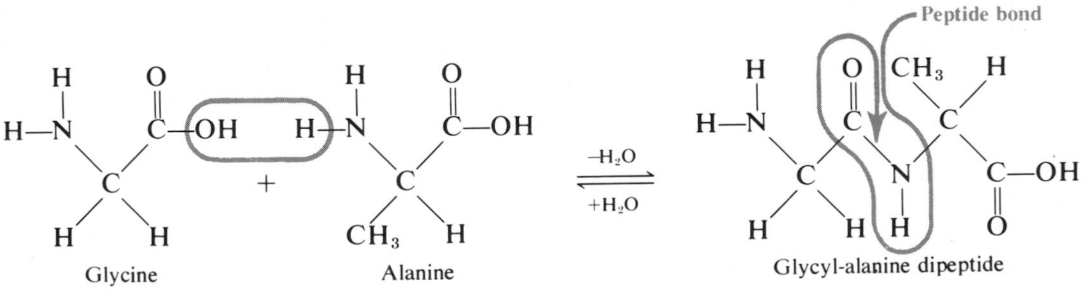

Figure 2.29
Amino acids have "zwitterion" properties (groups with positive and negative charges are present simultaneously) and can neutralize added H^+ ions in acid solution or OH^- ions in basic solution.

Figure 2.30
Amide linkage formation between adjacent amino acids in a peptide occurs as the carboxyl group of one amino acid joins with the amino group of its neighbor amino acid in a dehydration reaction.

glycogen. Other proteins may be composed of two or more kinds of polypeptides; the hormone insulin is an example (Fig. 2.33). While proteins may have molecular weights ranging into the millions of daltons, these very large molecules usually consist of individual polypeptides with molecular weights in an average range of 15,000 to 100,000 daltons. Insulin is an unusually small protein, with a molecular weight of about 6000 daltons, yet it is constituted of two polypeptide chains. The enzyme ribonuclease is larger, with a molecular weight of 13,700 daltons, but contains only one polypeptide chain of 124 amino acids. Insulin and ribonuclease were the first two proteins whose amino acid composition and exact sequence were described in the 1950s.

Protein Structure

Each kind of polypeptide consists of a unique sequence of amino acids, the primary structure. This **primary structure** is important for at least two reasons. First, the primary structure determines the three-dimensional conformation of the protein and in that way, the cellular role of the molecule. Second, the primary structure of a protein is a co-linear translation of the sequence of nucleotides in DNA and therefore provides crucial information about genetic input to protein synthesis.

These two features of primary structure can be illustrated by examination of one of the polypeptide chains in hemoglobin. The kinds and sequence of

the 146 amino acids of the β-chain of hemoglobin in humans are identical in normal and sickle-cell mutant varieties of the molecule except for one residue. The glutamic acid residue at position number 6 in non-mutant forms is replaced by a valine residue in people with the blood disease of sickle-cell anemia (Fig. 2.34). This mutant variant of hemoglobin is determined by a recessive gene. Even though the other 145 amino acids remain unaltered in the β-chain and the other polypeptides are also identical, the one mutant residue in sickle-cell hemoglobin causes severe effects. The substitution of this amino acid causes changes in the interactions between amino acids in the molecule and alters the shape of the protein. The change in molecular shape causes hampered binding between the heme group in the protein and molecular oxygen in the bloodstream. It is virtually certain that sickle cell anemia is caused by a mutational change in *one* nucleotide codon which leads to the substitution of *one* amino acid, and the crippling effects of sickle-cell anemia.

This sort of drastic effect from one amino acid substitution is not typical of most proteins. For example, as many as 40 amino acids out of 104 may be different in yeast and human *cytochrome c,* a respiratory protein, yet the protein functions equally well in both species. The amino acid substitutions in this case do not involve the critical regions of the molecule that are responsible for cytochrome *c* function in respiration. These two examples illustrate the significance of the three-dimensional conformation of pro-

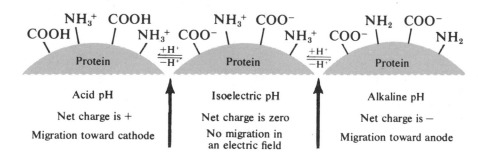

Figure 2.31
Migration of proteins in an electrical field depends on the net charge of the molecule. The net charge is zero at the isoelectric pH, even though the number of positive and negative charges is at the maximum for the molecule.

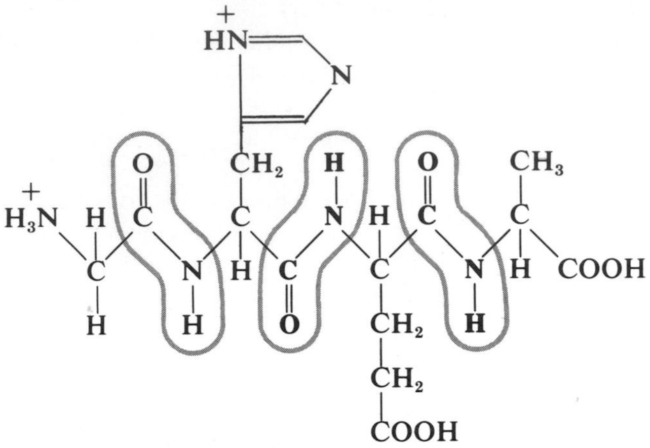

glycyl—histidyl—glutamyl—alanine
(at pH 7)

Figure 2.32
The amide link or peptide bond is the same throughout the length of a peptide whether there are two or two hundred amino acids in the molecule.

teins in their physiological activity, and point out that protein shape depends on interactions between some of the amino acid residues but not necessarily all the amino acids in a molecule.

The standard dimensions of the backbone of a polypeptide chain are determined by the bond lengths and bond angles (Fig. 2.35). The restrictions imposed by the zigzag, rigid polypeptide backbone, whose amide groups are planar, lead to restrictions on the manner in which this linear chain can fold into three-dimensional structures. From physical assays it is clear that proteins are rigid, compact molecules of relatively short length and must, therefore, be folded since the molecule is much shorter than we would expect from the lengths of the constituent polypeptides. The interactions between neighbor amino acid residues contribute to the **secondary structure** of polypeptide chains; interactions between residues at some distance from each other in a chain contribute to **tertiary structure.** The three-dimensional character of

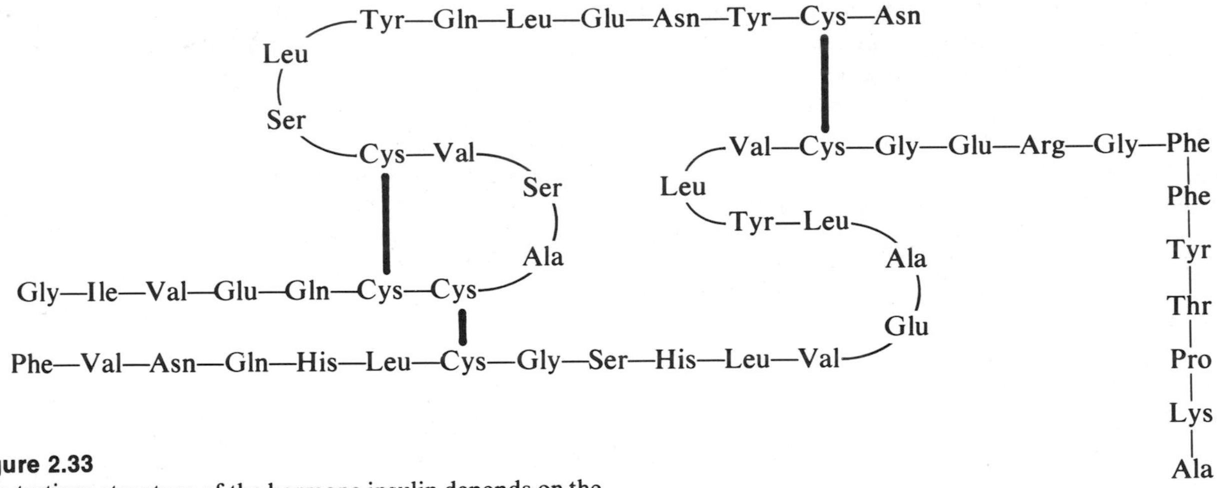

Figure 2.33
The tertiary structure of the hormone insulin depends on the formation of disulfide (—S—S—) bridges that form at specific regions by interaction between sulfhydryl (—SH) groups of amino acids at some distance from one another in one or both polypeptide chains of the protein.

Normal hemoglobin

glutamic glutamic
valine—histidine—leucine—threonine—proline—acid — acid—

Sickle-cell hemoglobin

glutamic
valine—histidine—leucine—threonine—proline—**valine**—acid—

↑
Amino acid
no. 6

Figure 2.34
The only difference in primary structure between normal and sickle-cell hemoglobin is the particular amino acid at position number 6 in the β polypeptide chain composed of a linear sequence of 146 amino acids.

a polypeptide results from both these spatial considerations.

Fibrous proteins exhibit much greater regularity of secondary structure, leading to highly ordered tertiary structures of these principal building materials in organisms. Similar secondary structures also occur in globular proteins, such as enzymes and antibodies, but there is usually a mixture of ordered secondary structure and unordered **random coil** regions in the polypeptide, all of which contributes to enhanced folding and produces a globular rather than a fibrous shape (Fig. 2.36).

A principal mode of secondary structure, called the **α-helix,** was first postulated by Linus Pauling and Robert Corey in 1951 (Fig. 2.37). This is the kind of secondary structure typical of α-keratin and of many globular proteins in at least some portion of their polypeptide chain length. The α-helix is extremely stable in the watery environment of the cell since each peptide linkage of the chain participates in hydrogen bonding. Hydrogen bonds are about 30 times weaker than covalent bonds, however, which permits greater flexibility in molecule modulation in chemical reactions. The α-helix shows a regularly repeated structure because the angles of rotation about the

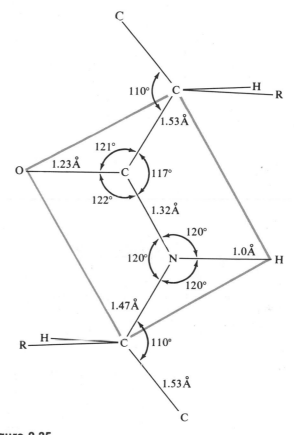

Figure 2.35
Bond angles and bond lengths of the peptide linkage in a polypeptide chain are shown. The plane is somewhat rigid because there are some double-bond characteristics to the C — N atoms that lie at its center.

bond formed by the α-carbon and the carbon of one amino acid group and the same α-carbon — nitrogen bond of the adjacent amino acid are essentially invariant. The chain falls into a helix because there are the same twists at every α-carbon in the polypeptide backbone. The invariance of this type of secondary structure is offset by regions of unordered random coiling and by secondary structure of the **pleated sheet** type (see Fig. 2.36).

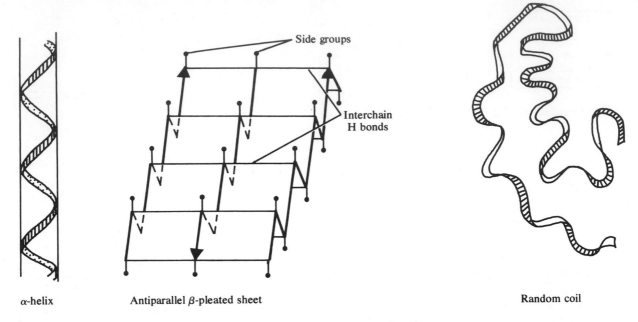

α-helix   Antiparallel β-pleated sheet   Random coil

Figure 2.36

Bonding between all adjacent residues lends great regularity of secondary structure to the α-helix and β-pleated sheet, whereas a random coil region in a protein is produced by lack of bonding or irregular bonding between neighboring residues. All three types of secondary structure may occur in regions of globular protein molecules, but α-helix and β-pleated sheet arrangements are typically found in fibrous proteins.

The particular tertiary structure assumed by polypeptides depends on the nature and locations of the amino acid side-groups. When the amino acid cysteine is present, it can form covalent cross-links with other cysteine residues in the same or an adjacent polypeptide chain of a protein molecule. These cross-links lead to disruption of the regular α-helix formation in the region of the cysteine residues and to folding of polypeptides in particular tertiary structures, as in the hormone insulin (see Fig. 2.33).

Tertiary structures, which result from interactions between amino acid side-groups, vary according to the pH of the medium and the aqueous or nonaqueous nature of the surroundings, as well as binding with metals and other constituents that are present. In the aqueous cellular medium the hydrophobic, nonpolar side-groups tend to be clustered in the interior of the globular protein, while the polar groups are on the outside surface. **Hydrophobic** or **apolar interactions** thus serve to sequester nonpolar groups and hold the molecule together in a watery medium. Various kinds of ionic interactions characterize associations among the charged and uncharged polar side-groups on the outside of the protein. The variety of side-groups and interactions involving them contribute to protein shape and to the particular physiological roles carried out by molecules of specific conformations. We will discuss some of these features in relation to enzyme activities in Chapter 3.

When two or more polypeptides are present in a protein molecule, their spatial organization imposes a **quaternary structure** on the compound (Fig. 2.38). Disruption of quaternary structure or disorganization of the molecule leads to dysfunction. Polypeptide

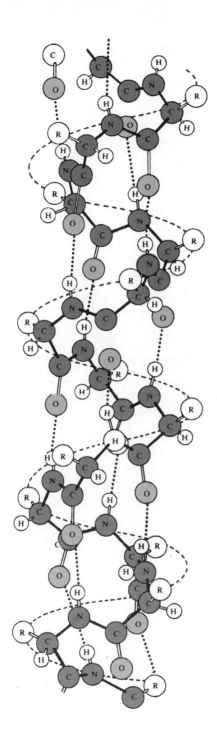

subunits in a particular protein may be identical or different or some combination of the two. The hemoglobin protein contains two α-chains that are slightly different from each other and two β-chains that also differ somewhat from each other. The tetramer of four polypeptide chains constitutes the active hemoglobin molecule. Studies of protein subunits are in active progress in many laboratories. We will consider some of these studies in Chapters 3 and 9, in relation to the construction and activity of certain enzymes of carbohydrate metabolism.

The ways in which the variety of structural and regulatory proteins influence the spectrum of cellular activities are discussed in all the chapters of this book. The uniqueness of living chemistry is derived from the protein constituents of cells. The stunning diversity of cellular activities and the fine controls that regulate these activities are due to the properties of proteins. No study of the biology of the cell can be fully entertained without including a consideration of the behavior of the proteins that underwrite the structure or activity under investigation.

NUCLEOTIDES AND NUCLEIC ACIDS

Mononucleotides are involved in two major cellular functions: (1) they are monomeric units from which polymers of DNA and RNA are constructed, and (2) they act as agents for transfer of hydrogens and electrons in a variety of metabolic interactions. A mononucleotide is made up of three components: a nitrogenous base, a pentose sugar, and a phosphate residue derived from phosphoric acid. When the phosphate group is removed, the remaining two components constitute a **nucleoside**. **Nucleotides** are therefore called **nucleoside phosphates**, and are more

Figure 2.37
Hydrogen bonding between each peptide linkage of a polypeptide chain leads to the highly ordered secondary structure of the α-helix conformation.

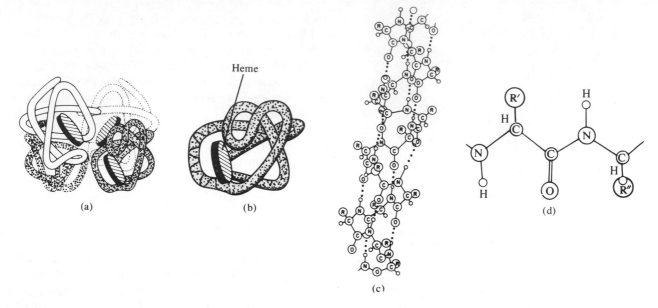

Figure 2.38
The four levels of structural organization in proteins as exemplified by the hemoglobin molecule: (a) *quaternary,* aggregation of two or more polypeptide chains (four in hemoglobin); (b) *tertiary,* folding of a polypeptide chain in space produces a globular three-dimensional shape; (c) *secondary,* neighboring interactions as in the α-helix portion of a polypeptide chain; and (d) *primary,* the sequence of amino acids joined by peptide bonds shown here.

specifically designated as mono-, di-, and triphosphates depending on the number of residues in the molecule. The various synonyms and conventions used to refer to nucleosides and nucleotides are given in Table 2.4.

The nitrogenous bases commonly found in nucleic acids and their nucleotide building blocks are derivatives of the heterocyclic compounds **purine** and **pyrimidine** (Fig. 2.39). The commonly occurring purines **adenine** and **guanine** are found in both DNA and RNA, as is the pyrimidine compound **cytosine.** The

Table 2.4 Nomenclature of nucleic acids and their constituent units

BASE	NUCLEOSIDE	NUCLEOTIDE	NUCLEIC ACID
Purines:			
Adenine	Adenosine	Adenylic acid	RNA
	Deoxyadenosine	Deoxyadenylic acid	DNA
Guanine	Guanosine	Guanylic acid	RNA
	Deoxyguanosine	Deoxyguanylic acid	DNA
Pyrimidines:			
Cytosine	Cytidine	Cytidylic acid	RNA
	Deoxycytidine	Deoxycytidylic acid	DNA
Thymine	Thymidine	Thymidylic acid	DNA
Uracil	Uridine	Uridylic acid	RNA

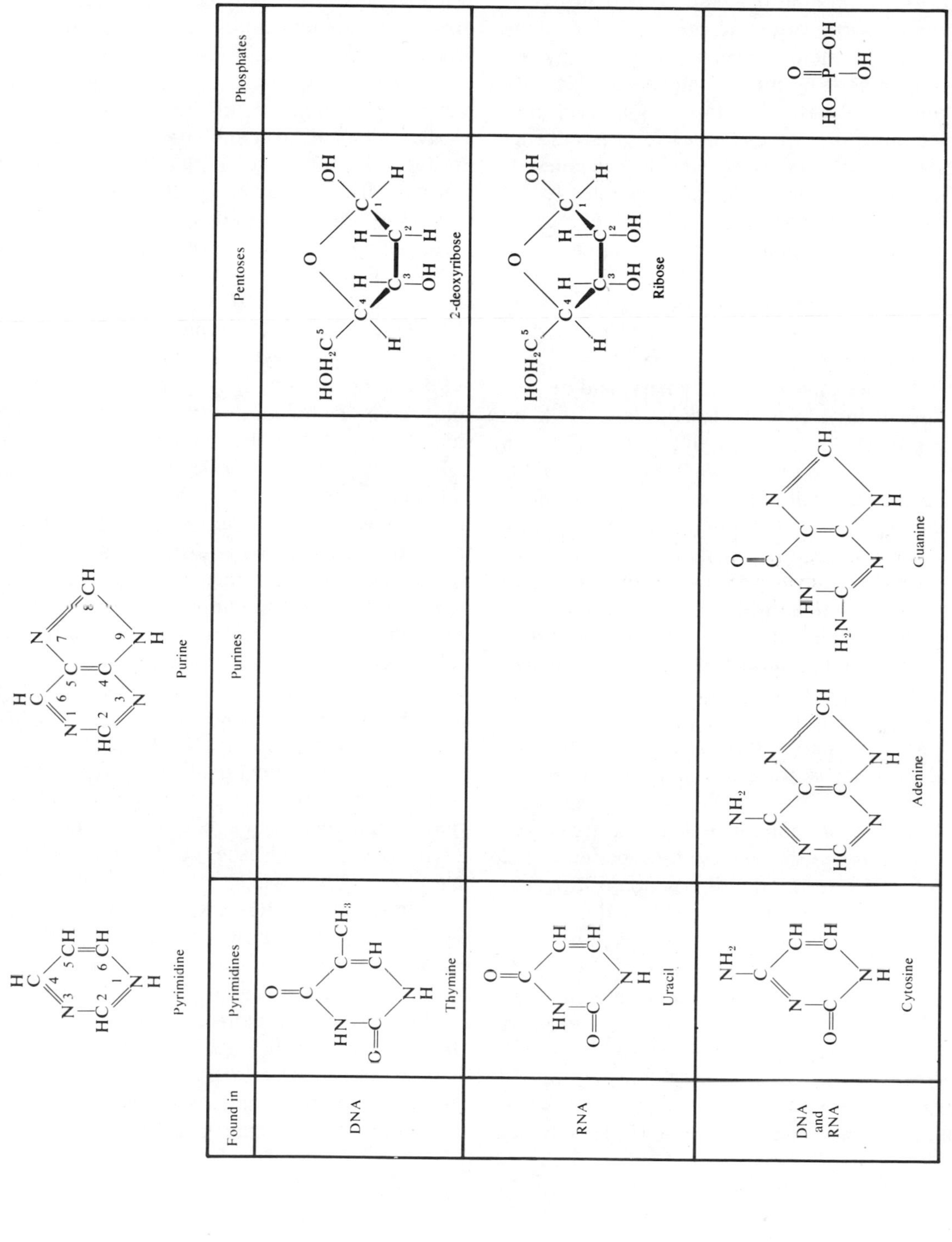

Figure 2.39
The building blocks of DNA and RNA.

second kind of pyrimidine in DNA is **thymine,** while its demethylated form, **uracil,** occurs in RNA. Since each kind of nucleic acid contains one unique pyrimidine, it is convenient to study synthesis and activity of DNA and RNA using isotopically-labeled precursors containing one or the other of these nitrogenous bases. Usually the nucleosides **uridine** or **thymidine,** or their nucleotide forms, are added to the biological system under study.

Nucleosides may include either D-**ribose** (ribonucleosides) or **2-deoxy-D-ribose** (deoxyribonucleosides). The furanose form of the sugar is bonded by β-glycosidic linkage to nitrogen atom-1 of pyrimidine or to nitrogen atom-9 of the purine residue. Ribose occurs in RNA monomers and polymers, and deoxyribose is found in DNA molecules and building blocks. The presence or absence of a hydroxyl group on carbon atom-2 of the sugar leads to profound differences in stabilities, pairing potential, and functions of DNA and RNA. Nucleotides have a phosphoric acid group esterified to one of the free hydroxyls of the sugar. Free hydroxyls occur at positions 3′ and 5′ in deoxyribose and at positions 2′, 3′, and 5′ in ribose. All these types of nucleotides are found in nature, but 5′-ribonucleotides and 5′-deoxyribonucleotides occur most commonly. The nucleoside 5′-diphosphates and 5′-triphosphates generally occur as complexes with divalent cations such as Mg^{2+} and Ca^{2+}, but the former is more prevalent in cells.

Polynucleotides of both the DNA and RNA varieties are built from mononucleotides that are linked covalently via *phosphodiester bridges* between the 3′ position of one unit and the 5′ position of the next (Fig. 2.40). Since there is no restriction on the vertical sequence of adjacent mononucleotides linked by these 3′,5′-phosphodiester bridges in either DNA or RNA, a considerable variety of molecules is possible even though only 4 kinds of mononucleotide occur in RNA (rAMP, rGMP, rCMP, and rUMP) and in DNA (dAMP, dGMP, dCMP, and dTMP). The theoretical variety is calculated as 4^n, where 4 is the number of different kinds of nucleotides, and n is the number of monomers in the polymer. For a molecule made of only 75 monomeric units, as in some of the smallest RNAs, there may be 4^{75} different arrangements of the constituent units, and each arrangement theoretically constitutes a molecule of different specificity. Where an average gene may include about 500 nucleotides in a DNA sequence, 4^{500} different sequences are theoretically possible and therefore, there may be that many different and specific genes. The astronomically high numbers of possible genes underlie the uniqueness of genetic materials and can easily account for all past and present life forms.

Synthesis of DNA and RNA requires specific enzymes called **polymerases,** energy sources such as ATP (adenosine 5′-triphosphate), Mg^{2+} ions, and all four kinds of nucleotides in their *tri*phosphate forms. Spontaneous assemblies involving any and all available precursors can produce short oligomer or longer polymer fragments, but faithful *replication* of genetic material requires a nucleic acid **template** that provides the molecular guideline for the synthesis of new molecules that are identical to the originals. DNA acts as a template for the synthesis of new DNA and for the manufacture of cellular RNAs, the latter process being referred to as **transcription. DNA-dependent DNA polymerases** catalyze DNA synthesis while **DNA-dependent RNA polymerases** catalyze the transcription of RNA from DNA templates. The qualifying term "DNA-dependent" (or "RNA-dependent") indicates the nature of the template used by the polymerase synthesizing nucleic acid. Except for certain RNA-containing viruses, **RNA-dependent RNA polymerases** are unknown.

In some of the RNA viruses and in certain kinds of eukaryotic cells, however, there is an **RNA-dependent DNA polymerase** (also called **"reverse transcriptase"**). In these systems the RNA template is first copied into DNA molecules by this enzyme, after which the newly formed DNA molecules guide the manufacture of new RNA strands. The RNA tumor viruses containing this polymerase package their new genetic RNA strands in the mature particle after

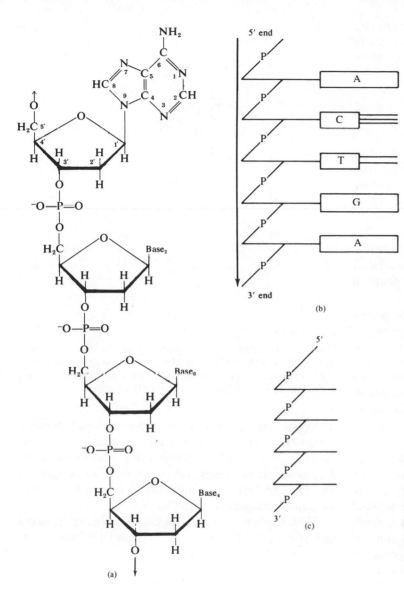

(a)

(b)

(c)

Figure 2.40
Regions of nucleic acid showing the 3',5'-phosphodiester links between adjacent nucleotides in a polynucleotide chain as (a) conventional formulae; (b) a shorthand notation showing whole mononucleotide units in a chain; and (c) a more abbreviated shorthand notation that emphasizes the 5' to 3' direction of one of the antiparallel chains in a duplex molecule or of a single-stranded molecule.

a somewhat circuitous synthesis route. Other RNA viruses, however, copy new RNA from existing genetic RNA without mediation of a step requiring a DNA template.

The discovery of the reverse transcriptase system led to a modification of the genetic dogma which stated that the flow of genetic information proceeded by irreversible steps from gene to intermediate to product of gene expression, or **DNA → RNA → protein.** In systems with reverse transcriptase, the notation must read DNA $\rightleftarrows$ RNA $\rightarrow$ protein. The underlying genetic themes of cellular biology are explored more thoroughly in Chapter 5 and are referred to often in many other chapters in this book.

WATER AND pH

Life depends on water. From 60 to 95 percent of the cell consists of water, and even dormant seeds and spores contain from 10 to 30 percent water. A number of important properties distinguish water from most other compounds and lead to the unique suitability for water as the medium of cellular activities. Because water molecules have a high degree of electrical polarity, they exert strong intermolecular attractions that lead to tight packing and to orienting effects in the liquid state (Fig. 2.41).

Water is a very good solvent for many organic and inorganic compounds, even when some of these are in the solid, crystalline state. Solubility of organic molecules is enhanced when groups that have the capacity to form hydrogen bonds with water are present or added. A hydrocarbon is insoluble in water, but becomes more soluble when a hydroxyl group is added to form an alcohol. Carboxyl, amino, phosphate, and some other groups increase the solubility of a compound because hydrogen bonds can form between the oxygen or other electronegative atoms of the compound and the hydrogens of water, or the oxygens of water and the hydrogens of the soluble substance (Fig. 2.42). Water is an excellent solvent for electrically charged substances, such as ions of salts, since water molecules have regions of positive charge that are precisely separated from centers of negative charge.

Water can enhance the dissociation of substances such as weak acids or bases, that already exist in partially dissociated form. In addition, water itself can undergo slight dissociation to ionized components. Although conventionally expressed as $H_2O \rightleftarrows H^+ + OH^-$, naked protons (nuclei of hydrogen atoms, or

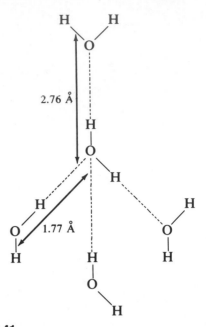

Figure 2.41
Hydrogen bonding around a water molecule in ice, showing the bond lengths involving the central oxygen atom and the oxygens of water molecules with hydrogens in two different regions of the tetrahedral arrangement.

hydrogen ions) do not occur among the small number of ionized components of water. Instead, water dissociates into H_3O^+ (hydronium ions) and OH^- ions. For practical purposes, however, this can be ignored and H^+ and OH^- ions are usually referred to as dissociation products of water.

Dissociation of water is an equilibrium process and at constant temperature this can be expressed by

$$K_{eq} = \frac{[H^+][OH^-]}{[H_2O]} \qquad [2.1]$$

where K_{eq} is the equilibrium constant and concentrations of water molecules and their ionized components are presented in moles per liter (signified by enclosure in brackets [ ]). Since the molar concentration of water in pure water is 55.5 M (number of grams of water in a liter divided by grams molecular weight

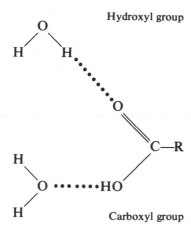

Hydroxyl group

Carboxyl group

Amino group

Figure 2.42
Hydrogen bonding between H and O atoms of H_2O with electronegative and electropositive atoms, respectively, of three kinds of residues in molecules.

of water, or 1000/18), and H^+ and OH^- ion concentrations are very low (1×10^{-7} M at 25°C), we can simplify the equilibrium constant to $55.5 \times K_{eq} =$ [H^+][OH^-]. The term $55.5 \times K_{eq}$ is called the **ion product of water** or the constant K_w.

$$K_w = [H^+][OH^-] \qquad [2.2]$$

K_w is the basis for the **pH scale,** which is a means of designating the actual concentration of H^+ ions (and thus of OH^- ions as well) in any aqueous solution in the acidity range between 1.0 M H^+ and 1.0 M OH^- (Table 2.5). The term **pH** is defined as

$$pH = -\log_{10}[H^+] \qquad [2.3]$$

It is convenient to use a logarithmic scale in determining pH values because of the wide variations in H^+ ion concentrations. The negative logarithm to the base 10 is used so that a positive scale of readings can be obtained. In a precisely neutral solution at 25°C, [H^+] = [OH^-] = 1×10^{-7} M, and the pH of such a solution is

$$pH = -\log_{10}[H^+] = -\log_{10}(10^{-7}) = 7.0 \qquad [2.4]$$

The value of pH 7.0 for a neutral solution is thus derived from the ion product of water at 25°C and not from some arbitrary basis. In an acidic solution, the pH is less than 7.0 since the H^+ ion concentration is high, whereas in an alkaline solution, the pH is more than 7.0 because the solution has a low H^+ ion con-

Table 2.5 The pH scale and the molar concentrations of H^+ and OH^- ions

H^+ ions (M)	pH	OH^- ions (M)
1.0	0	10^{-14}
0.1	1	10^{-13}
0.01	2	10^{-12}
0.001	3	10^{-11}
0.0001	4	10^{-10}
10^{-5}	5	10^{-9}
10^{-6}	6	10^{-8}
10^{-7}	7	10^{-7}
10^{-8}	8	10^{-6}
10^{-9}	9	10^{-5}
10^{-10}	10	0.0001
10^{-11}	11	0.001
10^{-12}	12	0.01
10^{-13}	13	0.1
10^{-14}	14	1.0

centration. Because the pH scale is logarithmic, there is a tenfold difference between one pH unit and the next, a 100 times difference in H^+ ion concentration between any two whole pH units, a 1000 times difference for a span of three pH units, and so forth. Measurements of H^+ ion concentration are made rapidly and routinely by using a pH meter.

Cellular activities are extremely sensitive to even slight changes in internal pH, mostly because enzyme activity is affected by H^+ ion concentration. Enzymes have maximal activity at a characteristic pH, called the **optimum pH,** and their activities decline sharply above and below this optimum value. The striking effects of pH on enzyme activity almost certainly reflect electrical changes in surface groups of the enzymes, which contribute to altered shape of the proteins and lowered reactivity.

Variations in fractions of a pH unit may be damaging or even lethal to some cells. These fluctuations in pH are modulated by powerful buffering action of coupled **proton donors** (acids) and **proton acceptors** (bases) that are present in intracellular and extracellular fluids of living organisms. Buffered systems tend to resist changes in pH when H^+ and OH^- ions are added. A major intracellular buffer is the donor—acceptor pair $H_2PO_4^-$—HPO_4^{2-}, while H_2CO_3—HCO_3^- acts as the principal buffering system in blood plasma of vertebrate species. The pH of human blood plasma is closely regulated to about pH 7.40. Irreparable damage may occur if plasma pH falls below 7.0 or rises higher than 7.8. The difference between pH 7.4 and pH 7.8 in blood reflects a change in H^+ ion concentration of only $3 \times 10^{-8} M$. The small magnitude of this change emphasizes the importance of pH-regulating mechanisms as precise modulators of acidity and alkalinity of cellular fluids.

BIOENERGETICS

Energy is broadly defined as the capacity to do work. There are different kinds or states of energy (potential, kinetic, electrical, radiant, thermal, and so on) and different kinds of work (chemical work, mechanical work, and osmotic work, for example). One form of energy can be transformed into another form, and the energy can then be applied to do work. For example, thermal energy of steam can be transformed into mechanical energy by a steam engine, and the mechanical energy can be used to perform mechanical work, which is driven by the energy input. The application, movement, and transformation of energy ultimately underlies all physical and chemical processes. The area of physical science that deals with exchanges of energy in collections of matter is known as **thermodynamics,** a term handed down from the earliest of these studies when heat was the focus of measurements. The equivalent term, applied more specifically to the study of energy transformations in living systems, is **bioenergetics.**

Thermodynamic analysis distinguishes the collection of matter, or **system,** from the **surroundings,** or all other matter in the universe apart from the system under study. The total energy content of the system in its **initial state,** before the process begins and in its **final state** of equilibrium can be measured in some cases. It is easier, however, to measure the *difference* in energy content between the initial and final states of a system, that is, the amount of energy exchanged between the system and its surroundings as the process takes place. Energy may be absorbed from or released to the surroundings, but the total energy of the universe remains constant.

The principle of the conservation of energy is defined by the **First Law** of thermodynamics: energy can be neither created nor destroyed. Since transformations of energy can take place, the First Law further implies that there is a quantitative correspondence between kinds of energy. This correspondence has been confirmed by many kinds of physical measurements.

Heat is a universal form in which energy can be transferred, and virtually every physical or chemical event is accompanied by exchange of thermal energy between the system and its surroundings. In an exothermic process, heat is lost to the surroundings, whereas in an **endothermic** process heat is absorbed by the system from its surroundings. The energy content of organic compounds can be measured as the heat of combustion (kilocalories of heat released to the surroundings as 1 mole of a substance is burned completely at the expense of molecular oxygen). The energy release (hence, initial energy content) during the combustion of fat is much greater than during the combustion of either carbohydrates or amino acids of proteins and more oxygen is required to burn fat to carbon dioxide because fat molecules are more fully hydrogenated (reduced).

Although heat serves quite well as a useful form of energy for performing work in many manufactured devices, it is not useful for biological work. Heat can do work only if there is a temperature differential through which it can act, and living cells have little or no temperature differences between their parts. Living cells utilize **free energy** to perform work.

Free Energy

The First Law states that energy is conserved, and the **Second Law** of thermodynamics defines the *direction* of a process. The Second Law states that all physical and chemical change proceeds in such a direction that the **entropy** of the universe (system + surroundings) increases to the maximum possible, at which point equilibrium exists: entropy never decreases. Entropy is generally defined as randomness or disorder and is expressed in entropy units or calories per mole degree. The universe proceeds toward *entropic doom,* or maximum chaos, according to the Second Law.

In biological systems, under the prevailing conditions of constant temperature, pressure, and volume, the quantitative relationship between change in entropy during a process and change in total energy of a system can be expressed as free energy. The change in free energy, or ΔG, is important in biochemical reactions because it is easily measured and can be used to predict the direction and equilibrium state of these

reactions. Assuming constant temperature, pressure, and volume, the First and Second Laws have been combined in a simple equation

$$\Delta G = \Delta E - T \Delta S \qquad [2.5]$$

where ΔG is the free-energy change, ΔE is the total energy change of the system, T is the absolute temperature (°Kelvin), and ΔS is the change in energy of the universe expressed in entropy units. If the equation is rearranged to form

$$\Delta E = \Delta G + T \Delta S \qquad [2.6]$$

then we see that for spontaneous reactions ΔE is the sum of the term $T \Delta S$ which is always positive and ΔG which is always a negative quantity. We can thus define ΔG as that fraction of the total energy of a system that is available to do work as the system proceeds toward equilibrium at constant temperature, pressure, and volume. Just as entropy proceeds toward a maximum, free energy declines to a minimum during an event, and predictions can be made from these readily measured free-energy changes. It is equally clear that living systems grow at the expense of their surroundings, and in terms of energetics, the increase in order of a growing organism leads to increasing disorder in its surroundings.

Measurement of Free Energy

In a reaction such as $A \rightarrow B$, which proceeds toward equilibrium, or $A \leftrightarrows B$, the equilibrium constant, K_{eq}, is determined from the concentration of the reaction product divided by the concentration of the initial reactant

$$K_{eq} = \frac{[B]}{[A]} \qquad [2.7]$$

In reactions that have more components, such as $aA + bB \rightarrow cC + dD$ which proceeds to $aA + bB \rightleftarrows cC + dD$ (where a, b, c, and d are the number of molecules or concentration of A, B, C, and D, respectively), the equilibrium constant is the product of the concentration of all the reaction products divided by the product of all reactants, at equilibrium, or

$$K_{eq} = \frac{[C]^c[D]^d}{[A]^a[B]^b} \qquad [2.8]$$

These concepts then lead us to predict that the free-energy change of a chemical reaction must be some mathematical function of its equilibrium constant, as shown by the following series of equations. When the reaction is at equilibrium there is minimum free energy and no further change can take place, that is, ΔG is 0.0. The relationship of ΔG to ΔG^0 is

$$\Delta G = \Delta G^0 + RT \ln K_{eq} \qquad [2.9]$$

where R is the gas constant (1.987 cal/mole °), T is the absolute temperature, $\ln K_{eq}$ is the natural logarithm of the equilibrium constant, and ΔG^0 is the **standard free-energy change,** or a thermodynamic constant representing the difference between the standard free energy of the reactants and the standard free energy of the products. From this equation we can derive the relationship between the standard free-energy change and the equilibrium constant

$$\Delta G^0 = -RT \ln K_{eq} \qquad [2.10]$$

or,

$$\Delta G^0 = -2.303 \, RT \log_{10} K_{eq} \qquad [2.11]$$

When the reaction occurs under standard conditions of temperature (usually 25°C), pressure (1 atmosphere), and concentration (1.0 M), then the superscript 0 is used (as in ΔG^o). The reference state in biological systems is pH 7.0 (pH 0.0 is used in physical systems), and this is indicated by a superscript prime mark (as in $\Delta G^{0\prime}$).

The relationship between equilibrium constant and standard free-energy change is shown in Table 2.6. When K_{eq} is more than 1.0, $\Delta G^{0\prime}$ is a negative value and the reaction proceeds with a decline in free energy. When K_{eq} is less than 1.0, $\Delta G^{0\prime}$ is a positive value and energy must be put into the system for the

Table 2.6 Relationship between the equilibrium constant and the standard free-energy change at 25°C and pH 7.0

K'_{eq}	$\Delta G^{0\prime}$ (kcal/mole)
0.001	+4.09
0.01	+2.73
0.1	+1.36
1.0	0
10.0	−1.36
100.0	−2.73
1000.0	−4.09

process to proceed. Reactions with a negative standard free-energy change are called **exergonic.** those with a positive standard free-energy change are **endergonic.** Exergonic reactions run "downhill" and release energy to the surroundings; endergonic reactions run "uphill" and require energy input to proceed.

The standard free-energy change, or $\Delta G^{0\prime}$, is rarely realized in living cells since 1.0 M concentrations generally do not exist. The actual free-energy change, or $\Delta G'$, is measured and varies according to the concentrations of reactants and products. But, for purposes of quantitative comparisons and maximum predictive value in analyzing biochemical reactions, the standard free-energy change is used to identify reactions and events. Since the equilibrium constant can be measured analytically, the K'_{eq} of a reaction permits us to calculate the standard free-energy change, or $\Delta G^{0\prime}$, of any chemical reaction according to the magnitude of its K'_{eq}.

The following example will illustrate these points: During the breakdown of glycogen to lactic acid in the cell, glucose 1-phosphate is converted to glucose 6-phosphate in the presence of a specific enzyme. If enzyme is added to 0.020 M glucose 1-phosphate in a medium at pH 7.0 and 25°C, chemical analysis at equilibrium shows that the concentration of glucose 6-phosphate has risen from zero to 0.019 M while glucose 1-phosphate has decreased to 0.001 M. The equilibrium constant is then

$$K'_{eq} = \frac{[\text{glucose 6-phosphate}]}{[\text{glucose 1-phosphate}]} = \frac{0.019}{0.001} = 19 \qquad [2.12]$$

and the standard free-energy change can be calculated as

$$\begin{aligned} \Delta G^{0'} &= -RT \ln K'_{eq} \\ &= -1.987 \times 298 \times \ln 19 \\ &= -1.987 \times 298 \times 2.303 \times \log_{10} 19 \qquad [2.13] \\ &= -1363 \times 1.28 \\ \Delta G^{0'} &= -1745 \text{ cal/mole, or } -1.745 \text{ kcal/mole} \end{aligned}$$

If the concentration of each compound were maintained at 1.0 M concentration at 25°C (298°K) and pH 7.0, then a decline in free energy of 1.745 kcal would accompany the conversion of 1 mole of glucose 1-phosphate to 1 mole of glucose 6-phosphate at equilibrium under these standard conditions.

The actual concentrations of reactant(s) and product(s) often are different when measured in living systems than when measured in a test tube analysis of the same reaction. If the *in vivo* concentrations of glucose 6-phosphate and glucose 1-phosphate were found to be 0.003 M and 0.015 M, respectively, then the actual free-energy change would be

$$\begin{aligned} \Delta G' &= \Delta G^{0'} + RT \ln \frac{[0.015]}{[0.003]} \\ &= -1745 + (1.987 \times 298 \times 2.303 \log_{10} 5) \qquad [2.14] \\ &= -1745 + 960 \\ \Delta G' &= -785 \text{ cal/mole or } -0.785 \text{ kcal/mole} \end{aligned}$$

Even though the actual free-energy change under cellular conditions may be different from the calculated standard free-energy change for a reaction, quantitative comparisons of the energetics of chemical reactions make it mandatory to consider standard free-energy changes in the interests of consistency.

In additon to the problem of equating free-energy change in living cells to standard free-energy changes, there is the additional difficulty of relating the **open system,** which exists in cells, to the **closed system** normally accepted as the classical thermodynamic situation. In a closed system there is no exchange of matter between the system and its surroundings at equilibrium, while an open system does exchange energy and matter with its surroundings.

Furthermore, cells do not exist in states of thermodynamic equilibrium. Many interacting events usually take place at any one time, and thermodynamically possible reactions may occur slowly or not at all unless catalyzed by enzymes in biological systems. Living cells exist in different **steady states** in which the rate of input equals the rate of output of matter at each given moment, but a different pair of rates may lead to a different steady state from moment to moment.

Classical thermodynamic theory has been modified to accommodate analysis of the nonequilibrium open systems that characterize living organisms. The significant features of open systems existing in steady states are:

1. These systems can perform work precisely because they do not attain equilibrium; equilibrium is a state of no work.
2. Only systems that are away from equilibrium can be subject to control and regulation of their activities.
3. The rate of entropy production is minimal in the steady state of an open system.

The steady state is thus the more *orderly* state and succumbs least to entropic doom. In other words, the outcome of the Second Law is minimized because the maintenance of a steady state produces entropy at a minimum rate.

The ATP System and the Flow of Energy

The transfer of energy in biological systems occurs through links between energy-yielding and energy-requiring reactions mediated by substances that donate and accept chemical-bond energy or electrons. The major link in transfer of chemical energy is the ATP-ADP system, first described in 1940 by Fritz

Lipman. Energy released from fuel molecules is recovered by phosphorylation of ADP to yield ATP in a reaction coupled to fuel breakdown. The energy-rich ATP formed in this way then transfers its energy to cellular activities that require energy to proceed. Examples of such activities are biosynthesis, active transport of molecules and ions against a concentration gradient contraction, and so forth.

The transfer of a "high-energy phosphate bond" from ATP energizes the biochemical reaction or event and simultaneously regenerates ADP for another cycle of phosphorylation. The transfers actually involve **phosphoryl** rather than **phosphate** groups, that

$$
\text{is, } -\overset{\displaystyle O^-}{\underset{\displaystyle O^-}{P}}\!\!=\!O \text{ rather than } -O-\overset{\displaystyle O^-}{\underset{\displaystyle O^-}{P}}\!\!=\!O, \text{ but it is common}
$$

to speak of phosphate-bond transfer nonetheless.

Bond energy refers to the *difference* in energy content between reactants and products and can thus be designated by $\Delta G^{0\prime}$, which stipulates the difference between standard free-energy of reactants in the initial state and standard free-energy of products in the final equilibrium state under idealized conditions. In this sense then, phosphoryl bond energy reflects the relative free energy content of reactants and products and does not refer to energy localized to the chemical bond itself. These cautions should be kept in mind because it may be misleading when discussions of the subject refer to bond-energy, the widely accepted colloquial expression for this feature of energetics.

An examination of standard free-energy change values for the hydrolysis of a number of important phosphate compounds reveals a gradation rather than a sharp division between "high-energy" and "low-energy" molecules (Table 2.7). More critical than this inaccuracy in the theory of energy transfer is the observation, from these data, that such compounds occupy relative positions on the thermodynamic scale, in relation to transfer of phosphate-bond energy

Table 2.7 Standard free energy of hydrolysis of some phosphorylated compounds

COMPOUND	$\Delta G^{0\prime}$ (kcal/mole)	DIRECTION OF TRANSFER OF PHOSPHORYL GROUP
Phosphoenolpyruvate	−14.8	
1,3-diphosphoglycerate	−11.8	
Phosphocreatine	−10.3	
Acetyl phosphate	−10.1	
Phosphoarginine	− 7.7	
ATP	− 7.3	
Glucose 1-phosphate	− 5.0	
Fructose 6-phosphate	− 3.8	
Glucose 6-phosphate	− 3.3	
3-phosphoglycerate	− 2.4	
Glycerol 3-phosphate	− 2.2	

from one reactant to another in metabolism. ATP has an intermediate value and is thus significant as the thermodynamically intermediate carrier of phosphoryl groups which it takes from higher energy phosphate compounds and gives to acceptor molecules lower on the thermodynamic scale.

Two classes of phosphorylated compounds have substantially more negative (higher) standard free-energy change than ATP. These compounds are often referred to as **high-energy phosphates.** One class includes phosphate compounds produced during enzyme-catalyzed breakdown of fuel molecules; the second class includes compounds that act as storage reservoirs of phosphate-bond energy. Two of the most important members of the first class of breakdown molecules are 1,3-diphosphoglycerate and phosphoenolpyruvate, compounds which form during fermentation of glucose. In the cell, one phosphoryl group is transferred from 1,3-diphosphoglycerate to ADP, forming ATP along with 3-glycerophosphate. Specific enzymes catalyze this reaction as well as the formation of phosphoenolpyruvate and its donation of a phosphoryl group to ADP to form ATP (see Chapter 7).

The second class of high-energy phosphate compounds includes **phosphagens,** or reservoirs of phosphate-bond energy, such as **phosphocreatine** of most vertebrates and **phosphoarginine** found in many invertebrates. These molecules are derivatives of guanine in which the N atom is bonded directly to the phosphorus atom. High-energy phosphate groups are designated as $\sim P$, so that phosphocreatine may be symbolized as $Cr\sim P$, ATP as $A—R—P\sim P\sim P$, ADP as $A—R—P\sim P$, and so forth. These notations also indicate that the phosphoryl group bonded to ribose in ADP or ATP has a low energy value. Only the terminal phosphoryl group of ADP releases substantial energy on being removed, but both terminal phosphates are high-energy in ATP. A phosphoryl group transfer would occur as follows

$$A—R—\overset{\overset{O^-}{|}}{\underset{\underset{O}{||}}{P}}—O\sim\overset{\overset{O^-}{|}}{\underset{\underset{O}{||}}{P}}—O\sim\overset{\overset{O^-}{|}}{\underset{\underset{O}{||}}{P}}—O^- + ROH \rightarrow$$

ATP

[2.15]

$$A—R—\overset{\overset{O^-}{|}}{\underset{\underset{O}{||}}{P}}—O\cdot\cdot\overset{\overset{O^-}{|}}{\underset{\underset{O}{||}}{P}}\quad O\ |\ R\quad O—\overset{\overset{O^-}{|}}{\underset{\underset{O}{||}}{P}}—O^-$$

ADP

Hydrolysis of ATP or of ADP yields -7.3 kcal/mole, while hydrolysis of adenosine monophosphate (AMP) yields considerably less energy

$$\Delta G^{0\prime}$$

	$\Delta G^{0\prime}$	
$ATP + HOH \rightarrow ADP + H_3PO_4$	-7.3 kcal	[2.16]
$ADP + HOH \rightarrow AMP + H_3PO_4$	-7.3 kcal	[2.17]
$AMP + HOH \rightarrow$ adenosine $+ H_3PO_4$	-3.4 kcal	[2.18]

The bonds between adjacent phosphate groups are acid anhydride linkages, whereas an ester linkage joins phosphate to ribose in AMP, ADP, and ATP molecules (Fig. 2.43). The standard free-energy change calculated for ATP varies according to temperature, concentration, and other factors. Although a value of -7.3 kcal for $\Delta G^{0\prime}$ has been calculated for ATP in some test systems, an average value of -7.0 kcal is often used. Values as high as -12.5 kcal are believed to be more typical of living cells, but undoubtedly $\Delta G'$ varies among cells and from one time to another.

ATP as a Common Intermediate in Biological Energy Transfer

There are two unique features of ATP in relation to energy flow in cells:

1. The standard free-energy change in the hydrolysis of ATP is intermediate between phosphate compounds of high potential and those of low potential (see Table 2.7).
2. ATP and ADP participate in almost all phosphate transfers in metabolism.

Because ATP occupies an intermediate position in the thermodynamic scale, it connects reactions involving high-potential and low-potential compounds. ATP is formed when ADP accepts phosphate groups from high-potential compounds. Subsequently ATP donates its terminal phosphoryl group to particular phosphate-acceptors (for example, glucose, glycerol, and others). These acceptors have a higher energy potential once they have been transformed, and reactions continue in the general direction dictated by the standard free energies of the products and reactants.

Because ATP and ADP take part in nearly all phosphate transfers, ATP serves as a general energy carrier in the cell. All sets of enzymes which function in this process catalyze transfer of phosphate from high-energy compounds to ATP, and from ATP to lower-energy compounds. No enzymes which transfer phosphate groups directly from some high-energy component to low-energy substances have been discovered (Fig. 2.44).

ATP participates as a **common intermediate** in reactions that deliver energy and in those which receive energy. In such sequential reactions, the product

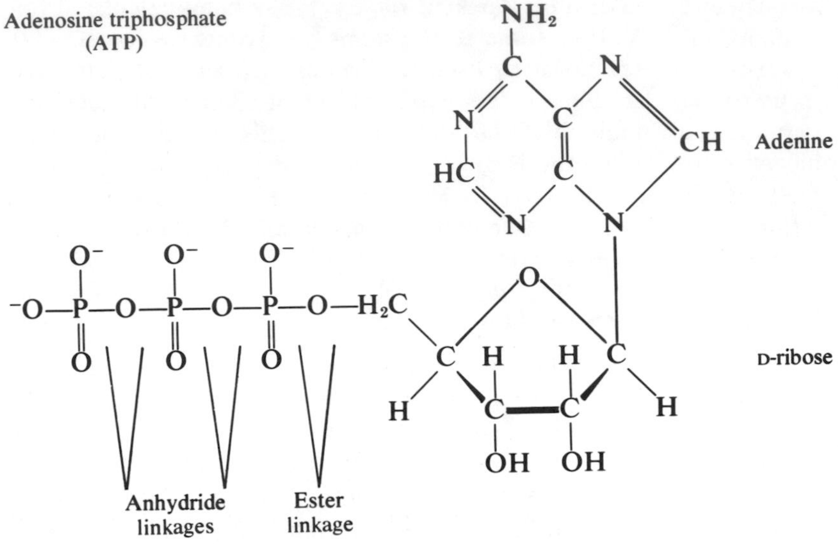

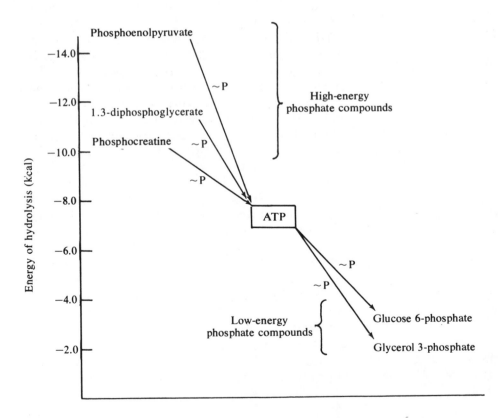

Adenosine triphosphate
(ATP)
NH_2
Adenine
D-ribose
Anhydride
linkages
Ester
linkage

Figure 2.43
Location of the anhydride and ester linkages in ATP.

Phosphoenolpyruvate
1.3-diphosphoglycerate
Phosphocreatine
~P
High-energy
phosphate compounds
ATP
Low-energy
phosphate compounds
Glucose 6-phosphate
Glycerol 3-phosphate
Energy of hydrolysis (kcal)
−14.0
−12.0
−10.0
−8.0
−6.0
−4.0
−2.0

Figure 2.44
Energy transfer from compounds of high free-energy content to those of low free-energy content occurs through mediation of ATP, whose free-energy level is intermediate between those two groups of molecules. (Adapted with permission from Albert L. Lehninger, *Bioenergetics: The Molecular Basis of Biological Energy Transformations,* Second Edition, copyright © 1971, W. A. Benjamin, Inc., Menlo Park, California.)

of one reaction becomes the substrate of the next, and so on, through a number of steps. For example

$$X{\sim}P + ADP \rightarrow X + ATP \qquad [2.19]$$

$$ATP + Y \rightarrow ADP + Y{-}P \qquad [2.20]$$

In this series of reactions, ADP accepts a phosphoryl group from a higher-energy phosphate compound $(X{\sim}P)$ and is thus converted to ATP. ATP then acts as a substrate in the second reaction by transferring its terminal phosphoryl group to energize compound Y, forming Y—P, and ADP is regenerated to cycle again. These reactions can proceed independently of each other since each is complete, but ATP participates in both cases and links the phosphoryl transfer events of the two reactions. Another way in which such reactions are often depicted is

$$[2.21]$$

In this format it is clear that the dephosphorylation of $X{\sim}P$ is *coupled with* the acceptance of the phosphoryl group by ADP to form ATP. In the second reaction the formation of Y—P is coupled with ADP formation by the transfer of the phosphoryl group from ATP to compound Y.

If these two reactions occurred in a biochemical pathway in which the energy transfers depended on reaction [2.19] preceding reaction [2.20], linked by an ADP−ATP cycle, then we might indicate the relationship as

$$[2.22]$$

In this example, **coupled reactions** are involved in transfer of phosphoryl-bond energy from a higher-potential to a lower-potential compound via a common ADP−ATP cyclic link.

Oxidation-Reductions and Redox Potentials

The transfer of energy or of electrons by coupled reactions is a unique feature of biological chemistry. Reactions involving loss of electrons, or **oxidations,** are coupled with **reduction** reactions in which electrons are gained. An oxidizing agent, or **oxidant,** loses electrons to a reducing agent, or **reductant,** which accepts electrons in coupled **oxidation-reductions,** or **redox reactions.**

The ability to lose or accept electrons varies among oxidants and reductants. The tendency to lose electrons can be quantitatively compared and expressed as a positive or negative numerical value in a **redox series,** arranged according to the oxidation-reduction potential of various substances. These values are obtained from measurements of electrode potential made against the standard of hydrogen, according to the reaction

$$H_2 \rightleftarrows 2H^+ + 2e^- \qquad [2.23]$$

In this reaction molecular hydrogen is in equilibrium with its oxidation products, hydrogen ions (protons) and electrons, under standard conditions.

When electrons of a substance are donated to the hydrogen electrode, the potential has a negative value (<0); when the substance accepts electrons its potential registers as positive (>0). The **standard electrode potential** of a substance, or E_0, is its potential relative to a hydrogen electrode, expressed in volts (V). Stronger oxidizing agents have a higher positive potential, or a greater affinity for electrons, than weaker oxidants or than reductants, which have negative potential.

The redox series represents a range of increasing electron affinity, going from negative to positive E_0 values (Table 2.8). The redox series has great predictive value for determining which reactions are theoretically possible, just as the ΔG^0 series permits predictions for direction of energy flow and thermodynamically possible reactions. An oxidant can be reduced by a substance with a lower E_0 value than its own, and conversely, a reductant can be oxidized by a substance with a higher E_0 value than its own.

Table 2.8 Standard redox potentials of biochemical systems at pH 7.0 and 25–37°C*

REDUCTANT	OXIDANT	E_0' (volts)
Pyruvate	Acetate	−0.70
H_2	$2H^+$	−0.42
NADH	NAD^+	−0.32
Lactate	Pyruvate	−0.19
Cytochrome c (reduced)	Cytochrome c (oxidized)	+0.26
H_2O	$\frac{1}{2}O_2$	+0.82

* Systems having more negative standard reduction potential than the H_2–$2H^+$ couple have a greater tendency to lose e^- than H_2. Those with more positive potential than the H_2–$2H^+$ couple have a lesser tendency to lose e^- than H_2.

In coupled oxidation-reductions, therefore, electron transfer occurs spontaneously in predictable directions according to E_0 values. The symbol E_0 refers to measurements at pH = 0.0, while E_0' is used to describe potentials measured at any other pH, most often at a pH of 7.0.

Standard redox potentials for biochemical systems, at the physiological pH of 7.0, are usually determined by potentiometric titrations in which an oxidizing agent is added to the completely reduced form of a system. The difference in potential is then determined at various stages of oxidation by reference to a standard electrode. The process can be carried out in reverse, by adding reducing agent to the completely oxidized form of a substance. The inflection point of the titration curve is the point at which 50 percent of the reaction is complete, and this point is taken as the standard redox potential, or E_0' (Fig. 2.45).

The standard free-energy change ($\Delta G^{0'}$) is related to difference in redox potentials ($\Delta E_0'$) between the oxidation and reduction reactions, according to

$$\Delta G^{0'} = -nF\Delta E_0' \qquad [2.24]$$

where n refers to the number of electrons that move, F is the Faraday (23,040 cal/volt), and $\Delta E_0'$ is the redox potential difference (in volts). The interconverti-

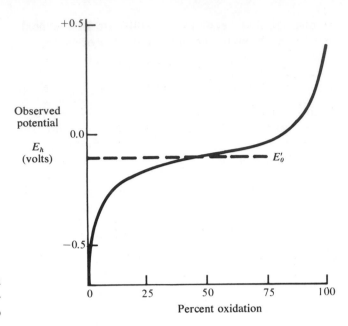

Figure 2.45
Titration curve of a biological substance, showing the point that determines the standard electrical potential (E_0') of that substance.

bility of $\Delta G^{0'}$ and $\Delta E_0'$ is illustrated in the following hypothetical scheme for the reaction $AH_2 + B \rightleftarrows BH_2 + A$

$$AH_2 \rightleftarrows A + 2H^+ + 2e^- \qquad E_0' = 0.20V \qquad [2.25]$$

$$B + 2H^+ + 2e^- \rightleftarrows BH_2 \qquad E_0' = -0.10V \qquad [2.26]$$

Reaction [2.25] proceeds by being coupled to reaction [2.26], so that

$$\Delta E_0' = 0.20V - (-0.10V) = 0.30V \qquad [2.27]$$

and according to equation [2.24]

$$\Delta G^{0'} = -2 \times 23,040 \text{ cal/V} \times 0.30V$$
$$= -13,824 \text{ cal or } -13.8 \text{ kcal} \qquad [2.28]$$

Energy Transfer by Electron Carriers

Transfer of electrons or of hydrogens containing these electrons is an alternative means for energy transfer that involves **redox couples,** or compounds that occur in both the oxidized and reduced forms. The standard reduction potential E_0' of a redox couple is compared with the E_0' of the reduced-substrate—oxidized-substrate couple with which it interacts. From this comparison one can predict the direction and equilibrium composition of the oxidation-reduction system (Table 2.9).

The neutral term **reducing equivalent** is often used to refer to an electron or a hydrogen atom which is transferred, because either electrons, or hydrogens, or both may be involved in transfers. Since hydrogens may be involved, a dehydrogenation reaction is viewed as an oxidation, and a reaction in which hydrogens are added is viewed as a reduction.

Three principal groups of electron-transferring substances are of major importance in cell chemistry: **pyridine nucleotide** couples, **riboflavin** derivative couples, and **iron-porphyrin** couples. These systems generally function as parts of enzymes and are thus considered to be **coenzyme** or **prosthetic group** components bound to the protein portion of the enzyme.

The pyridine nucleotide **nicotinamide adenine dinucleotide,** or **NAD,** and its relative **NADP** exist as NAD^+-NADH and $NADP^+-NADPH$ couples which interact with substrate couples according to the general reaction

reduced substrate + NAD^+ (or $NADP^+$)

$$\mathbin{\updownarrow} \qquad [2.29]$$

oxidized substrate + NADH (or NADPH) + H^+

The NAD^+-NADH and $NADP^+-NADPH$ systems can transfer electrons from one substrate couple to another since they can act as common intermediates shared by two different pyridine-linked reactions, each catalyzed by a specific enzyme.

In the breakdown of glucose during glycolysis, for example, glyceraldehyde 3-phosphate is oxidized by pyruvate in reactions sharing NADH as a common intermediate.

glyceraldehyde 3-phosphate + P_i + NAD^+

$$\mathbin{\updownarrow} \qquad [2.30]$$

1,3-diphosphoglycerate + NADH + H^+

$\underline{NADH + H^+ + \text{pyruvate} \rightleftarrows NAD^+ + \text{lactate}}$ $\quad [2.31]$
sum:

glyceraldehyde 3-phosphate + P_i + pyruvate

$$\mathbin{\updownarrow} \qquad [2.32]$$

1,3-diphosphoglycerate + lactate

Each NAD^+ or $NADP^+$ transfers two electrons, one of which is part of a hydrogen atom and the remaining proton occurs as free H^+ ion in the medium. Essentially, two reducing equivalents are transferred from the substrate to NAD^+ or $NADP^+$. One of these equivalents appears in NADH or NADPH as a hydrogen atom, the other as an electron. Two-electron transfers usually involve structural changes in the carrier molecules, and relatively large free-energy changes due to the redistribution of energy among reaction components resulting from chemical-bond

Table 2.9 Redox potentials of redox couples and substrate couples

REDOX COUPLE (red/ox)	E_0'	SUBSTRATE COUPLE (red/ox)	E_0'
$NADH/NAD^+ + H^+$	−0.32	α-ketoglutarate/succinate	−0.67
$FMNH_2/FMN$	−0.12	Ethanol/acetaldehyde	−0.20
Cytochrome c, red/ox	+0.26	Lactate/pyruvate	−0.19
Cytochrome a, red/ox	+0.29	Malate/oxaloacetate	−0.17
Fe^{2+}/Fe^{3+} (nonprotein)	+0.77	Succinate/fumarate	+0.03
		Hydrogen peroxide/oxygen	+0.30
		Water/oxygen	+0.82

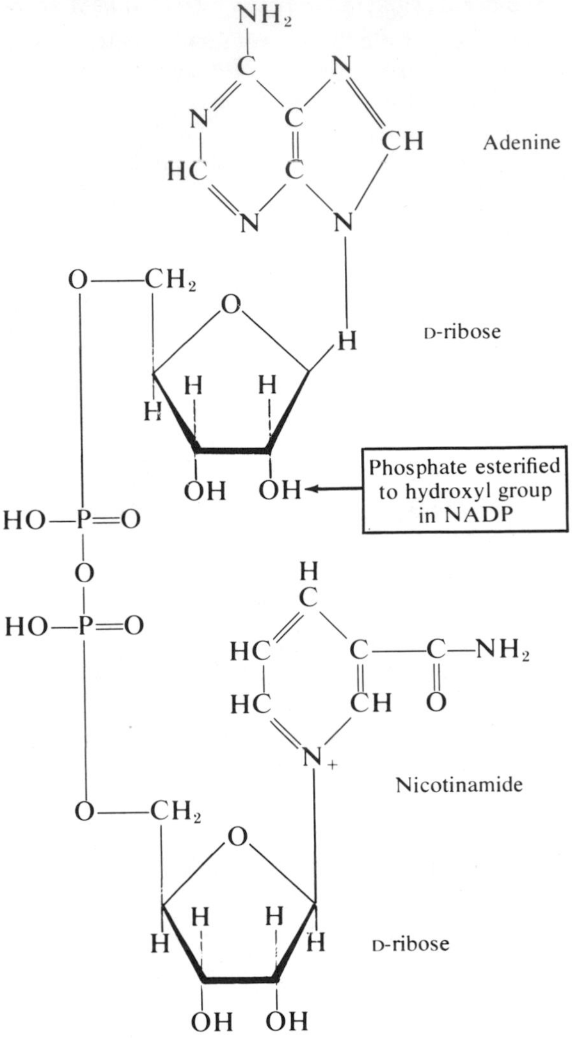

NAD (nicotinamide adenine dinucleotide)

Figure 2.46
The energy-transferring electron carrier NAD. The pyridine ring in the nicotinamide portion of the molecule undergoes oxidation and reduction.

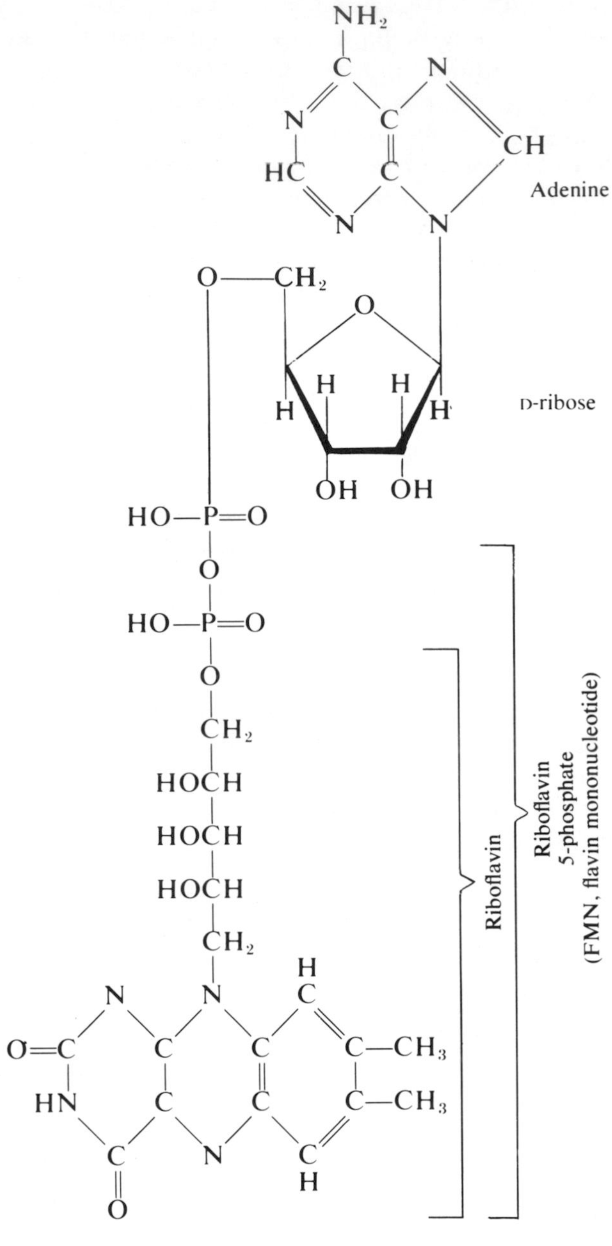

Flavin adenine dinucleotide (FAD)

Figure 2.47
FAD, the coenzyme portion of electron-carrier flavoprotein enzymes.

changes. In most cases the pyridine nucleotide is reversibly bound to the enzyme protein and thus acts more like a substrate than a prosthetic group or coenzyme. Some enzymes have NAD^+ or $NADP^+$ prosthetic groups covalently bonded to the protein portion of the catalytic molecule. NAD^+-linked enzymes generally are involved in breakdown of fuel molecules, while $NADP^+$-linked enzymes are more commonly involved in biosynthesis reactions. In both NAD^+ and $NADP^+$ it is the pyridine ring that undergoes alteration during oxidation-reduction (Fig. 2.46).

Flavoproteins often act as electron-carrier enzymes in mixed reactions where electron flow is from a two-electron donor to a one-electron acceptor, or vice versa. The electron-carrier component is a riboflavin derivative bound to a protein as a coenzyme. The commoner derivative is **flavin adenine dinucleotide (FAD),** but **riboflavin 5′-phosphate** (also called **flavin adenine mononucleotide,** or **FMN)** also acts as an electron-carrier component in certain enzymes. Two hydrogens are added to FAD or FMN when either is reduced in coupled reactions where substrate is oxidized (Fig. 2.47). The $FAD-FADH_2$ system is part of the enzyme succinic dehydrogenase, an enzyme of the mitochondrial respiration pathways. Flavoprotein enzymes are important catalysts in photosynthetic reactions as well as in oxygen-consuming processes such as respiration.

Many compounds transfer one electron at a time; these usually are metal-containing compounds, and iron is the commonest of these metals. It is the valency state of the metal that changes in oxidation-reduction, not the structure of the molecule. Electron carriers containing a **heme** (iron-porphyrin) group, such as **cytochromes,** transfer single electrons between the Fe^{2+} and Fe^{3+} valency states. The cytochromes act sequentially to transfer electrons from flavoproteins to molecular oxygen in the respiratory chain of aerobic cells, and are usually bound very tightly to the plasma membrane of prokaryotes or the inner membrane of the mitochondrion in eukaryotes. In fact, these components are a part of the membrane structure (see Chapter 7).

Both electron transfer and phosphoryl transfer contribute to cellular energetics. In general, greater energy release can be achieved by electron transfer than by phosphoryl transfer, but each of these modes fulfills a particular function in the overall cellular economy.

SUGGESTED READING

Books, Monographs, and Symposia

Carter, L. C. 1973. *Guide to Cellular Energetics.* San Francisco: W. H. Freeman.

Dickerson, R. E., and Geis, I. 1969. *The Structure and Action of Proteins.* Menlo Park, Calif.: W. A. Benjamin.

DuPraw, E. J. 1972. *The Biosciences: Cell and Molecular Biology.* Stanford: Cell and Molecular Biology Council.

Lehninger, A. L. 1971. *Bioenergetics.* Menlo Park, Calif.: W. A. Benjamin.

Stryker, L. 1975. *Biochemistry.* San Francisco: W. H. Freeman.

Articles and Reviews

Albersheim, P. 1975. The walls of growing plant cells. *Scientific American* 232(4):80–95.

Doty, P. 1957. Proteins. *Scientific American* 197(3):173–184.

Lambert, J. B. The shapes of organic molecules. *Scientific American* 222(1):58–70.

Pauling, L., Corey, R. B., and Hayward, R. 1954. The structure of protein molecules. *Scientific American* 191(1):51–59.

Perutz, M. F. 1964. The hemoglobin molecule. *Scientific American* 211(5):64–76.

Chapter 3

Enzymes: Catalysts of Life

Catalysts are substances that modulate reaction rates without altering the equilibrium point of the reaction. The unique catalysts of the living world are **enzymes,** all of which appear to be globular proteins. Nonliving systems do not have protein catalysts, but substances such as iron atoms or hydrogen ions may be catalytic components in both living and nonliving systems. Enzymes are uniquely suited to the processes of cellular chemistry because these proteins possess both catalytic and regulatory properties. Organic chemical reactions are notoriously slow under the mild conditions of temperature and pressure of the average cell.

Enzymes not only accelerate these reaction rates, but they also exert a fine control over the actual rate at which a reaction will proceed. Enzymes themselves are subject to regulation, sometimes by external factors and sometimes by particular qualities of their own.

Since there are about 3000 different enzymes in an average cell, regulation of enzyme activity and regulation of rates of reactions in progress are important aspects of orderliness that distinguish the open system of the cell from the disordered surroundings. The intricate meshwork of activities taking place at different times in a cell can undergo profound and rapid changes under the catalytic direction of enzymes. The distinctiveness of the chemistry of different kinds of cells or of different compartments of the cell can be traced to differential enzyme content and activity. These features, in turn, are the outcome of differential gene action and of regulation over the activity of enzymes that are produced from genetic instructions in certain cells at certain times.

The concept of catalysis was formulated in the 1850s, but it was not until the 1930s that the idea that biological catalysts were proteins was accepted. The first enzyme to be purified in crystalline form was *urease,* which catalyzes the hydrolysis of urea to carbon dioxide and ammonia end products.

$$\begin{array}{c} H_2N \\ \diagdown \\ C{=}O + H_2O \rightarrow CO_2 + 2NH_3 \\ \diagup \\ H_2N \end{array} \qquad [3.1]$$

$$\Delta G^{0\prime} = -13.8 \text{ kcal/mole}$$

Although crystalline urease was reported in 1926 by J. B. Sumner, it was not considered sufficiently convincing evidence to warrant the generalization that enzymes were proteins. There was strong opposition to this concept by respected and leading chemists. Their opposition finally was overcome within the following 10 years as other enzymes were purified or crystallized and were shown to be proteins. Well over 100 different enzymes have since been crystallized, many more have been purified but not in crystalline form, and the complete structure of several enzymes has been elucidated.

Enzyme nomenclature has been formalized by international agreement, but common or trivial names still remain in general use because of their familiarity and also because they are less cumbersome than the formal names. Six classes of enzymes have been designated, and specific categories have been recognized within each of these classes (Table 3.1). The suffix **-ase** identifies an enzyme protein. This suffix is added onto the name of a substrate or a reaction catalyzed by the designated enzyme, and the enzyme is thus named more informatively. For example, *succinic dehydrogenase* is a dehydrogenating enzyme that catalyzes the oxidation of succinic acid in the mitochondrial respiratory system.

In some cases the older, trivial name has been retained because it is very familiar and extensive reference to the older name appears in biochemical and biological literature. Enzymes such as pepsin, trypsin, chymotrypsin, and papain are protein-digesting enzymes whose older designations remain in wide use today, even though the suffix *-ase* is not a part of the name of the catalyst.

ENZYME ACTIVITY

All the complexity of cellular biochemistry is based ultimately upon simple chemical reactions that are catalyzed at each step by a specific enzyme. Enzyme-catalyzed reactions follow the same basic rules of chemistry as any other chemical reaction that may take place in our world. The hydrolysis of urea can take place spontaneously in the absence of any catalyst, but the same reaction leading to the same end products and energy release occurs when a catalyst is present. If a source of H^+ ions is provided, the reaction rate is speeded up by the inorganic ion catalyst. The rate of hydrolysis is even faster if the enzyme urease is provided to the system. In all three cases of urea hydrolysis, the only difference is in the rate of the reaction, not in the end products, the energy balance sheet, the direction of the reaction, or the point at which equilibrium is reached. In order for urea to react with water there must be sufficient energy to form an activated complex (Fig. 3.1). The activation energy barrier is reduced to the greatest extent in an enzyme-catalyzed system and so the reaction occurs at an accelerated rate when urease is present as compared to the rates when H^+ ions are present or in the total absence of a catalyst (Fig. 3.2).

The main function of an enzyme is to lower the

Table 3.1 International system of enzyme classification*

ENZYME CLASS NAME	TYPE OF REACTION CATALYZED	CODE NUMBER**
Oxido-reductases	Oxidation-reduction reactions	1.
	Acting on $-\overset{\mid}{C}H-OH$	1.1
	Acting on $-\overset{\mid}{C}=O$	1.2
	Acting on NADH; NADPH	1.6
Transferases	Transfer of functional groups	2.
	One-carbon groups	2.1
	Phosphate groups	2.7
	Sulfur-containing groups	2.8
Hydrolases	Hydrolysis reactions	3.
	Esters	3.1
	Peptide bonds	3.4
Lyases	Addition to double bonds	4.
	$-\overset{\mid}{C}=\overset{\mid}{C}-$	4.1
	$-\overset{\mid}{C}=O$	4.2
Isomerases	Isomerization reactions	5.
	Racemases	5.1
Ligases	Formation of bonds, with ATP cleavage	6.
	C—O	6.1
	C—N	6.3
	C—C	6.4

* Only some examples of specific reaction types are shown in column 2.
** A specific enzyme is identified by two additional numbers. For example, hexokinase is the trivial name of the enzyme catalyzing the reaction: ATP + glucose → glucose 6-phosphate + ADP. Its formal name is ATP:hexose phosphotransferase. Its Enzyme Commission code number is EC 2.7.1.1.

activation energy barrier to a chemical reaction. With or without a catalyst a reaction will proceed spontaneously in the direction of lower energy potential (higher ΔG). Reactions proceed in this way because molecules are in constant motion by virtue of possessing kinetic energy. Some molecules have more energy than others, and there is a statistical probability of some molecules overcoming the energy barrier at all times. If the temperature is raised, thereby adding thermal energy to the system, more molecules are energized and reaction rates will accelerate. Cells experience little change in temperature however.

Enzymes act to lower the activation energy barrier and thus increase the probability of molecules passing this barrier and proceeding to completion of the reaction. The enzyme combines with the **substrate** (molecule changed in the reaction) in temporary association, and since the enzyme-substrate complex has lower energy requirement, more substrate molecules can pass beyond the energy of activation barrier. Some enzymes can process millions of substrate molecules every second, repeatedly dissociating from the temporary enzyme-substrate complex. A relatively few enzyme molecules, therefore, can handle many millions of substrate molecules. High rates of reactions are thus encouraged and are aided by the *specificity* of enzyme affinity for particular substrates. Inorganic catalysts such as iron or H^+ ions are much less specific

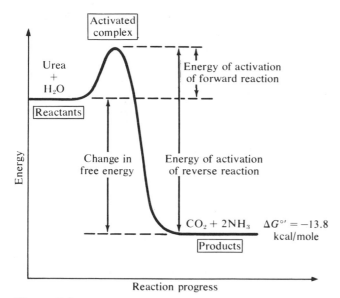

Figure 3.1
Diagram showing the energy relations of a reaction in which urea is hydrolyzed to form carbon dioxide and ammonia, and the reverse reaction.

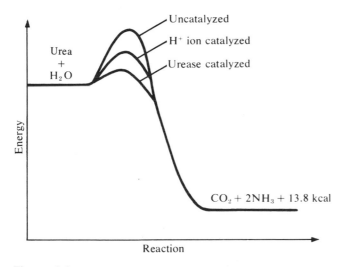

Figure 3.2
Relative effects on reducing the activation energy barrier in hydrolysis of urea when catalyzed by H^+ ions or by the enzyme urease.

or are even nonspecific. Since a particular enzyme has a highly specific affinity for a particular substrate, it is highly *efficient* in catalyzing the particular reaction in question. Each enzyme handles its own share of the workload, and many reactions can proceed simultaneously in the confines of an organelle or a cell because of enzyme specificity and efficiency.

Some enzymes rely only on their protein structure for activity while others require a nonprotein **cofactor,** which may be a complex organic molecule, a metal ion, or both. Intact protein−cofactor complexes are called **holoenzymes.** When the cofactor is removed, the protein remaining is called the **apoenzyme,** and the cofactor component itself is a **coenzyme.** If a coenzyme is covalently bound or otherwise tightly associated with the apoenzyme, it may be referred to as a **prosthetic group.** The heme cofactor of cytochrome enzymes in the respiratory electron transport system is this sort of prosthetic group and it is difficult to dissociate from the apoenzyme portion.

Other coenzymes are readily dissociated from apoenzyme and act almost like substrates in the chemical reaction. The coenzyme undergoes no permanent change during the reaction and is thus a catalytic component in the true sense of the definition. Coenzymes such as NAD^+ and FAD^+ usually function as carriers of electrons. Other kinds of coenzymes carry specific atoms or functional groups of the substrate involved in the reaction.

Most enzymes work only within narrow limits of pH, known as their **optimum pH** for the reaction. The large majority of enzymes have pH optima well within the physiological range of most cells, about pH 6.0 to 7.0. Some enzymes, such as pepsin in the stomach, have pH optima of 1.5–2.0 for particular substrates, while others work best at alkaline pH values near pH 10.0 in some systems. The same enzyme usually shows a different pH optimum for different substrates, if it has the capacity to act on more than one substrate.

Like any chemical reaction, increasing temperature of the reactants will lead to increase in reaction rate for enzyme-catalyzed systems. But because enzymes are proteins, they are subject to **denaturation** (unfolding of tertiary structure), with resulting inactivation. Many enzymes are inactivated at 45°C, and most are rapidly denatured at 55°C and above. Organisms that live at unusually high temperatures, such as thermophilic bacterial and blue-green algal species that inhabit hot springs, obviously possess enzymes with high-temperature tolerance. These cell types are often useful in providing enzymes and other protein components that can be analyzed in experiments conducted at temperatures that would denature the average set of proteins.

Substrate Specificity

The mechanism of enzyme action first proposed was described as a "lock and key" relationship by Emil Fischer in 1894. This analogy was based on spatial and conformational relationships between enzyme and substrate that were so specific that exquisite discriminations could be made by enzymes between almost identical molecules. Different enzymes catalyze reactions involving the optical isomers of the same compound, different enzymes hydrolyze β-galactose and β-glucose linkages even though the only apparent difference is in the position of the hydroxyl group on carbon atom 4 of these sugars, and many similar examples.

Some enzymes, however, are specific for a whole group of molecules rather than one specific compound. For example, trypsin and chymotrypsin disrupt peptide linkages in many proteins, phosphatases hydrolyze various esters of phosphoric acid, and so forth. Even in these cases, however, the substrates share a similar type of molecular site that is attacked by the particular enzyme (Fig. 3.3).

Specificity is believed to be determined by the catalytic site, or **active site,** of the enzyme which combines with or adsorbs onto the substrate molecule. The active site of an enzyme contains the specific functional groups that can bind the substrate or its relevant portion and then bring about the catalytic

event. From enzymes whose amino acid sequence is known and whose three-dimensional conformation has been constructed (lysozyme, ribonuclease, and chymotrypsin, for example), it is clear that the critical side-groups on the protein that participate in catalysis are not necessarily very close together along the length of the protein chain. Folding brings these groups into closer proximity (Fig. 3.4). When enzyme protein is denatured, unfolding separates the component side-groups of the active site, or prevents their coordinated participation in a chemical reaction. Enzyme inactivation is caused by these gross events of denaturation, but it may also occur when even one side-group is displaced or substituted in some cases.

The substrate molecule is oriented to the enzyme by the active site(s) of the enzyme, and the process of forming or breaking chemical bonds is enhanced by enzyme—substrate complexing (Fig. 3.5). Physical interaction between enzyme and substrate molecules leads to "induced fit" through distortion of the molecular surfaces (Fig. 3.6). Although the three-dimensional shape and organization of only a handful of enzymes have been studied, it has been suggested that the active site of an enzyme is in a crevice, a shallow depression, or a pit in the surface of the protein. The geometry of the active site is closely related to the conformation of the particular substrate molecule and to the way in which this molecule is processed in the catalytic reaction. An example of two of these three situations may help to illustrate the concept.

Ribonuclease and some others are crevice enzymes. Ribonuclease acts on RNA chains to cut bonds which are located in relatively exposed positions. The RNA substrate chain appears to fit into the cleft containing the active site of the enzyme, very much like a thread that can be snipped somewhere in the middle of its length by a pair of scissors.

Carboxypeptidase A, like chymotrypsin, has its active site in a pit. In this case, the enzyme can fit up against the substrate molecule and cut away at the bond near the end of the substrate chain. Since poly-

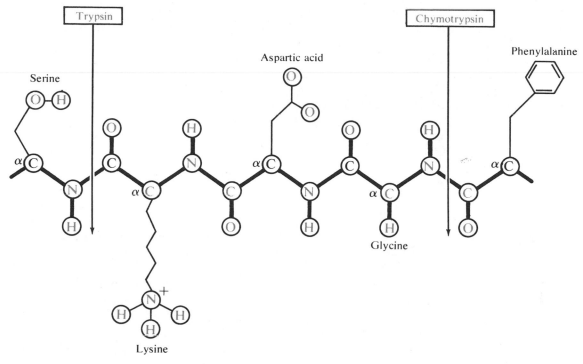

Figure 3.3
Trypsin digests peptide bonds that follow either lysine or arginine, which are positively charged side chains, while chymotrypsin acts only at peptide bonds that follow large, hydrophobic side chains, such as phenylalanine. These and other proteases act on the same peptide bonds, but each of the enzymes cleaves only specific links depending on the side chains projecting from the polypeptide "backbone."

peptides in fibrous proteins are tightly packed, exposed strands are rare, and it would be difficult to cut the chain except for the occasional exposed ends. A crevice enzyme would be ineffective in this case, whereas if the active site were contained in a pit the end of a polypeptide chain could be accommodated and brought into appropriate juxtaposition to the active site of the enzyme (Fig. 3.7). The shape of the region of the enzyme which contains the active site seems to correspond to the nature of the substrate molecule and the way in which the enzyme cuts the molecule. Ribonuclease has a cleft that accommodates an exposed long-chain substrate molecule which is then cut in the middle, while chymotrypsin chews away at the exposed ends of polypeptide chain substrate molecules and has effective access and contact with these ends through its pit conformation in the region of the active site.

The active site is a relatively small portion of the rather large enzyme molecule. Furthermore, almost all substrate molecules are considerably smaller than the enzymes with which they interact. These two features indicate that only a small portion of the surface of the enzyme actually engages in catalysis. The remainder of the enzyme surface is used to bind molecules involved in the regulation of enzyme action and for the association of the subunits of which the individual enzyme is composed.

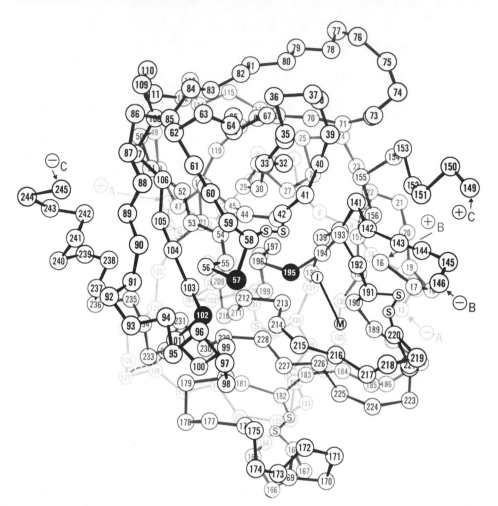

Figure 3.4
Folding of the main chain of α-chymotrypsin brings distant amino acid residues (numbered within the circles according to primary sequence of chymotrypsinogen) into proximity at the active site of the enzyme. The amino acids at positions 102 (aspartate), 57 (histidine), and 195 (serine) are catalytically important and are shown as black circles. (Reprinted by permission from *The Structure and Action of Proteins,* by R. E. Dickerson and I. Geis. W. A. Benjamin, Inc., Menlo Park, California. Copyright © 1969 by R. E. Dickerson and I. Geis.)

Inhibition of Enzyme Activity

Studies of enzyme inhibition have provided valuable information about a number of catalyst characteristics and the nature of the catalytic reaction. Enzymes may be inhibited by various chemicals in various ways, but the two basic types of inhibition have been classified as **reversible** and **irreversible.** Reversible inhibition does not alter the functional portions of the enzyme molecule and inhibiting effects can be removed by increasing the concentration of the substrate or by some other appropriate remedy. The enzyme molecule itself is altered in situations of irreversible inhibition. Examples of these two categories will illustrate the general principles.

There are two types of reversible inhibitions, competitive and noncompetitive. **Competitive inhibition** can be reversed by addition of substrate so that the ratio of inhibitor to substrate concentrations reaches an appropriate level which reduces or cancels the effect of the inhibitor. The absolute concentrations are unimportant: it is the ratio of the two substances that matters.

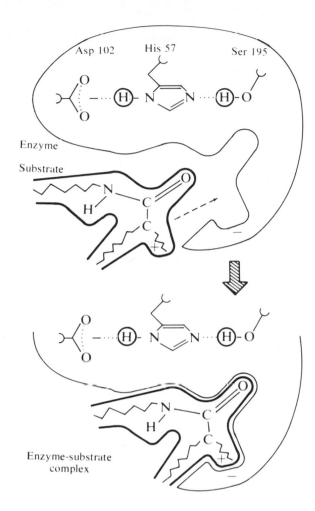

Figure 3.5
The complexing of the enzyme chymotrypsin with its peptide substrate occurs at the active site of the enzyme. Three catalytically important amino acid residues in the enzyme are shown (see Fig. 3.4).

A classical example of a competitive inhibitor is malonic acid, which inhibits the catalytic action of the respiratory enzyme succinic dehydrogenase in the oxidation of succinic acid to fumaric acid (Fig. 3.8). In this and other cases of competitive inhibition, the substrate and inhibitor compete for the same portion of the active site of the enzyme. The molecules are sufficiently similar that either type can form a complex with the enzyme, but there is enough difference so that the inhibitor—enzyme complex is inactive, whereas the substrate—enzyme complex engages in catalytic activity. When enough substrate is added to increase its proportion in relation to the inhibitor concentration, inhibition is reversed and succinic dehydrogenase activity resumes. Another example of competitive inhibition involves the inhibition of p-aminobenzoic acid contribution to bacterial growth by the drug sulfanilamide. When sulfanilamide is present in relatively high proportion to substrate con-

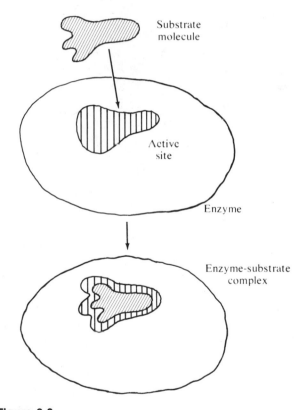

Figure 3.6
Interaction between substrate and enzyme molecules is facilitated by a conformational change at the active site of the enzyme, a phenomenon referred to as "induced fit."

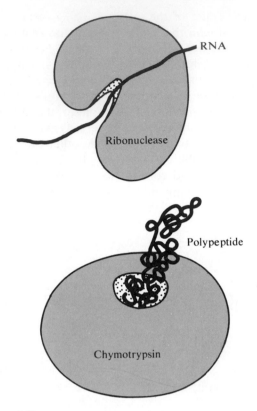

Figure 3.7
Diagram illustrating the relationship between active site geometry and conformation of the substrate in relation to the way the substrate is processed. Ribonuclease, a crevice enzyme, snips exposed regions in the middle of the RNA chain, whereas chymotrypsin, a pit enzyme, attacks the accessible ends of the polypeptide chain that is its substrate.

centration, competitive inhibition occurs because these molecules are very similar, a characteristic feature of competitive inhibitor—substrate—enzyme interactions (Fig. 3.9).

Reversible **noncompetitive inhibition** is distinguished by the fact that increasing the proportion of substrate to inhibitor does not cancel the effect of the inhibitor. By analysis of the pattern of effects on enzyme activity in the presence of increasing proportions of substrate, this type of inhibition pattern can

be separated from the competitive sort. The enzyme is not altered in either case, but noncompetitive inhibitors are believed to bind to some part of the enzyme other than the active site. In some cases, the noncompetitive inhibitor binds to a metal ion that is essential for enzyme activity. The chelating agent ethylenediamine tetraacetate (EDTA) binds ions like Mg^{2+} reversibly and inhibits those enzymes dependent on the cation for activity (Fig. 3.10). Other noncompetitive inhibitors may bind reversibly to sulfhydryl (—SH) groups and affect catalytic activity of enzymes whose —SH groups are required to maintain three-dimensional conformation or contribute to catalysis in some other way. In any case, removal of the agent restores enzyme activity.

Irreversible inhibition occurs in systems in which the inhibitor modifies the enzyme and reduces or prevents its catalytic action. Various nerve poisons act in this way, including compounds that are contained in a number of insecticides. Some inhibitors

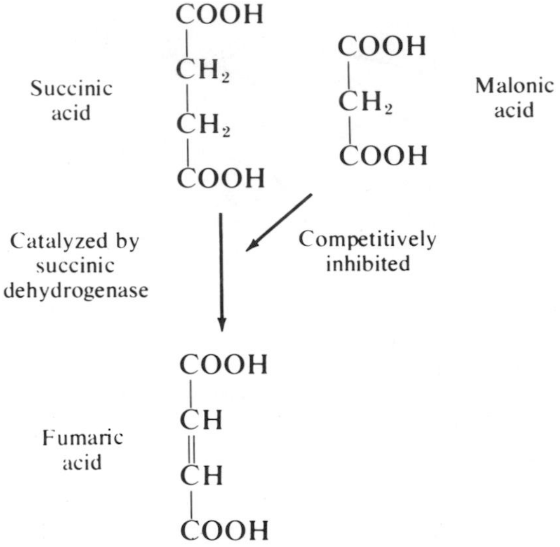

Succinic acid

$$COOH$$
$$|$$
$$CH_2$$
$$|$$
$$CH_2$$
$$|$$
$$COOH$$

Malonic acid

$$COOH$$
$$|$$
$$CH_2$$
$$|$$
$$COOH$$

Catalyzed by succinic dehydrogenase

Competitively inhibited

Fumaric acid

$$COOH$$
$$|$$
$$CH$$
$$||$$
$$CH$$
$$|$$
$$COOH$$

Figure 3.8
Malonic acid competitively inhibits the reaction producing fumaric acid by competing with succinic acid for the enzyme succinic dehydrogenase.

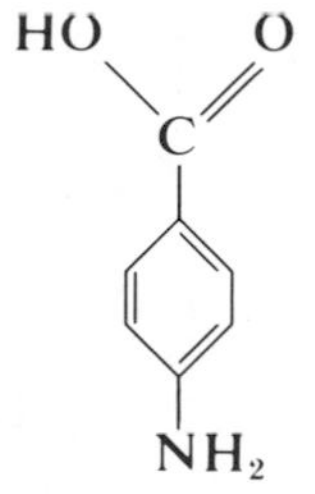

p-aminobenzoic acid

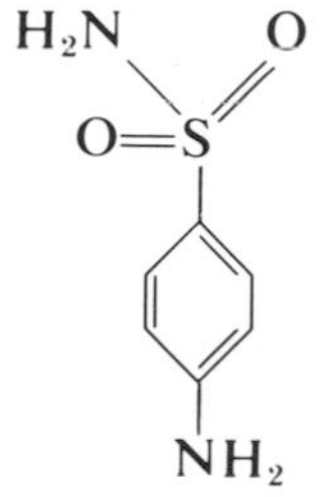

Sulfanilamide

Figure 3.9
Sulfanilamide is a competitive inhibitor in reactions involving p-aminobenzoic acid as a substrate.

have been shown to bind covalently to the enzyme and form a stable but inactive complex. Other inhibitors cause sufficient chemical alteration of the enzyme so that it loses its activity and does not regain function, even after the inhibitor has been removed.

Self-Regulation

Considerable interest has developed recently in the ability of enzymes to regulate the overall rate of reactions in which they are involved. The end product of a reaction sequence will often inhibit the formation of the initial product catalyzed by the first enzyme of the series. In such **feedback inhibition** phenomena, the rate of the entire reaction sequence is determined by the concentration of the end product. The first enzyme in the pathway is usually inhibited by the end product (Fig. 3.11). This enzyme is called a **regulatory enzyme** and the inhibitory metabolite is an **effector** or **modulator** molecule. Since the inhibitor and the substrate of the regulatory enzyme are not structurally similar, the term **allostery** ("different structure") has been coined to describe the modification of an enzymatic reaction by a compound that is different in shape from the true substrate. Considering the specificity of fit between the substrate molecule and the active site of an enzyme, **allosteric modulators** must bind to the enzyme at some other site than its catalytic region.

The allosteric modulator may be positive and enhance the catalytic activity of an enzyme, as well as being negative and inhibiting enzyme action. For example, the synthesis of the amino acid L-isoleucine from the amino acid L-threonine occurs in a sequence of five enzymatic steps. Accumulation of the isoleucine end product leads to feedback inhibition of the regulatory enzyme L-threonine deaminase. The same regulatory enzyme can be activated by the amino acid L-valine. In this system then, isoleucine is a negative allosteric modulator, while valine is a positive allosteric modulator of threonine deaminase. These observations have led to the proposal that the regulatory enzyme has two allosteric regulatory sites in addition to its active site that complexes with the substrate threonine.

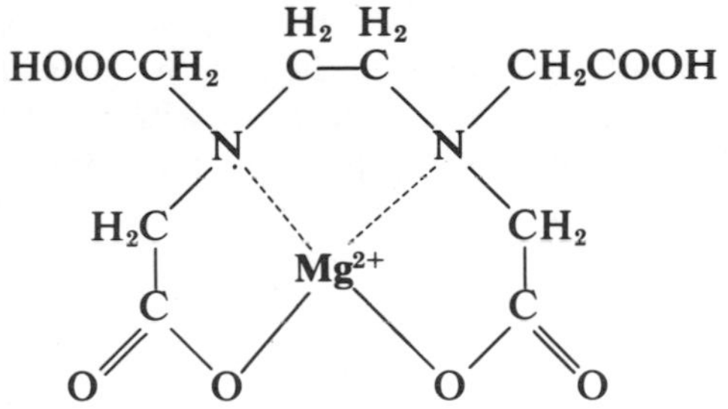

Figure 3.10
Chelation, or sequestering, of Mg^{2+} ions by EDTA removes the ions required for activity of some enzymes, resulting in noncompetitive inhibition of enzyme action.

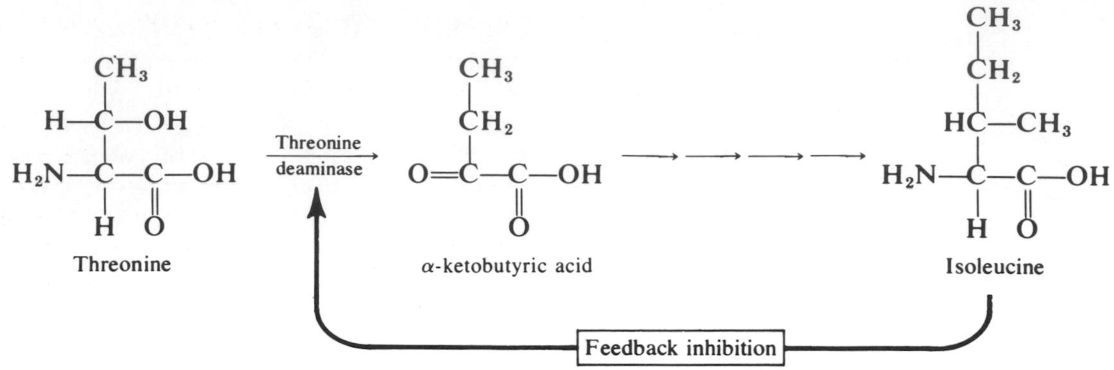

Figure 3.11
The feedback inhibition of activity of the regulatory enzyme threonine deaminase by the allosteric modulator L-iso- leucine, the endproduct of the five-step reaction sequence.

The action of the allosteric modulator has been shown to be due to an induced change in the conformational shape of the enzyme. Since all known regulatory enzymes are large proteins composed of more than one polypeptide subunit, considerable study has been directed toward understanding the interactions between the allosteric modulators and subunits of the enzyme. One hypothesis states that binding between modulator and one subunit of the enzyme leads to conformational changes in all enzyme subunits in unison. An alternative hypothesis postulates that modulator binding leads to perturbations in only part of the enzyme shape and leaves other subunits entirely unchanged in conformation. In either case the modulator molecules are not consumed in the reaction with the enzyme. These molecules participate in regulating enzyme activity via conformational changes but are otherwise still available in the cell for metabolism. Inhibitory modulators distort the enzyme shape; activators restore the proper conformation of the enzyme. Modulator molecules do not undergo chemical modification during binding with the enzyme protein. Once released from the complex with the enzyme, the modulator returns to the metabolic pool in the cell. Using the same set of components, enzymes and

modulators, the cell regulates its metabolism through this system of enzyme protein flexibility that constitutes a system of "on-off" switching of catalysis.

Another way in which enzymes regulate their activity in relation to the pool of metabolites in the cell has been called **cooperativity.** The binding patterns of most enzymes and substrates fall along a hyperbolic curve indicating an increase in enzyme activity as concentration of substrate bound to the enzyme increases (Fig. 3.12). Other binding patterns show an increase in enzyme activity level taking place to the same extent but with much less substrate required. This is an enzyme with a pattern of **positive cooperativity.** An enzyme that does not reach its maximum activity level even at saturating concentrations of substrate is one that exhibits **negative cooperativity.** These features of regulatory enzyme activity have also been explained on the basis of conformational flexibility of the protein upon binding with substrate molecules. In positive cooperativity binding of the substrate to the active site of one subunit of the enzyme leads to a conformational change that makes the next subunit more receptive to the substrate, and it therefore binds more readily. A kind of "domino" effect leads to more efficient enzyme—substrate interaction and heightened

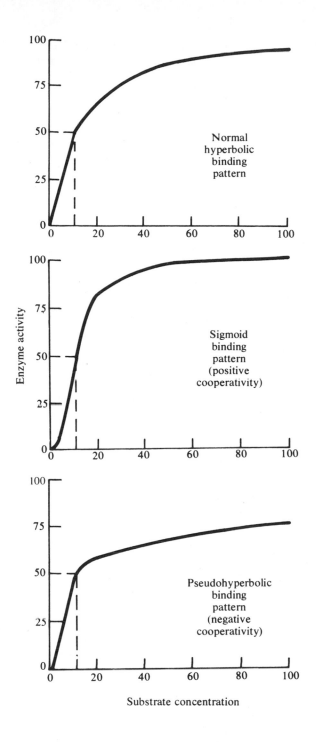

(a)

(b)

(c)

enzyme activity levels off at relatively lower substrate concentrations than in noncooperative systems. In the case of negative cooperativity, it has been assumed that binding of the substrate to the enzyme induces a conformational change in the protein which makes additional binding to substrate less likely. Activity levels remain lower in this case, even when very high concentrations of substrate exist in the cell.

The regulation of enzyme activity by protein cooperativity helps explain the different degrees of sensitivity of enzymes to the actual concentration of substrates in the cell. Positively cooperative enzymes are extremely sensitive to minor fluctuations in substrate concentration, whereas negatively cooperative enzymes appear rather insensitive to such fluctuations. Each type of cooperativity provides an advantage in particular catalytic situations, especially in relation to the amount of available substrate at any one time in the cell.

Systems involving activators and inhibitors (positive and negative modulators) are different but complementary to systems involving positive and negative cooperativity. Activators turn on enzymes and inhibitors turn them off. Cooperativity increases or decreases the sensitivity of these enzymes to the environmental fluctuations in concentration of modulator molecules. Substrates, inhibitors, and activators may show noncooperativity, positive cooperativity, or negative cooperativity on the basis of the binding pattern displayed. Activators and inhibitors play a major role in regulation, but cooperativity adds a

Figure 3.12
Binding patterns of enzyme with substrate indicate whether or not the enzyme is self-regulating: (a) Standard increase in enzyme activity in relation to substrate concentration produces a hyperbolic curve representing the binding pattern of enzyme with substrate for an enzyme that is not self-regulatory; (b) a self-regulating enzyme with properties of positive cooperativity shows activity increases requiring less substrate than in a; and (c) a self-regulating enzyme with properties of negative cooperativity does not reach the same high level of enzyme activity as in a, even at saturating concentrations of the substrate.

further dimension of fine-tuning of the catalytic system. All or most of these regulations depend on the ability of an enzyme to undergo changes in shape, some of which involve a displacement only a few angstroms in extent.

REGULATION OF ENZYME SYNTHESIS

Regulation of enzyme activity provides a fine control over cell metabolism, and is effective at the level of the enzyme protein itself. The important but coarser controls over catalysis are those determining the *kind* of enzyme and the *amount* of enzyme synthesized in the cell. These aspects of regulation are influenced by the genes. The gene codes for enzyme protein, thus determining the kind of enzyme that can be synthesized in particular cells with the appropriate nucleotide sequences. If the gene for the enzyme threonine deaminase does not exist in a species, that protein will not be manufactured. If the gene is mutant and either fails to direct synthesis of the enzyme or directs synthesis of a defective enzyme, the catalyst is not available for its specified reaction. If the gene

is present, the enzyme may be synthesized but the amount of enzyme manufactured is controlled by processes that take place at the level of the gene itself. Amounts of enzyme protein generally are regulated by transcriptional controls, that is, by control over manufacture of the messenger RNA that provides the directions for making the protein according to coded specifications in the gene from which the messenger was transcribed initially (Fig. 3.13).

The more explicit discussion of gene control studies in relation to enzyme synthesis will be found in Chapter 5. Here we will sketch in some general features of control systems as they relate more directly to the enzymes themselves. Three general categories of enzymes have been described on the basis of cellular conditions for their synthesis. Enzymes that are found in essentially constant amounts under various growth conditions are called **constitutive**. These enzymes apparently are formed at rates that bear little relation to concentrations of metabolites with which they interact. Some of the enzymes of glycolysis, or sugar breakdown, are constitutive types that are found in essentially similar amounts regardless of the concentrations of their substrates

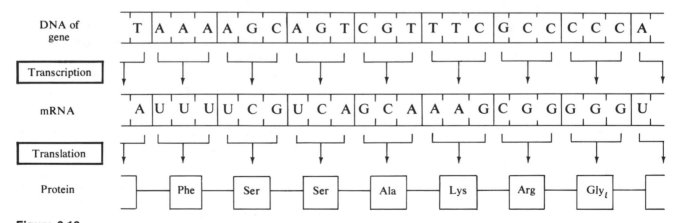

Figure 3.13
The *kind* of enzyme synthesized in the cell depends on the genetic blueprint in DNA. The *amount* of enzyme synthesized depends on controls over transcription of genetic information into messenger RNA and translation of that message into the primary structure of the enzyme protein.

in the medium. However, many enzymes fluctuate rapidly and substantially in response to the presence and absence of substrates. These are **inducible** and **repressible** enzymes, the other two categories described according to conditions for their synthesis.

Inducible enzymes are formed in response to the introduction of the substrate on which they act. For example, when glucose is not available, the cell will begin to elaborate enzymes needed to metabolize an alternate energy source that is available. If lactose or some other β-galactoside is present and glucose is not, the cell begins to synthesize the enzyme β-galactosidase. Only a few molecules of this enzyme usually occur in glucose-grown cells, but up to 5000 molecules of the enzyme can be found in an *E. coli* cell within minutes of the addition of lactose to the glucose-free medium. When lactose is removed or depleted, synthesis of β-galactosidase stops abruptly (Fig. 3.14).

The synthesis of repressible enzymes, on the other hand, stops when the product of the reaction is present. For example, the amino acid tryptophan is required in relatively small amounts since it is an in-

frequent constituent of proteins. When tryptophan is synthesized in excess of its need in protein synthesis, further manufacture of this amino acid stops because the enzymes in the reaction pathway are repressed (Fig. 3.15). The actual synthesis of new enzymes stops. This process is called **endproduct repression** of enzyme synthesis since the end product of a reaction sequence usually shuts down the system. In addition, existing enzymes in the cell are

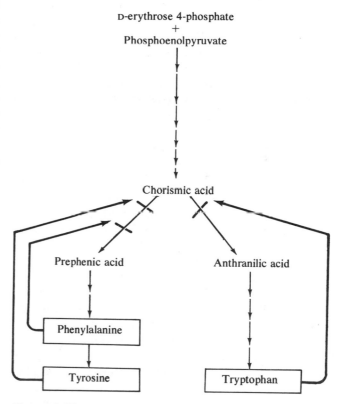

Figure 3.15
Cessation of tryptophan synthesis occurs by repression of one enzyme active early in the pathway, a phenomenon called endproduct repression of enzyme synthesis. The repressible enzymes usually act after a branch point in a larger biosynthetic sequence, as shown here affecting synthesis of repressible enzymes leading to three different amino acid endproducts via branching pathways after chorismic acid formation.

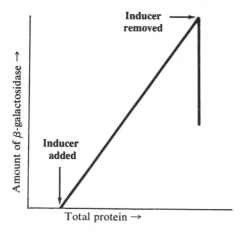

Figure 3.14
Synthesis of the inducible enzyme β-galactosidase begins on addition of the inducer substrate and declines very rapidly when substrate is depleted or removed from the cell.

inhibited from further activity. This process is called **feedback inhibition** of enzyme activity since the usual inhibitor is the final product of the reaction pathway. Repression of enzyme synthesis and inhibition of existing enzyme activity shut down catalysis when substrate is added.

Inducible enzymes generally catalyze breakdown reactions. They are usually found in trace amounts but may exist in higher concentrations when their substrates are present. **Repressible enzymes** usually catalyze biosynthetic reactions. They are present and active in growing cells in relation to the amount of substrate being produced and utilized in biosynthesis. If enzymes are inducible or repressible, their responses to substrates are referred to as **enzyme induction** and **enzyme repression,** or induction and repression of enzyme synthesis.

Negative Control of Transcription

Enzymes will be synthesized if messenger RNA has been transcribed from the coded genes. Control over manufacture of the messengers will therefore determine whether or not RNA transcripts are available for translation into proteins. Messenger transcription is turned "off" and "on" according to the interactions between **repressors** and **effectors,** or **inducers.** Repressors are protein products of regulatory genes. Effectors or inducers are the substrates or metabolites involved in a catalytic pathway. In the β-galactosidase system, lactose is the effector; in the threonine deaminase system, threonine is the effector.

Induction of enzyme synthesis will not take place in the absence of the effector. Repressor protein, manufactured at all times from information in regulatory genes, blocks the transcription of the inducible enzyme. When the effector is present, however, it combines with the repressor and leads to detachment of the molecular block from the DNA. Once the repressor is removed, RNA polymerase is free to move along the gene and transcribe its information into messenger RNA. The messenger can then be trans-lated into enzyme during synthesis of the protein along ribosomes. In the case of β-galactosidase, repressor product of the regulatory gene for this system blocks transcription. When lactose is present, this inducer binds with repressor and the block to transcription is released from the DNA site (Fig. 3.16).

Repression of enzyme synthesis will not take place in the absence of the effector. In this case the repressor product of the regulatory gene for the system cannot prevent transcription all by itself. When the effector is also present, it binds to the repressor and the repressor-effector complex is then able to bind to DNA and block transcription of the messenger RNA. The effector in these systems is usually referred to as a **co-repressor.** Threonine deaminase synthesis continues in the presence of repressor, but absence of the co-repressor threonine. When threonine occurs in excess, it combines with the repressor. The repressor—co-repressor binds to the DNA and blocks further transcription, effectively halting further synthesis of the enzyme (Fig. 3.17).

Both induction and repression of enzyme synthesis are negative control systems. In each case, the synthesis of enzyme takes place only when the repressor is removed from its blocking site on DNA. The two systems are only different sides of the same coin.

Cyclic AMP and Positive Control of Transcription

Negative control systems of transcription appear to be fairly common, at least in bacteria. Some examples of positive control have also been studied, although less is known about them than about negative controls. While repressor proteins modulate negative control, **inducer proteins** have been implicated in positive control over enzyme synthesis. One of the better known systems of positive control involves the effector molecule **cyclic AMP,** or cyclic adenosine 3'5'-monophosphate (Fig. 3.18), and the inducer protein called **CRP,** or cyclic AMP receptor protein. This system is present in both bacteria and

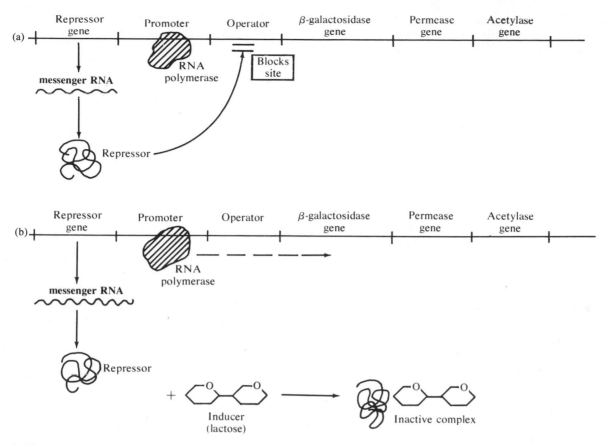

Figure 3.16

Negative control of inducible enzyme synthesis occurs at the level of transcription: (a) Repressor protein binds to the operator site of a gene cluster and prevents formation of messenger RNA transcripts by RNA polymerase; (b) when repressor is removed from the operator site after inactiva- tion by the inducer (the substrate of the enzyme to be synthesized), the way is opened to transcription of messenger RNA that will be translated into inducible enzyme protein.

eukaryotic cells, but we will briefly discuss the bacterial studies since these are simpler to interpret and understand.

When cells such as *E. coli* are provided with glucose as a source of carbon and energy, the rates of growth and total protein synthesis are faster than with many other substrates. Synthesis of inducible enzymes, however, is very slow at the same time that overall growth is most rapid. When both glucose and lactose are present in the growth medium, *E. coli* still synthesizes β-galactosidase at relatively low rates. The rate of β-galactosidase synthesis can be increased if lactose is the only substrate present. It also can be increased to similar levels if cyclic AMP is added to the medium containing glucose alone or glucose plus lactose. These and other observations

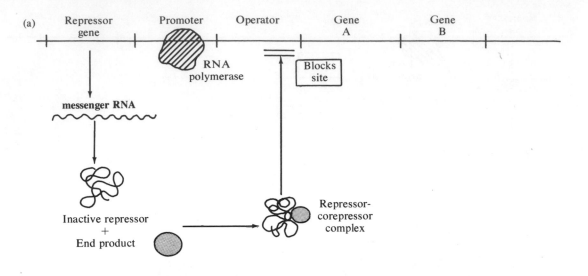

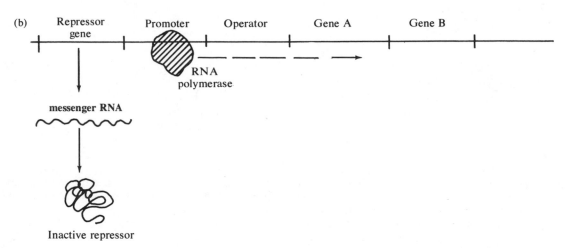

Figure 3.17

Negative control of repressible enzyme synthesis occurs at the level of transcription: (a) The end-product metabolite of a reaction sequence acts as a co-repressor, activating the binding of repressor protein to the operator site of a gene cluster and preventing transcription by RNA polymerase; (b) removal of the endproduct leads to transcription of messenger RNA for repressible enzyme, since the repressor now cannot bind to the operator site.

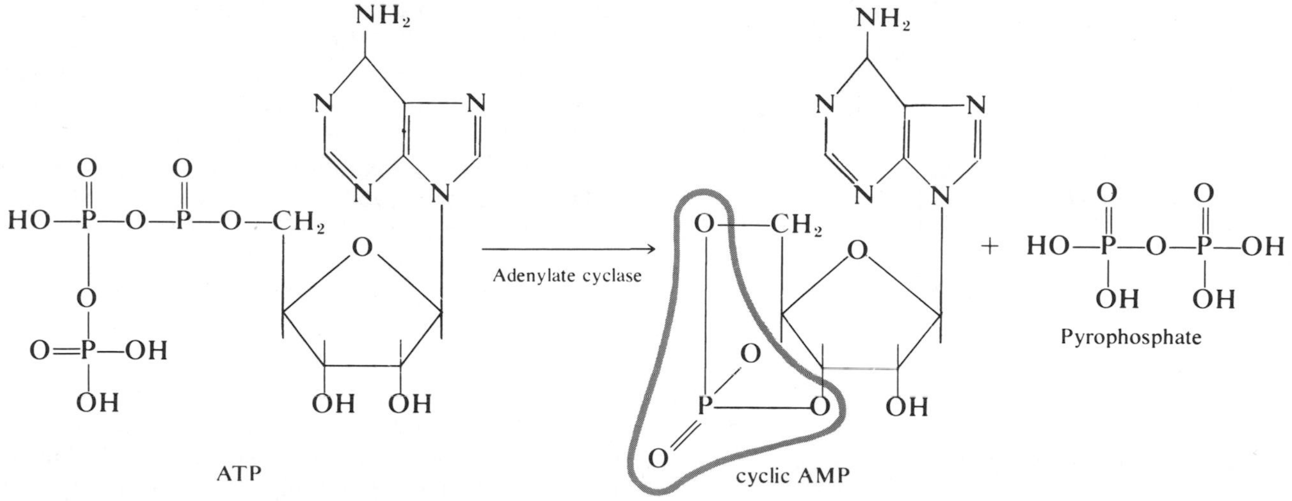

Figure 3.18
Formation of cyclic AMP from ATP.

indicate that cyclic AMP influences the rate of β-galactosidase synthesis (Fig. 3.19). This cyclic nucleotide has a similar effect on some other inducible enzymes, too. The response is specifically stimulated by cyclic AMP, since other adenine nucleotides and other cyclic nucleotides cannot substitute for cyclic AMP in inducing enzyme synthesis under similar conditions.

Evidence from a number of studies reported by Ira Pasten, Robert Perlman, and others showed that cyclic AMP influenced β-galactosidase synthesis by increasing the transcription of messenger RNA at the lactose gene sites. Cyclic AMP did not exert its effect on β-galactosidase through the negative control system of repressor product of the regulatory gene for the lactose system. This fact was demonstrated in a study using mutants with specific defects in this repressor system. Such mutants responded equally well to added cyclic AMP as did normal cells; both were stimulated to synthesize β-galactosidase. Since the repressor was not functional in the mutant, its response to cyclic AMP must have been due to some other control system. However, cyclic AMP—CRP

must act in conjunction with the negative control of the β-galactosidase repressor system since these proteins bind to different sites within the lactose gene group. We will discuss some of these features further in Chapter 5, after laying a proper basis for interpreting the genetic evidence.

ISOENZYMES AND MULTIENZYME SYSTEMS

Regulation of catalyzed reactions operates at many levels. Two features of enzyme systems that have been shown to influence regulation are their presence in interacting groups in a particular location in the cell, and the existence of more than one molecular form for the same enzyme species.

Isozymes

Multiple molecular forms of a single species of enzyme, differing principally in their net electrical charge, are known as isoenzymes or **isozymes.** Isozymes of an enzyme catalyze the same chemical reaction but have different reaction-rate properties

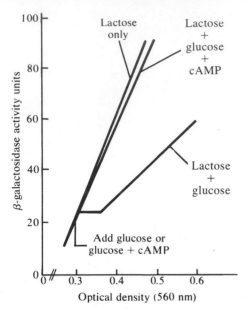

Figure 3.19

The rate of β-galactosidase synthesis, observed as enzyme activity, is high when lactose is the only substrate but is reduced when glucose is present along with lactose. The rate of enzyme synthesis can be increased to its original levels if cyclic AMP is present in the lactose + glucose medium. This observation indicates that cyclic AMP participates in positive control of inducible enzyme synthesis since the rate of synthesis increases when cyclic AMP is present. In negative control systems, a substance must be removed for the rate of enzyme synthesis to increase.

and can be distinguished by appropriate activity assays. According to current information, different proportions of a common set of subunits combine to form the isozyme varieties of a particular enzyme.

One of the best studied of these isozyme groups is lactate dehydrogenase from rat tissues. There are five isozymes, each of which is made up of four subunit polypeptide chains arranged in a tetramer quaternary structure (Fig. 3.20). The individual polypeptides have a molecular weight of 33,500 daltons, while the whole enzyme protein has a molecular weight of 134,000 daltons. Only two *kinds* of polypeptide sub-

units occur in an enzyme molecule. These are called the M chain and the H chain.

Lactate dehydrogenase occurs in all or most cells in its five isozymic forms. M_4 is made up of 4 M chains exclusively, H_4 has 4 H chains in the tetramer, and the other possible combinations are M_3H, M_2H_2, and MH_3. These isozymic forms do not necessarily occur in equal proportions in all cells. In fact, the muscle tissue contains predominantly the M_4 isozyme, while heart tissue has been found to contain the H_4 isozyme predominantly. These locations led to the naming of the subunits as M (muscle) and H (heart).

The relative proportions of the five isozymes depends on random combinations of subunits, but some cells produce more of one kind of subunit than the other. The M and H subunits are translations of two different genes. If these genes are equally active in a cell, then M and H subunits may be produced in equal proportions and all five isozymes will occur in

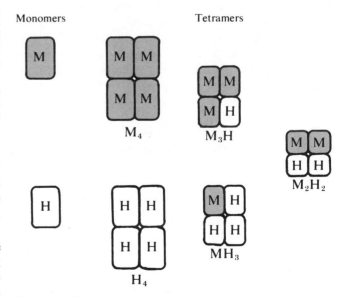

Figure 3.20

Functional lactate dehydrogenase consists of tetramers of the M and H monomer polypeptide units, in any of five possible isozyme combinations.

amounts that correspond to random combinations according to a simple arithmetical prediction. If one gene is more active than the other, then the ratio of isozyme types will reflect the greater abundance of one subunit over the other. The high amounts of M_4 and M_3H in white skeletal muscle and the high proportions of MH_3 and H_4 isozymes in tissues that have higher aerobic respiratory activities are probably explained in this way.

Tissues such as white skeletal muscle are very active in glycolysis, during which pyruvate is reduced to lactate by lactate dehydrogenase. The MH_3 and H_4 isozymes have a much greater affinity for pyruvate as an electron acceptor, so that NADH is reoxidized with pyruvate to produce lactate. Isozymes like M_4 and M_3H have much lower reactivity with pyruvate so that NADH is more readily handled by the aerobic respiratory system of the mitochondrion. The glycolytic versus aerobic pattern of different cells and tissues are thus aided by different isozymic forms of the same enzyme. Regulation of pyruvate and NADH oxidation-reduction in relation to cellular activities is genetically programmed. Isozyme synthesis is genetically programmed in this case, and regulation of these reactions varies according to cell differentiation, or with the stage of cell and tissue development. Regulation of metabolism by isozymic forms of an enzyme extends the influence of cellular controls beyond the stage of simple chemical reactions.

Multienzyme Systems

Multienzyme systems are sequential chains of catalysts involved in reactions in which the product of one reaction becomes the substrate of the next in the sequence. There are three levels of complexity of molecular organization that have been recognized for multienzyme systems. The simplest involves individual enzymes in suspension in the cytosol, existing as independent molecular entities that are not physically associated at any time during their action. The small substrate molecules move rapidly among

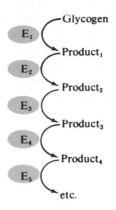

Figure 3.21
Degradation of a substrate such as glycogen occurs through sequential action of a system of enzymes not physically associated with one another in the cell. Intermediates freely diffuse among the enzyme molecules.

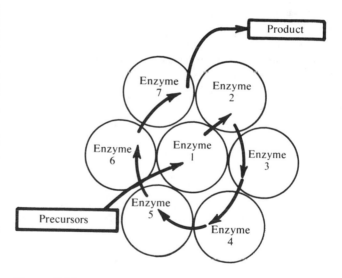

Figure 3.22
Mode of action and organization of the fatty acid synthetase enzyme complex. The reaction intermediates remain sequestered within the physically associated enzymes of the cluster during steps that process precursors into the final fatty acid product, which is then released from the complex.

the enzyme molecules due to high rates of diffusion. Enzymes of the glycolysis sequence provide an example of such a simple system (Fig. 3.21).

Groups of individual enzymes that are physically associated and function together are known as **enzyme complexes.** The individual enzymes are inactive if separated, but when present in a tightly bound cluster, the entire complex functions to guide the synthesis of a molecule from smaller precursors. One well known system of this kind is the fatty acid synthetase complex that catalyzes synthesis of fatty acids from two-carbon precursors (Fig. 3.22). The metabolite intermediates remain sequestered within the enzyme cluster and are released only after the completed fatty acid has been formed by coordinated action of the seven enzymes in this particular complex.

The most intricate and highly organized multienzyme systems are those associated with cellular structures, especially membranes and ribosomes. In the case of components of the electron transport chain of aerobic respiration, the enzymes are an integral part of the membrane structure. In prokaryotes they are part of the surrounding plasma membrane, and in eukaryotes they are incorporated within the structure of the inner membrane of the mitochondrion (see Chapter 7). These multienzyme systems are particularly difficult to study by conventional biochemical methods because their catalytic activities require an intact membrane structure.

SUGGESTED READINGS

Books, Monographs, and Symposia

Dickerson, R. E., and Geis, I. 1969. *The Structure and Action of Proteins.* Menlo Park, Calif.: W. A. Benjamin.

Lehninger, A. L. 1970. *Biochemistry.* New York: Worth.

Roller, A. 1974. *Discovering the Basis of Life: An Introduction to Molecular Biology.* New York: McGraw-Hill.

Articles and Reviews

Changeux, J. 1965. The control of biochemical reactions. *Scientific American* 212(4):36–45.

Dickerson, R. E. 1972. The structure and history of an ancient protein. *Scientific American* 226(4):58–72.

Koshland, D. E., Jr. 1973. Protein shape and biological control. *Scientific American* 229(4):52–64.

Pastan, I. 1972. Cyclic AMP. *Scientific American* 227(2):97–105.

Phillips, D. C. 1966. The three-dimensional structure of an enzyme molecule. *Scientific American* 215(5):78–90.

Stroud, R. M. 1974. A family of protein-cutting enzymes. *Scientific American* 231(1):74–88.

Chapter 4

Membrane Structure and Functions

All cellular life must maintain membrane integrity to survive and function amidst diverse and potentially disruptive surroundings. The plasmalemma is capable of rapid self-sealing if punctured, but any extensive damage that is not repaired will lead to cell death. This cellular membrane is a dynamic barrier separating the internal order of cell functions from surrounding disorder, since it not only sequesters living material but also regulates the movement of solutes, solvents, and particles into and out of the cell.

All organelles as well as the cell itself have *selectively permeable* membranes that allow some substances to move through them at different rates from other substances. In addition to monitoring the diffusion of solutes, membranes and their component molecules participate in moving solutes across the membrane against a diffusion gradient. This process is usually accomplished at the expense of energy manufactured within the membrane itself. The domain of membrane biology has long been claimed by cell physiologists, but biochemists, biophysicists, and cell biologists have recently staked their claims to certain features of investigation in this mine of activities. We will examine some of these studies now.

STRUCTURAL MODELS

During the 1890s there were enough physiological studies of various membranes for E. Overton to propose that surface membranes basically were lipid barriers. Lipid-soluble substances penetrated membranes much more readily than water-soluble materials. This observation was most easily interpreted by postulating a lipid membrane in which lipid-soluble components could move through easily. Using plasma membranes isolated from red blood cells (erythrocytes), chemical studies indeed showed that lipids were a constituent of this membrane. Very little membrane is present in mature erythrocytes other than the plasmalemma so the analysis was relatively specific for the plasmalemma of these cells. By measuring the surface area of these cells and calculating the amount of lipid per cell, it was concluded that the lipid must occur in a layer that was just two molecules thick. Such a bimolecular layer could account for the fact that there was twice as much lipid per cell than was needed to make a plasma membrane only one molecule thick, based on measured cell surface area.

Danielli and Davson Model

The surface properties of a postulated lipid film as a cell membrane were not the same as the properties of the living cell surface however, so there was a problem in proposing a simple lipid bilayer as a model of the cell membrane. This difficulty was disposed of by James Danielli and Hugh Davson who postulated that a coating of protein over the lipid bilayer could account for the observed differences in the properties of the cell membrane and a film of pure lipid. Their model of membrane structure, proposed in 1935, was the first significant attempt to describe the membrane in molecular terms and to relate the structure to observed biological and chemical properties.

Danielli and Davson proposed a model in which two layers of phospholipid molecules were arranged so that their hydrophobic fatty acid chains faced each other in the interior of the membrane. The hydrophilic portion of each phospholipid layer faced the outer borders of the membrane and were coated with globular proteins. In effect, the proteins sandwiched the phospholipid bilayer (Fig. 4.1). Hydrophobic properties of the membrane were explained on the basis of the lipid center, and hydrophilic properties were accounted for by the protein coatings attached to the phosphate residues of the phospholipid molecules. Davson and Danielli paid more attention to the lipid than to the protein portions of the membrane in relation to membrane functions; this was a reflection of the extent of our knowledge about the two classes of organic compounds in the 1930s.

The Robertson Unit Membrane Model

J. David Robertson studied the membranes of the cell from electron micrographs of sectioned materials. The preparations involved the usual steps of killing cells in fixing fluids such as solutions of osmium tetroxide or potassium permanganate, dehydrating in solvents such as acetone and alcohol, and embedding in plastics before sectioning. The earlier photographs were of comparatively poor quality, but by 1960 there had been enough improvement in electron microscopy to provide clear images of a three-layered membrane about 75 Å thick (Fig. 4.2).

Building on the Danielli-Davson membrane model, Robertson proposed a more detailed structure based partly on electron microscopy and partly on functional criteria. He postulated a single layer of protein molecules over the inner and outer surfaces of the lipid bilayer, with extended (β-sheet) rather than globular protein conformations (Fig. 4.3). This model fit the measurements of the three layers of a membrane that could be seen in electron micrographs, allowing 20 Å each for the two protein coatings and 35 Å for the interior lipid bilayer. Extended protein

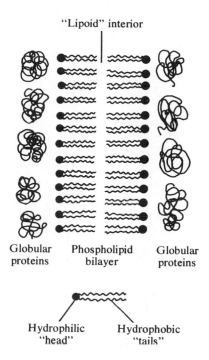

"Lipoid" interior

Globular proteins   Phospholipid bilayer   Globular proteins

Hydrophilic "head"   Hydrophobic "tails"

Figure 4.1
Model of membrane structure according to Danielli and Davson.

conformation fit the 20 Å dimension fairly well, whereas globular proteins would generally be larger than 20 Å in diameter. Because Robertson found the same dark-light-dark staining pattern and a uniform thickness of 75 Å for all membranes he examined, he proposed that eukaryotic and prokaryotic membranes had a common structural plan, or **unit membrane.** The plasmalemma was a single membrane and mitochondria were two-membrane systems, but in each case, an individual membrane conformed to the unit membrane on the basis of its thickness and staining properties.

Robertson's model was widely approved, but there was a continuing undercurrent of dissatisfaction with the model, mostly because these dynamic structures had been pictured as static and uniform elements. Other points of dissatisfaction involved questions about the presence of artifacts following the harsh procedures used to prepare materials for electron microscopy, and the difficulties in reconciling the common unit membrane structure with the variety of membrane functions and specificities. Variety in function is usually accompanied by some variety in structure in living systems, but Robertson's unit membrane was postulated to be essentially the same in all parts of all cells.

These initial difficulties in accepting the model became even more troublesome as additional studies showed that membranes varied between 50 and 100 Å in thickness, with particular membranes showing a characteristic thickness. It was also difficult to understand how active globular proteins could remain active, if they assumed an extended conformation over the lipid bilayer. Change in protein shape usually leads to profound change in protein activity, and the unfolding of globular proteins usually leads to inactivation. From a chemical standpoint, the model did not provide a basis for explaining why some proteins were difficult to extract from membranes while others dissociated readily from the lipid components.

The unit membrane did account for a number of observed properties of cell membranes, however:

1. The densely packed lipid bilayer easily accounted for the presence of 40 percent lipid, by weight, in membranes.
2. Hydrocarbons are poor electrical conductors, so the continuous hydrocarbon phase of the natural membrane would explain the known high electrical resistance of membranes.
3. High permeability of natural membranes to nonpolar molecules could be explained by their ready solubility in the nonpolar lipid phase, and at the same time account for relative impermeability to small ions which do not dissolve readily in this medium.

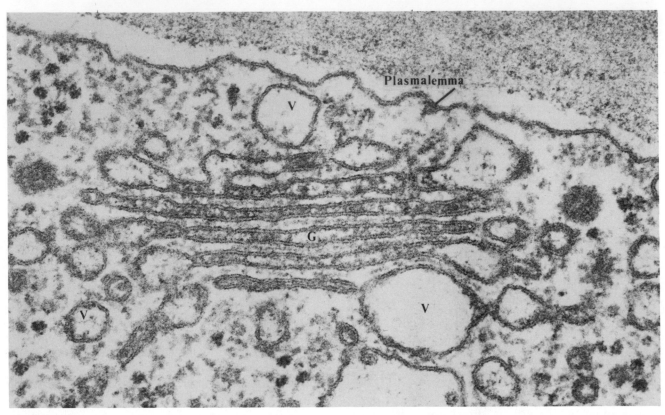

Figure 4.2
Electron micrograph of a thin section through a root cell of *Potamogeton natans* showing the typical dark-light-dark staining pattern of the plasmalemma. Membranes of the Golgi apparatus (G) and its associated secretion vesicles (V) also show the trilaminate staining pattern of a "unit" membrane. × 151,000. (Courtesy of M. C. Ledbetter)

4. Phospholipids spontaneously form bilayer systems *in vitro* when added to an aqueous environment, and there is no requirement for work input to maintain this minimum-energy conformation of the artificial membrane.

5. The model accounted for the three-layered staining pattern of fixed membranes seen with the electron microscope.

Another positive feature of the unit membrane model was the asymmetry of the structure. Robertson based this in part on differences in appearance of the inner and outer stained layers of the unit membrane, and it fit well with known cellular membranes that contained saccharides and conjugated proteins on one surface and not on the other. Current references to "unit" membranes have come to signify the three-layered stained image and not necessarily acceptance of Robertson's proposed model of the membrane.

The Fluid Mosaic Model

Objections to the unit membrane model increased during the 1960s and this led to reexaminations of lipid—protein interactions and to new models. Studies

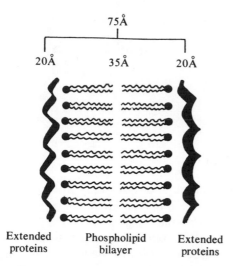

75Å

20Å   35Å   20Å

Extended   Phospholipid   Extended
proteins   bilayer   proteins

Figure 4.3
Model of the "unit" membrane structure according to interpretations of electron micrographs made by Robertson.

of mitochondrial and chloroplast membranes especially underscored the differences between observed features of membranes and the uniformity that was required by the unit membrane concept. Mitochondrial and chloroplast membranes contain displays of particulate units in or on the membrane. The plasmalemma did not present the same appearance as mitochondrial or chloroplast membranes. It seemed that different models might be needed to describe different functional membrane types. This unsuitable approach became unnecessary when a mosaic membrane model was proposed. This model could accommodate functionally distinct membranes if different protein or lipoprotein subunits were arranged within the lipid framework of any membrane.

David Green suggested several possible models for a mosaic membrane, and together with Ronald Capaldi has more recently proposed a model in which the basic membrane structure emerges from protein—protein interactions, with patches of lipid bilayer dispersed among the proteins (Fig. 4.4). Since proteins may account for as much as 75 percent by weight

of some membranes, Green and Capaldi emphasized the protein more than the lipid constituents, although both are essential to membrane structure and function (Table 4.1).

In contrast with the Green-Capaldi model, which requires some restriction on protein · movements through the thickness of a membrane, S. J. Singer and Garth Nicolson described a **fluid mosaic model** in which the lipid bilayer is the cementing framework of the membrane, and attached or embedded proteins interact with each other and with the lipid but retain the capacity to move laterally in the fluid lipid phase (Fig. 4.5). The *predominant* interactions which hold the membrane together and account for its activities are viewed differently by the two groups. Green and Capaldi stress protein—protein interactions, whereas Singer and Nicolson suggest that protein—protein, lipid—protein, and lipid—lipid interactions contribute to membrane structure and dynamics.

Singer and Nicolson consider two kinds of noncovalent interactions to be predominant: **hydrophobic** and **hydrophilic**. The stable structure can exist in the minimum free-energy state if both hydrophobic and

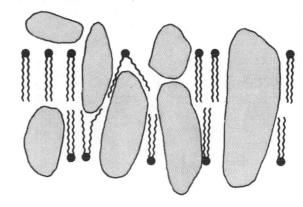

Figure 4.4
Model of a fluid mosaic membrane structure proposed by Green and Capaldi. In this model protein-protein interactions predominate. Proteins are distributed at the surfaces and within the patches of phospholipid bilayer, extending from one surface to the other or only partway through the membrane thickness.

Table 4.1 Composition of mammalian membrane preparations*

COMPONENT	MYELIN	ERYTHROCYTE PLASMA MEMBRANE	LIVER PLASMA MEMBRANE	HEART MITOCHONDRIA
Proteins	22	60	60	76
Total lipids	78	40	40	24
Phospholipids	33	24	26	22
Glycolipids	22	trace	0	trace
Cholesterol	17	9	13	1
Other lipids	6	7	1	1

* Values taken from various sources, expressed in percentage, by weight.

hydrophilic interactions are maximized. Hydrophobic and hydrophilic interactions are maximized in the aqueous environment inside and outside the cell if the nonpolar fatty acid chains of the phospholipids and the nonpolar amino acid residues of the proteins are kept from contact with water to the greatest possible extent and at the same time, the polar and ionic groups of the membrane proteins, lipids, and carbohydrates are in contact with water (Fig. 4.6). The Robertson unit membrane model does not meet these requirements since the covering protein layers keep the lipid polar groups away from contact with water while the nonpolar residues of these proteins are exposed to water instead of being sequestered

away from the aqueous solvent. Both phospholipids and proteins are **amphipathic** molecules, having both polar and nonpolar regions in the same molecule (see Chapter 2), but these regions are not distinguished or accounted for in the unit membrane. Since neither the hydrophobic nor the hydrophilic interactions can be maximized in such a model, Singer has rejected it as being thermodynamically unstable and thus unlikely to exist in the cell.

Both the Singer-Nicolson and the Green-Capaldi models call for two broad categories of proteins: **peripheral** (or extrinsic) and **integral** (or intrinsic).

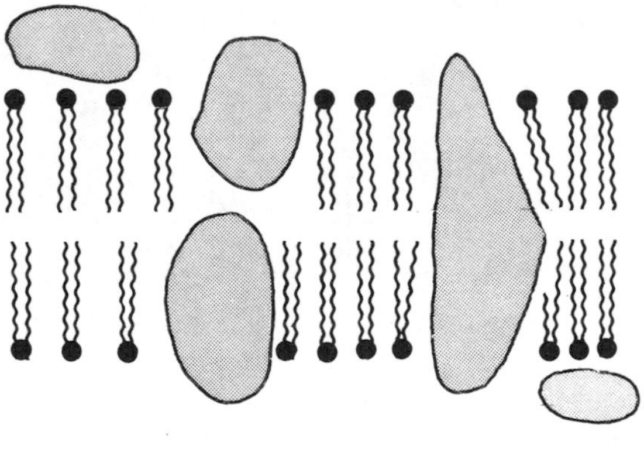

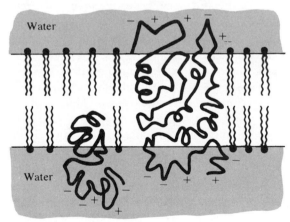

Figure 4.6
Hydrophilic interactions occur among the charged residues of amphipathic membrane proteins, the water phase, and the polar heads of amphipathic phospholipids. Hydrophobic interactions occur between uncharged protein residues and nonpolar tails of the membrane phospholipid bilayer.

Figure 4.5
The fluid mosaic model of membrane structure proposed by Singer and Nicolson. The phospholipid bilayer is a cementing framework with attached peripheral proteins or integral proteins embedded within the bilayer.

Most of the proteins of more active membranes are believed to be integral; they are tenaciously held in place by strong hydrophobic or hydrophilic interactions (or both) and thus are difficult to remove from membranes except by harsh and disruptive methods of extraction. Peripheral proteins are located more superficially at the membrane surface and are thus more easily separated during extraction. Some integral proteins span the entire thickness of the membrane and protrude at one or both surfaces from the lipid bilayer, while others project from only one surface with the remainder being embedded within the lipid bilayer. No protein is believed to be entirely embedded in lipid.

The widely differing activities and specificities of the various kinds of membranes in cells are assumed to reflect different kinds of proteins that occur throughout the membrane or in some localized regions of the structure. Differences in the content of enzyme proteins have been studied particularly in mitochondria and chloroplasts. In both organelles the enzymes of electron transport and ATP formation are situated within the inner membrane and do not occur in the outer envelope of the organelle.

The photosynthetic complexes involved in the capture and conversion of light energy are also tightly bound within the innermost chloroplast membrane system. These enzyme systems must be integral proteins since they can be extracted from organelle inner membranes only with great difficulty after relatively harsh treatment. In prokaryotic blue-green algae and photosynthesizing bacteria the photosynthetic enzymes are localized to infoldings of the plasmalemma, but the plasmalemma at the cell surface does not have these particular proteins in its structure. The continuous sheet of membrane in these cell types shows distinctly localized groups of enzymes involved in photosynthesis. Immunological specificities clearly involve surface membrane proteins that are located in particular regions of the membrane and that differ from one cell type to another according to antigen-antibody tests.

Observed differences in membrane thickness can be related to the presence of substantial amounts of peripheral proteins in thicker membranes and to predominance of integral proteins in thinner membranes such as those of the mitochondrion. Measurements of mitochondrial membrane show it to be a little over 50 Å, which is not much thicker than the 50 Å-thick lipid bilayer alone. This membrane has very few peripheral proteins so that its major thickness is due to the lipid, with a little bit of protruding integral protein added on.

Membranes of most systems are asymmetrical, having different functional proteins on the outer and inner surfaces. The plasmalemma of erythrocytes, and probably of other cell types as well, have saccharide residues of glycolipid and glycoprotein molecules exclusively on the outer surface and are therefore obviously asymmetrical membranes. The mitochondrial inner membrane is definitely asymmetrical since some respiratory components are found only on the outer surface and some proteins occur on the inner surface exclusively. Various respiratory proteins of this membrane have been shown to be situated within the membrane interior. This "sidedness" of membranes is considered by some investigators to be an important structural aspect in transport of substances across the membrane and in lipid-protein interactions required for certain enzyme activities.

The lipid matrix of the membrane has the consistency of a light oil, according to several lines of physical evidence. Lipids and proteins have been shown to move laterally within this fluid layer. The lateral, or translational, motion within the plane of the membrane provides one way in which membrane proteins can interact with one another and with lipids. This interaction would be a required feature in certain metabolic interactions that are associated with the preservation of an intact membrane system to maintain functional activity. If the interacting components are separated during extraction, they may lose their capacity to establish physical contacts at specified

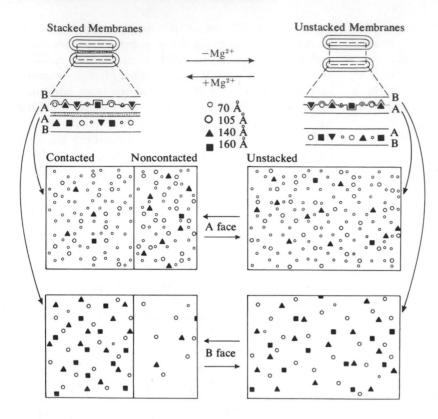

Figure 4.7
Summary diagram illustrating lateral particle movement during thylakoid unstacking in wild type and thylakoid stacking in mutant *Chlamydomonas* chloroplast structure. Four types of particles were identified within the membrane. Some remained with one fracture face, and others remained with the alternative fracture face in freeze-fractured membrane preparations. (From Ojakian, G. K., and P. Satir, 1974, *Proc. Natl. Acad. Sci.* **71**:2052–2056, Fig. 9.)

times and places for a metabolic reaction to occur.

Photosynthetic membranes within chloroplasts become inactive or show altered enzymatic activity if the tight joining between individual membranes is interrupted. Examination of the protein particles inside these membranes showed that "stacked" membranes contained a different distribution of certain particles than "unstacked" membranes (Fig. 4.7). There was no change in the numbers of particles per unit area of membrane, only a change in their distribution within the membrane thickness. This kind of information provides support for the fluid mosaic model of the membrane as proposed by Singer and Nicolson. It also shows that membrane function is related to the spatial arrangement of components as well as to the kinds of components that are present. Relating membrane structure to membrane functions will require more than chemical analysis of the protein and lipid constituents of different cell and organelle types. The geometry of membrane organization must play an important role in underwriting function.

Membrane junctions within the chloroplast, between different organelles in the cell, and between

Figure 4.8
Thin sections through a portion of a chloroplast: (a) leaf mesophyll cell of timothy grass (*Phleum pratense*). There is considerable density at the junctions between adjacent stacked membranes, but no dense material is evident at the surfaces of unstacked membranes within the common matrix of the organelle. × 80,000. (Photograph by W. P. Wergin, courtesy of E. H. Newcomb) (b) Leaf cell of *Elodea canadensis,* an aquatic flowering plant. The dense junctions between the stacked membranes are even clearer in a higher magnification view. × 135,000. (Courtesy of M. C. Ledbetter)

(a)

(b)

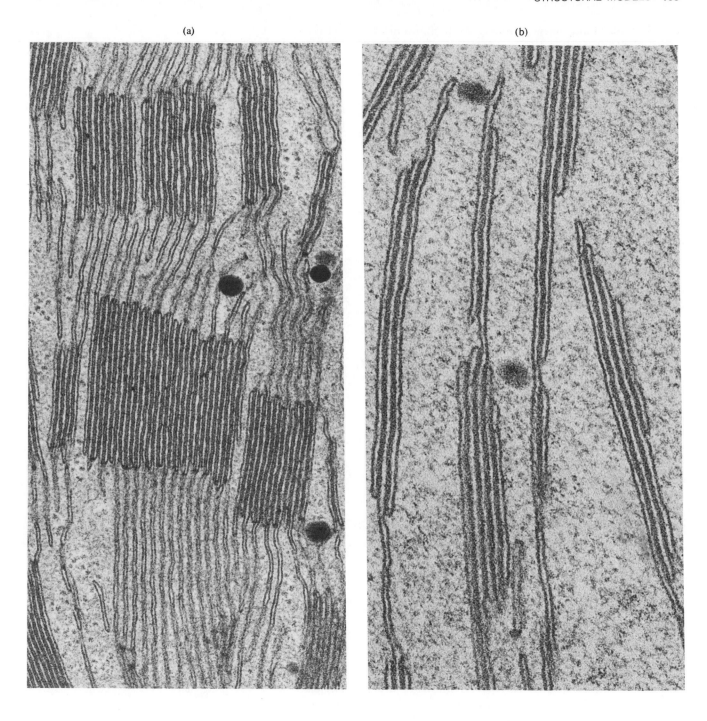

adjacent cells have been proposed as facilitating exchange of substances. These junctions apparently form as a result of molecular redistributions within the apposed membranes, which leads to tight coupling between previously separate structures (Fig. 4.8).

MOVEMENT OF SUBSTANCES ACROSS MEMBRANES

Substances move through membranes *selectively* in the living cell. Once the membrane has been damaged or the cell has been killed, molecules move freely across the membrane and may reach equilibrium. Equilibrium is rarely achieved in living systems, so that the dynamic state of the living membrane acts as a regulatory barrier for entry and exit of molecules and particles. There are three general routes by which substances cross the membrane barrier: (1) **free diffusion** along a gradient going from higher to lower concentration of the molecule; (2) assisted passage, or **transport,** across the membrane either by a process of **facilitated diffusion** in which the substance moves as expected in the direction of its lower concentration, or by energy-expending **active transport** against a concentration gradient; and (3) enclosure in membranous vesicles to enter the cell by the process called **endocytosis** or be expelled from the cell in the reverse process of **exocytosis.**

Free Diffusion

According to a large amount of evidence, many substances move through membranes at rates of free diffusion that are directly proportional to their solubility in lipid (Fig. 4.9). Water molecules are a notable exception to this rule since they freely diffuse through membranes regularly and rapidly. It has been suggested that membranes contain 8–10 Å-wide pores that are lined by hydrophilic residues (Fig. 4.10). Such openings would be large enough for water molecules but would hardly accommodate other water-soluble substances to pass through the narrow

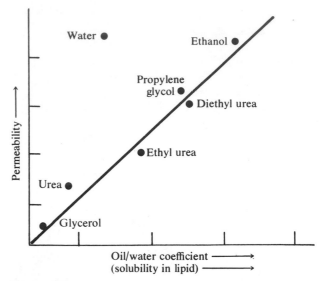

Figure 4.9
Graph showing that the rate of free diffusion of chemicals through an oil/water phase system (analogous to a membrane) is directly proportional to the solubility of the chemical in lipid. Water molecules are an exception to this rule.

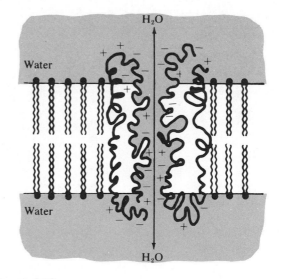

Figure 4.10
Illustration of the concept that narrow pores perforate the membrane allowing rapid diffusion of water molecules through their interactions with hydrophilic residues of adjacent membrane proteins.

passageway. The suggestion is quite logical to explain diffusion of water through membranes, but there is little direct evidence to support the concept.

Free diffusion along a concentration gradient could be maintained for entry and exit of metabolites and waste products as long as substances were changed or washed away to keep one end of the gradient lower in concentration. If metabolites are chemically changed on entering the cell, then the concentration of the metabolite remains high outside the cell and is kept low inside the cell if it is not retained in its original form. Similarly, wastes leaving the cell could continue to move from the higher concentration inside as long as these substances were washed away or removed somehow from the immediate vicinity of the cell (Fig. 4.11).

The idea is attractive because no particular apparatus is required to maintain diffusion, and no energy expenditure is needed to keep the system operating. The major difficulty, of course, is that most biologically important molecules of cellular metabolism are insoluble in lipid and would be confronted by the lipid barrier regardless of the favorable state of the concentration gradient. Important lipid-insoluble substances, such as sugars, move through membranes selectively and efficiently. Simple diffusion cannot explain these movements and membrane transport mechanisms have been widely accepted instead as explanations for selective movement of many biologically significant solutes across the lipid layer.

Transport Mechanisms

Essential lipid-insoluble metabolites, such as sugars and amino acids, enter and leave the cell or its compartments through processes involving reversible combination with membrane proteins. These **carriers** are integral proteins that form part of the structure of the membrane. Carriers are highly specific. Each is assumed to have a characteristic binding site which will bind one particular kind of molecule but not another that may be almost identical. The binding between carrier and metabolite is transient. After the bound molecule has been translocated to the other side of the membrane the carrier is freed and may recycle to assist other molecules of this substance to get across the membrane. The relative solubility of the metabolite in lipid is not significant since its interaction is with protein rather than lipid constituents of the membrane.

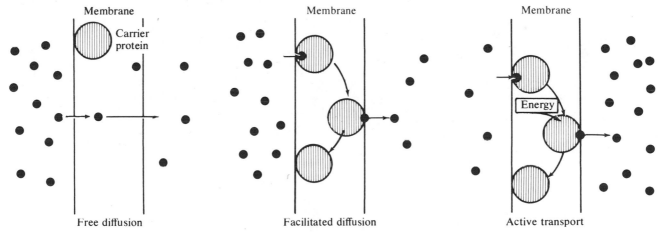

Figure 4.11
Three modes of movement across the membrane by metabolites. The relative concentrations of metabolite are indicated by the number of dots on either side of the membrane.

Certain carriers are called **permeases,** in recognition of their resemblance to enzymes in certain respects. In particular, permeases accelerate transport, provide selectivity to transport, and recycle unchanged at the conclusion of the transport event. These properties are enzymelike, but permeases and other carriers differ in one very important respect from enzymes. Permeases may alter the equilibrium point of a chemical reaction, sometimes to a large extent, whereas enzymes only change the rate at which equilibrium is reached but do not alter the point of equilibrium itself.

Permeases may assist molecules across the membrane in the usual direction of the concentration gradient, from high to lower amounts of the metabolite, or against this usual "downhill" direction. At least two carrier-mediated transport mechanisms must exist, therefore, one that sponsors facilitated diffusion and another that aids energy-requiring active transport in the "uphill" direction of a gradient. In facilitated diffusion, the carrier helps the molecule along the direction in which that molecule ordinarily would move. The diffusion is facilitated because the molecule cannot penetrate the selectively permeable membrane on its own by free diffusion movement. In active transport, a substance can continue to be accumulated in a region where it already exists in higher concentration only if energy from metabolism is continually supplied to the carrier system. Metabolic energy drives active transport.

Biochemical, physiological, and genetic studies have provided support for the existence of permeases. Relatively few permeases have been identified specifically or characterized from purified preparations. However, experimental results of various kinds have been interpreted most readily according to the permease concept. There are indications that regulation of membrane transport by permeases depends on the amount of carrier protein, the kinds of carriers produced, and the rate of carrier activity. In these respects the regulation of transport resembles regulation of metabolism by enzymes. In addition, both systems are genetically determined.

ACTIVE TRANSPORT AND ION PUMPS. Cells can accumulate substances in excess of expected amounts in at least three ways:

1. The substance can be precipitated from solution once it is inside the cell, effectively reducing the concentration of solute in water.
2. The molecule can be chemically changed after it has gone through the membrane, thus reducing the concentration of the specific molecule involved in the concentration gradient.
3. The transport of a metabolite can be coupled directly to a second reaction that is energetically favorable to driving the transport reaction "uphill."

The accumulation of Ca^{2+} ions in cisternae of muscle endoplasmic reticulum operates by way of precipitation. Calcium phosphate is the precipitate in this case, and its formation maintains a favorable gradient of calcium ions. Sugars may accumulate inside the cell by a combination of facilitated diffusion and phosphorylation events, particularly in bacterial cells. If the entering sugar is phosphorylated immediately, then the inside concentration of the nonphosphorylated sugar remains low, and additional molecules will be transported into the cell along the usual "downhill" direction of the concentration gradient. Both of these situations differ from active transport in that their direction of metabolite movement is toward the region of lower concentration, whereas active transport leads to continued accumulation of a substance in the region where it occurs in its highest concentration. The end result is very similar in all these cases, namely, a situation exists by which a substance accumulates inside the cell or some cell compartment in great amount.

One unifying concept that has developed in recent years explains active transport on the basis of pumping actions. In particular, the active pumping of one substance out of a cell or compartment provides the driving force for active transport of various other substances inward. The pumping in of these metabolites is thus considered to be coupled to the outward transport of some one other material, but in a manner that provides the energy required to drive the needed materials inward, in the "uphill" direction of their concentration gradients. The pump is an economical as well as a simple system since the outward movement of one kind of substance helps to drive in many kinds of metabolites.

The solutes that are most actively pumped into cells are K^+ ions, sugars, and amino acids. The driving force for this inward transport is believed to be a Na^+ gradient across the membrane, created by active transport of Na^+ ions pumped out of the cell. The external Na^+ ion concentration remains high and the internal ion concentration remains low as Na^+ continues to be transported outside the cell. The energy required to pump Na^+ ions out of the cell is provided by ATP, which is hydrolyzed by a Mg^+-activated ATPase believed to be situated within the membrane. The free-energy difference between ATP and its ADP and inorganic phosphate products of hydrolysis is made available for active transport in the operation of the **primary sodium pump.** This pump acts in animal cells but apparently is not a feature of plant cells or of bacteria, neither of which requires Na^+ ions for its metabolism. A H^+ ion pump is active in bacterial cells; the pump in plant systems is not well understood.

Two distinct types of Na^+ pump have been described for animal cells. In one, the outward pumping of Na^+ is tightly linked to the inward transport of K^+ ions. Since Na^+ and K^+ are exchanged in a compulsory way, outward movement of Na^+ is always accompanied by inward movement of K^+. This type of sodium pump is called the **sodium/potassium exchange pump,** or the **coupled neutral pump** (Fig. 4.12). Inward transport of K^+ does not necessarily accompany outward extrusion of Na^+ in the **electrogenic sodium pump** (Fig. 4.13). The electrogenic pump has been so named because a gradient of electrochemical potential may be generated when exit of Na^+ is not compensated by one-to-one entry of K^+. In most cells there is an accumulation of K^+ that exceeds the loss of Na^+ (or H^+ in bacteria), which is one of the

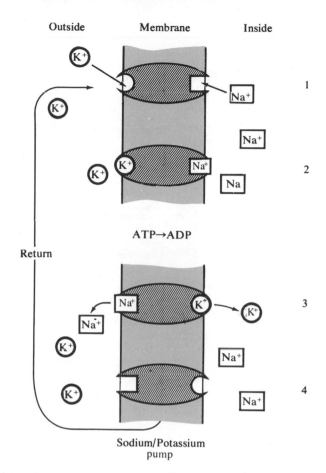

Figure 4.12
Diagram illustrating the action of a sodium pump of the *coupled neutral* type. Inward transport of K^+ ions is accompanied by compulsory outward transport of Na^+ ions via carrier proteins situated within the membrane. The sequence of events is shown in steps 1–4.

indications that the electrogenic pump rather than the neutral pump is operative. Nerve and muscle cells particularly are known to have the coupled neutral pump.

A high intracellular concentration of K^+, regardless of the external concentrations of Na^+ and K^+, is especially needed by aerobic cells of all species. Two vital processes that require a high concentration

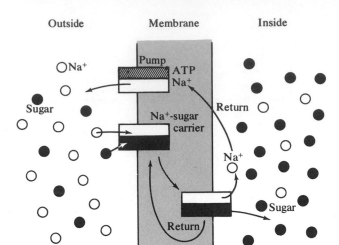

Figure 4.13
Diagram illustrating the action of an electrogenic sodium pump through which sugars and other metabolites enter the cell along an electrical gradient generated by the energy-requiring extrusion of Na^+ ions.

of K^+ are protein synthesis at the ribosomes and one of the critical enzymatic steps in glucose processing during glycolysis. The high internal K^+ concentration must be balanced by loss of some cation, such as Na^+ or H^+, or there would be excessive swelling that would cause the cell to burst by creating a condition of high internal osmotic pressure.

Active transport of amino acids into cells is another consequence of the action of an electrogenic Na^+ pump. Extrusion of Na^+ from the cell generates a gradient of lower internal and higher external concentration of Na^+. The energy inherent in this gradient is believed to provide the driving force for transport of amino acids into the cell in the "uphill" direction, leading to accumulation of these essential compounds. The Na^+ gradient itself is formed at the expense of ATP. Specific carrier protein systems assist amino acids across the cell membrane in these active transport events.

The capacity to accumulate sugars, like amino acids, is another feature of the cellular electrogenic

Na^+ pump activity. Both sugar and amino acid active transport are accomplished under conditions in which the outside concentration of Na^+ is kept high enough to create an appropriate gradient whose energy drives metabolites into the cell from very dilute outside solutions of these substances. The accumulation of sugars is therefore coupled to Na^+ extrusion, as with amino acids, and is also assisted by specific active-transport proteins (see Fig. 4.12).

TRANSLOCATION ACROSS THE MEMBRANE. The carrier proteins assist hydrophilic molecules across a membrane thickness of 60 to 100 Å. The metabolites are considerably smaller than 60 Å, so it is important to know how these molecules are translocated across this relatively large distance by the carriers. Several alternatives have been proposed, but two have been studied more intensively than the other possibilities.

One alternative hypothesis postulates that the carrier binds with the hydrophilic molecule and then the entire transport protein rotates across the membrane and delivers its bound metabolite to the other

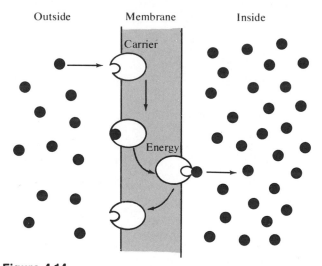

Figure 4.14
Diagram illustrating the carrier mechanism concept for translocation of hydrophilic molecules across the membrane barrier.

side (Fig. 4.14). The second alternative proposes that the carrier is fixed in place within the structure of the membrane, and that the carrier molecule undergoes a conformational change that translocates the binding site across the membrane and the bound metabolite along with it at the same time (Fig. 4.15). Once the metabolite has been translocated the binding site is freed and restored to its original conformation, ready to bind another hydrophilic molecule in another transport event. This second alternative has been referred to as the **fixed-pore mechanism.** The first alternative is known as the **carrier mechanism.**

From an energetic standpoint the carrier mechanism would require considerable energy expenditure to support rotations of integral carrier proteins across the span of the membrane. Fixed-pore mechanisms, on the other hand, have a lower energy cost and can be related more directly to membrane structure. Integral proteins with both hydrophilic and hydrophobic residues could be envisioned existing along the pore channels, with their hydrophilic portions lining the opening through the membrane while their

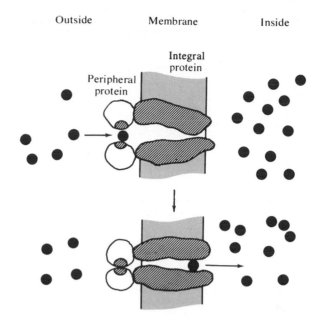

Figure 4.16
Participation by peripheral proteins in the translocation of solutes across the membrane according to the fixed-pore mechanism. This mechanism uses the Singer and Nicolson structural model of the fluid mosaic membrane.

hydrophobic regions were within the lipid matrix. The bound hydrophilic solute molecule would be translocated across when some energy-yielding process caused a conformational change in the carrier. Peripheral proteins also could aid in translocating solutes through fixed pores in active transport processes, as suggested by S. J. Singer and others (Fig. 4.16). Although still incomplete, the weight of available evidence is more in favor of the fixed-pore than of the carrier mechanism.

Transport by Vesicle Formation

Most cell membranes can enclose materials in **vesicles** and bring the substances into the cell in this way, or package materials for discharge from the cell in a reverse process. The process is called **endocytosis** when materials are brought in, and **exocytosis**

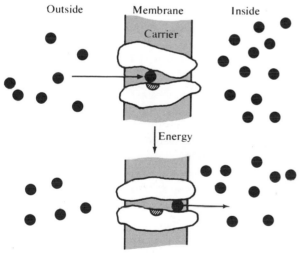

Figure 4.15
Diagram illustrating the conformational change of carrier protein according to the fixed-pore mechanism for translocation of hydrophilic molecules across the membrane.

when vesicle discharge takes place (Fig. 4.17). Endocytosis has features that are analogous to active transport. For example, substances enter along an "uphill" concentration gradient, and energy is required to support the process. Endocytosis will stop if poisons that stop energy production in the cell are added, and the process can be stimulated by the addition of ATP to cells in suspension.

Various cell secretions are discharged from the cell by exocytosis. The mucus droplets of intestinal goblet cells, digestive enzyme precursors in granule packages from pancreatic cells, and other cell secretions leave the cells in which they are produced by this general pathway (Fig. 4.18). Although more difficult to see by light microscopy, numerous endocytotic infolded regions can be seen along the plasma

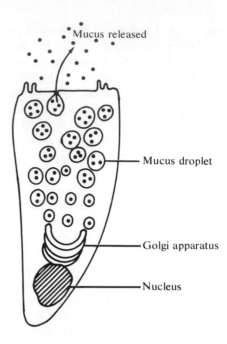

Figure 4.18
Release of mucus, by exocytosis, from mucus droplets formed at the Golgi apparatus within the goblet cell of the intestine.

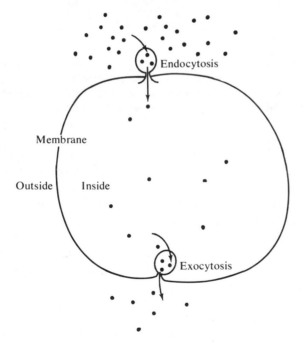

Figure 4.17
Entry by endocytosis and exit by exocytosis transport molecules into and out of the cell across the membrane barrier. The substances are enclosed in a membranous vesicle that fuses with the plasma membrane before entry into or exit from the cell.

membrane in suitable electron micrographs (Fig. 4.19).

Endocytosis and exocytosis play essential roles in **lysosome** digestive activities within the cell. Incoming vesicles fuse with lysosomes, and the lysosomal enzymes proceed to digest the material contained in the endocytotic structures. The digested substances and occasional undigested residues then are discharged from the cell by exocytosis. These processes are discussed at greater length in Chapter 10, along with cellular events associated with the release of secretions packaged within the **Golgi apparatus** in various cell types.

At the present time relatively little information has been obtained on the way in which molecular organization of the membrane can be related to vesicle formation and substance transport. Part of this phenomenon certainly is related to membrane fluidity and repair, but membrane biogenesis must also be

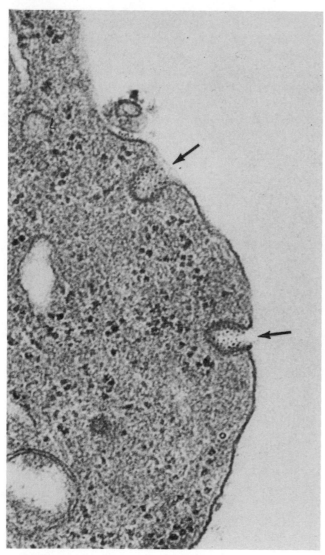

Figure 4.19
Electron micrograph showing invagination of endocytic vesicles at the plasma membrane of an erythroblast from guinea pig bone marrow. Dense ferritin particles adhere to these specialized areas and are carried into the cytoplasm as vesicles pinch off from the plasma membrane. Ferritin uptake by this mechanism is the normal pathway by which these cells obtain iron for hemoglobin synthesis. × 69,000. (Courtesy of D. W. Fawcett, from Fawcett, D. W., 1965, *J. Histochem. Cytochem.* **13**:75–91.)

considered. We can anticipate more rapid progress in the next few years as experimental studies become focussed more precisely on relationships between models of membrane structure and the dynamic properties of functional membranes in living cells.

INTERCELLULAR COMMUNICATION THROUGH JUNCTIONS

Rapid and effective exchanges of ions and metabolites is characteristic of some kinds of cells in communities of living tissues and in cell cultures. These cell-to-cell communications may operate at close range and require direct physical contact between interacting cells in a group. Both the chemical and the physical expressions of communication vary according to cell type and stage of growth and development. The chemical exchanges may include ions or organic metabolites and regulatory molecules. The physical basis for ionic and metabolic cooperation may range from the formation of actual **cytoplasmic bridges** to localized membrane **junctions** that may vary in extent from a few angstroms to micrometers of distance. Cell junctions are specialized regions of close-range contact between adjacent cells. They are associated with a differentiation of the contributing cell surface membranes and intercellular matrix, or they may only involve the matrix material. Junctional contacts range from fused membranes to areas separated by a space as wide as 200 Å.

Three major types of junctional membrane localizations have been recognized by electron microscopy: **gap junctions, tight junctions,** and **septate junctions.** In each case there is a recognizable membrane region which is physically distinguishable from nonjunctional regions elsewhere on the membrane surface (Fig. 4.20).

Gap junctions are seen as localized regions of 7-layered structure when seen in thin-sections by electron microscopy. They were clearly seen and described in 1967 as areas formed from two "unit" membranes with a space or "gap" between the ad-

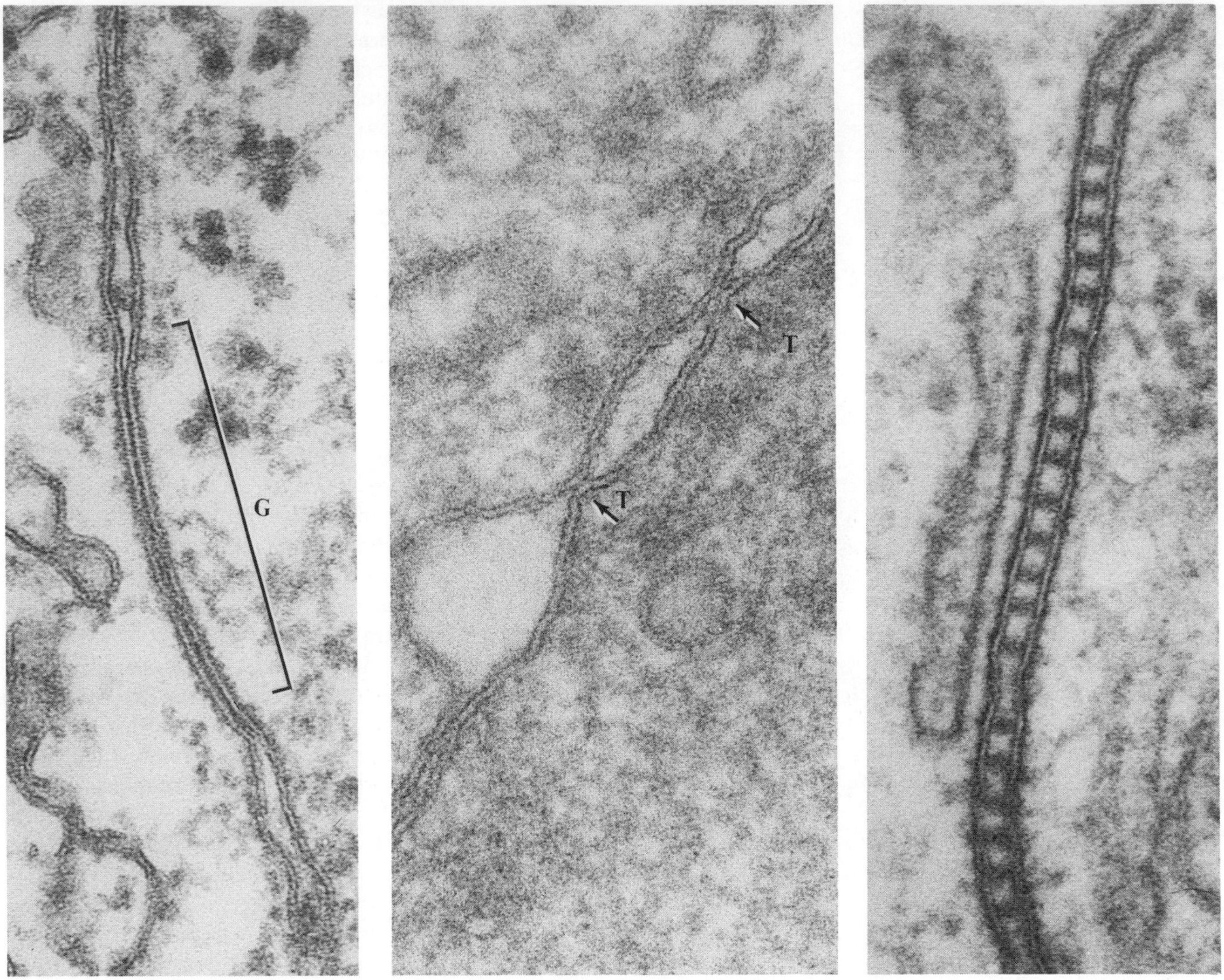

(a) (b) (c)

Figure 4.20
Thin-section appearance of junctions between adjacent animal cells: (a) gap junction between hamster fibroblasts. The two closely apposed junctional membranes are separated by a 20-40 Å space, or "gap." × 210,000. (b) Tight junctions (T) formed by fusion of adjacent plasma membrane areas in epithelial cells of rat small intestine. × 247,500. (c) Invertebrate septate junction. Transverse image of the septate junction between two molluscan ciliated epithelial cells. A periodic arrangement of electron-dense bars, or septa, is present within the intercellular space between the two lateral plasma membranes. × 216,000. (All photographs courtesy of N. B. Gilula. Refer to Gilula, N. B., 1974, *Cell Communications* [ed. R. P. Cox], John Wiley & Sons, pp. 1–29.)

jacent plasmalemmas. The typical 75 Å thickness of each plasmalemma plus the 20–40 Å gap produces an average thickness of 170–190 Å for gap junctions in vertebrate and invertebrate animals. They are commonly observed in excitable and nonexcitable cell types, but they are absent from types such as skeletal muscle fibers and circulating blood cells. Tight junctions represent true membrane fusions. There is no space between the fused membrane width of 140–150 Å. Unlike the widely distributed gap junctions, tight junctions are characteristic of vertebrate tissues such as epithelium. Septate junctions have only been found in invertebrate tissues. They are the largest junctional types, and they sometimes extend for micrometers in length, forming a belt surrounding the apical or basal regions of a cell.

All three kinds of junctions can be described in somewhat greater detail from freeze-fracture preparations for electron microscopy (Fig. 4.21). When the membrane is laid open by fracturing, internal surface textures become visible. The membrane surface nearer the cytoplasm-bordering part of the plasmalemma is the **A fracture face**, and the **B fracture face** is closer to the extracellular matrix around the cell. Face A in gap junction regions shows plaques of relatively small size which are distinguished by a regular latticework of homogeneous particles. Face B is the complementary surface showing pits or depressions that match the orderly arrays of face A particles. The internal pattern of bordering nonjunctional membrane consists of random arrays of different-sized particles.

The A fracture face of tight junctions has a series of ridges in a meshwork pattern, while the B fracture face is a complementary meshwork of grooves. The ridges and grooves are interwoven and represent sites of true membrane fusion. This pattern varies in different epithelial tissues, from "very tight" to "leaky" systems according to permeability criteria. In general, tight junctions exclude movements of large molecules between cells, but "leaky" epithelia are relatively more permeable to such molecules than "tight" or "very tight" tissues.

In freeze-fracture preparations of septate junctions in invertebrate tissues, there are parallel rows of particles exposed on the A face that exactly correspond to the septa between cells in the junction area. The B face contains depressions that are complementary to A face particles. The two adjacent membranes are joined by the septate or ladderlike intermembrane region, which is similar in width in various tissues and species. The spacing between septa, however, often varies from one sample to another.

The particle arrays are not artifacts of the preparative method since the same systems of orderly subunit arrangement can be seen after negative staining of membranes (Fig. 4.22).

Ion and Metabolite Exchange

If molecules are tagged with radioactive isotopes or a fluorescent component, they can be traced as they move from the extracellular medium into the cell and from one cell into another. Radioisotope-labeled reagents can be detected in the intercellular space of the gap junction, showing that the "gap" can serve as a channel through which substances move and penetrate the cell. Molecules of the general size of metabolites and regulatory substances can be injected into one cell and can then be located later in the adjacent cell if there is gap junction communication (Fig. 4.23). Cells which lack gap junctions or which have been dissociated into separate units do not receive the added molecules injected into the test cell in the system. Movement of ions between cells takes place, as detected by electrical measurements using microelectrodes.

Septate junctions also serve as communication links between joined cells, since tracer substances readily penetrate the differentiated region of the membranes. This kind of continuum provided by septate and gap junctions is not typical for tight junctions. The "tighter" the tight junction the more impermeable it is to movement of large molecules from cell to cell. Tight junctions are permeability barriers between cells since the intercellular space is

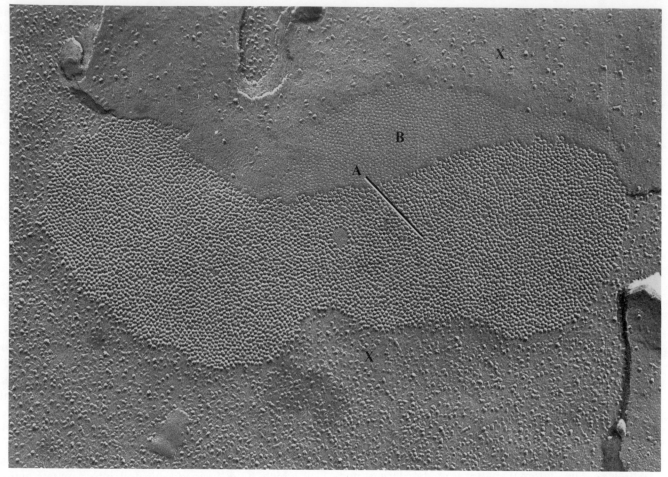

(a)

occluded. The tighter the occlusion, the more impermeable the cell is to penetration by molecules from a neighboring cell.

Because of their wide distribution among multicellular animals and their participation in cellular continuities, more attention has been directed toward study of gap junctions. Gap junctions can be "unzipped" by placing cells in hypertonic sucrose solutions. The osmotic sensitivity of gap junctions is not correlated with changes within the membrane however, since freeze-fracture faces are unchanged after

the junction has been disrupted. Joined cells can also be separated by treatments with protein-digesting enzymes. In such cases the entire gap junction remains part of only one of the two cells. This situation is indicated by the presence of junction areas that are 150–190 Å thick, located on the surface of a separated cell.

According to preliminary chemical studies, gap junctions contain some neutral lipid, some phospholipid, and a few kinds of proteins. Since all these constituents are usually found in membranes gen-

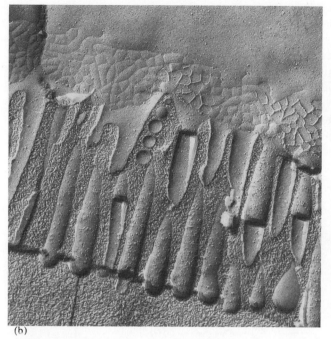

(b)

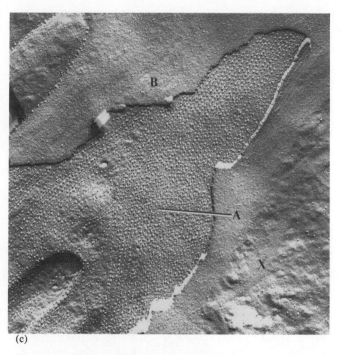
(c)

Figure 4.21

Freeze-fracture appearance of junctions between adjacent animal cells: (a) gap junction from mouse liver. This unique membrane differentiation is characterized by a polygonal arrangement of membrane particles on the fracture face closest to the cytoplasm (A) and complementary pits or depressions on the fracture face closest to the external border of the plasma membrane (B). Note that the gap junction occurs as a plaquelike region that is segregated from regions of nonjunctional plasma membrane (X). The nonjunctional membrane fracture faces are characterized by a random distribution of heterogeneously sized particles. × 108,000. (b) Tight junction between epithelial cells of rat small intestine. The meshwork arrangement of ridges and grooves represent sites of true membrane fusion. The arrangement of anastomosing ridges (facing the cytoplasm) and grooves (facing the extracellular surface) is responsible for the occlusion properties of the tight junction. The fracture faces are exposed when the plasma membrane is split open during freeze-fracture preparation. × 47,000. (c) Septate junction from molluscan ciliated epithelium. Two complementary fracture faces are exposed in the junctional region. The innermost fracture face (A) contains parallel rows of membrane particles that correspond to the arrangement of intercellular septa seen in thin sections. The outermost fracture face (B) contains an arrangement of linear depressions or grooves that complement the particle rows of fracture face A. The particles in the nonjunctional membrane regions (X) are randomly arranged. × 45,000. (All photographs courtesy of N. B. Gilula. Photographs a and c from Gilula, N. B., 1974, *Cell Communications* [ed. R. P. Cox], John Wiley & Sons, pp. 1–29. Photograph b from Friend, D. S., and N. B. Gilula, 1972, *J. Cell Biol.* **52**:758.)

erally, there is little known to distinguish gap junctions chemically from nonjunctional membrane regions. Some evidence from cytochemical staining tests shows that ATPase activity product is deposited in the gap junction area. If this information can be confirmed by biochemical assays, it would show that ATPase activity was a membrane-associated function of the gap junction. Since tests using inhibitors of ATP synthesis also abolish ion and molecule movements between joined cells, an energy requirement is

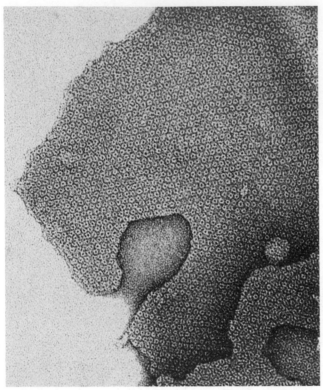

Figure 4.22
Negatively stained gap junction from isolated rat liver plasma membranes. An electron-dense spot is present in the center of all or most of the polygonal subunits in the lattice of the gap junction. × 240,000. (Courtesy of N. B. Gilula)

believed to be necessary for intercellular movement of substances across the junction. ATPase activity in this area would, therefore, be expected, but clear-cut evidence has yet to be obtained.

One kind of system suited to study of metabolite exchanges in communicating cells is a culture made up of wild-type and mutant human fibroblasts (connective tissue components). Certain established lines of mutant cells cannot convert purine precursors into nucleic acids because they lack the required enzyme. If these mutants are mixed with wild-types containing the necessary enzyme, both the mutant and the wild-

type cells are able to incorporate radioisotope-labeled purine precursors into nucleic acid polymers.

This phenomenon has been called **metabolic cooperation,** and it is dependent on formation of junctional contacts between the two kinds of cells. If gap junctions are disrupted, communication between cells stops and the mutant can no longer utilize purines for nucleic acid synthesis. The ability of cells to form gap junctions is inherited. Some kinds of cells do not form junctions, and are considered to be "noncommuni-

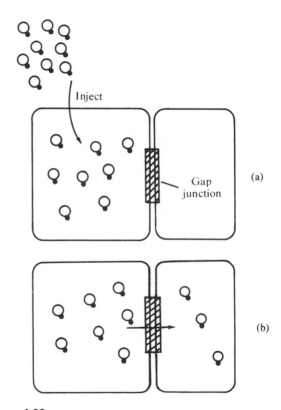

Figure 4.23
Diagram illustrating the movement of tagged molecules (a) into a cell and (b) from one cell to its neighbor through a communicating gap junction. Presence and movement of the molecules can be assayed during an experiment by locating and measuring the fluorescent or radioactive tag added to the molecules that are injected into the cell at the start of the experiment.

cating" types. For example, when wild-type cells are mixed with lymphocytes from patients lacking the enzyme for purine incorporation, the two kinds of lymphocytes do not establish contact and the mutant cells remain unable to carry out the particular biosyntheses. When fibroblasts are mixed together, however, both genetic kinds of cells are able to make nucleic acids from the available precursor purines. In fibroblasts, the wild-types and mutants form gap junctions, and metabolic cooperation can take place.

There are mutant cell lines which lack gap junctions, so we may consider the capacity to be a genetic variable within a cell type as well as between different kinds of cells in the organism. When wild-type and junctional mutants are mixed together in culture, contacts are only made between wild-type cells and not between wild-type and mutant or two mutant cells. Mutants of this kind should prove useful in studying cell-to-cell communication mechanisms.

Intercellular Adhesion

There is a class of intercellular contacts called **desmosomes** which act primarily as sites of intercellular adhesion and as anchoring sites for filamentous structures (Fig. 4.24). These differentiated regions are widely distributed among vertebrate and invertebrate tissues and may vary in form in different species or groups of cells. Desmosomes are not considered to be permeability sites since cell-to-cell permeability remains unaffected if desmosomes are disrupted by treatment with proteolytic enzymes or by selective removal of divalent cations such as calcium and magnesium. Disruption usually is evidenced by disintegration of the condensed intercellular material that occupies the 250–350 Å-wide space between adjacent plasmalemmas of a desmosome area.

Because desmosomes are especially abundant in tissues such as cardiac muscle or outer layers of skin that are subjected to severe mechanical stresses, they are believed to act in maintaining cell-to-cell

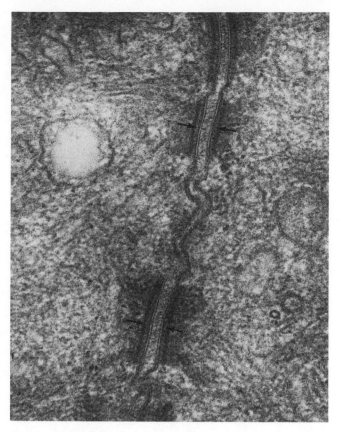

Figure 4.24
Desmosomes in rat intestinal epithelium. The desmosomes occur as symmetrical plaques between two adjacent cells in thin-section. Their characteristics include (1) a wide intercellular space containing dense material; (2) two parallel cell membranes; (3) a dense plaque associated with the cytoplasmic surface (at arrows); and (4) cytoplasmic microfilaments that converge on the dense plaque. × 156,000. (Courtesy of N. B. Gilula, from Gilula, N. B., 1974, *Cell Communications* [ed. R. P. Cox], John Wiley & Sons, pp. 1–29, Fig. 12.)

adhesion. Filaments are inserted into desmosomal plaques in several kinds of epithelia and other tissues, which indicates that desmosomes act as anchoring sites for these motility components within the cell.

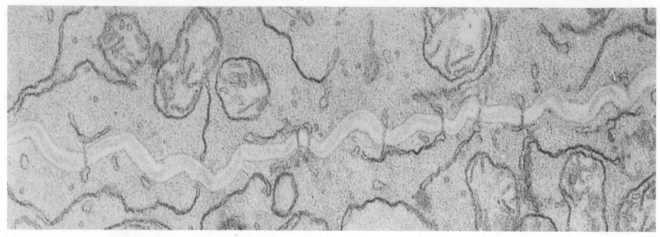

(a)

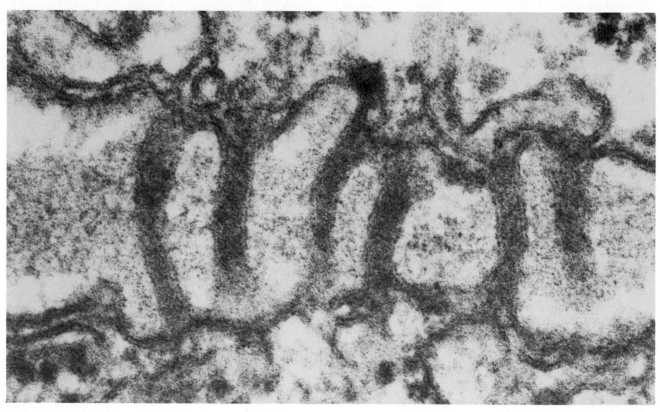

(b)

Cells and Cell Systems

One of the central postulates of the cell theory is that individual cells are separated units of function. It is quite possible, however, that interconnected cell systems rather than individual cells are functional units in the multicellular organism. At least we can say that cell-to-cell contacts provide the multicellular organism with a physical framework for metabolic cooperation and control of metabolic activities. Direct cell contacts are known to be involved in regulating growth and differentiation, and in development of the embryo. Cell junctions provide channels for ionic and metabolite exchanges and therefore provide ways for cells to circumvent the permeability barrier imposed by the plasma membrane. Interconnected cells can act coordinately in responses to stimuli, and order can be produced through these means for communication. Cell junctions are present during times of active growth and differentiation in animal embryos, and are not present earlier or later in development of the embryo. Orderliness of development appears to be enhanced by cell contacts at appropriate times.

Multicellular organisms would have a considerable evolutionary advantage if intercellular communications formed between interacting cells at critical times in their activities. Cell junctions or cytoplasmic channels provide selective communication networks between particular cells at different times in their history or during different stages of activity. Direct junctional contact between plasma membranes is an obvious device in animals but cannot function in multicellular fungus or plant systems where a thick and rigid cell wall surrounds each cell in the organism. Cytoplasmic channels in plants, called **plasmodesmata,** provide the communication links through pores in the walls (Fig. 4.25). Cytoplasmic channels are also a prominent feature of animal embryos during development, and they may provide a more rapid and coarser pathway for metabolite exchanges during rapid growth.

Certain kinds of cancerous cell growths are characterized by a failure of cell junctions to form. This defect may lead to uncontrolled growth if regulatory substances are transmitted slowly or erratically between cells in the group. This abnormality cannot be the only basis for malignant growths since many tumor cell populations do maintain regular junctional contacts. Indeed, it may not be the basis for any cancerous growths, and perhaps uncontrolled growth may be better explained by an altered response to regulatory substances, but the role of cell contacts in general cell communication remains a subject of intensive study. If we can learn how normal cells communicate and interact, we will then be in a better position to understand derangements in growth which take place in certain cells or cell communities.

Figure 4.25

Plasmodesmata in plant root cell thin sections: (a) permanganate-fixed cell of timothy grass (*Phleum pratense*) showing membranous elements extending across the cell wall and continuous between the two cells. × 31,000. (Photograph by W. Ridge) (b) High magnification view of *Potamogeton natans* cells fixed in osmium tetroxide. The plasmodesmata perforate the walls between adjacent cells. × 263,000. (Courtesy of M. C. Ledbetter)

SUGGESTED READING

Books, Monographs, and Symposia

Finean, J. B., Coleman, R., and Michell, R. H. 1974. *Membranes and Their Cellular Functions*. New York: John Wiley.

Packer, L., ed. 1974. *Biomembranes: Architecture, Biogenesis, Bioenergetics, and Differentiation*. Academic Press, New York.

Articles and Reviews

Branton, D. 1966. Fracture faces of frozen membranes. *Proceedings of the National Academy of Sciences* 55:1048–1056.

Branton, D. 1969. Membrane structure. *Annual Reviews of Plant Physiology* 20:209–238.

Bretscher, M. S. 1973. Membrane structure: Some general principles. *Science* 181:622–629.

Capaldi, R. A. 1974. A dynamic model of cell membranes. *Scientific American* 230(3):26–33.

Cox, R. P., Krauss, M. R., Balis, M. E., and Dancis, J. 1974. Metabolic cooperation in cell cultures: A model for cell-to-cell communication. In *Cell Communication,* ed. R. P. Cox, pp. 67–95. New York: John Wiley.

Finean, J. B. 1972. The development of ideas on membrane structure. *Sub-Cellular Biochemistry* 1:363–373.

Fox, C. F. 1972. The structure of cell membranes. *Scientific American* 226(2):30–38.

Gilula, N. B. 1974. Junctions between cells. In *Cell Communication,* ed. R. P. Cox, pp. 1–29. New York: John Wiley.

Hokin, L. E., and Hokin, M. R. 1965. The chemistry of cell membranes. *Scientific American* 213(4):78–86.

Loewenstein, W. R. 1970. Intercellular communication. *Scientific American* 222(5):78–86.

Sharon, N. 1974. Glycoproteins. *Scientific American* 230(5):78–86.

Singer, S. J. 1974. The molecular organization of membranes. *Annual Reviews of Biochemistry* 43:805–833.

Singer, S. J., and Nicolson, G. L. 1972. The fluid mosaic model of the structure of cell membranes. *Science* 175:720–731.

Steck, T. L. 1974. The organization of proteins in the human red blood cell membrane. A review. *Journal of Cell Biology* 62:1–19.

Tamm, S. L., and Tamm, S. 1974. Direct evidence for fluid membranes. *Proceedings of the National Academy of Sciences* 71:4589–4593.

Tooze, J., ed. 1973. The external surfaces of cells in culture. In *The Molecular Biology of Tumour Viruses,* pp. 173–268. New York: Cold Spring Harbor Laboratory.

Chapter 5

The Genetic System

The potential for expression and development of structure and function in any virus or cellular organism is encoded in the genetic material, which is usually DNA but which may be RNA in certain viruses. The coded instructions are eventually translated into proteins, and proteins, in turn, form all other chemical constituents of protoplasm through enzyme-catalyzed reactions. The organization of chemical components into recognizable structures is mediated by physical and chemical conditions of cells and their surroundings and is regulated by both coarse and fine controls over synthesis and organization into functional molecular assemblies.

DNA: THE GENETIC MATERIAL

One of the first important lines of experimental evidence which showed that DNA was the genetic material was reported in 1944 by O. T. Avery, C. M. MacLeod, and M. McCarty. They demonstrated that highly purified DNA extracted from one genetic strain of pneumonia-causing bacteria could **transform** another strain so that it was genetically altered. The receptor of the DNA extract assumed the genetic properties of the donor bacterial strain and transmitted its altered genetic traits to all descendant generations (Fig. 5.1). Neither protein nor RNA extracts could genetically transform a receptor pneumococcus strain. These results did not exactly overwhelm the scientific world at the time, mostly because follow-up experiments of the same general type could not be done very easily with other organisms. There was also a reluctance to accept DNA, rather than proteins, as the most probable genetic chemical material. But a great deal of experimental study was under way in various laboratories, and by 1952 there was a more favorable climate for the consideration of DNA as the prime candidate for the stuff of which genes are made.

In 1952 a set of important experiments was re-

ported by A. D. Hershey and M. Chase using the T2 bacteriophage of *Escherichia coli*. The viral hereditary material enters the host bacterial cell and directs the synthesis of new progeny viruses. The host cell dies in the process, releasing the new viruses for other rounds of infection. Since the T2 virus is made up exclusively of almost equal amounts of DNA and protein, it provided a useful subject for investigation of the chemical composition of the hereditary material.

A new method for biological studies, labeling with radioactive isotopes, had recently been developed, and Hershey and Chase designed their experiments around this technique (Fig. 5.2). Viruses formed in infected cells grown in media containing ^{35}S-labeled sulfate incorporated ^{35}S in their proteins. The amino acids cysteine and methionine contain S atoms, but DNA has no S in its molecular structure. When viruses were produced in infected cells grown

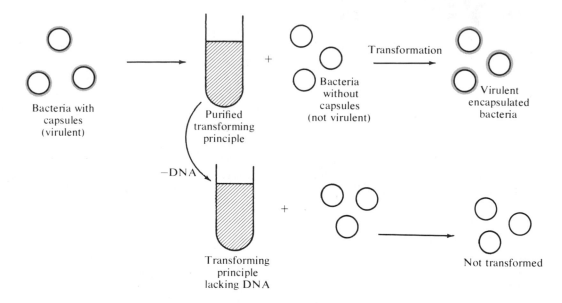

Figure 5.1
Diagram illustrating the results of a bacterial transformation experiment. DNA in the extract of transforming principle from donor bacteria is the agent responsible for transformation of an avirulent receptor strain to a virulent strain like the donor cells. No capsule surrounds avirulent cells in these pneumococcus bacteria.

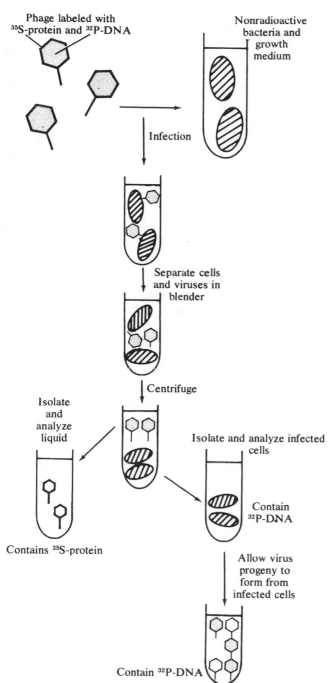

Phage labeled with
^{35}S-protein and ^{32}P-DNA

Nonradioactive
bacteria and
growth
medium

Infection

Separate cells
and viruses in
blender

Centrifuge

Isolate
and
analyze
liquid

Isolate and analyze infected
cells

Contain
^{32}P-DNA

Contains ^{35}S-protein

Allow virus
progeny to
form from
infected cells

Contain ^{32}P-DNA

in ^{32}P-labeled media containing phosphates, their DNA became radioactively labeled with ^{32}P. Viral proteins contain no phosphorus. In this way, the DNA and the protein could be distinctively labeled and the location of each kind of chemical could be followed during viral infection.

When ^{35}S-protein-containing viruses were presented to *E. coli*, the labeled protein remained outside the host cells for the most part. When ^{32}P-DNA-containing viruses were used, however, ^{32}P-DNA was found within the host cells. Since new viruses must be formed according to the genetic instructions of the parent viruses, only the part of the parents that actually entered the host cells could contain the genetic information. DNA, rather than protein, was the substance that entered the host, so DNA was the genetic material. Other experiments in this series verified the major results and strengthened the hypothesis that DNA, rather than protein, was the genetic material.

By 1953 when James D. Watson and Francis H. C. Crick proposed a molecular model for DNA, most investigators immediately recognized that the DNA molecule was admirably suited to playing a genetic role. Watson and Crick pointed out important features of the DNA molecular model that could explain mutation and replication, two fundamental qualities of genetic material. They suggested that molecules of DNA could be replicated precisely through the synthesis of new complementary partners for each of the parental strands in the double helix structure.

Figure 5.2
Diagrammatic illustration of the main features in the experiments of Hershey and Chase. T2 phages carrying radioactively labeled protein or DNA were allowed to infect *E. coli*. After infection was initiated, the viruses were mechanically separated from their infected host cells. The whole mixture was then centrifuged to separate and recover the viruses and *E. coli*. Only labeled viral DNA was found in the infected host cells, and only labeled viral DNA was later recovered in progeny viruses formed in the isolated bacterial cells allowed to complete the infection cycle.

Their suggestions concerning the suitability of DNA as a genetic molecule had a profound impact on the final acceptance of DNA as the genetic material of life.

In subsequent years it was shown that DNA contained encoded information for protein synthesis. The information in DNA was carried to the sites of protein synthesis by an intermediary called **messenger RNA (mRNA).** Two other major kinds of RNA had been identified earlier as **ribosomal RNA (rRNA)** and **transfer RNA (tRNA).** All three types of RNA participated in synthesis of protein from coded instructions contained in DNA molecular structure (Fig. 5.3). A molecule of mRNA was required to move from its origin at the DNA template to the regions of active protein synthesis. A mRNA molecule must have a nucleotide sequence that accurately reflects its complementarity to the sequence of DNA from which it is transcribed. The existence of such messengers was first postulated in 1961 by Sidney Brenner and colleagues, and they have been found in all genetic systems investigated so far. Purified mRNA species have been isolated and shown to direct synthesis of specific protein molecules both *in vivo* and *in vitro*.

The **transcription** of DNA information into molecules of mRNA is followed by **translation** of the encoded information into proteins. The expression of genetic information is subject to **regulation.** Gene action, or transcription, can be turned on and off in different cells at different times. Control over gene action results in synthesis of different proteins in genetically identical cells. Such cells may have entirely different phenotypes (appearance). Each cell in our bodies contains the same set of genes and, therefore, the same potential for development. But the total genetic potential is not expressed in each of these cells. Development and differentiation of cells and organisms proceed as a consequence of gene action, but the regulation of gene expression provides a primary basis for the uniqueness of cells in an organism or a population.

The Double Helix

Each of the two long polynucleotide chains in the duplex DNA molecule is unbranched. The double helical molecule is a remarkably stable conformation as a result of two sets of forces: (1) hydrogen bonding between all the bases along the length of the chain pair, and (2) hydrophobic interactions between stacked bases in the vertical sequence of the individual chains (Fig. 5.4). The backbone of each strand in the duplex is formed by bridges between the phosphate at the $5'$ position of one nucleotide and the hydroxyl at the $3'$ position of the sugar in the next nucleotide. These **$3',5'$-phosphodiester bridges** lend a considerable stiffness to the polynucleotide. The two chains of the duplex are antiparallel, one chain running in the $3'$ to $5'$ direction and the other following the $5'$ to $3'$ direction (Fig. 5.5).

The adenine—thymine and guanine—cytosine base pairs are of precisely the same size and shape, a fact that led directly to the model proposed by Watson and Crick in 1953. The duplex, therefore, has a constant diameter of 20 Å along its entire length and most importantly, any sequence of base-pairs is possible within a two-stranded molecule. Given one strand of DNA, the complementary second strand of the duplex is predictable in base sequence and its backbone

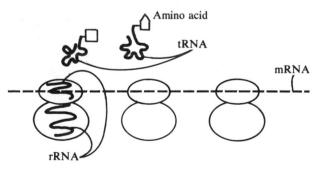

Figure 5.3
Amino acids are brought by transfer RNA to the messenger RNA copy of coded DNA instructions. Synthesis of the polypeptide takes place along messenger RNA situated at the ribosomes, which provide ribosomal RNA and proteins for the biosynthesis reactions.

(a)

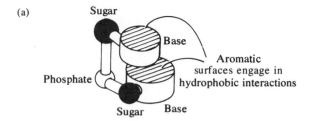

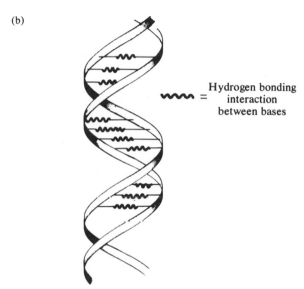

Figure 5.4

Figure 5.4
Stability of the duplex DNA molecule results from (a) hydrophobic interactions between aromatic surfaces of bases stacked vertically in each DNA strand (with permission of M. Levitt) and (b) hydrogen bonding interactions between bases in the complementary strands of the double helix molecule.

orientation. The base sequence is exactly *complementary* in the two chains, and their backbones are *antiparallel*. These built-in features of DNA provide the basis for precise replication, as first noted, in part, by Watson and Crick. They were aware of the complementary base-pairing importance, but the antiparallel orientation of the two chains was not discovered until a later time.

Denaturation and Renaturation of DNA

DENATURATION. A highly significant feature of the DNA double helix is the relative ease with which the component strands are separated and rejoined. When native (undenatured) DNA is exposed to high temperatures or to titration with acid or alkali, the two strands unwind and separate. This denaturation, or **melting** behavior, is the result of hydrogen bond disruption between the paired bases. Since G−C pairs are triple-bonded while A−T pairs are double-bonded, molecules with a higher mole percent of G−C are more stable structures and they require higher temperature or pH to melt. Acid is not usually used in denaturation studies, since it affects purine bonds, but alkaline conditions or high temperature are employed routinely in melting studies or in extraction of separated strands from a sample of native DNA.

One of the simplest ways to monitor DNA melting is by observing the change in absorbance using a spectrophotometer set at 260 nanometer (nm) wavelength (Fig. 5.6). DNA absorption of light energy is maximum for this particular wavelength of the spectrum, but the individual bases are responsible for the absorption property. When bases occur in a duplex DNA molecule, their absolute absorbance is lowered because of their packing into the double helix. When duplex DNA is denatured, the bases in single strands absorb more energy even though there has been no change in the concentration of DNA in the solution. The same amount of single strands has higher absorbance than its equivalent in duplex molecules. This increase in absorbance at the same concentration of the chemical is called the **hyperchromic effect** (Fig. 5.7). Thermal melting causes the bases to "unstack" and thus permits more light to be absorbed by the unhindered bases than can be realized in the duplex DNA conformation. Since single-stranded DNA does not show this hyperchromic effect, the phenomenon can be used to distinguish single- or double-stranded DNA in an unknown sample.

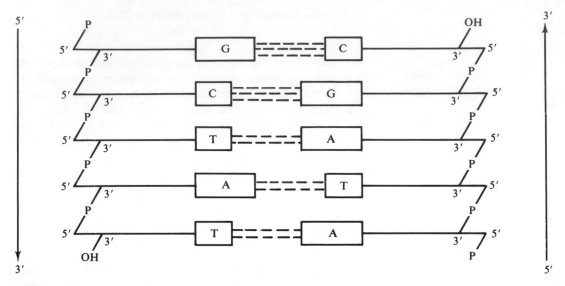

Figure 5.5
Diagrammatic illustration of the antiparallel orientation of
the two sugar-phosphate "backbone" strands of duplex
DNA molecules.

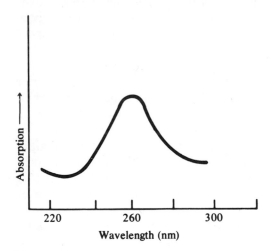

Figure 5.6
When duplex DNA in solution is exposed to different wavelengths of radiation, the wavelength of maximum absorbance
is found to be 260 nm.

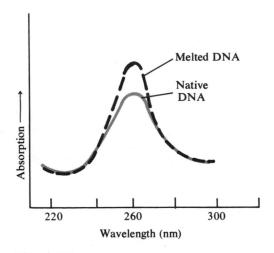

Figure 5.7
When melted DNA in solution is scanned at different wavelengths, there is a higher absorbance of the solution at
260 nm than is found with the same concentration of duplex
DNA molecules. This increase in absorbance is called
hyperchromicity.

Chain unwinding during thermal denaturation begins in regions high in A—T base pairs and moves progressively to regions of increasing G—C content. Carefully controlled heating of DNA in solution can be used to determine melting curves which show the increase in absorbance during the transition from duplex to single-stranded molecules (Fig. 5.8). There is a sharp transition in absorbance as duplex strands go to single strands, and the **midpoint melting temperature, T_m,** (the point at which the transition is half completed) is characteristic of a particular source of DNA. The T_m is directly proportional to the G—C content of the DNA in question and therefore can be used to determine this feature of DNA in different species or different parts of the same cell. Mitochondrial and chloroplast DNAs usually can be distinguished from nuclear DNA of a species on the basis of G—C content according to this method or some other quantitative procedures (Fig. 5.9). When a mixture of DNAs is centrifuged to equilibrium in density gradients of cesium chloride, the DNAs can be identified according to their buoyant density in cesium chloride (Fig. 5.10).

Separated single strands of duplex DNA can be recovered from alkaline CsCl density gradients. These complementary strands have different buoyant densities and reach different equilibrium positions in the gradients (Fig. 5.11). The two strands are identified as **"heavy"** and **"light,"** in relation to their buoyant densities.

In some experimental designs it is important to recognize and identify the two strands in relation to some function of DNA. For example, do both strands

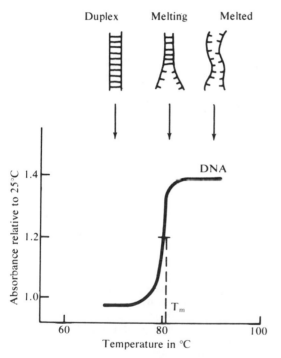

Figure 5.8
Melting curve for DNA in solution. The increase in absorbance is measured at different temperatures and plotted according to the absorption at 260 nm at 25°C. The midpoint melting temperature of the transition from duplex to single-stranded molecules (shown in diagrams at the top of the figure), where 50 percent denaturation has occurred, is identified as the T_m of the particular DNA sample or source. No further increase in absorption occurs when all the DNA is melted.

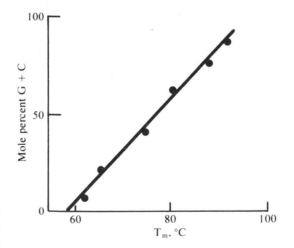

Figure 5.9
Graph showing that there is a direct correlation between the G + C content of a DNA sample and its T_m. Each dot represents a different specific DNA from various species. The straight-line relationship is illustrated.

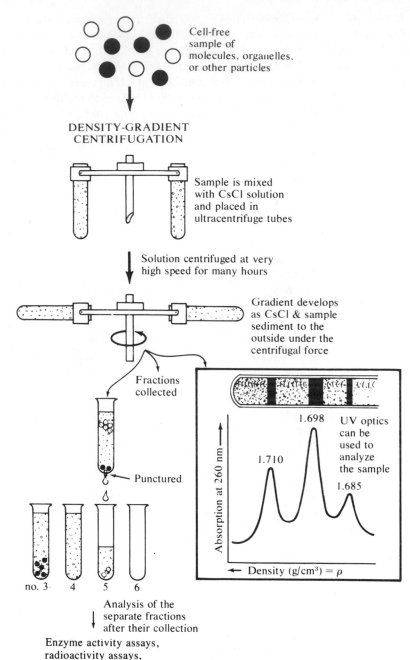

Figure 5.10
Diagram illustrating the protocol for
equilibrium density gradient centrifugation.
After materials have been centrifuged to
their equilibrium positions in the gradient,
fractions may be collected dropwise from
a hole punctured in the bottom of the tube
(or may be withdrawn in regular amounts
beginning at the top of the tube). Fractions
may also be photographed using ultraviolet
optics. Densitometer tracings may then be
made to show the positions and amounts of
DNA or RNA in the gradient (inset
diagram). The buoyant density of the solute
(ρ) is expressed in g/cm³, and reflects the
density of the solute relative to the density
of CsCl or other gradient material.

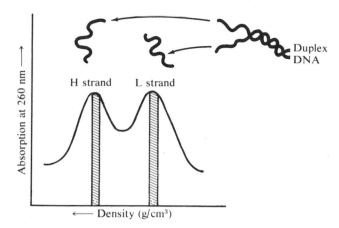

Figure 5.11
When duplex DNA is centrifuged to equilibrium in alkaline CsCl density gradients, the two strands of the duplex molecule separate and each reaches its own equilibrium position in the gradient. The heavy (H) strand occupies a region of higher CsCl density than the complementary light (L) strand of the DNA double helix.

transcribe RNA or is only one strand the template for messenger or other RNAs? This question was studied in several mitochondrial systems by hybridizing RNA with the heavy and the light single strands of melted mitochondrial DNA. Ribosomal RNA and most of the transfer RNAs hybridized specifically with the heavy strand, but a few transfer RNA types did hybridize exclusively with the light DNA strand. The heavy strand contains most of the coded information in the mitochondrion, but some information appears to be contained in the light strand as well. These kinds of studies help to establish the differences between the two strands of the duplex in gene expression, and therefore provide more detailed information on total informational content of a genetic system.

The location of regions rich in A−T sequences can be accomplished in some cases by **denaturation mapping.** Since A−T regions melt sooner than G−C regions in duplex DNA, a sample can be subjected to a temperature just high enough to melt A−T but not

G−C regions. When DNA is examined with the electron microscope after such treatment, the intact G−C-rich stretches are distinguished from the melted A−T-rich portions of the duplex molecule (Fig. 5.12). This method is crudely similar to chemical determination of base sequence in DNA, a process beset with technical difficulties. Chemical analysis of even the smallest DNAs would require identification of 5000

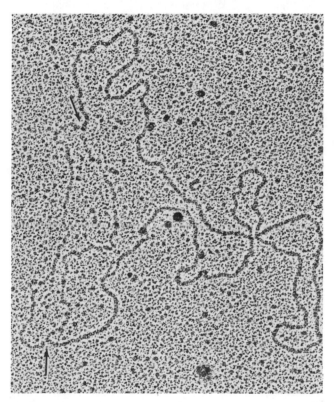

Figure 5.12
Electron micrograph of a platinum-shadowed, duplex mitochondrial DNA molecule from an egg of *Drosophila melanogaster.* The preparation was heated to 40°C for 10 minutes in 0.05 *M* sodium phosphate and 10 percent formaldehyde to partially denature the molecules. Approximately one-fourth of the 6.1 μm-long circular duplex has denatured (at arrows), due to the region being rich in adenine and thymine. × 85,700. (Courtesy of D. R. Wolstenholme and C. M.-R. Fauron)

to 6000 base pairs in a molecule about 2 μm long. Denaturation mapping at least provides some useful general information about sequences of A–T and G–C in DNA molecules of a reasonable size for such studies, many of which are much longer than 2 μm.

RENATURATION. When melted DNA is incubated at a temperature about 25°C below its T_m, the two separated strands begin to reassociate, or **reanneal,** and reconstitute the original duplex combination. Renaturation can be monitored in several ways, including decrease in absorbance at 260 nm, as measured in a spectrophotometer. This is the reverse of the hyperchromic effect described earlier in this chapter.

The precision of reannealing can be demonstrated for shorter-length DNAs, using the electron microscope. If one chain from wild-type virus DNA is allowed to reanneal with the complementary chain from a mutant strain that is lacking at least 100 nucleotide residues, a loop can be seen at this particular site on the reannealed duplex (Fig. 5.13). The intact strand has no matching bases in the deleted region of the mutant strand and loops out around this deleted site in the reformed double helix. **Deletion mapping** of DNA has proven to be as useful a tool in genetic analysis of some viruses as it was an important method of mapping genes in eukaryotic chromosomes during the 1930s.

The kinetics of renaturation provide an accurate measure of the **size of the genome** (or, the full complement of genes in the organism) and of the **complexity of the DNA** (or, the amount of unique versus repeated sequences of nucleotides). Renaturation is conducted under standardized conditions of temperature, salt concentration of the medium, and size of sheared fragments of the single strands of DNA. The rate of reannealing can be measured from the ratio of reassociated to single-strand fragments remaining at intervals during the renaturation process (Fig. 5.14).

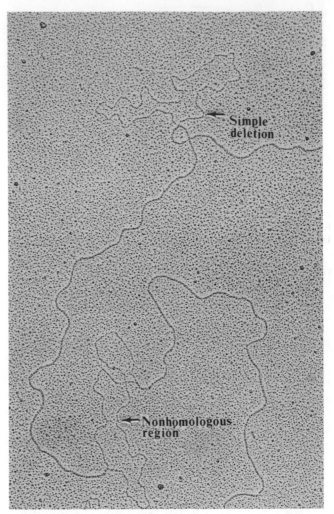

Figure 5.13
Electron micrograph of a platinum-shadowed heteroduplex DNA molecule made by annealing melted single strands from two different strains of phage lambda. The DNA of one strain has a deleted region plus another deletion into which a short piece of nonhomologous DNA has been substituted for the native longer region. The region of the simple deletion can be detected by a single-stranded loop, and the unpaired, nonhomologous region is evident by an opened area with one single strand longer than the other. (Courtesy of B. Westmoreland and H. Ris, from Westmoreland, B. *et al.,* 1969, *Science* **163**:1343–1348.)

(a)

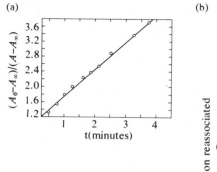

(b)

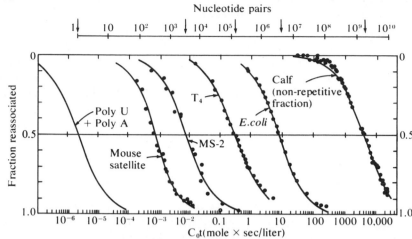

Figure 5.14

Kinetics of DNA renaturation measured by light absorption of a preparation at intervals during reannealing of short, single-stranded pieces: (a) The genome size (expressed as molecular weight) and genome complexity (molecular weight of nonrepeating nucleotides) are derived from the renaturation rate constant, which is inversely proportional to molecular weight. (b) An alternative method is the measurement of the fraction of reassociated strands in the preparation, which is plotted according to the C_0t value (concentration of DNA in moles of nucleotides per liter times the number of seconds of reaction time). Genome size and complexity is expressed in numbers of nucleotide pairs (shown in the upper nomographic scale for various DNAs). (Reproduced with permission from Britten, R. J., and D. E. Kohne, *Science* **161**:529–540, 9 August 1968. Copyright © 1968 by the American Association for the Advancement of Science.)

If the DNA includes only unique sequences, then the rate of renaturation will be slower than a comparable sample with repetitious DNA sequences. Since complementary fragments must "find each other" to reanneal, it will take longer for each unique sequence to "find" the unique fragment that is its complement. On the other hand, if DNA contains many repeated segments, then there will be many fragments with the same nucleotide sequence, and renaturation will proceed at a faster rate.

By tests like this one, it has been possible to describe eukaryotic DNA and to determine the percentage of unique and repeated nucleotide sequences. These studies in turn have opened new avenues of investigation about the functional role of repeated DNA and its significance in evolution. These ques-

tions will be considered in greater detail in other chapters.

REPLICATION OF DNA

In a recent monograph, Arthur Kornberg set forth six basic rules of DNA replication:

1. Replication occurs by a semiconservative process.
2. Replication proceeds in one or both directions from a given point, usually in both directions.
3. Replication starts at a unique point, or origin, and there may be one or more origins in a DNA molecule.

4. Both strands replicate by addition of nucleotide monomers in the 5′ to 3′ direction.
5. Replication occurs in short, discontinuous pulses which produce short fragments that later are joined to the main body of the molecule.
6. A short segment of RNA primer is required to initiate DNA polymerase action in lengthening the polynucleotide chain.

We will discuss some of the evidence for these rules of replication briefly in this section and deal, at greater length, with certain topics in later chapters.

The Semiconservative Model of Replication

Continuity of life requires conservation of genetic information from generation to generation. The explanation lies in the mechanism by which new DNA (set of genes) is manufactured so that identical replicate molecules are synthesized accurately in every cycle. One compelling feature of the Watson-Crick model for DNA was that it could form the basis of a model of replication which would be guided by the property of complementary base pairing inherent in the structure of the molecule itself. Watson and Crick suggested that the two strands of the duplex could each serve as a template for synthesis of a new complementary partner, thus producing two identical molecules from the original duplex. Since only one strand of the old duplex is conserved in each new duplex, this mode of replication has been called **semi-conservative;** two other theoretically possible mechanisms are the **dispersive** and the **conservative** modes (Fig. 5.15).

In 1958, Matthew Meselson and Franklin Stahl

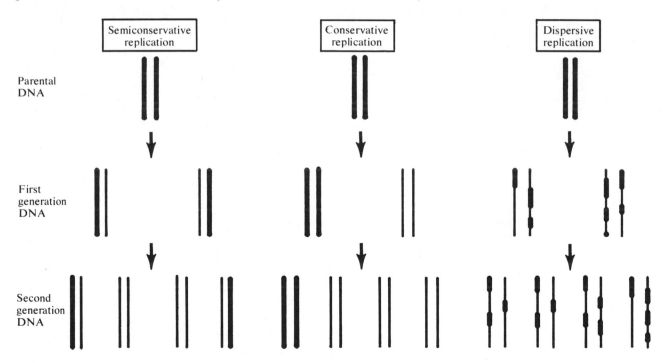

Figure 5.15
The distribution of first and second generation progeny molecules follows different predictions according to the three modes of DNA replication set out as hypotheses before 1958.

provided strong evidence which supported the semi-conservative mode and worked against the other two possibilities. Using the method of **equilibrium density gradient centrifugation** to separate DNA in cesium chloride gradients (see Fig. 5.10), Meselson and Stahl designed their experiments to obtain data that could support only one of the theoretical possibilities and simultaneously exclude the other two. These experiments are models of excellent experimental design, and the same design has been employed successfully in a few other cases since 1958.

The *E. coli* cultures to be analyzed had previously been grown in media containing the heavy nitrogen isotope ^{15}N, so that the parent DNA was uniformly labeled at time-zero. These ^{15}N-labeled cells were transferred afterward to media containing the ordinary ^{14}N isotope. In the first generation after cell and DNA doubling, DNA was extracted and centrifuged to identify the kinds of isotope-marked molecules that had been synthesized.

If replication was conservative, then two distinct DNA populations would be identified after centrifugation: half the molecules would be heavy and half would be light. The heavy molecules ($^{15}N - {}^{15}N$) would be the conserved parental duplexes formed in the original ^{15}N-medium. The light molecules ($^{14}N - {}^{14}N$) would be the newly synthesized duplex DNA manufactured from ^{14}N-precursors in the medium in which the first generation cells had been produced (Fig. 5.16).

If replication were semiconservative, it was predicted that all first-generation DNA molecules would be hybrid $^{15}N - {}^{14}N$, that is, all would be half-heavy. If replication was dispersive, the population of DNA molecules would contain varying amounts of $^{15}N -$ and $^{14}N - DNA$, depending on which original ^{15}N-pieces and how much of the original had been incorporated into the new molecules along with new ^{14}N-labeled sequences.

Sedimentation of extracted and purified DNA was carried out using high-speed centrifugation and a CsCl density gradient. At equilibrium the DNA molecules would settle in a specific region of the density gradient that corresponded to the buoyant density of DNA in CsCl. The $^{15}N - {}^{15}N$, $^{15}N - {}^{14}N$, and $^{14}N - {}^{14}N$ duplexes each settle in a distinct region that does not overlap with the others. The contents of the centrifuge tubes were photographed and interpreted (Fig. 5.17). All DNA molecules from first-generation cells were half-heavy $^{15}N - {}^{14}N$, an observation which supported the semiconservative mode of replication.

These interpretations were confirmed by examining purified DNA extracted at various times during growth. The distribution of labeled DNA always followed the predictions based on semiconservative replication. For example, semiconservatively replicated DNA from the second-generation cells would be predicted to consist of 50 percent half-heavy ($^{15}N - {}^{14}N$) and 50 percent light ($^{14}N - {}^{14}N$) molecules. With conservative replication, on the other hand, second-generation molecules would be expected to be 25 percent heavy ($^{15}N - {}^{15}N$) and 75 percent light ($^{14}N - {}^{14}N$), and no half-heavy molecules at all since parental duplexes are conserved $^{15}N - {}^{15}N$ molecules and all new DNA is synthesized only from ^{14}N precursors. Dispersively replicated molecules would be quite variable. Semiconservative replication is the mode of DNA synthesis, a fact that has been verified many times by independent experiments since 1958.

Experiments of the Meselson-Stahl design have also been conducted in studies of organelle DNA replication. One highly successful study which was reported in 1967 by K.-S. Chiang and N. Sueoka showed that chloroplast DNA in unicellular *Chlamydomonas reinhardii* replicated semiconservatively (Fig. 5.18). The distribution of ^{15}N and ^{14}N in heavy, half-heavy, and light chloroplast DNA paralleled the predictions made according to the semiconservative model and was consistent over a number of replication cycles. Similar evidence has been difficult to obtain for mitochondrial DNA because of technical problems. One recurring problem is that pools of labeled precursors remain, rather than being exhausted at the end of each round of replication. If

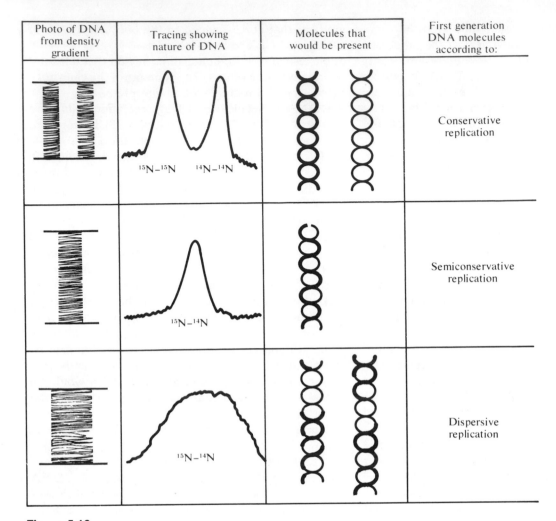

Photo of DNA from density gradient	Tracing showing nature of DNA	Molecules that would be present	First generation DNA molecules according to:
	$^{15}N-^{15}N$ $^{14}N-^{14}N$		Conservative replication
	$^{15}N-^{14}N$		Semiconservative replication
	$^{15}N-^{14}N$		Dispersive replication

Figure 5.16
Three predictions for the distribution of ^{15}N-labeled parental DNA strands. Each prediction was made from one of the hypothesized modes of DNA replication. Each set of predictions is unique. A correct prediction can be distinguished experimentally by the labeling found in first generation DNA molecules banded during centrifugation to equilibrium positions in CsCl density gradients (see Fig. 5.10 for reference to the methods).

there is "leftover" heavy precursor that can be used after the transfer to ^{14}N-medium, then the results are distorted in succeeding generations of replicated molecules. However, new methods have been developed which permit direct observations of replicating DNA molecules using the electron microscope. These recent studies have provided important information on the way in which semiconservative replication proceeds in mitochondrial DNA systems, as we will see shortly.

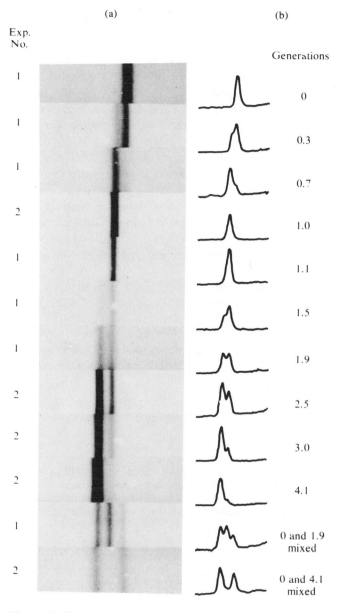

(a)

Exp. No.

(b)

Generations

1   0

1   0.3

1   0.7

2   1.0

1   1.1

1   1.5

1   1.9

2   2.5

2   3.0

2   4.1

1   0 and 1.9 mixed

2   0 and 4.1 mixed

Figure 5.17
Some of the results of the Meselson and Stahl 1958 experiment showing (a) ultraviolet absorption photographs of DNA bands resulting from density gradient centrifugation of bacterial preparations sampled at different times after development in ^{14}N-labeled medium of ^{15}N-labeled cells, and (b) densitometer tracings of the DNA bands shown in the adjacent photographs. The bottom photograph and tracing, which show fully unlabeled and labeled DNA from a mixed preparation of known content, serve as a reference for DNA density. (From Meselson, M., and F. W. Stahl, 1958, *Proc. Natl. Acad. Sci.* **44**:675.)

A variation on the Meselson-Stahl experimental design was exploited by J. H. Taylor to examine the mode of replication in whole eukaryotic chromosomes. Cells were grown in media containing thymidine which was labeled with the radioactive isotope **tritium** (^{3}H). Fully-labeled cells were transferred to a medium containing unlabeled thymidine (^{1}H—thymidine) and were allowed to undergo one cycle of doubling. These cells were then examined for distribution of silver grains in light microscope autoradiographs (Fig. 5.19). The pattern conformed to predictions based on semiconservative replication, since each daughter chromosome was half-labeled. On a conservative scheme half the chromosomes would have had all the label and half the chromosomes would have had none (Fig. 5.20).

Direction of Replication

The end products of a replication event are semiconserved duplexes. But, how does replication lead to this end result of the process? The first experimental evidence showing that replication proceeded in a specific direction was provided in 1963 by John Cairns. *E. coli* cultures were grown in ^{3}H—thymidine for one 30-minute generation to label the DNA entirely. Afterward they were allowed to grow in the same medium for varying intervals during the second 30-minute cycle before they were removed for autoradiography. The replicating regions of the large molecule would thus have twice the label density as regions that had not yet been replicated (Fig. 5.21). The DNA molecules were seen to be circular, verifying the earlier assumption, based on genetic mapping studies, that the genome was circular. The molecules

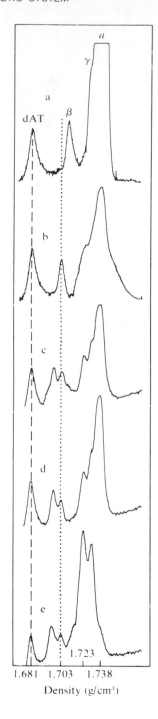

looked like the Greek letter *theta* and have been referred to as Θ forms for this reason.

These forms were interpreted as showing that replication had proceeded in one direction, beginning at a fixed starting point or origin, until the entire circle had been replicated to produce two semiconserved daughter duplexes (Fig. 5.22). The two parts of the molecule with denser tritium label were measured to be the same length, while the third portion of the theta figure was the segment of different length. This unequal-length segment had a lower density of radioactive label and was therefore the part that had not yet been replicated before removal from the medium. The exact length of the unequal segment varied according to the amount of replication that had taken place, beginning at the point of origin and proceeding to the growing fork.

The same Θ figure could have arisen if replication had proceeded in *both* directions away from the point of origin (Fig. 5.23). Later studies using *E. coli* showed this to be true. DNAs from most viruses, prokaryotes, and eukaryotes, analyzed in a variety

Figure 5.18
Some results of the 1967 Chiang and Sueoka experiment analyzing replication of chloroplast DNA in *Chlamydomonas*. Cells were fully labeled with ^{15}N and were then transferred to ^{14}N-containing media at time zero. Samples of the culture were taken at different times during the experiment and were analyzed after density gradient centrifugation in CsCl. Molecules shown in the densitometer tracings of the DNA bands are fully labeled at time a, all are half-labeled at time b, half are half-labeled and half are unlabeled at time c, and there is a predictable decrease in proportion of half-labeled DNA at times d and e. These results are consistent with predictions based on semiconservative replication of DNA. (From Chiang, K.-S., and N. Sueoka, 1967, *Proc. Natl. Acad. Sci.* **57**:1506–1513.)

Figure 5.19
Protocol for autoradiography and a photograph showing the localization of silver grains over those chromosome regions that contain ^3H-thymidine-labeled DNA. (Photograph courtesy of P. B. Moens, from Moens, P. B., 1966, *Chromosoma* **19**:277–285.)

Radioactive
substance

Organisms
or
cells

Incubate
(allow radioactivity
to be incorporated)

Stop the incorporation, fix the organisms,
separate them from fluid, wash, and place
on microscope slide coated with adhesive

In the darkroom

Cover specimen
with thin coating
of special photographic
emulsion

Place slide in light-tight
container for several days;
during this time the
radioactivity "exposes"
the emulsion directly above it

Developer   Rinse   Acid-fix   Wash

Stain if necessary

Dry

Examine under microscope

Before development

Emulsion
Organism
Adhesive
Slide

Radioactivity

Silver grains

After development

Unexposed silver halide has been
removed, while exposed and developed
silver grains remain in the layer
of transparent gelatin; these grains
are superimposed upon the source
of radioactivity

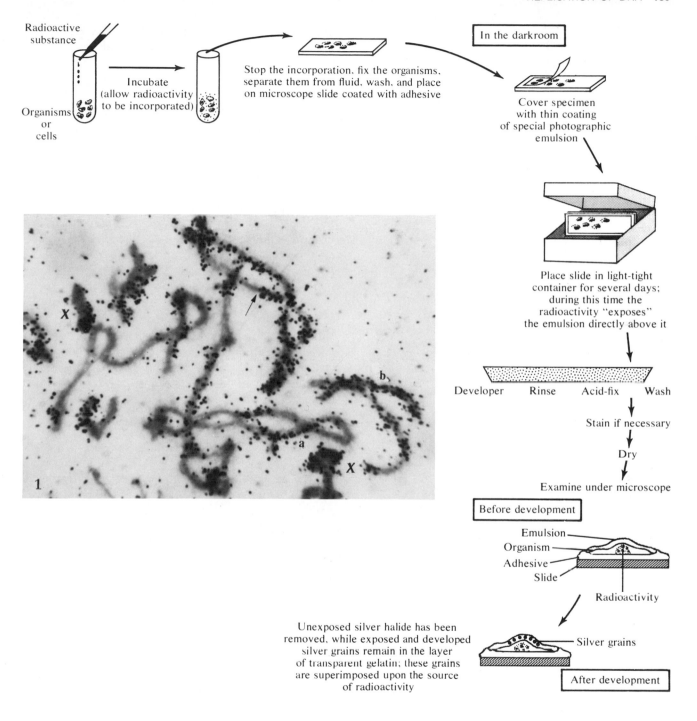

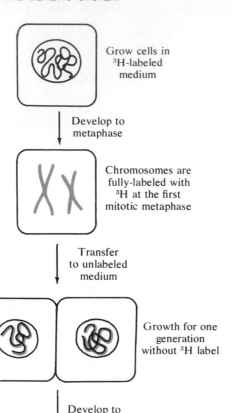

Grow cells in ³H-labeled medium

Develop to metaphase

Chromosomes are fully-labeled with ³H at the first mitotic metaphase

Transfer to unlabeled medium

Growth for one generation without ³H label

Develop to metaphase

Chromosomes are half-labeled at the next metaphase following transfer.

They are *not* fully-labeled and unlabeled, as predicted by conservative replication

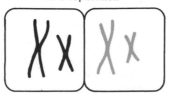

of ways in different laboratories, have also been shown to replicate in this way. Bidirectional replication is the commoner pattern.

The two patterns can be distinguished in autoradiographs if the regions between the two forks of the theta are examined for density of label. In a molecule replicating unidirectionally the denser label will be found only in the region of the growing fork. If replication proceeds in both directions, then there will be denser label at *both* forks and less label in between the forks.

In addition to autoradiographic evidence of bidirectional replication in most systems, independent confirming evidence has been obtained from genetic analysis and from electron microscopy. Molecules of eukaryotic chromosomal DNA show multiple origins, in the form of "eyes" or loops along the molecule, while most viral and bacterial DNA contain only one origin point (Fig. 5.24). Replication may begin at the same fixed origin in prokaryotes and viruses while an earlier round is still in progress. In other words, replication need not go to completion before another cycle of replication is initiated on the same molecule at the one fixed origin.

Addition of Nucleotide Monomers

If the same enzymatic mechanisms are responsible for synthesis of both new strands, then chain growth cannot be simultaneous and synchronous for the two strands at the growing fork. It was predicted that (1) newly added monomers for both strands would occur only at the 3′ end of the replicating chains, and (2) an asymmetric single-strand would be found in the duplex in the region where the opposite strand was being replicated. Chemical and enzymatic analysis verified the first prediction, and electron microscopy verified

Figure 5.20
Diagrammatic illustration of the autoradiographic evidence showing semiconservative distribution of labeled parental DNA in the first generation chromosome complement of daughter cells.

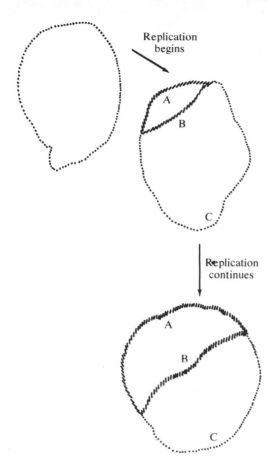

Figure 5.21
Drawing illustrating autoradiographs of *E. coli* circular DNA during replication. The labeled molecules continue to replicate in ³H-thymidine-containing medium. A double thickness of silver grains identifies the newly replicated regions of the molecule. The lengths of the replicated A and B segments are equal, and the unreplicated segment is different in length from the other two.

the second. In phage lambda (λ), DNA the size of such a single-stranded unit was measured to be up to 0.4 μm, which is equivalent to about 1000 base-pairs.

Single-strand regions have been identified in replicating mitochondrial DNA from animal cells, from centrifugation and electron microscopic studies. In the 5 μm-long circles of animal mitochondrial DNA

there is a novel replication pattern in which replication begins along the light or L-strand of the duplex, causing the heavy or H-strand to be displaced. In the displaced- or **D-loop** molecules it is possible to isolate the newly formed L-strand component, if the hydrogen bonds that bind the segment to the replicating DNA are broken. The identity of the isolated fragments can be confirmed by centrifugation analysis and by hybridizing the newly synthesized L-strands with complementary L-strands from duplex DNA. Replication of the parental H-strand is initiated later. The process therefore takes place by a staggered copying:

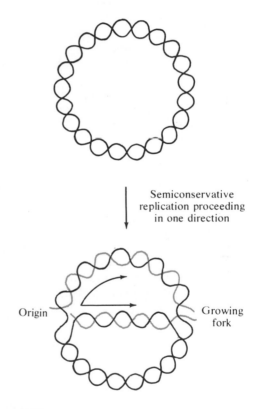

Figure 5.22
Interpretation of the autoradiographic evidence from photographs of replicating *E. coli* DNA (as shown in Fig. 5.21 drawings), if unidirectional replication is responsible for the observed pattern of molecule labeling.

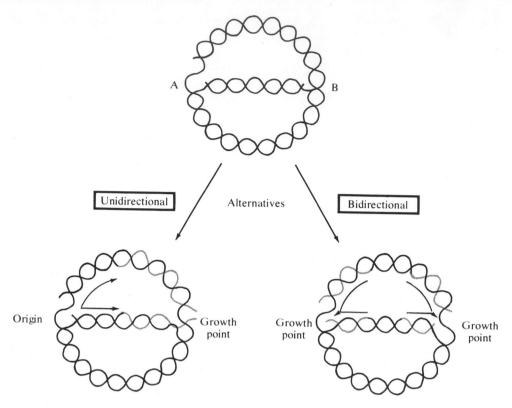

Figure 5.23
Bidirectional replication of θ-form DNA molecules can be distinguished from unidirectional replication according to the pattern of labeling (gray strands) in the replicated regions of the molecule.

the L-strand is copied first and the H-strand is copied later, with each proceeding unidirectionally from the 5' to the 3' end of their respective chains (Fig. 5.25).

Synthesis of Short Pieces

R. Okazaki first showed that the most recently synthesized DNA occurred in short pieces that were later incorporated into the high-molecular-weight DNA chains. When growing cultures were provided with labeled precursor in brief pulses at low temperature, almost all the precursor label could be found in DNA fragments. These short pieces sedimented in density gradients where stretches containing 1000 to 2000 nucleotide residues would be expected to equilibrate (equal to a sedimentation value of 10S). Mutants which lacked enzymes necessary to extend the DNA chain or to join the short fragments were particularly useful because all their newly synthesized DNA occurred as short pieces. This finding suggested that synthesis proceeded along both chains of the duplex in brief, discontinuous pulses. The suggestion was verified by showing that the fragments annealed to both strands of parental *E. coli* DNA to an equal extent, in the hybridization tests.

These **replication (Okazaki) fragments** were employed to verify that nucleotide monomers were added in the 5' → 3' direction along both chains. The most recently added monomers regularly were found at the 3' end of fragments for both antiparallel chains of the double helix.

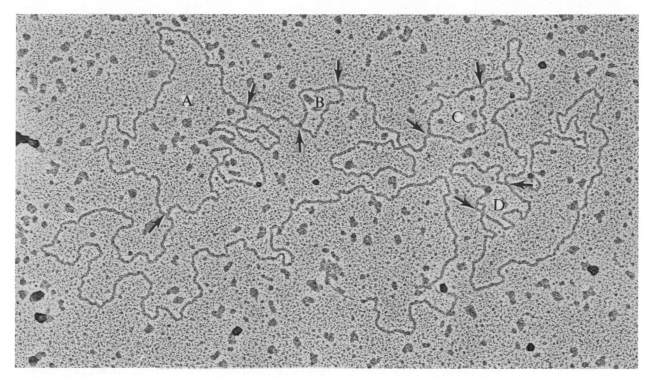

Figure 5.24
Replicating regions of eukaryotic nuclear DNA molecules resemble "eyes," or "bubbles," along the length of the duplex molecule. Four regions of replication (denoted by A, B, C, D) are evident in this molecule from *Drosphila* *melanogaster,* as well as unreplicated areas (between pairs of arrows). × 51,000. (Courtesy of D. R. Wolstenholme, from Wolstenholme, D. R., 1973, *Chromosoma* **43**:1–18, Fig. 6.)

With the discovery that DNA synthesis proceeded in short steps, it became clear that chain growth required some phosphodiester joining action to link the replication fragments to each other and to the growing chain. Other processing steps also had to be accounted for, including excision of the RNA primer nucleotides that initiate each fragment and addition of deoxyribonucleotides to fill in these gaps on excising the ribonucleotide residues. We will examine these steps next.

Linking the Short Pieces

A **DNA ligase** appears to be the enzyme responsible for joining the replication fragments to the main DNA chain. The existence of this enzyme had been postulated earlier on the basis of recombination events, and it has also been shown to be essential in repair synthesis of damaged DNA regions. Since the first demonstration of its existence in 1967, the ligase has been purified and employed in studies of the mechanism of its action, as well as in the synthesis and manipulation of DNA made in the laboratory.

E. coli mutants deficient in ligase activity have been useful in confirming how the enzyme functions in phosphodiester bridge formation and in establishing that DNA synthesis proceeds by formation of short Okazaki fragments. These fragments accumulate in ligase-deficient mutants and can be examined for

(a)

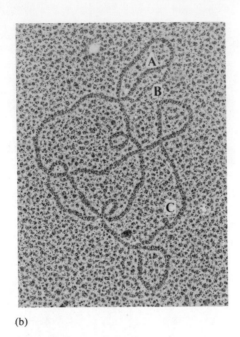

(b)

Figure 5.25
Replication of mitochondrial DNA: (a) diagram illustrating the staggered copying of the parental DNA strands during D-loop synthesis of mouse mitochondrial DNA. Heavy strands are shown thicker than light complementary strands, and new strands are shown by dashed lines. (b) Replicating mitochondrial DNA molecule from regenerating rat liver. In this early D-loop stage the single strand is indicated as line B, the region of parental template from which the new single strand is copied is indicated as line A, and line C represents the unreplicated portion of the parental duplex. × 65,900. (Courtesy of D. R. Wolstenholme, from Wolstenholme, D. R. *et al.*, 1973, *Cold Spring Harbor Sympos. Quant. Biol.* **38**:267–280, Fig. 2a.)

presence of precursor label in the most recently added 3' positions.

Initiation of Replication by RNA Primer

Since the discovery in 1970 of the requirement for **RNA primer** in DNA synthesis by A. Kornberg, R. Okazaki, and colleagues, a substantial amount of information has been collected. It had been known for a long time that no known DNA polymerase could start a new DNA chain, although the polymerases could catalyze chain growth. Initiation of chain synthesis occurs as a brief transcriptional event catalyzed by a **DNA-directed RNA polymerase.** The priming pieces of RNA must later be excised from the chain

by some nuclease enzyme that specifically recognizes the region where the 3' primer terminus is covalently bonded to the 5' deoxyribonucleotide position (Fig. 5.26). After the RNA primer is excised, the gap must be filled in with deoxyribonucleotides by DNA polymerase action, thus reconstituting the continuous DNA chain that continues to grow in this way along both strands in the 5' → 3' direction.

Preliminary information about the nature of RNA and DNA polymerases made it quite reasonable to postulate chain initiation by an RNA primer, but specific support for the event was first reported by A. Kornberg using the phage M13 system. M13 DNA exists as a single-stranded circle in mature viruses, but

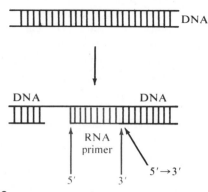

Figure 5.26
Diagram showing that initiation of synthesis of a segment of a complementary DNA strand requires a priming piece of RNA (gray segment). Synthesis proceeds in the 5' to 3' direction along each DNA strand.

it is converted to a duplex replicating form when the virus infects host cells. The duplex circles multiply during the infection cycle, and single-stranded DNA is finally packaged into mature virus particles.

If the drug *rifampicin* is added during the infection process, the single-stranded M13 DNA is not converted to the duplex form. If rifampicin is added about 5 minutes after infection (about the time the converted duplexes begin to synthesize new duplex DNA in the host cell), new duplex DNA is not produced. Since it is known that rifampicin specifically inhibits *initiation* of RNA chains by RNA polymerase, these data support the view that DNA synthesis is initiated by RNA primer formation. Inhibitors of protein synthesis had no effect on DNA synthesis in this system, further supporting initiation as the target of rifampicin inhibition.

The M13 phage studies were rapidly extended to other systems, and it was found that RNA involvement in DNA chain initiation was a widespread feature in prokaryotic and eukaryotic species. It remains to be established that all DNA replicating systems possess this feature, however. New questions were asked as soon as the RNA primer evidence had been obtained. For example, what signals the start of the RNA sequence at each replication-fragment

synthesis? What determines the size of the RNA segment (about 10 percent of the length of a replication fragment) and its specific termination? What system is responsible for excising the 5' → 3' RNA segment? Some answers to these and other questions are now in sight.

Unwinding the Double Helix

It had been postulated for a long time that some protein(s) must be present to facilitate unwinding of the duplex DNA in advance of the replicating fork. The first **"unwinding" protein** to be isolated and characterized was the protein product of gene 32 in *E. coli* phage T4, reported in 1970 by B. Alberts and L. Frey. Other unwinding proteins have since been recognized in eukaryotic species.

The gene 32 protein product converts a duplex DNA to single strands at temperatures 40° below the T_m for this DNA because tight binding occurs between the protein and single-stranded DNA and little binding with duplex DNA or RNA takes place. The protein binds to A−T-rich low-melting regions of DNA first, and the energy of this binding drives the melting to completion. The unwinding protein binds to the sugar-phosphate backbone so that no sequence specificity, other than for low-melting regions rich in A−T pairs, is involved. The elongated shape of the protein molecule permits it to bind tightly to about 10-nucleotide-long stretches. Because protein−protein interactions are highly favored, protein molecules bind cooperatively and thus line up adjacent to one another rather than at isolated sites on the DNA (Fig. 5.27). The gene 32 protein is not catalytic. Its structural function is indicated by the proportions of the binding reaction. If single-stranded circular DNA is incubated with enough gene 32 protein to coat one-third of the molecule, electron micrographs reveal molecules in which one-third of the length is extended and protein-covered while the remaining two-thirds of the molecule are bare and collapsed.

Unwinding proteins not only melt duplex DNA by preferential binding to single-stranded sugar-phosphate backbone, but they also facilitate base-pair

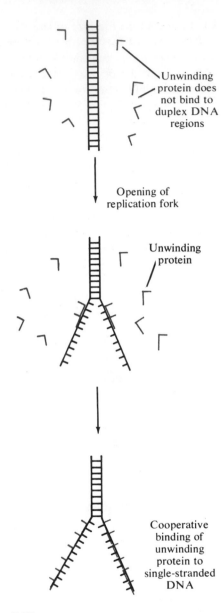

Unwinding protein does not bind to duplex DNA regions

Opening of replication fork

Unwinding protein

Cooperative binding of unwinding protein to single-stranded DNA

Figure 5.27
Illustration of the action and effect of unwinding protein (gray) at the replication fork of duplex DNA.

alignment between single strands during DNA replication and recombination. Phage T4 has an absolute requirement for gene 32 protein product for its DNA replication. The relation of the unwinding protein to gene recombination will be discussed again in Chapter 13.

Enzymes Involved in DNA Replication

The first known *E. coli* DNA polymerase, **DNA polymerase I** (the Kornberg enzyme), was not generally accepted as the primary replicating enzyme because its behavior had a number of peculiar aspects *in vitro*. For example, branched DNA chains were synthesized in the presence of DNA polymerase I *in vitro,* whereas unbranched DNA is the only form synthesized *in vivo*. In 1969 a polymerase I-deficient mutant was isolated by John Cairns. This *polA⁻* mutant synthesized DNA at normal rates despite very low (virtually undetectable) amounts of polymerase I, and a search was instituted for other polymerizing enzymes, using this mutant strain as an experimental system.

Polymerase I activity is so high in wild-type cells that it masks activities of other enzymes involved in nucleotide incorporation, but in *polA⁻* mutants, **DNA polymerase II** and then **DNA polymerase III** were found. Another polymerase mutant, *polB⁻* was found to lack DNA polymerase II, but it still synthesized DNA at normal rates. DNA polymerase II did not seem to be the primary replicase either. Mutants deficient in DNA polymerase III, *polC⁻* strains, are under study, and it appears that this might be the major replicating enzyme in *E. coli*. Since no mutant has been found to be totally lacking in DNA polymerase I, this enzyme must also fulfill some vital function in DNA chain growth. A series of temperature-sensitive *dna* mutants have also been isolated. Their gene products are soluble proteins and enzymes that are required for DNA synthesis, but many details remain to be established.

According to Kornberg and others, DNA polymerase III accounts for a large part of chain elongation. DNA polymerase I probably fills in the gaps between Okazaki fragments and may also remove segments of RNA primer since the enzyme can cleave nucleotides from both the $5' \rightarrow 3'$ and the $3' \rightarrow 5'$ directions. DNA polymerases II and III, however, only can remove $3' \rightarrow 5'$ pieces but not $5' \rightarrow 3'$ sections of DNA. The manner in which some of these proteins act at the replicating fork has been tentatively suggested in several studies (Fig. 5.28).

Certain tumor-causing RNA viruses have been shown to specify an **RNA-directed DNA polymerase** which catalyzes DNA synthesis from RNA templates. The new DNA then serves as the molecule from which messenger RNA is transcribed, and these transcripts are translated into viral protein. The cycle appears to involve formation of RNA—DNA hybrid

molecules. This synthesis is directed by DNA polymerase acting on the virus RNA template and leads to subsequent formation of DNA duplexes that become incorporated into the host cell chromosome (Fig. 5.29). These DNA duplexes presumably form after the DNA strand has been separated from the RNA—DNA hybrid molecule.

The enzyme has been called a **"reverse transcriptase"** because it directs the formation of DNA from RNA, which is the reverse of the usual transcription format in which RNA molecules are transcribed from DNA molecules. There are many similarities between the RNA-directed and DNA-directed DNA polymerases, but the reverse transcriptase shows a great efficiency and preference for RNA templates. The enzyme is capable of using DNA as well as RNA

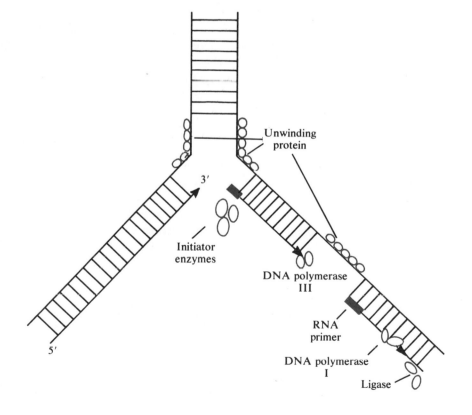

Figure 5.28
Summary diagram of the requirements and events relative to replication of duplex DNA molecules. The initiating pieces of RNA primer are the shaded areas. (From *DNA Synthesis* by Arthur Kornberg. W. H. Freeman and Company. Copyright © 1974.)

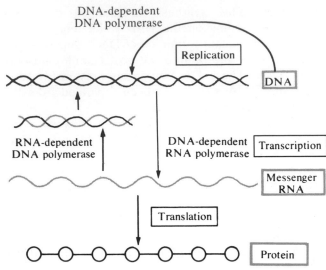

Figure 5.29
Diagram illustrating the usual events of replication, transcription, and translation in RNA viruses. The distinctive step of making DNA copies of viral genetic RNA under the guidance of an RNA-dependent DNA polymerase ("reverse transcriptase") is also shown.

as template and of directing duplex DNA synthesis from conventional DNA strands.

The viral enzyme has been isolated from human leukemic cells and from brain tumors, and thus holds considerable medical interest. Normal animal cells also appear to contain RNA-directed DNA polymerases, but there are some differences in enzyme cofactor requirements and in the inability of the enzymes to utilize natural RNAs as templates. Future investigations hold considerable promise for understanding the relationship of polymerase activities to both pathological and normal events during cell and tissue differentiation.

INFORMATION STORAGE AND FLOW

DNA has two main functions: replication and storage of the genetic information that specifies the characteristics of cells and organisms. DNA acts as the template for its own replication, as discussed earlier. Its coded information is transcribed into RNA from which proteins are translated. Each activity requires particular enzymes that catalyze the reaction steps, sources of energy for biosynthesis, and precursor molecules. One theme that underlies all of these events is the specific pairing of complementary bases: guanine with cytosine and adenine with thymine or uracil. Another feature of DNA which aids in both replication and information transfer is the relative ease with which the two strands of the duplex can unwind. The unwinding opens up accessible regions along which new complementary sequences can be aligned during polymerization of partner DNA or transcript RNA strands.

Transcription requires:

1. Template DNA.
2. RNA polymerase which catalyzes the addition of mononucleotides to the free 3'-hydroxyl end of the growing RNA chain.
3. ATP as an energy source.
4. All four precursors in their nucleoside triphosphate forms.

Translation is an exceedingly complex set of processes (see Chapter 6). In addition to RNA messengers, translation into polypeptides requires transfer RNAs, amino acids in an activated form, enzymes that serve both to convert amino acids to an activated form and then to join these residues to specific transfer RNA carrying molecules, and an assortment of other factors and enzymes. These components cooperatively lead to joining of amino acids into polypeptide chains specified by genetic information (Fig. 5.30).

Complementary base pairing through recognition of one nucleotide by its partner has been firmly established in various ways. One method which is very useful in demonstrating that some nucleic acid sequence is the complement of another is the hybridization between single strands of DNA or RNA in DNA−DNA, DNA−RNA, and RNA−RNA com-

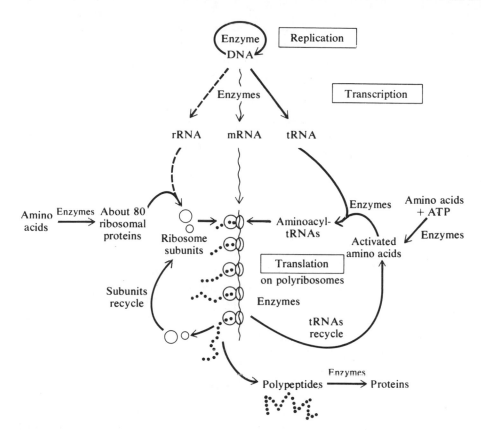

Figure 5.30
Summary scheme showing information flow from coded DNA to the final protein products of gene action.

binations (Fig. 5.31). From these tests and others we know that base pairing or recognition is the basis for specificity in interactions between DNA and DNA in replication, DNA and RNA in transcription, and RNA and RNA in translation (Fig. 5.32). The genetic code provides important insights to our understanding of the operation of the genetic informational and transfer systems.

Deciphering the Genetic Code

The nature of the information storage system was opened to study once it was realized that DNA nucleotide sequence could vary enormously. As Watson and Crick pointed out, the number of theoretically possible genes or nucleotide sequences fully satisfies the diversity required for all past and present life. It was clear that groups of nucleotides were part of a coded repertory of information that specified the protein products of gene action. With 20 kinds of amino acids as units in translation, it was presumed that triplets of nucleotides must serve as individual codewords for these amino acids. If only two nucleotides specified an amino acid, then only 16 unique codewords could be formed from the four kinds of bases in all possible pairwise combinations ($4^2 = 16$) and this number of codewords seemed too small. But, nucleotide triplets could form 64 (4^3) unique permutations and could thus easily form codes for the 20 naturally-occurring amino acids. A triplet code was favored above any other, although alternatives were proposed at various times in the 1950s.

The first triplet codeword, or **codon,** to be identified

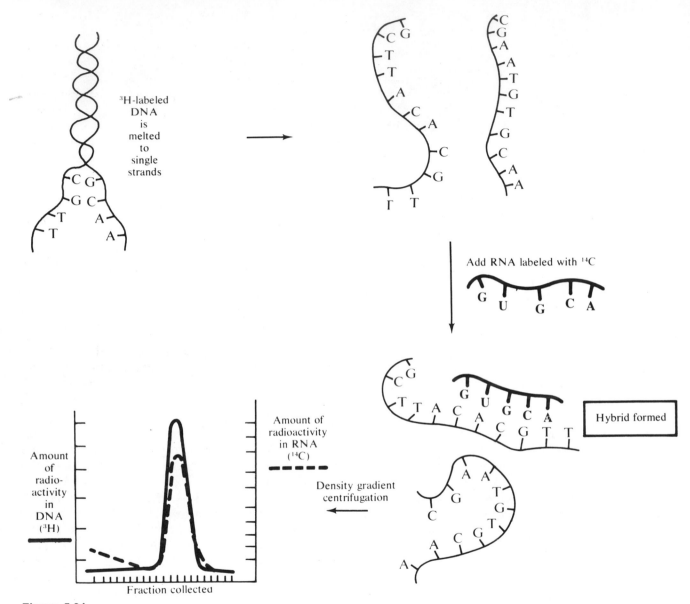

Figure 5.31
Protocol for DNA-RNA hybridization and the detection of complementary nucleotide sequences in radioactively labeled molecules collected after density gradient centrifugation.

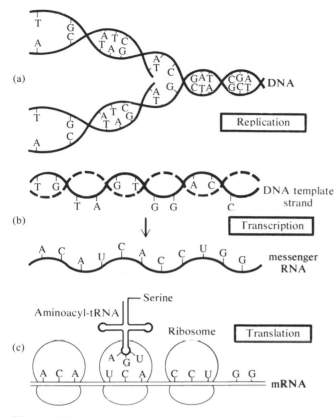

Figure 5.32
Complementarity is the underlying theme for recognition between nucleotides during (a) replication, (b) transcription, and (c) translation of genetic information.

transcription of an AAA codon in the DNA template (Fig. 5.33).

In 1961, Marshall Nirenberg and Henry Matthaei showed that artificial messenger RNA made up only of uracils (polyuridylic acid or **poly U**) could direct the formation of a polypeptide containing only phenylalanine residues. The *in vitro* test system contained a mixture of 18 kinds of amino acids, as well as ribosomes, transfer RNAs and activating enzymes in cell extract, an energy source, and Mg^{2+} ions, along with poly U. This assortment was set up in 18 different tubes with only one of the amino acids radioactively labeled in each tube. The job of finding the specific amino acid corresponding to UUU codons was much simpler because only one tube would end up with radioactive polypeptides and the others could be ruled out. In the presence of poly U, only the tube containing radioactively labeled phenylalanine had labeled polypeptides. When the labeled polypeptide was hydrolyzed, it was confirmed that phenylalanine was the only amino acid in the polymer. Subsequent studies using the poly C codon showed that this codon directed formation of polyproline exclusively. When poly A was used as the artificial messenger, only polylysine was formed. These studies demonstrated that CCC was the RNA codon for proline and AAA was the RNA codon for lysine.

Between 1961 and 1964 Nirenberg, Ochoa, and others reported various studies in which artificial messengers made up of nucleotides in various combinations and proportions were used. About 50 codons were identified after statistical analyses of the experimental results. The composition of the codons could be established in this way, but the sequence of the bases could not. For example, two Us and one G were found to be involved in coding for leucine, valine, and cysteine, since these three amino acids were incorporated into polypeptides when poly UG messengers were used in which there were twice as many uracils as guanines in the RNA. There was no way, however, to determine from these tests that UUG was the codon for leucine, GUU for valine, and

was UUU which specified the amino acid phenylalanine. The genetic code is usually described in terms of the codons in messenger RNA transcripts rather than directly in DNA codewords. This convention arose from the fact that almost all experimental studies have employed RNA transcripts involved in protein translation, because it is easier than working with the DNA. Furthermore, translation takes place in the presence of the transcript, even if DNA is omitted from the test system. We refer all information ultimately to DNA, but discuss the RNA relation to protein more directly. The UUU codon in RNA is a

Second nucleotide

First Nucleotide	A or U	G or C	T or A	C or G	Third Nucleotide
A or U	AAA *UUU* / AAG *UUC* Phenylalanine / AAT *UUA* / AAC *UUG* Leucine	AGA *UCU* / AGG *UCC* Serine / AGT *UCA* / AGC *UCG*	ATA *UAU* / ATG *UAC* Tyrosine / ATT *UAA* / ATC *UAG* "Stop"	ACA *UGU* / ACG *UGC* Cysteine / ACT *UGA* "Stop" / ACC *UGG* Tryptophan	A or U / G or C / T or A / C or G
G or C	GAA *CUU* / GAG *CUC* / GAT *CUA* / GAC *CUG* Leucine	GGA *CCU* / GGG *CCC* / GGT *CCA* Proline / GGC *CCG*	GTA *CAU* / GTG *CAC* Histidine / GTT *CAA* / GTC *CAG* Glutamine	GCA *CGU* / GCG *CGC* / GCT *CGA* Arginine / GCC *CGG*	A or U / G or C / T or A / C or G
T or A	TAA *AUU* / TAG *AUC* Isoleucine / TAT *AUA* / TAC *AUG* Methionine	TGA *ACU* / TGG *ACC* / TGT *ACA* Threonine / TGC *ACG*	TTA *AAU* / TTG *AAC* Asparagine / TTT *AAA* / TTC *AAG* Lysine	TCA *AGU* / TCG *AGC* Serine / TCT *AGA* / TCC *AGG* Arginine	A or U / G or C / T or A / C or G
C or G	CAA *GUU* / CAG *GUC* / CAT *GUA* Valine / CAC *GUG*	CGA *GCU* / CGG *GCC* / CGT *GCA* Alanine / CGC *GCG*	CTA *GAU* / CTG *GAC* Aspartic acid / CTT *GAA* / CTC *GAG* Glutamic acid	CCA *GGU* / CCG *GGC* / CCT *GGA* Glycine / CCC *GGG*	A or U / G or C / T or A / C or G

Figure 5.33
The genetic code. The DNA codons are shown in roman and the complementary RNA codons are in italics. The 20 amino acids are specified by 61 of the 64 triplet codons. The remaining 3 codons are "punctuation marks" that signal the end of a genetic message in a sequence of nucleotides.

UGU for cysteine. These studies did lead the way, however, and they demonstrated one important generalization: the genetic code is *degenerate,* that is, one amino acid is specified by more than one codon. These studies, therefore, showed that all or most of the 64 possible codons were part of the dictionary that spelled out amino acids in proteins.

The sequence of bases in each codon was opened to analysis in 1964 with the development of a *binding assay* by Nirenberg and P. Leder. They discovered two important features of the experimental system which permitted them to assay individual codon triplets. First, if they omitted guanosine triphosphate from the system, they could prevent the joining of individual amino acids into polymers, while still allowing transfer RNA—activated amino acid complexes to bind to messenger RNA. Second, they could get the same efficiency of tRNA—amino acid binding if they used messenger fragments only three nucleo-

tides long or longer pieces of poly U. This second feature provided support for the hypothesis that codons were triplets of nucleotides, and also opened up possibilities for studies using different nucleotide-triplets whose base sequence and composition were *known.* To decipher the code, the researchers looked for specific binding of radioactively labeled amino acid—tRNA complexes to specific triplet RNA codons. This approach opened the way to rapid solution of the codon dictionary and what had appeared an impossible and lengthy task a few years earlier turned out to require only a relatively simple set of defined tests.

A different experimental approach was taken by H.G. Khorana and his associates in deciphering the meanings of triplet codons. Their studies provided important independent evidence in support of the interpretations based on the binding assays. Khorana synthesized RNA polymers by stepwise additions of known doublet or triplet nucleotide fragments so that the final molecule contained alternating stretches of these two or three-nucleotide-long inserted pieces. A polymer such as UGUGUGUGUG could be made' from repeated UG dinucleotides, but the alternating triplet codons that are generated include UGU and GUG (Fig. 5.34). This molecule could code for two amino acids, which would appear in an alternating sequence in the polypeptide translation. When a polypeptide synthesized from this RNA ·is shown to contain alternative cysteine and valine residues, the assignments are UGU = cysteine and GUG = valine.

These experimental results, together with information from binding assays, led to the assignment of 61 codons to specific amino acids. By 1967 the remaining 3 codons had been deciphered to be termination signals for the stop point of a gene or transcript RNA. It has also been discovered that the codon specifying methionine (AUG) also acts as a codon which signals the initiation of a protein chain. Another codon that can signal the start of protein synthesis is GUG, the RNA codon that also specifies valine under

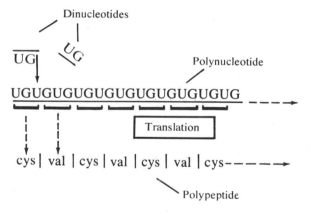

Figure 5.34
Diagram illustrating the principle of an experiment designed to decipher the genetic code, as conducted by Khorana. Triplets of nucleotides spell out the amino acids cysteine and valine in an alternating sequence that corresponds to the alternating sequence of triplet codons in the artificial poly UG messenger RNA guiding translation.

appropriate conditions. These codons will be discussed further in Chapter 6.

General Features of the Genetic Code

The genetic code is degenerate but consistent; that is, more than one codon specifies most of the amino acids, but each codon is translated as only one amino acid and no other. Two of the 20 amino acids, tryptophan and methionine, are specified by one codon each. The other amino acids are specified by 2 to 6 different codons.

Examination of the codons shows that most amino acids are coded by triplets having identical bases in positions one and two but varying in the third base. Exceptions to this rule are tryptophan and methionine which have one codon each, and leucine, serine, and arginine which have six codons each. Because the third base often varies in codons specifying the same amino acid, Crick has discussed the relationship of the "wobble" nature of the third base position to evolution of the genetic code. He has suggested that only the first two bases in the primeval code may have had a coding function and that the third base assumed a coding specificity only later in biological evolution. The codon must have been made up of three bases from the beginning, but the third base may have acted differently in the past. Certain amino acids are thought to be more ancient than others because they occur commonly in many proteins and in proteins known to have been in existence for over a billion years. By itself, however, the idea of a change in the triplet codon from specificity due to only the first two bases to specificity due to all three bases is still only an idea and there is little evidence on which to base judgment.

The code is commaless. Translational readout continues without interruption from one codon to the next until a termination codon is reached, after which the finished protein is released. Translation proceeds along a reading frame, beginning at some specified point of origin and ending at one of the three punctuating terminator codons (UAA, UAG, or UGA). In a series of experiments where single bases were added or deleted in wild-type and mutant phage, Crick demonstrated that a garbled translation occurred in the region of the gene that had been modified (Fig. 5.35). The reading frame could be restored by appropriate additions or deletions of other bases. Such "frameshift" mutants showed there was a direction to translational readout, that one added or missing base could lead to synthesis of a nonfunctional protein, and that the code was almost certainly triplet. The latter point was interesting at the time it was made in the 1960s.

The code is essentially universal. Nirenberg and colleagues showed that amino acid−tRNA complexes from an amphibian (toad) and a mammalian (guinea pig) species could bind to messenger RNA from *E. coli* just as well as such complexes from *E. coli*. These three widely different species thus contained transfer RNAs that could recognize the same codons in messenger RNA from a single source. Since the test was a binding assay, it further showed that the amino acids carried by tRNA to the messenger were very much the same in all three species. The same codewords also seem to be used for initiation and termination of polypeptide chain synthesis in prokaryotes and eukaryotes. Some differences have been found in the way the initiating amino acid is modified,

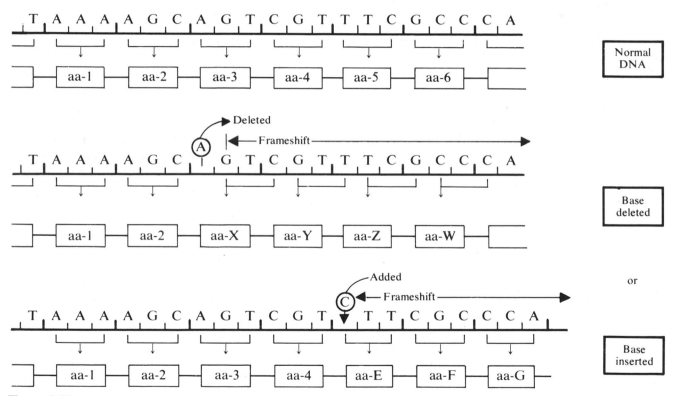

Figure 5.35
Diagram illustrating the results of experiments with T4 phage in which bases were added to or deleted from DNA. Garbled results of translation occurred after introducing these alterations into the reading frame of the genetic message.

but a methionine-carrying tRNA recognizes and binds to the AUG initiator codon in cells and virus protein-synthesizing systems.

The gene and polypeptide are co-linear. The linear sequence of nucleotides corresponds to the linear sequence of amino acids in the polypeptide for which it codes. Two elegant lines of evidence have clearly demonstrated **co-linearity,** although it had been presumed to be the correct model in earlier years. Charles Yanofsky and associates showed that amino acid replacements in one chain of the enzyme tryptophan synthetase of *E. coli* coincided in location with the position of the responsible change in DNA. High-resolution genetic mapping of the gene provided the basis for comparison with the amino acid sequence in the enzyme protein (Fig. 5.36). In each of the twenty or more mutants studied, there was a precise corre-

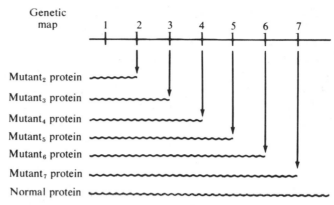

Figure 5.37
Demonstration of co-linearity of gene and protein in T4 phage. The known position of an inserted termination codon in a mutant DNA sequence corresponded to the observed site of premature termination in the protein chain translated from the altered DNAs.

spondence between the mutant codon and the amino acid replacement.

In a different set of studies using other methods, Sidney Brenner, Crick, and others showed the co-linearity between a T4 phage gene and the protein product that forms part of the head of the virus. In these mutants, terminator codons were present in place of the usual codons specifying amino acids. The head protein chains therefore terminated protein synthesis prematurely in the mutants compared to protein produced in wild-type phage (Fig. 5.37). In each mutant the length of the protein chain corresponded to the position of the terminator codon in the mutant gene, again showing co-linearity between gene and protein.

Some known amino acids are not included in the genetic code. In each of these cases, it has been shown that these uncoded amino acids are either products of degradation of some larger molecule, or else they are residues that become modified after another amino acid has been inserted into the growing protein. The protein collagen contains an un-

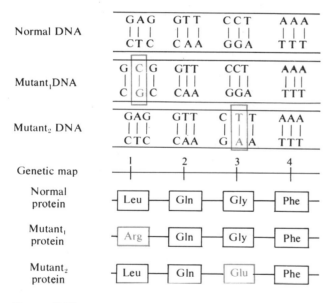

Figure 5.36
Demonstration of co-linearity between gene and protein in *E. coli.* Alteration of specific codons in DNA lead to specific alterations of amino acids in corresponding locations in the proteins made by the mutant cells.

usually high amount of the amino acid hydroxyproline which is not included in the genetic code. During collagen synthesis, proline, a coded amino acid, is inserted into the growing polypeptide; hydroxylation of proline takes place afterward. All the 20 amino acids specified by the genetic code are inserted directly into the polypeptide as it is being synthesized.

Regulatory and Structural Genes

During the 1940s and 1950s there was increasing support for the **one gene—one enzyme** concept. Genetic studies showed that a certain reaction would be catalyzed if the wild-type allele of the gene was functioning, but the reaction would not take place in strains with a mutant allele. The concept was generalized and variously expressed as the **one gene—one protein** or the **one gene—one polypeptide** concept as more was learned about the relationship between a gene and its product. The underlying theme remained: each gene specified some protein—either a functional enzyme or a globin molecule. Hemoglobin, myoglobin, immunoglobins, and other nonenzymatic proteins were included in the repertory of gene products. These ideas were changed profoundly by the studies of François Jacob, Jacques Monod, and their associates. In 1961 they proposed a new concept which stated that functionally different kinds of genes participated in control of the expression of coded information specifying the synthesis of enzymes and other proteins. The interplay of regulatory and structural genes was spelled out in the **operon hypothesis** of gene action.

INDUCIBLE AND REPRESSIBLE ENZYMES. A brief introduction to this topic was given earlier in Chapter 3, but it may be helpful to describe some pertinent background once more. Synthesis of many enzymes may either be induced in the presence of substrate or repressed when the metabolite is in the medium. Both induction and repression of enzyme synthesis are alike in that the amount of enzyme present fluctuates rapidly and substantially in response to some metabolite in the medium. **Inducible enzymes** occur in only

trace amounts most of the time, however, and are synthesized quickly when the substrate is provided. **Repressible enzymes** are present and active when their substrates occur in low amounts.

When *E. coli* uses glucose for growth, little of the enzymes necessary for utilization of lactose or some other sugars are present. If lactose is provided in the culture medium instead of glucose, the bacteria begin to elaborate β-galactosidase almost immediately. The enzyme concentration may rise from only a few molecules to about 5000 within minutes. Removal or depletion of lactose in the system causes an abrupt stop in β-galactosidase synthesis, and the enzyme concentration is diluted down to its usual low levels during subsequent generations of growth and division.

The genes specifying the enzymes involved in utilization of lactose, the *lac* gene system, were studied by Jacob and Monod. The principles derived from these experiments could be applied to other inducible enzyme systems and were used to explain repressible enzyme synthesis as well. These principles led to a totally new view of genes and gene regulation, and they also provided significant insights into modulations of metabolism that underlie the thrifty economy of cell activity.

THE OPERON CONCEPT. Jacob and Monod studied the effects of mutations on the activities of three genes responsible for proteins needed to hydrolyze lactose in *E. coli*. Genetic mapping showed that the three genes (*z*, *y*, and *a*) were next to one another in that order. The *z* gene codes for the enzyme *β-galactosidase*, which splits lactose and other β-galactosides to glucose and galactose. The *y* gene specifies *β-galactoside permease*, a transport protein found in the cell membrane which assists in the uptake of β-galactosides from the medium into the cell. Gene *a* codes for *β-galactoside transacetylase*. This enzyme catalyzes the transfer of an acetyl group from acetyl CoA to β-galactosides, but its role in lactose processing is uncertain. Mutations in either the *z* or *y* genes produce mutants that cannot use lactose. The z^- mutants cannot metabolize lactose and the y^- mutants cannot take up this substrate from the medium. Mutants that are a^- have modified β-galactoside transacetylase, but this does not affect the ability of the cells to take up and hydrolyze lactose. Most studies of this gene system have therefore examined β-galactosidase activity as an index of gene action and its regulation (Fig. 5.38).

Earlier studies of the *lac* system had shown that molecules resembling lactose could also induce enzyme synthesis, even though these alternative inducers were not acted upon by the enzymes they induced. It had also been shown that *any* inducer was equally effective or ineffective on all three proteins of the *lac* system. The relative amounts of the three proteins coded by the *z y a* cluster increased and decreased coordinately under different conditions for growth. All these observations indicated that some common controlling element governed coordinate synthesis of the three proteins, probably by interaction with the inducer. It was unlikely that these three genes themselves were the agents responsible for regulating synthesis of their proteins since their action was coordinate.

Based on these and other considerations, Jacob and Monod postulated the existence of two classes of genes: **structural** and **regulatory**. Structural genes coded for the enzyme proteins, and regulator genes were responsible for turning the structural genes "on" and "off" so that proteins were or were not synthesized in the cell. Jacob and Monod predicted that structural gene mutations would affect only one of the three proteins, depending on whether the change was in the *z, y,* or *a* locus. The other two proteins would be unaffected in a one-locus gene mutant.

Mutations in regulator genes, however, should affect all three inducible proteins equally and coordinately. The effects of regulator gene mutation would be related to the inducer and would result in failure to induce enzyme synthesis even when lactose was present, or in enzyme synthesis which would take place in both the presence and absence of the inducer. Normal or wild-type regulatory systems would sponsor inducible enzyme synthesis only in the presence of lactose or a related metabolite. Jacob and Monod found regulatory mutants showing all these predicted effects. Structural gene mutants also were found, as predicted.

The regulator gene *i* was mapped at some distance away from the structural gene cluster. Two mutants were found: i^-, in which the three inducible proteins were synthesized whether or not lactose was present, and i^s, in which protein synthesis was uninducible even if lactose was available. Putting together various kinds of information, Jacob and Monod proposed that the product of the regulator gene was a **repressor.** The repressor interacted with the inducer (lactose) and regulated enzyme synthesis at the level of transcription (Fig. 5.39). When the inducer was absent, the repressor blocked transcription of the structural genes. If messenger RNA was not transcribed, then enzyme translations could not be manufactured. When lactose was present, this inducer combined with repressor and prevented the repressor from blocking transcription of the *z y a* structural genes. With the inducer present, transcription was turned on; with the inducer absent, transcription was turned off. In the i^- mutant, synthesis presumably was turned on all the time because the repressor was defective or missing. The i^s regulator mutant presumably made a repressor that could not be removed from its blocking site even if inducer was present.

Where does the repressor act? It blocks transcription of all three genes coordinately, so it must be at

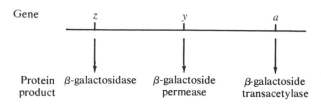

Figure 5.38
Genetic map showing the genes that specify enzymes in the *lac* system of *E. coli.*

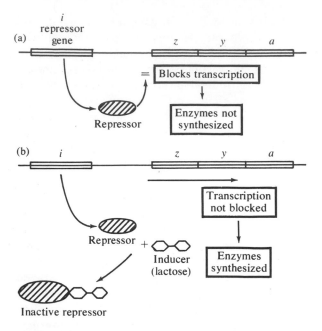

Figure 5.39
Diagram illustrating the interaction between the repressor product of the regulator gene i and transcription of the structural genes that specify the three inducible enzymes of the *E. coli lac* operon in the (a) absence and (b) presence of the inducer metabolite lactose.

one side of the z y a cluster. It was shown by mapping studies of another kind of regulatory mutant that an **operator** site was just to the left of the z gene. The repressor acts at the operator site, which may be a gene or a site on a gene. The operator mutant, o^c, produced *lac* proteins all the time, just as the i^- mutant did. The o^c and i^- mutants were distinguished by mapping studies of the z y a cluster. They were also distinguished in genetic tests using specially constructed strains that were diploid for the *lac* region but haploid everywhere else in the genome. In such partial diploids, the i^+ wild-type allele acted as a dominant to the i^- allele, whereas o^+ acted as a recessive to its o^c allele. The operator mutant is apparently incapable of binding with the repressor and is therefore turned "on" all the time. The regulator mutants make altered repressors and are always turned "on"

(i^-) or "off" (i^s), depending on the alteration in the repressor molecule.

The regulatory and structural genes of the interacting system were parts of an **operon.** The regulator gene product was a repressor, the structural gene product was an enzyme or other protein of cell metabolism, and the operator served as the recognition site for the repressor. Induction of enzyme synthesis took place only when transcription of the structural genes could proceed. Repressor blocking was relieved when the inducer combined with the repressor and removed it from the operator site. With the operator unblocked, transcription was "on." Messenger RNAs were then made available for translation into proteins at the ribosomes in the cell cytoplasm.

The same operon model serves equally well to explain repression of enzyme synthesis. In this case the enzyme is synthesized in the absence of a metabolite but is repressed when the metabolite is present. Since the metabolite is usually an end product of the biosynthetic sequence, the phenomenon had been referred to as **endproduct repression** of enzyme synthesis. According to the operon concept, the repressor product of the regulator gene cannot block the operator site unless the repressor is in combination with the inducer metabolite. The inducer in these systems is called a **co-repressor** because it contributes to blocking the operator when in combination with repressor (Fig. 5.40).

In the tryptophan synthesizing pathway, repressor produced by the regulator gene cannot block transcription of the messengers for enzymes of the pathway. But, when repressor combines with tryptophan (co-repressor), the operator site is blocked by the repressor—co-repressor complex. The operon concept led to the understanding that induction and repression of enzyme synthesis were different expressions of the same kind of **negative control system.** In each case enzyme synthesis stops when appropriate conditions exist in the cell.

REPRESSOR PROTEINS. A number of refinements have been introduced into the operon concept since 1961,

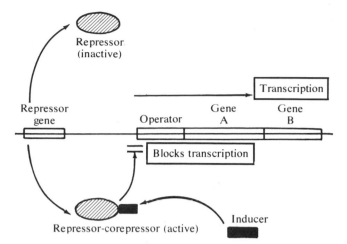

Figure 5.40
Diagram illustrating the negative control of repressible enzyme synthesis in prokaryotes. Transcription of mRNA is blocked when co-repressor binds with repressor at the operator site of the operon. Transcription proceeds when the co-repressor metabolite is absent since repressor alone cannot bind to the operator in such a system.

and vital new information has been obtained. Although Jacob and Monod had suggested that repressors were protein products of regulator genes, direct confirmation was not obtained until 1966. Using different methods and systems, Walter Gilbert and Mark Ptashne independently isolated repressors. It was shown that these were proteins and that the repressor combined specifically with the inducer in the inducible enzyme systems which were studied. Gilbert examined the *E. coli lac* operon repressor and Ptashne described the repressor governing synthesis of phage lambda proteins early in its infection of *E. coli* host cells.

The repressor protein bound specifically to DNA containing the genes regulated by the repressor. Even more specifically, the repressor bound to the operator site of the operon. In combination with inducer, the repressor detached from the operator DNA. The repressor protein must have at least two binding sites, one that recognizes operator DNA and another that binds the inducer. Recent experimental evidence reported by Benno Müller-Hill and associates led them to suggest that a very short sequence of amino acids near one end of the repressor protein is the specific DNA-binding region. This short piece may fit into the wide groove of the DNA double helix, binding to both the phosphates of the backbone and the bases. The repressor conformation which allows binding with DNA may be altered when the inducer binds to another part of the repressor protein. Inducer binding, then, would lead to the repressor detaching from the operator DNA, unblocking transcription.

PROMOTERS AND OPERATORS. When Gilbert's and Ptashne's studies were published, the way was opened to a more detailed exploration of the regulatory components of transcriptional control over protein synthesis. One important outcome of studies in the late 1960s was the identification of the **promoter** site of the operon. The promoter is on the other side of the operator, which is sandwiched between the promoter site and the structural genes. Like the operator, the promoter is a recognition site on DNA. Neither operator nor promoter is transcribed, but both are regions to which components of the system bind specifically and they influence transcription of structural genes and, presumably, regulator genes which code for repressor proteins.

The promoter is the site recognized by RNA polymerase which binds and catalyzes polymerization of mononucleotides into RNA transcripts. Since the operator sits between the promoter and the structural genes of the operon, the polymerase has access to structural genes only when the operator site is not blocked by the repressor (Fig. 5.41). Once the repressor is removed, inducible-enzyme transcript-RNAs can be made as the polymerase moves across the operator and onto the structural gene sequences. During the 1970s it became clear that the promoter was more complex than a mere recognition site for RNA polymerase. Although this topic is still under active investigation, some additional properties of the promoter have been described. Since these properties are concerned with a widespread system of **positive**

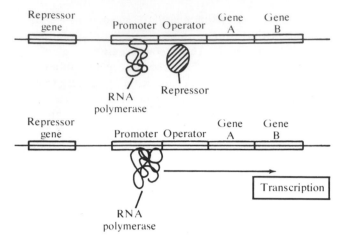

Figure 5.41
Relationship of the promoter site to the other components of the operon. RNA polymerase binds to the promoter DNA. The polymerase can then direct transcription of messenger RNA from the structural gene sequences. Transcription occurs only if its progress is not blocked by repressor bound at the adjacent operator site.

control over inducible enzyme synthesis, we will examine the story briefly. Some general discussion of the positive control system was presented earlier in Chapter 3.

As described in Chapter 3, synthesis of inducible enzymes such as β-galactosidase takes place at low rates if glucose is present alone or in combination with the inducing substrate, such as lactose. The rate of enzyme synthesis can be increased to maximum levels if lactose is the only carbon source or if cyclic AMP is added to a medium containing glucose alone or together with lactose. Since cyclic AMP can mimic the effects of lactose alone in the β-galactosidase system, the role of cyclic AMP in regulating inducible enzyme synthesis has been studied intensively.

Cells of *E. coli* that are grown in glucose media have lowered levels of cyclic AMP. Adding this nucleotide to cultures using glucose therefore raises the level of cyclic AMP and leads to enzyme synthesis. Since a protein is required to effect cyclic

AMP control, the relationship of cyclic AMP and the cyclic AMP receptor protein (CRP) to transcription were determined from studies of suitable mutants. The three kinds of mutants were (1) mutants with defective *adenyl cyclase,* which catalyzes cyclic AMP formation from ATP; (2) mutants with abnormal receptor protein but which are otherwise normal; and (3) mutants for the regulator, operator, or promoter regions of the *lac* operon.

Adenyl cyclase-deficient mutants could be restored to normal function if cyclic AMP was added to the medium. This experiment showed that cyclic nucleotide was required in cell growth and metabolism. Function in mutants with defective CRP was not restored by addition of cyclic AMP. This experiment showed that protein was needed in the system, a need that was subsequently verified using purified CRP in test systems. Both cyclic AMP and CRP act at the level of transcription, since promoter mutants showed decreased sensitivity to cyclic AMP or were totally unresponsive to cyclic AMP if most of the promoter region was deleted in the operon. Mutants with altered regulator or operator activities showed no changes in their responses and activities in relation to cyclic AMP or CRP. These regulatory components were therefore ruled out, but promoter interaction with cyclic AMP and CRP proved to be critical.

Using promoter mutants, it was shown that cyclic AMP and CRP must influence transcription of messengers for structural gene proteins. Further support for this interpretation came from two kinds of studies, among others. (1) If a wild-type system was inhibited from *initiating* RNA transcription, the response was the same as it was when cyclic AMP was absent. Inhibitors of RNA chain *elongation* had no such effect. Since the initiation inhibitor mimicked the responses obtained with cyclic AMP, it was reasoned that cyclic AMP affected initiation at the promoter site of the *lac* operon. (2) Using purified CRP, investigators demonstrated that the protein bound specifically to DNA if cyclic AMP was present. In the presence of cyclic AMP, therefore, the receptor protein binds to

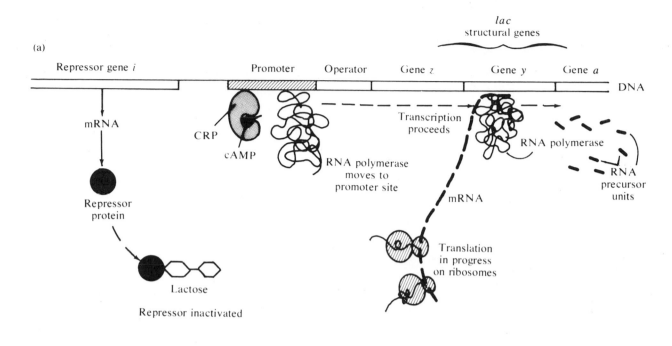

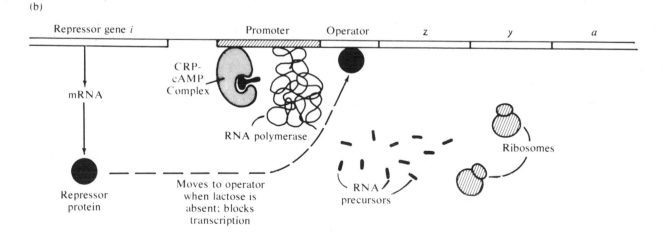

Figure 5.42

The interacting components of gene expression of the *lac* operon in *E. coli*: (a) Transcription proceeds, directed by RNA polymerase, when stimulated by the presence of CRP—cyclic AMP at the promoter site; (b) even with CRP—cAMP and other components present, transcription may be blocked by the repressor of inducible enzyme mRNA transcription, since the repressor is active in the absence of its inducing metabolite lactose.

the promoter site of the operon and stimulates transcription by RNA polymerase. The stimulation occurs at the initiation of transcription and not during RNA chain growth itself (Fig. 5.42).

Positive and negative controls over inducible enzyme synthesis must be cooperative since each has a particular effect. The negative control system is more specific since its response is elicited by specific inducers, such as lactose. A positive control system, such as the cyclic AMP—CRP system, operates over a broader range. Various inducible enzymes are synthesized when CRP binds to promoter regions in the presence of cyclic AMP. The interactions among these systems lead to much finer levels of control over metabolism than either alone could exert.

Promoters must have at least two sites for recognition and binding. One site accommodates cyclic AMP receptor protein, and another is the place to which RNA polymerase binds. These observations immediately show that promoters are not just simple DNA sites of recognition. Many studies are now in progress which we expect to reveal more about the promoter and other regulatory components of the genetic system. Together with studies of enzyme activity and its regulations, we are coming closer every day to an understanding of the incredibly intricate machinery that regulates the thousands of reactions which an average cell is capable of carrying out.

Control systems in eukaryotic cells are less well known, mostly because of the greater complexity of their chromosomes and the levels of interaction among cell compartments. Less is known about the genetic basis for controls in eukaryotes, but a number of controls must be similar to those in prokaryotes. The genetic material of eukaryotic cells is DNA, messengers are transcribed, and proteins are translated on similar ribosomal machineries. Superimposed on all this is the complexity of compartmentation within the eukaryotic cell and the specialization of different cell types in a multicellular organism. The molecular basis for development and differentiation depend in large part on genetic controls. Some of these topics will be discussed in subsequent chapters.

SUGGESTED READING

Books, Monographs, and Symposia

Cold Spring Harbor Symposia on Quantitative Biology. 1968. *Replication of DNA in Micro-organisms*, vol. 33. New York: Cold Spring Harbor Laboratory.

Kornberg, A. 1974. *DNA Synthesis*. San Francisco: W. H. Freeman.

Raacke, I. D. 1971. *Molecular Biology of DNA and RNA. An Analysis of Research Papers*. St. Louis: Mosby.

Watson, J. D. 1975. *Molecular Biology of the Gene*. Reading, Mass.: W. A. Benjamin.

Articles and Reviews

Alberts, B. M., and Frey, L. 1970. T4 bacteriophage gene 32: A new structural protein in the replication and recombination of DNA. *Nature* 227:1313–1318.

Borst, P., and Aaij, C. 1969. Identification of the heavy strand of rat-liver mitochondrial DNA as the messenger strand. *Biochemical and Biophysical Research Communications* 34:358–364.

Cairns, J. 1966. The bacterial chromosome. *Scientific American* 214(1):36–44.

Crick, F. H. C. 1966. The genetic code: III. *Scientific American* 215(4):55–62.

Dickson, R. C., Abelson, J., Barnes, W. M., and Reznikoff, W. S. 1975. Genetic regulation: The *lac* control region. *Science* 187:27–35.

Ingram, V. M. 1958. How do genes act? *Scientific American* 198(1):68.

Jacob, F., and Monod, J. 1961. Genetic regulatory mechanisms in the synthesis of proteins. *Journal of Molecular Biology* 3:318–356.

Kornberg, A. 1968. The synthesis of DNA. *Scientific American* 219(4):64–78.

Kriegstein, H. J., and Hogness, D. S. 1974. Mechanism of DNA replication in *Drosophila* chromosomes: Structure of replication forks and evidence for bidirectionality. *Proceedings National Academy of Sciences* 71:135–139.

Meselson, M., and Stahl, F. W. 1958. The replication of

DNA in *E. coli. Proceedings of the National Academy of Sciences* 44:671–682.

Nirenberg, M. W. 1963. The genetic code: II. *Scientific American* 208(3):80–94.

Nirenberg, M. W., and Matthaei, J. H. 1961. The dependence of cell-free protein synthesis in *E. coli* upon naturally occurring or synthetic polynucleotides. *Proceedings of the National Academy of Sciences* 47:1588–1602.

Pastan, I. 1972. Cyclic AMP. *Scientific American* 227(2):97–105.

Ptashne, M., and Gilbert, W. 1970. Genetic repressors. *Scientific American* 222(6):36–44.

Spiegelman, S. 1964. Hybrid nucleic acids. *Scientific American* 210(5):48–56.

Stent, G. S. 1972. Prematurity and uniqueness in scientific discovery. *Scientific American* 227(6):84–93.

Temin, H. M. 1972. RNA-directed DNA synthesis. *Scientific American* 226(1):24–33.

Watson, J. D., and Crick, F. H. C. 1953. Molecular structure of nucleic acids: A structure for deoxyribonucleic acid. *Nature* 171:737–738.

Westmoreland, B., Szybalsi, W., and Ris, H. 1969. Mapping of deletions and substitutions in heteroduplex DNA molecules of bacteriophage lambda by electron microscopy. *Science* 163:1343–1348.

Yanofsky, C. 1967. Gene structure and protein structure. *Scientific American* 216(5):80–94.

Part
two

Cytoplasmic
Compartments
of the Cell

Chapter 6

Ribosomes and Protein Synthesis

Living cells manufacture proteins that have many kinds of functions. Structural proteins, regulatory proteins, catalytic proteins, and many other types are constantly formed as cells grow, reproduce, and repair themselves. Proteins are not permanent parts of cells; they must be replenished during the life of the cell. Synthesis of new proteins takes place on the surface of ribosomes, particles so small they cannot be seen with the ordinary light microscope. Every living cell contains ribosomes during all or part of its existence. During these times, a cell engages in expression of its genes through two separable sets of events: (1) **transcription** of coded genetic information, and (2) subsequent **translation** of the RNA transcripts into polypeptide chains. Ribosomes provide the surface on which protein synthesis takes place, and they actively coordinate and catalytically assist the processes of **initiation, elongation,** and **termination** of polypeptide molecules being synthesized. Ribosomes are the activity center for many interacting processes and molecules required for synthesis of every protein in the cell.

Prokaryotic cells contain abundant ribosomes distributed throughout the cytoplasm. In eukaryotic

167

cells, however, ribosomes may occur free in the cytoplasm or they may be bound to various cellular membranes. Ribosomes also are present in mitochondria and chloroplasts. These organelle ribosomes are readily distinguishable from cytoplasmic ribosomes in the same eukaryotic cell. Whether in the cytosol or in organelles, ribosomes may be bound loosely or firmly to specific membranes. One of the striking ribosomal displays in eukaryotes is the system known as **rough endoplasmic reticulum** (Fig. 6.1). Ribosomes are bound to the outer surface of the membrane sheets of the endoplasmic reticulum. Where ribosomes are absent from these membranes, the system is called **smooth endoplasmic reticulum.**

In addition to endoplasmic reticulum, ribosomes also are found attached to the outer membrane of the nucleus, on the side facing the cytoplasm. Protein synthesis takes place on all these ribosomes in all their locations in the cell. Only the acellular viruses lack ribosomes. Viruses therefore cannot synthesize their own structural and enzyme proteins. Instead, they subvert the ribosomal machinery of the host cell so that host ribosomes translate virus genetic information.

GENERAL FEATURES OF RIBOSOMES

Ribosomes are made up of two different subunits which act in concert during protein synthesis. Neither subunit can act alone in guiding protein synthesis, but each performs a set of functions that is complementary to the other's. The entire functional **ribosome monomer** can be characterized by physical and chemical characteristics, and the two kinds of subunits are also distinguishable in these ways.

Monomers and Subunits

The commonest way to identify ribosome monomers and their constituent subunits is by their sedimentation properties. Under standard conditions of centrifugation, monomers and subunits settle in particular regions of a density gradient according to size, shape, and other features of the particles. The particles are described by their **S value,** which is their sedimentation constant expressed in **Svedberg units.** Ribosomes from *E. coli* are 70S and these particles are frequently used as a reference standard for studying and describing monomers or subunits from other species and sources. The subunits of *E. coli* ribosomes are 50S and 30S particles. Since sedimentation behavior of particles and molecules is influenced by shape as well as size and molecular weight, the individual subunit values are higher than one might otherwise expect. The combined values for the two subunits are always higher than the S value of the monomer particle.

Ribosomes of prokaryotes other than *E. coli* are also 70S. On the other hand, ribosomes from cytoplasm of eukaryotic cells are 80S. The subunits of eukaryotic 80S cytoplasmic ribosomes are 60S and 40S. Common reference standards for eukaryotic 80S monomers and their subunits are particles from yeast and rat liver cells. Since S values can vary because of slight changes in conditions for centrifugation, it is often useful to include some standard reference particle whose S value is firmly established and accepted.

Mitochondrial and chloroplast ribosomes are also made up of two subunits. Mitochondrial ribosomes range from 55S in multicellular animals to 80S in some of the protozoa and fungi. Chloroplast ribosomes appear to be 70S in all green cells that have been studied to date. Further discussion of organelle ribosomes will be found in Chapter 9. In this chapter the cytoplasmic ribosomes will be the focus of discussion.

Ribosome dimensions have been reported differently depending on preparation methods used in electron microscopy of the particles. In general, dried ribosomes in fractions of isolated particles extracted from prokaryotes measure about 170 Å wide and 200 Å long. Eukaryote cytoplasmic ribosomes are larger. Absolute dimensions vary according to the methods

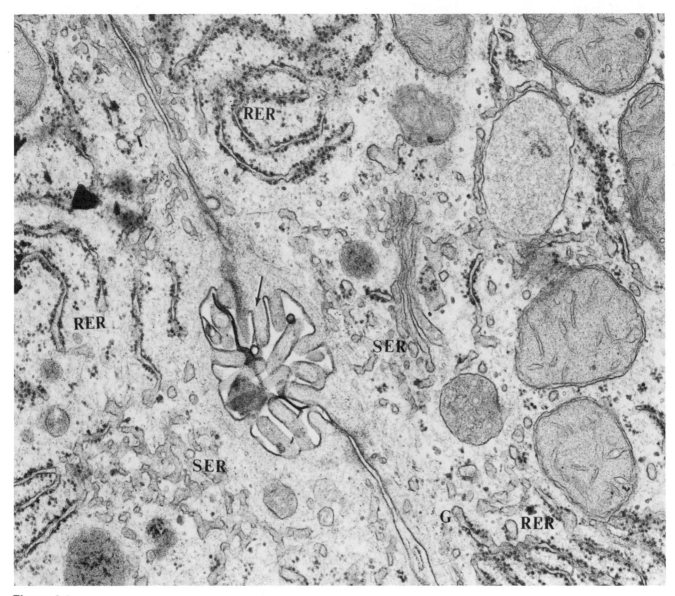

Figure 6.1
Thin section of rat liver cells showing rough (RER) and smooth (SER) endoplasmic reticulum. In addition to mitochondrial profiles, there are microvilli (at arrow) of a bile canaliculus where the adjacent cell membranes come together. × 40,000. (Courtesy of H. H. Mollenhauer)

used for preparation, and they have been described as somewhat flattened (250 Å wide by 200 Å long) or somewhat elongated (250 Å wide by 275 Å long). The larger size of eukaryotic ribosomes has been established by several methods, including electron microscopy and S value determinations. The size difference between prokaryote and eukaryote ribosomes has been found consistently, even if absolute measurements vary. Another difference that has been noted from electron micrographs is the presence of a visible cleft between the two subunits of the 70S prokaryotic ribosome, but an opaque spot rather than a cleft is seen in 80S eukaryotic monomers (Fig. 6.2).

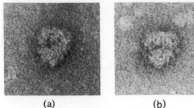 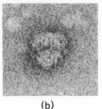 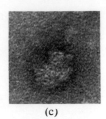

(a)	(b)	(c)

Figure 6.2
Negatively stained 80S cytoplasmic ribosomes from *Tetrahymena:* (a) the principal *dorsal* view showing the dense spot on the left and the small subunit at the top. (b) Rare *frontal* view. (c) Rare *lateral* view. × 400,000. (Courtesy of J.-J. Curgy, from Curgy, J.-J. *et al.,* 1974, *J. Cell Biol.* **60**:628–640.)

Ribosomal RNA and Protein

Ribosomes are made up of some 50 to 80 different proteins and from 3 to 4 kinds of RNA molecules. Each subunit has a unique and different set of these two kinds of constituents (Table 6.1). Prokaryotic ribosomal proteins are far better characterized than eukaryotic ones. In fact, there is considerable uncertainty about the actual number of proteins in eukaryotic ribosomes whereas bacterial ribosomes have been described more precisely. The small subunit of *E. coli* ribosomes contains 21 proteins, and another 32–34 proteins are part of the large subunit. This seemingly simple particle has more than 50 different proteins, while the larger but apparently similar ribosomes from eukaryotic cells may have as many as 80 different proteins.

While the number of proteins and even their percentage of the total molecular weight of ribosome monomer and subunits is still somewhat uncertain, especially for eukaryotes, ribosomal RNA is relatively well characterized in both prokaryotes and eukaryotes.

Table 6.1 Some characteristics of cytoplasmic ribosomes

SOURCE	INTACT RIBOSOME	RIBOSOME SUBUNITS	rRNA IN SUBUNIT	NO. PROTEINS IN SUBUNIT
Prokaryotes	70S	30S	16S	21
		50S	23S, 5S	32–34
Eukaryotes	80S	40S		~30
		60S		~50
Animals		40S	18S	
		60S	28S, 5S, 5.5S	
Plants		40S	18S	
		60S	25–26S, 5S, 5.5S	
Fungi		40S	18S	
		60S	25–26S, 5S, 5.5S	
Protozoa (and some other protists)		40S	18S	
		60S	25–26S, 5S, 5.5S	

There is only one RNA molecule in the small subunit of all ribosomes. Like ribosome monomers and subunits, the RNAs of these particles routinely are described and discussed according to their S values. The prokaryote small subunit has one 16S RNA molecule, and the eukaryote small subunit has one 18S RNA molecule. The RNA is a single-stranded polynucleotide, but there are short duplex regions due to local hydrogen bonding between complementary bases, such as guanines and cytosines. On the basis of its molecular weight, the number of nucleotides in the RNA chain can be estimated. Assuming an average molecular weight for one nucleotide to be 330 daltons, there are about 1650 nucleotides in prokaryotic 16S RNA and about 2100 nucleotides in 18S eukaryotic RNA. The molecular weights have been verified by several independent methods, but the nucleotide numbers are estimated arithmetically for these and larger molecules.

The large subunits not only are more complex in protein content, but also in their RNA components. In prokaryotes the 50S subunit contains one 23S and one 5S RNA. The 23S RNA chain has been estimated to have 3300 nucleotides in its single-strand length. The 5S RNA has been determined specifically in some cases where the entire nucleotide sequence has been worked out. The 5S RNA of *E. coli* contains 120 nucleotides, while 115–116 nucleotides occur in the single strand of 5S RNA of the bacterium *Bacillus subtilis*. Some prokaryotes among the photosynthetic bacteria and blue-green algae groups contain a 23S RNA only during early stages of ribosome subunit formation. In these species it has been found that the initial 23S molecule is cleaved to form two separate fragments before the particle matures. The utility and significance of this variation is not known.

Eukaryote 60S subunit RNAs are different from prokaryotic homologues, and they also differ according to species groups. The 60S subunit of multicellular animals includes one 28S RNA molecule, while plants, protozoa, and at least some fungi have a 25–26S RNA in the large ribosomal subunit. All these eukaryotes contain one 5S RNA, just like the prokaryotic species. A third kind of RNA is also present in eukaryote 60S subunits, which has no counterpart in prokaryotic particles. This third kind of RNA is intimately associated with the 28S RNA molecule, and has therefore been referred to as 28S-associated ribosomal RNA (28S-A rRNA). Its sedimentation value is about 5.5S.

Significant differences are superimposed on the general similarities in these features between prokaryotic and eukaryotic ribosomes. Prokaryotic monomers and subunits are smaller, have smaller major RNA molecules, and probably have fewer proteins than their eukaryotic counterparts. Both kinds of ribosomes contain 5S RNA in their larger subunit, but eukaryotic subunits have a third kind of RNA not found in prokaryotic 50S particles. Totally different proteins have been shown to occur in prokaryotic and eukaryotic ribosomes.

RIBOSOME FORMATION

Each constituent of the ribosome is an expression of genetic coded information. Mature **ribosomal RNA (rRNA)** molecules are *transcripts* of DNA information, however, while ribosomal proteins are themselves *translations* of transcribed messenger RNAs of their respective genes. According to several lines of evidence, ribosomal RNA transcripts are *never* translated into proteins. Directions for synthesis of proteins and RNAs of the ribosome are encoded in different genes, all of which are transcribed, but only transcripts for proteins are carried to the final step of gene expression by translation into proteins. As far as we know there is nothing unusual about the synthesis of the ribosomal proteins. Formation and maturation of ribosomal RNA, however, has been studied extensively. We will examine some of these events next.

Processing of Ribosomal RNA

We know a great deal about rRNA formation in eukaryotes. Most of the work has been done with amphibian and mammalian species, and especially valuable insights have come from studies by Robert Perry, Sheldon Penman, and others. When ribosomes are isolated and purified in cell-free preparations, it is a relatively simple matter to separate the protein from RNA and see what RNAs are present in each of the subunits or in whole monomers (Fig. 6.3). If RNA is extracted from whole cells, however, additional kinds of RNA fractions are found after centrifugation in sucrose density gradients (Fig. 6.4). Two of these

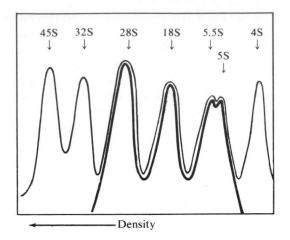

Figure 6.4
Idealized density gradient profiles of the discrete forms of RNA in whole cell extracts, including 4S tRNA. (Messenger RNAs have been omitted for purposes of simplicity.) For comparison, the RNAs isolated from purified ribosomes are indicated by a heavier line.

RNAs can be identified by appropriate tests as **messenger RNA (mRNA)** and **transfer RNA (tRNA)**. Messenger RNAs do not form a discrete fraction in one part of the gradient, indicating they are of various sizes. The difference in size of different mRNAs is expected for transcripts of genes of different lengths which specify proteins of different lengths. More specific tests, of course, are usually carried out. The tRNA fraction settles in the gradient as a group of 4S molecules. Since tRNAs are molecules that carry amino acids to the ribosomes for assembly into proteins, 4S tRNA can be identified by specific interactions with amino acids. Of the remaining RNA fractions, the 28S, 18S, 5S, and 5.5S molecules are the same as RNAs isolated from purified ribosome preparations, and these are therefore mature rRNA. Two other RNA peaks in density gradients from actively growing cells remain to be identified. Studies of these 45S and 32S RNA fractions have shown their relationship to synthesis of ribosomal RNAs of the mature particles.

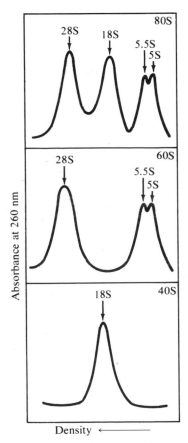

Figure 6.3
Idealized density gradient profiles showing the ribosomal RNAs isolated from 80S monomers and the particular rRNAs of the 60S and 40S subunits of these monomers.

Both 45S and 32S RNA have methyl groups bound to some of their nucleotides. Similar methylation patterns are also typical of 28S and 18S rRNA. This pattern suggests some relationship between all these RNAs, and one of the more important possibilities is that 45S and 32S are precursors of the final 28S and 18S products. One way to test this possibility is to supply the cells with radioactively labeled RNA nucleotide and follow the label over a period of time. Precursor RNAs will then be labeled first and RNA products will receive the radioactive label later on. In a typical experiment of this sort, ^{3}H-uridine is "pulsed" into the culture for a brief time, perhaps 10 minutes or less. Cells can then be placed in a nonradioactive medium and studied at different times afterward to see what happens to the tritiated uridine incorporated into RNA chains. Cells are removed from the culture at specified intervals so their RNA can be isolated, purified, and checked for radioactivity (Fig. 6.5).

The first RNA to be labeled is 45S during the brief "pulse" period. After transfer to nonradioactive medium no further labeled 45S is formed. Instead, the label later appears in 32S and 18S RNA, and after this the ^{3}H radioactivity is found in 28S rRNA. The sequence that was reconstructed showed that 45S RNA is transcribed from genes and is then processed into 28S and 18S rRNA of ribosomes. Since 18S appears first it must be a cleavage fragment of the 45S molecule, and 32S RNA must be the remainder of this 45S RNA chain. Later, 32S is processed still more to form the 28S rRNA of the large subunit of ribosomes. Further support for this scheme comes from several independent kinds of information, including a quantitative correlation between amounts of 45S RNA and its maturation products.[1]

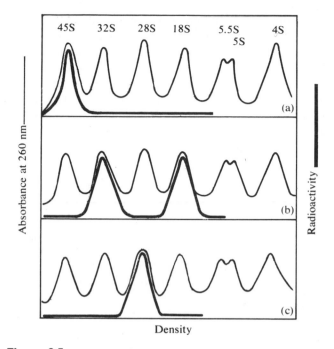

Figure 6.5
Idealized illustrations of experiments using "pulse" labeling with ^{3}H-uridine followed by a "chase" with unlabeled uridine. The time sequence in which the ^{3}H label (heavier line) appears in different RNA fractions is shown, as it is observed from RNA samples centrifuged to equilibrium in density gradients: (a) 45S RNA is labeled first; (b) the label is transferred to 32S and 18S RNA later in the "chase" period; and (c) label appears last in the 28S RNA fraction. The sequence from precursor (45S RNA) to products (18S and 28S RNAs) can be reconstructed from such data.

---

[1] Although it is a technicality, it is pertinent to note that the absolute S value of precursor ribosomal RNA varies according to the evolutionary ranking of the animal species. Human cells contain the 45S precursor, but amphibia have 40S and insects have 38S RNA precursor molecules. We will refer only to the values found for human and other mammalian RNA in the interests of simplicity.

Experiments such as those just described are not quite so simple. For example, other kinds of RNA are also being made during the time the cells are in radioactive media. These can be distinguished from the RNAs involved in the synthesis of the two major ribosomal RNA species by additional tests and experimental systems. All these studies confirm the basic sequence described above and show that messengers, 5.5S, 5S, and 4S RNAs are not involved in forming the 28S and 18S rRNA molecules.

Since ribosomal RNAs are transcribed from DNA

in chromosomes, it is reasonable to assume that these RNA molecules are manufactured in the nucleus. The details of rRNA formation in the nucleus have been followed extensively. We will consider these studies in a later section of this chapter. The combined approaches of biochemistry, genetics, and electron microscopy have led to a fascinating story of nuclear involvement in ribosome formation.

RIBOSOMAL RNA FORMATION IN BACTERIA. Until recently it was believed that rRNA synthesis in prokaryotes was less complex than the eukaryotic system. Using brief pulses of radioactive uridine to follow rRNA formation, presumptive precursor molecules were found which were slightly larger than mature 23S, 16S, and 5S RNAs of ribosomes. The evidence continued to accumulate from biochemical studies, such as comparisons between mature RNA composition and that of the possible precursor molecules. In 1969, Carl Woese and associates provided the first experimental evidence showing that a single RNA transcript was made from genes for 16S, 23S, and 5S ribosomal RNAs, in that order. Genetic studies showed that these three genes were very closely linked and in the sequence of 16S—23S—5S, at least in *E. coli* and *Bacillus subtilis*. Using inhibitors of RNA synthesis it was further shown that a single transcript was made first and was later converted to the three RNAs of mature ribosomes.

No actual precursor has been isolated at this time. Two technical problems have prevented isolation of precursor molecules:

1. The precursor is cleaved before the complete transcript is formed.
2. RNA synthesis is so rapid that any transient kind of molecule would exist only briefly and in small amounts at any one time.

At the conservative estimate of 50 nucleotides per second, it takes only 30 seconds to transcribe the 16S gene and an additional 60 seconds to transcribe the

neighboring 23S gene. A specific enzyme is presumed to cleave the 16S portion of the precursor very soon after transcription has extended into the 23S region of DNA. A second enzymatic scission separates the 23S from the residual 5S precursor. The three precursor molecules, in a ribonucleoprotein aggregate form, then undergo further processing to form the mature rRNA species in subunits (Fig. 6.6).

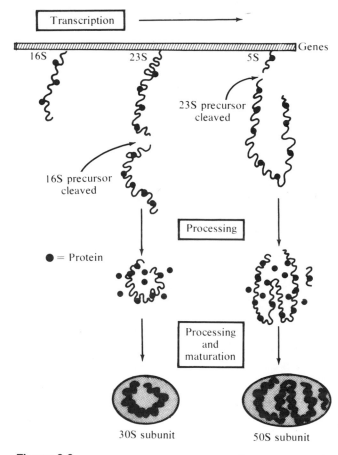

Figure 6.6
Flow diagram illustrating the formation of *E. coli* ribosome subunits from rRNA precursor. The proteins that associate with the rRNA molecules during processing are different from the proteins that finally assemble in the mature ribosome subunits.

COORDINATE SYNTHESIS OF RIBOSOMAL RNAs. In prokaryotes all three RNAs of the ribosome are transcribed in a coordinated sequence from closely-linked genes. The eukaryotic 45S precursor is transcribed from one gene, so formation of 28S and 18S rRNA must be on a 1:1 basis. The 5S RNA in eukaryotes is transcribed from completely different genes that are located elsewhere from the 45S rRNA genes in the chromosome set. Synthesis of all these RNAs must be coordinated in some way since 5S RNA occurs in a 1:1 ratio with the larger RNA molecules of the ribosome. It is known that inhibition of 45S precursor rRNA does not necessarily stop 5S RNA transcription, but the 5S RNA is degraded under these conditions. The 5S RNA can only be preserved if it is incorporated into the ribosomal subunit along with 28S rRNA. Little is known about the contributions made by 5S and 28S-associated ribosomal RNAs to ribosome formation and function. It is therefore difficult to understand the importance of the observed similarities and differences between prokaryotic and eukaryotic systems.

Maturation and packaging of rRNA in both prokaryotes and eukaryotes requires continued protein synthesis. Precursor RNA is synthesized even when protein formation is inhibited experimentally, but these precursors do not continue to be processed into mature ribosomes. Eukaryotes will continue to make 45S RNA under such conditions, but 18S rRNA does not accumulate and both 32S and 28S RNAs are degraded in the nucleus of the cell. A similar situation characterizes prokaryotic systems, so that new ribosomes are not made unless proteins as well as RNA are being synthesized. When syntheses proceed normally, proteins and RNAs are packaged into ribonucleoprotein complexes.

It was assumed at one time that these ribonucleoprotein complexes were the actual ribosome precursor particles. When the proteins of these complexes were compared with the specific proteins from mature ribosomes, however, it was shown that there was little resemblance between the two sets of molecules.

The function of the proteins which complex with newly-synthesized RNAs is not known at present. These proteins may protect the RNAs from degradation by the abundant ribonucleases in cells until they finally bind with the specific proteins of mature ribosome subunits. The precise location of mature ribosome subunit formation is uncertain. Since precursor aggregates are found in the eukaryotic nucleus while functional mature particles are in the cytoplasm, the final touches may be made either in the cytoplasm or in the boundary between the nucleus where subunits form initially and the cytoplasm where these particles assume their mature composition and function. Some important insights have been obtained from recent studies of disassembly and reassembly of proteins and RNAs of mature ribosomal subunits, particularly in *E. coli*.

Dissection of Bacterial Ribosome Subunits

Subunits of ribosome monomers can be collected in separate centrifugate fractions if the concentration of Mg^{2+} ions in the medium is adjusted appropriately. Subunits of bacterial ribosomes can be collected when there are less than 5–10 millimolar Mg^{2+} ions, while subunits from eukaryote cytoplasm dissociate from the monomers at a concentration of less than 1 millimolar Mg^{2+} in the medium. Various other conditions of the medium, such as pH and concentrations of some monovalent cations such as K^+, must also be regulated. Purified subunits can then be used *in vitro* to study their composition and manner of assembly or disassembly.

Studies of ribosome dissociation were begun after an observation in 1964 by Matthew Meselson. He found that a group of proteins split from residual cores of 30S and 50S subunits when these particles were centrifuged in CsCl. When subunits are exposed to CsCl or LiCl, a series of successive losses of proteins occurs. Each disruption of structure leads to losses of specific protein groups, rather than to a more gradual diminution. The first level of dissociation yields *split proteins* and *cores*. The 30S subunits yield particulate 23S cores and a group of split proteins called SP30; 50S subunits dissociate to 42S cores and SP50 split proteins. The core particles can be further dissociated to 16S rRNA and CP30 core proteins from the smaller subunit and to 23S and 5S rRNA plus CP50 core proteins from the larger subunit (Fig. 6.7). The individual split and core proteins from both subunits can be separated and identified by their pattern and position of migration in an electrical field.

The separated proteins and RNAs can be reassembled into 30S and 50S subunits under conditions that suggest the assembly is sponsored by hydrogen bonding and by hydrophobic interactions between the molecules. These reconstituted subunits can direct protein synthesis if they are placed in a suitable *in vitro* system containing messenger RNA, an energy source, and various amino acids and factors for polypeptide formation. This restoration of subunit function indicates that accurate assembly of the particles takes place in response to instructions that are contained within the structure of the molecular components themselves. It is a **self-assembly** process.

Assembly will only take place, however, if the core proteins are allowed to bind to rRNA first and the

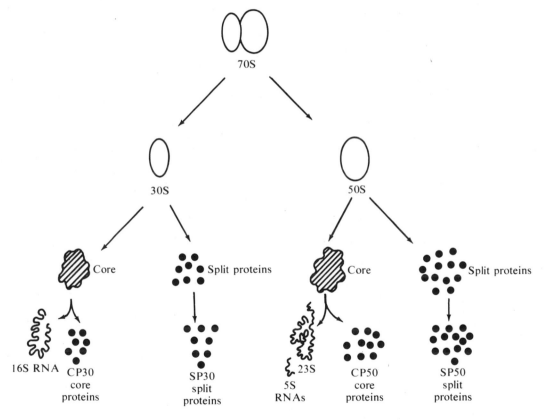

Figure 6.7
Flow diagram showing the procedure and results of ribosome disassembly in *E. coli*. Stepwise disruptive treatments lead to the recovery of the RNA and protein components of each subunit type. These isolated components may then be analyzed or used in other experiments.

split proteins are added after formation of 23S cores or 42S cores. Assembly of the whole 50S subunit is somewhat more complicated but follows essentially the same routes as assembly of the smaller 30S subunit. Leadership in these studies came from several laboratories, particularly from those of M. Nomura, C. Kurland, and others. Nomura's group showed that reconstitution of the 50S subunits required (1) higher temperature, (2) accurate addition of the 5S RNA molecule, and (3) careful regulation of the Mg^{2+} ion level during addition of the 5S RNA to the assembling subunit. One of the complications in 50S subunit assembly *in vitro* is due to the fact that reconstitutions are carried out in the absence of 30S subunits or their constituents. Studies of ribosome formation in living bacterial cells had shown that formation of active 50S subunits depended on synthesis of the 30S subunits in the same cells. Most studies have therefore concentrated on the simpler system of the 30S subunit.

ADDITION OF PROTEINS TO THE ASSEMBLING SUBUNIT. The 30S subunit of the *E. coli* ribosome includes 21 different proteins, compared to 32–34 unique proteins in the 50S subunit. Since all 21 proteins have been purified and identified, they can be added one at a time or in particular groups during subunit assembly. If one or more of these proteins is omitted from the reassembly system, it becomes possible to determine its contribution or necessity in subunit function during protein synthesis. These features of the assembly sequence have opened productive avenues for study of the contributions made by individual components to the activity of the whole particle.

Three general groups of subunit proteins have been recognized:

1. Those which are essential for assembly to begin.
2. Those which are essential for addition of the last group of proteins in a functional subunit.
3. Those which are dispensable for 30S subunit formation but required for subunit activity.

All these proteins are therefore needed to reconstitute an active 30S subunit, but the proteins assemble in a definite sequence and subunits with different kinds of defects are produced if specific proteins are omitted from the system. Six proteins are essential for assembly to begin. If any one of these is not included, subunit assembly stops at a certain stage and goes no further. If a protein from the second group is excluded, the final stage of formation stops just short of a 30S-size particle. Omission of one of the proteins from the third group does not stop formation of 30S particles, but these are found to be inactive when tested for protein synthesis *in vitro*.

When all 21 proteins of the 30S subunit are added in a defined sequence, an **assembly map** can be generated to show relationships among the constituents (Fig. 6.8). From this and other lines of information, assembly of the 30S subunit seems to occur as follows: (1) particular proteins assemble by binding to 16S RNA to form a small precursor particle; and (2) the remaining proteins then bind to these cores to form the active 30S subunit. It is uncertain whether this second group of proteins assembles in one or two steps after formation of the cores. Although some proteins assemble in relation to their final position in the finished subunit, it is not certain that the remainder become positioned inside or outside the particle, according to the time they are added to the core. Changes in final position of some proteins, at least, are believed to take place as a result of folding and other conformational changes during subunit assembly.

INTERACTION BETWEEN rRNA AND PROTEINS IN SUBUNITS. In order to see where the 6 essential proteins for core formation bind to the 16S RNA, binding studies have been done using 16S RNA broken into two unequal-size pieces. Five of the proteins bind specifically to the larger RNA piece, which contains the 5′ end of the polynucleotide chain (Fig. 6.9). The bound proteins are clustered within 900 nucleotides of this 5′ end. This binding pattern may be significant

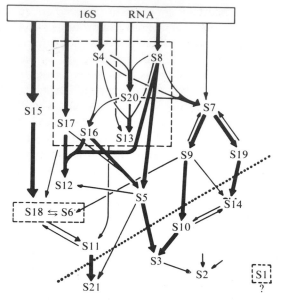

Figure 6.8

The assembly map of *E. coli* 30S ribosome subunit proteins. Arrows between proteins (numbered from S1 to S21) indicate the facilitating effect on binding of one protein on another. A thick arrow indicates a major effect; a thin arrow indicates a weaker effect. Various features of assembly are still uncertain. For example, it is not known which of the proteins enclosed in the box with a dashed outline facilitates the binding of protein S11. (From Held, W. *et al.,* 1974, *J. Biol. Chem.* **249**:3109, Fig. 6.)

Proteins from 30S subunits bind exclusively to 16S RNA, and proteins from 50S subunits bind specifically to 23S RNA. Functional subunits will not assemble if RNA and proteins are a mixture from both the 30S and 50S particles, but if RNA and proteins are taken from 30S subunits of widely different prokaryotic species, they will assemble into a functional 30S subunit. The same is true for mixtures of RNA and proteins derived from 50S subunits of different bacteria. There is an overall similarity among their subunit molecules, although RNAs and proteins differ greatly in their primary structure among bacteria. The similarity of subunit structure indicates a functional homology among these components, despite the differences in chemical sequence and composition. Further evidence for functional homology in ribosomes from various species of bacteria has come from studies of hybrid ribosomes. When 30S and 50S subunits have been taken from different species and placed into *in vitro* systems, the hybrid ribosome monomers participate properly in protein synthesis.

There is no homology between molecules or subunits from prokaryotes and eukaryotes. RNA from

in subunit formation since the 5′ end of the RNA is the first part of the transcript to peel off the DNA template. These proteins would not only protect RNA from digestion by ribonuclease, but they would also provide binding sites for other proteins in the growing subunit precursor. It is known that some of the proteins of the mature subunit are identical with some of the proteins of the nascent (growing) ribonucleoprotein aggregates. Many of the proteins of mature and precursor ribonucleoprotein particles, however, are known to be different. Whether or not the essential proteins that bind to 16S RNA *in vitro* are the same as the first proteins to bind to rRNA during its transcription *in vivo* remains to be determined.

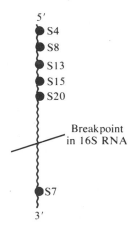

Figure 6.9

Binding of specific 30S ribosome subunit proteins to the 16S rRNA molecule. The 16S molecule has been cleaved into two unequal-sized pieces to facilitate the analysis and localizations.

ribosomes of yeast or rat cells cannot bind to ribosomal proteins derived from prokaryotes. When hybrid ribosomes are formed from small and large subunits taken from prokaryotic and eukaryotic cells, the monomers do not subsidize *in vitro* protein synthesis.

Interestingly, functional hybrid ribosomes have been formed using 30S and 50S subunits isolated from *E. coli* and from chloroplasts of various species. In addition to similarities in chloroplast and prokaryote 70S monomer and 30S and 50S subunits on the basis of particle size and RNA S-values and demonstrated functional homology in hybrid ribosome studies, a 5S RNA is found in the 50S subunit of chloroplasts as it is in prokaryotes (and eukaryotes). These striking similarities between prokaryote and chloroplast ribosomes do not seem to exist for mitochondrial ribosome comparisons. Ribosomes from mitochondria vary in monomer size from 55S to 80S, and subunits vary in size accordingly. Subunit exchanges between mitochondrial and prokaryotic ribosomes do not lead to functional hybrid ribosomes when tested *in vitro*, even if specific mitochondrial particles are selected from species with 70S monomers. On the basis of these and other observations, it would not be appropriate to consider all organelle ribosomes as "prokaryote-like".

RIBOSOMAL RNA

Chemistry and Structure of Ribosomal RNA

All kinds of RNA in ribosomes have the usual 4 bases (adenine, guanine, cytosine, and uracil) arranged in single-stranded molecules. A minor quantity of these nucleotides are modified, usually by methyl group substitutions. In *E. coli* the methyl groups are substituted in the nucleotide bases, whereas eukaryote rRNA has methylated ribose and unmethylated bases. Only the two major RNAs are methylated; neither prokaryotes nor eukaryotes have modified nucleosides in their 5S RNA.

Methylated nucleosides in prokaryotic rRNA are associated with specific polynucleotide sequences rather than being dispersed randomly in the molecules. Methylations are highly specific and these sequences appear to have been conserved during evolution, while other parts of the rRNA obviously have been profoundly changed. While these observations point to some essential contribution by methylated portions of rRNA, little has been discovered to correlate rRNA methylation and rRNA function. In fact, a mutant which is resistant to the antibiotic kasugamycin has been found to lack a particular methylating enzyme for 16S rRNA, yet these *E. coli* cells grow normally. Another *E. coli* mutant also grows normally despite an extra methylated nucleoside in the 23S RNA of its 50S subunit. It is difficult to interpret the importance of methylated rRNA on the basis of available information.

Evolutionary conservatism is also apparent since there is a fairly narrow range of variation in base composition of rRNA with the highly diversified DNA composition in the same species. The overall base composition of rRNA in different species is relatively similar, but the primary structure of the molecules is very different. Changes have taken place in nucleotide sequence but not in the percentages of the four kinds of bases in rRNA. Since there is similar base composition and functional homology among rRNAs from different prokaryotes, we infer that secondary and tertiary structures have been conserved in essentially similar conformations. Secondary and tertiary structures are far more important in rRNA—protein interactions than the precise base sequence of the RNA, as shown by subunit assembly studies.

Secondary structure of rRNA is believed to resemble a bewildering array of hairpin loops. This structural interpretation is based on chemical evidence which shows that about 60–70 percent of the rRNA molecules contain local groups of from 10 to 30 hydrogen-bonded nucleotides. Elucidation of the tertiary structure of these relatively large molecules is a long way away, however. If the sequence of 1650 nucleotides in 16S rRNA could be determined completely, then it should be possible to look for common regions in different 16S molecules. The identification of common regions could provide a clue to the basis for common function of 16S rRNA in different species. These sequences should stand out in otherwise very diverse rRNAs. This information would be of tremendous importance because, at the present time, we know almost nothing about the function(s) of ribosomal RNA. Analysis of its molecular architecture could provide valuable leads to interpretations of rRNA function in the overall contributions made by ribosomes to protein synthesis in the cells of every living species.

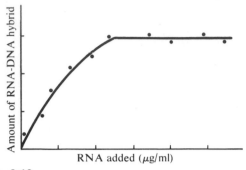

Figure 6.10
Idealized plot of the results of a DNA-RNA hybridization-saturation assay. DNA-RNA hybrid molecules continue to form until a plateau is reached at the point when the constant amount of added DNA has been saturated.

Molecular Hybridization

It is possible to obtain crude estimates of sequence homologies among species. Relatedness between ribosomal RNAs can be assessed from **molecular hybridization** of RNA and DNA using two different test systems. In **hybridization-saturation assays,** a constant amount of DNA from some source is annealed with increasing amounts of radioactively-labeled rRNA until a plateau is reached. No further labeled RNA enters into ribonuclease-resistant "true" hybrid molecules (Fig. 6.10). Once the complementary DNA sequences have been saturated with rRNA, the fraction of DNA which is complementary to added rRNA can be calculated.

In **hybridization-competition assays,** DNA from a given source is annealed with labeled rRNA from the same (homologous) source in the presence of a large molar excess of unlabeled rRNA from another (heterologous) source. If the two rRNAs have any sequence homologies, then the unlabeled rRNA competes with labeled rRNA for the same complementary DNA sequences (Fig. 6.11). In this assay the extent of the similarities between two rRNAs is measured by the difference in amounts of labeled, homologous rRNA which is hybridized to DNA in the presence and absence of competing heterologous rRNA. Re-

latedness between different ribosomal RNAs (or any RNAs) in saturation assays, however, is measured according to the amount of rRNA that hybridizes with heterologous DNA.

Hybridizations between RNA and RNA, RNA and DNA, and DNA and DNA have been carried out extensively in numerous cell biology and genetic studies. The method has provided a powerful tool for analyzing complex nucleic acid molecules that are too large for the chemical techniques of sequence analysis. During many of the chapters that follow we will refer repeatedly to information and insights gained from this incisive and specific technique. Complementarity underlies the specificity of interaction between all or parts of nucleic acid molecules, and complementarity implies relationship and homology. The probability that different nucleic acid molecules happen by chance to have complementary sequences is so low as to be ignored for all practical purposes. The assays described in this section are specific and reliable indicators of complementarity and homology.

Genes Specifying rRNA

Using hybridization-saturation assays it is possible to determine the amount of DNA that codes for ribosomal RNAs. This particular DNA is often

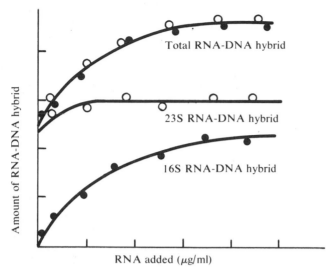

Figure 6.11
Idealized plot of the results of a DNA-RNA hybridization-competition assay. The 23S and 16S rRNAs compete for different sites on DNA, as seen from the amount of total hybrid formed relative to the separate DNA-23S RNA and DNA-16S RNA hybrid populations.

referred to as **ribosomal DNA (rDNA).** In *E. coli* the amount of rRNA that anneals with total cellular DNA is more than expected if only one gene determined the formation of one rRNA chain. Since 0.3 to 0.4 percent of total DNA anneals with rRNA at saturation, approximately 5 or 6 copies of each rRNA gene must be present in the *E. coli* genome. These calculations are relatively accurate since we know the length of these rRNA molecules and since each must be a transcript of complementary DNA with the same number of nucleotides as precursor rRNAs. *Mycoplasma* has about 20–25 percent the amount of DNA *E. coli* has, and according to hybridization-saturation assays, there seems to be only one gene for each of the rRNA types.

LOCATION OF rDNA IN E. COLI. It has not been possible to locate rRNA genes by the usual genetic mapping procedures because rRNA mutants have not been isolated for study. Various alternative approaches have been taken instead, such as constructing strains that are partially diploid and then determining which regions contain rDNA on the basis of which partial diploid strains produce extra amounts of rRNA. The measurements are carried out by DNA-RNA hybridization assays. Where the extra piece of chromosome leads to twice the production of rRNA, location on a map can be made in relation to other marker genes present in this same extra DNA fragment. On the basis of these and other studies, rDNA has been located in two regions of the *E. coli* map.

More specific information about the molecular organization of the rDNA regions in *E. coli* has shown that individual clusters of 16S – 23S – 5S genes occur in tandem and are separated from other clusters by stretches of DNA that are about 10 million daltons long. A similar situation has been found for *Bacillus subtilis* rDNA.

LOCATION OF rDNA IN EUKARYOTES. Beginning in the 1960s, a tremendous body of information has accumulated about eukaryotic rDNA. The earlier studies using *Drosophila* and the clawed toad *Xenopus laevis* clearly showed that rDNA was restricted to certain chromosomes, and to specific regions of these chromosomes. It is known that genes for ribosomal RNA exist in hundreds or thousands of copies in the nucleolar-organizing chromosomes of the nucleus. More specifically, rDNA is present in the particular region of these chromosomes to which the nucleolus is attached. The **nucleolar-organizing region (NOR)** is therefore a special and differentiated part of its chromosome (Fig. 6.12). At least one nucleolar-organizing chromosome is found in the haploid genome of eukaryotes, but more may occur in some species.

In the mid-1960s, Donald Brown and John Gurdon provided important evidence showing that ribosomal RNA was synthesized at the NOR in *Xenopus*. They studied a recessive mutant which lacked nucleoli, as

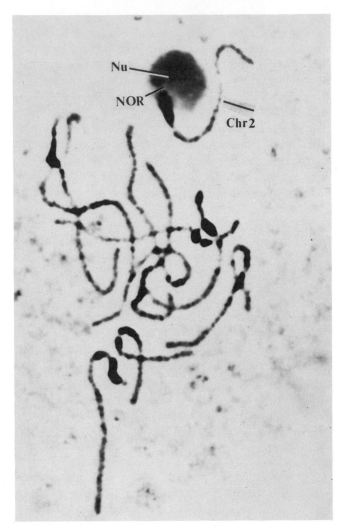

Figure 6.12
Pachytene stage of meiosis in castor bean (*Ricinus communis*). The main nucleolar-organizing chromosome is No. 2 in the haploid complement of ten chromosomes. The NOR is located at a secondary constriction near the end of the chromosome attached to the nucleolus (Nu). (Courtesy of G. Jelenkovic)

well as the homozygous wild-type with two nucleoli and the heterozygote which had one nucleolus. The wild-type and heterozygous strains could synthesize rRNA and manufacture new ribosomes during embryonic development and afterward. The anucleolate (lacking nucleoli) mutant could develop from the fertilized egg for a period of about two weeks, up to the time of embryo gastrulation. After this time, the mutants died. A continual supply of these mutants was obtained by interbreeding heterozygous animals. One-fourth of their progeny were the recessive anucleolate type, as expected in a simple case of Mendelian inheritance. In addition to lacking nucleoli, the mutant had shorter nucleolar-organizing chromosomes. This observation suggested that the entire nucleolar-organizing region was missing in the mutant nucleus.

When a radioactively labeled precursor was provided to the developing embryos, transfer RNA was synthesized but not ribosomal RNA. The mutant embryo survives for the first two weeks because very little new protein is synthesized during this time, and because there are enough ribosomes carried over from the fertilized egg of the heterozygous mother. When gastrulation begins, wild-type and heterozygous embryos undergo a burst of new protein synthesis, for which a burst of newly synthesized ribosomes are needed. Mutant embryos die because they cannot produce ribosomal RNA and, therefore, cannot make new ribosomes for synthesizing protein during growth and development.

The evidence from studies of anucleolate *Xenopus* indicated that ribosomal RNA was synthesized in or near the nucleolus and its chromosomal attachment site. In 1965, Frank Ritossa and Sol Spiegelman showed decisively that rRNA was made at the NOR of the chromosome. They constructed special strains of *Drosophila* which each contained either 1, 2, 3, or 4 nucleoli and NOR chromosome parts. Radioactively-labeled precursor was supplied and newly synthesized rRNA with label was isolated, purified, and hybridized with chromosomal DNA. The amount of rRNA that was synthesized exactly paralleled the number of NOR per strain (Fig. 6.13). In later studies it was shown that 5S rRNA and 4S transfer RNA did not increase, even when extra NOR segments were present in a strain. These RNAs are therefore transcribed from genes that are located somewhere else in the set of chromosomes and not at the NOR.

GENE AMPLIFICATION IN AMPHIBIAN OOCYTES. During the late 1960s and early 1970s, even more inter-

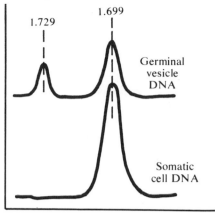

Figure 6.14
Densitometer tracings of *Xenopus* DNA showing the distinctive "satellite" rDNA in germinal vesicle preparations ($\rho = 1.729$) and the nuclear DNA ($\rho = 1.699$) common to both DNA sources.

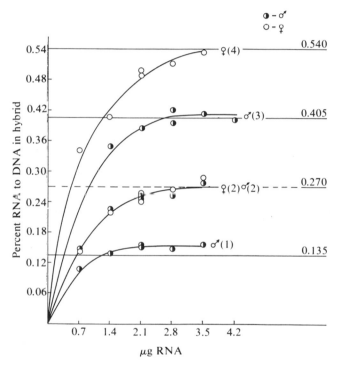

Figure 6.13
Saturation levels of DNA containing various dosages of nucleolar-organizing regions (indicated by the number in parentheses). The dashed horizontal line at 0.270 (percent RNA to DNA in the hybrid molecules) is the estimate for a dosage of 2 NOR, and the solid horizontal lines represent predicted plateaus for dosages of 1, 3, and 4 NOR, respectively. Numerical values of the plateaus are given on the right. (From Ritossa, F. M., and S. Spiegelman, 1965, *Proc. Natl. Acad. Sci.* **53**:742, Fig. 1.)

esting information came from molecular and cell biology studies of rRNA synthesis in amphibians such as *Xenopus*. Oocytes produced by these animals are extremely large, reaching a maximum size during the early prophase of the first meiotic division before egg formation. The nucleus enlarges greatly and is generally referred to as a "germinal vesicle" at this stage of development. During this time from 600 to 1600 individual nucleoli appear in the nucleus and become distributed near the periphery of the nucleus. Each of these numerous nucleoli contains DNA and ribosome precursor particles.

The nucleolar DNA is present in large amounts and can easily be seen as a separate "satellite" fraction, different from the rest of the DNA isolated from "germinal vesicles" of amphibian oocytes (Fig. 6.14). If the "satellite" DNA is hybridized with ribosomal RNA, it is clear that they are complementary. The "satellite" DNA, then, is ribosomal DNA that has been replicated and packaged in the many extra nucleoli. This rDNA is transcribed into rRNA and huge amounts of ribosomes are produced in oocytes at this stage of development.

While many copies of rDNA are replicated at this stage of ooctye development, the rest of the chromosomal DNA remains stable in amount (Fig. 6.15). This differential replication has been called **gene amplification**. Transcription of rRNA takes place in all nucleoli until the first metaphase of meiosis. At this time all further synthesis of RNA stops, the "germinal vesicle" membrane disappears, and so do all the nucleoli. The ribosomes manufactured in the oocyte sustain the cell before and after fertilization of the egg and also provide for the needs of the embryo during its first two weeks of existence.

According to measurements from DNA—RNA hybridization studies, there are about 450 copies of the 45S gene for rRNA in each NOR. This number is increased 600–1600 times during gene amplification in the early oocyte. Similar amplification occurs in other animal oocytes, but the extra rDNA copies are kept within the same number of nucleoli as in any other cell. Amphibians are unusual in producing extra nucleoli to house their extra rDNA in oocytes. Plant rDNA occurs in thousands of copies normally, and further amplification does not take place during oocyte differentiation.

TRANSCRIPTION OF rRNA. In the mid-1960s, Penman and others showed that rRNA was formed specifically in the nucleolus. The investigators separated cytoplasm and nuclei from whole cells and followed the time and place of incorporation of radioactively labeled precursor into newly synthesized rRNA (Table 6.2). The label appeared first in 45S rRNA in the nucleus. Later, 32S rRNA was found in the nucleus and labeled 18S rRNA in the cytoplasm as part of small subunits. Finally, labeled 28S rRNA was found in large subunits in the cytoplasm. This sequence indicated that precursor 45S and 32S rRNAs were made in the nucleus and processed into 28S and 18S rRNA in ribosomal subunits located in the cytoplasm.

More specific localization of rRNA synthesis was obtained in experiments in which the nucleus was separated into nucleolus and residual nucleoplasm and studied along with the cytoplasm of the cells. Results of these experiments showed that: (1) 45S rRNA is synthesized in the nucleolus; (2) it is processed there into the 32S intermediate, while the 18S rRNA fragment is rapidly transported to the cyto-

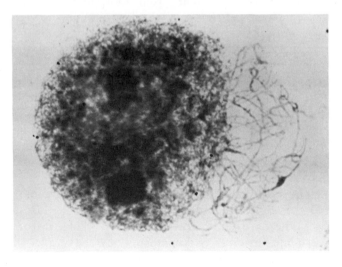

Figure 6.15
Light micrograph of an oocyte nucleus of a beetle (*Dytiscus marginalis*). A large cap of amplified rDNA is to the left, comprising approximately 90 percent of the nuclear DNA at this stage of meiosis. × 1,200. (Courtesy of J. G. Gall, from Gall, J. G., and J.-D. Rochaix, 1974, *Proc. Natl. Acad. Sci.* **71**:1819–1823, Fig. 1.)

Table 6.2 Types of labeled rRNA found in subcellular fractions collected after adding ^{3}H-uridine to the cell culture*

SUBCELLULAR FRACTION	TIME COLLECTED AFTER LABELING		
	10 MIN	30 MIN	60 MIN
Nuclei	45S	45S 32S	45S 32S 28S
Cytoplasm	—	18S	18S 28S

* Data adapted from Penman, S., Smith, I., Holtzman, E., and Greenberg, H. 1966. *Natl. Cancer Inst. Monograph* 23:489–511.

plasm; and (3) 28S rRNA is processed from 32S precursor in the nucleolus but moves slowly through the nucleoplasm out to the cytoplasm where it is eventually found in the large ribosomal subunits (Table 6.3). Supporting evidence was collected from other studies in which inhibitors were used to stop rRNA synthesis. The general scheme for rRNA formation seems to be essentially the same in the eukaryotic cells which have been analyzed (Fig. 6.16).

Direct visualization of rDNA—rRNA transcription complexes has been done by Oscar Miller, Barbara Hamkalo, and associates using the electron microscope. Transcribing rDNA has been isolated from nucleoli of eukaryotic cells and from *E. coli* (Fig. 6.17). The linear fibers of deoxynucleoprotein isolated from nucleoli of *Xenopus* are periodically coated with feathery regions of transcribing rRNA. Many separated transcribing areas are found on the same fiber and the regions in between are not active. Each transcribing segment measures 2–3 μm in length, which is the expected size for a single rDNA gene coding for 40S rRNA precursor in *Xenopus* and other amphibians. Each rDNA gene bears many transcripts along its length, with the newest growing rRNA chains at the "point" of the feathery region and earlier-started, longer rRNA chains at the terminus of the rDNA gene.

The linear fiber can be digested with deoxyribonuclease, evidence which shows it contains DNA.

Table 6.3 Types of labeled rRNA found in subcellular fractions collected after adding ³H-uridine to the cell culture*

SUBCELLULAR FRACTION	TIME COLLECTED AFTER LABELING		
	10 MIN	30 MIN	60 MIN
Nucleoli	45S	45S 32S	45S 32S
Nucleolar-free nucleoplasm	—	—	28S
Cytoplasm	—	18S	18S 28S

* Data adapted from Penman, S., Smith, I., Holtzman, E., and Greenberg, H. 1966. *Natl. Cancer Inst. Monograph* 23:489–511.

The growing chains in the "feathery" regions can be digested with ribonuclease, showing their RNA nature, and with a protein-digesting enzyme such as trypsin. The growing chains are actually ribonucleoproteins and not naked rRNA. These chains are precursors of mature ribosomal subunits which will undergo processing into the individual rRNAs and proteins, unique to each kind of subunit of the mature ribosome. About 100 fibrils of growing ribonucleoprotein exist per rDNA gene ("matrix unit").

Incorporation of radioactively labeled precursors also verifies the identifications of DNA in the "matrix units" and spacers in between, and the RNA composition of the growing fibrils. Similar evidence has been found for *E. coli* rDNA transcribing regions. These and other visual correlates of biochemical and biophysical evidence have greatly strengthened our understanding of ribosomal DNA and RNA in both prokaryotes and eukaryotes. We will discuss electron microscopy studies of translation complexes in the next section. These electron micrographs are equally as striking and informative as the ones of the transcription complexes just described.

A great deal is known about rRNA coding, synthesis, and packaging, but very little is understood about its function. It is thought that rRNA must contribute somehow to ribosomal participation in protein synthesis. As we will see in the next sections, most available information is about ribosomal constituents other than the rRNAs in protein synthesis. The functions of these RNAs remain important but unresolved problems for continued research.

PROTEIN SYNTHESIS

During the 1950s and early 1960s it was generally believed that proteins were synthesized on single, free ribosomes. This notion was shown to be incorrect by elegant studies reported in 1963 by Alexander Rich, Jonathan Warner, and their associates. These investigators demonstrated that polypeptide chains

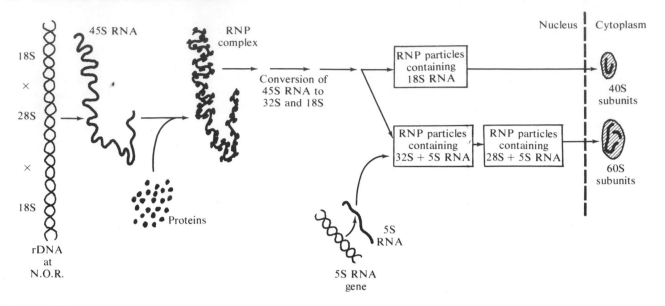

Figure 6.16
Flow diagram showing the components and sequences of synthesis and packaging of eukaryotic cytoplasmic ribosome subunits.

were made at groups of ribosomes called polyribosomes, or **polysomes,** rather than on free ribosome monomers. The polysome is a complex that includes a mRNA strand which holds together a variable number of ribosomes. The number of ribosomes is proportional to the length of the mRNA transcript coded for a polypeptide.

Rich used rabbit reticulocytes for these studies because these cells had a number of advantageous features. Mammalian reticulocytes have lost their nuclei at maturity, they have very little endoplasmic reticulum, and hemoglobin is almost the only protein they synthesize. This cell system is really an amplified hemoglobin-making structure, and experimental studies can be done without problems of identifying and separating many kinds of mRNA and proteins. Biochemical analysis using large amounts of cells is quite practical, and different components can be studied by density gradient centrifugation of ma-

terials isolated from cell-free lysates (broken cell contents).

When these lysates were examined after centrifugation, there was a region in the gradient that contained free ribosome monomers. Heavier particulate materials were present in the same gradients in addition. These particles were groups of ribosomes, with 5-ribosome aggregates being most common. Newly-synthesized protein was identified by incorporated radioactive amino acids, which were provided to the system at the beginning of an experiment. The peak of highest radioactivity was found to coincide with the pentamer (5-ribosome) aggregates in the gradient. This observation indicated that most protein was made at polysomes rather than at ribosome monomers (Fig. 6.18).

The particular polypeptide chain of the whole protein was known to contain 146 amino acids. A mRNA transcript of the gene which codes for this

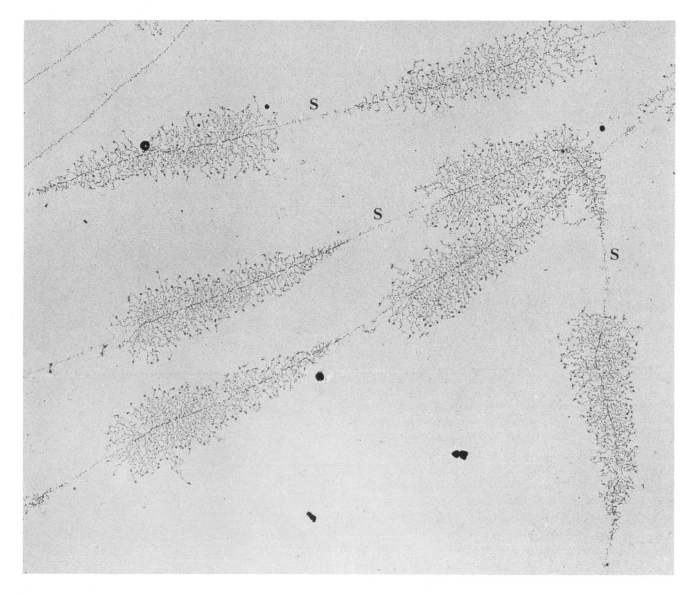

Figure 6.17
Electron micrograph of actively transcribing rDNA isolated from oocyte nucleoli of the spotted newt (*Triturus viridescens*), an amphibian species. The arrow-shaped "matrix units" are composed of fine fibrils of protein-complexed rRNA precursor molecules synthesized from the DNA template to which they are attached. The rRNA genes occur as tandem repeats separated by DNA "spacer" segments (S) of unknown function. × 28,000. (Courtesy of O. L. Miller, Jr., from Miller, O. L., Jr., and B. Beatty, 1969, *Science* **164**:955–957, Fig. 2.)

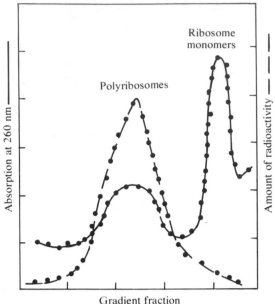

Figure 6.18
The peak of high radioactivity in the polyribosome fractions of the gradient indicate that new polypeptides were synthesized at these aggregates and not on single ribosome monomers.

translation complexes are not found in eukaryotic systems. Ribosomes are transported to the cytoplasm, as are mRNAs made in the eukaryotic nucleus. These structures and molecules arrive in the cytoplasm independently of each other and become associated only at the time translation is begun. The rabbit reticulocytes, therefore, contained a resident population of mRNAs and ribosomes. These structures persist and function for some time after the nucleus has disappeared in the mature cell.

Translation complexes in eukaryote cytoplasm have been isolated and studied in many species. It is virtually certain that polypeptides are synthesized

polypeptide should have a length of about 1500 Å, since each amino acid is coded by a triplet of nucleotides that is 10.2 Å long. At 10.2 Å per codon and 146 codons for the 146 amino acids, there should be $10.2 \times 146 =$ about 1500 Å for the messenger. When polysomes were examined by electron microscopy, the predominant pentamer group indeed measured about 1500 Å. A thin thread connected the five ribosomes in a polysome and seemed to be the mRNA, since it could be digested with ribonuclease (Fig. 6.19). Various additional lines of supporting evidence also were obtained.

Six years later Oscar Miller produced striking electron micrographs showing polysomes in *E. coli* (Fig. 6.20). In the bacterial system, transcription and translation take place concurrently. Ribosomes attach to mRNA while the messenger is still being transcribed from the DNA molecule. Such transcription—

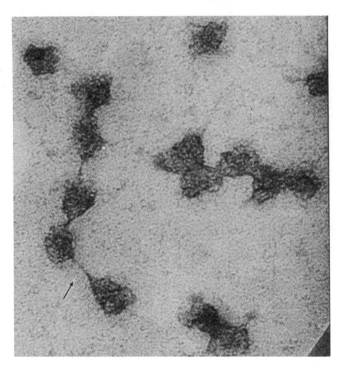

Figure 6.19
Polysomes from rabbit reticulocytes. Two pentamers are shown, each of the five ribosomes in the aggregates held together by a messenger RNA strand presumably coding for a hemoglobin polypeptide. × 382,000. (Reproduced with permission from Slayter, H. S. *et al.,* 1963, *J. Mol. Biol.* **7**:652–657, Plate IIIa.)

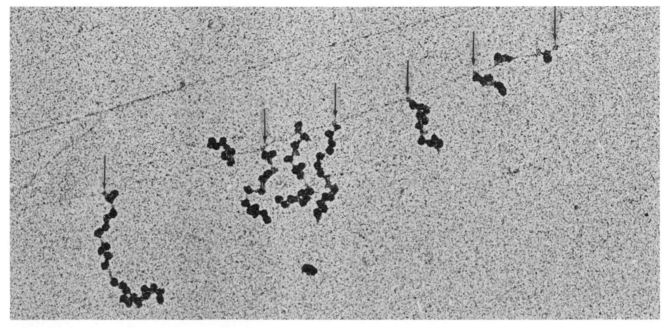

Figure 6.20
Electron micrograph of transcription-translation complexes isolated from *Escherichia coli*. The polysomes assemble while messenger RNA is being transcribed from active DNA template. Molecules of RNA polymerase (arrow), which catalyze synthesis of messenger RNA, are present on the DNA strand next to each mRNA being transcribed. × 115,300. (Courtesy of O. L. Miller, Jr., from Miller, O. L., Jr. *et al.,* 1970, *Science* **169**:392–395, Fig. 3.)

only from mRNAs that are associated with groups of ribosomes; that is, protein synthesis occurs at polysomes and not at free ribosome monomers. This observation is equally true for both prokaryotes and eukaryotes and for organelles such as mitochondria and chloroplasts. The size of the polysome generally corresponds to the length of the mRNA, which reflects the length of the protein which is translated from coded genetic information.

Polypeptide Synthesis

During formation of a polypeptide the separable events of **initiation, elongation,** and **termination** of the amino acid chain take place at the ribosome surface. The kinds and sequence of amino acids in the chain are determined by the coded mRNA associated with the polysome engaged in synthesis. Various enzymes and protein factors influence synthesis, and all three major classes of RNA (tRNA, mRNA, and rRNA) are required. Together, the interacting molecules and particles of the system lead to a linear translation of a linear informational sequence encoded in DNA. Co-linearity of gene and protein was discussed earlier in Chapter 5.

AMINO ACIDS AND TRANSFER RNA. If amino acids are to be linked together to form a specific polypeptide, there must be an energy source and a recognition mechanism. Energy is needed to sponsor the "uphill" reaction of amino acid additions as the polypeptide grows in length. If the correct amino acids are to be incorporated during synthesis, there must be some way in which the properly coded units recognize the codons of the mRNA during translation. It is also obvious that mRNA must be in an accessible conformation and that ribosome subunit surfaces also must be available for interactions to take place.

Energy for polypeptide chain growth is provided by hydrolysis of ATP, which occurs during activation of amino acids. When an amino acid is raised to a higher energy level after reacting with ATP, it has undergone one necessary step in preparation for protein synthesis. The second preparatory step involves the attachment of the activated amino acid to its tRNA carrier. Both activation of the amino acid and its attachment to tRNA are catalyzed by the same enzyme: an **aminoacyl-tRNA synthetase.** There is a different and specific synthetase for each of the 20 amino acids represented in the genetic code.

In the first of the two reactions, the carboxyl group of the amino acid interacts with the two terminal phosphoryl groups of ATP to yield an activated (higher energy level) amino acid complex which is called an amino acid-adenylate. In this reaction, the complex remains temporarily associated with the enzyme, so we can describe the events as

amino acid + ATP + enzyme →
  amino acid − AMP − enzyme + pyrophosphate

or   [6.1]

AA + ATP + enzyme → AA − AMP − enzyme + PP

In the second reaction, the enzyme portion of the AA − AMP − enzyme complex combines with a specific tRNA molecule and catalyzes the transfer of energy from AMP (adenosine monophosphate) to aminoacyl-tRNA. When this occurs, AMP is released into the medium and the freed enzyme can recycle to activate and attach another amino acid to another tRNA. The second reaction is

AA − AMP − enzyme + tRNA →
  AA − tRNA + AMP + enzyme   [6.2]

In the free state, each aminoacyl-tRNA synthetase presumably has one site for recognition of its particular amino acid and one site for binding with ATP. After formation of the amino acid-adenylate (AA − AMP) bond, the enzyme is believed to undergo a conformational change that unmasks another site at which the correct tRNA binds to the AA − AMP − enzyme complex.

Having identified the energy source, we are now confronted with the basis for recognition between mRNA codons and aminoacyl-tRNAs during translation. Is it the amino acid or its tRNA that recognizes the mRNA codon? A classic experiment, reported in the early 1960s, provided the answer to this question. It is the tRNA, not its amino acid, which recognizes and interacts with mRNA codons.

The amino acids cysteine and alanine are essentially identical except that cysteine has a sulfur atom and alanine does not. Each amino acid is coded by different triplet codons, and each is carried by different and specific tRNAs to these codons in the mRNA sequence. In this experiment, ^{14}C-labeled cysteine was enzymatically attached to its particular tRNA to form cysteine − tRNA^{cys}. Afterward, the complex was exposed to a nickel catalyst that removed the sulfur atom from cysteine but had no other effect on the complex. The catalyst treatment converted cysteine to alanine, but the newly-formed alanine was still attached to tRNA^{cys}. The hybrid alanine − tRNA^{cys} was supplied to an *in vitro* protein-synthesizing system. The critical observation was whether the complex behaved like alanine (which would show that the amino acid recognized the mRNA codons) or like cysteine (which would indicate recognition of mRNA by the tRNA^{cys} portion of the complex). The hybrid complex behaved just as if it were the original cysteine − tRNA^{cys} (Fig. 6.21). It was therefore established that amino acids are carried passively to the mRNA codon which is recognized by the region of tRNA called the **anticodon.**

A great many tRNAs from prokaryote and eukaryote systems have been completely sequenced since 1965. In that year Robert Holley reported the entire primary structure for the alanine-carrying tRNA from

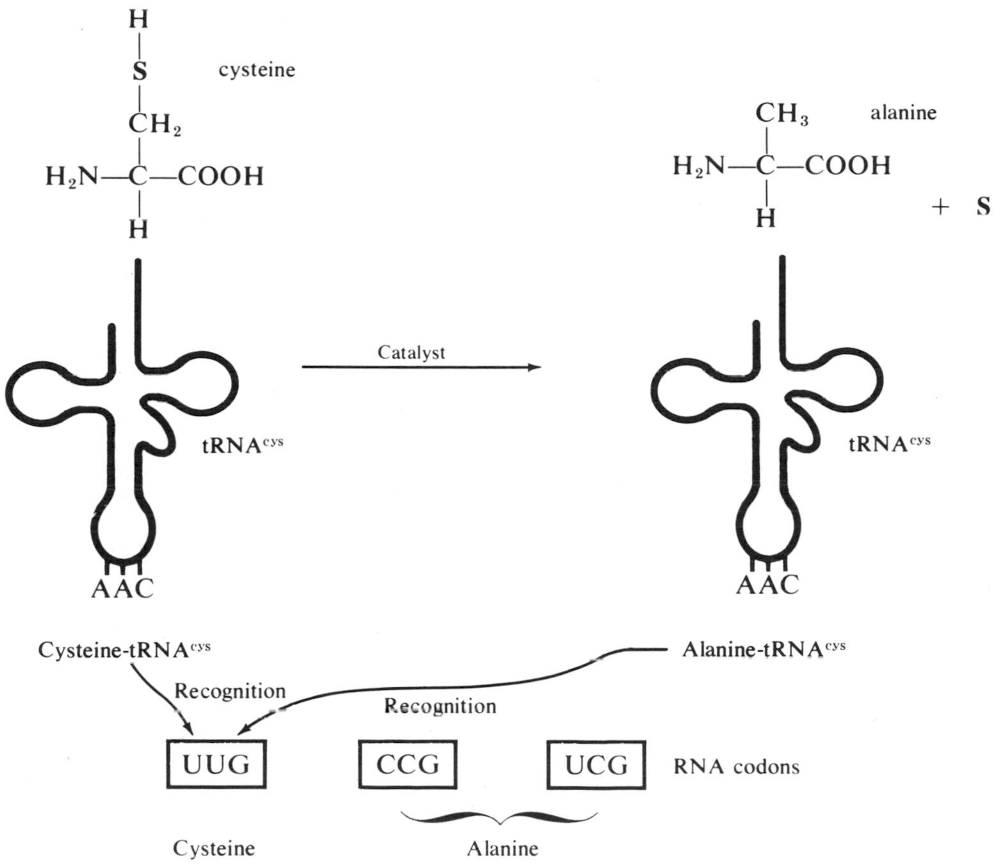

Figure 6.21
Diagrammatic illustration of the experiment showing that tRNA^{cys} recognized the mRNA codon for cysteine even though it was carrying alanine instead of its usual amino acid. Alanine-carrying tRNA^{cys} installed the alanine it was carrying into polypeptide sites meant for a cysteine residue.

yeast cytoplasm. There are 75–85 nucleotides in the single-stranded tRNAs from various sources. All these molecules share some features in common and it has been possible to construct a model of possible secondary structure of tRNA, based on known primary sequences (Fig. 6.22). The "stem" region of the molecule is created when 7 pairs of bases are held together by hydrogen bonding, while the remaining 4 bases of the segment project beyond the paired region. The projecting bases form the region of tRNA that accepts and binds to the amino acid. The anticodon region which interacts with mRNA is located elsewhere. According to the prevailing **clover leaf model** of secondary structure, the anticodon occurs within the "arm" farthest from the "stem" of the tRNA molecule. More recently it has been possible to construct reasonable models showing teritary structure of tRNA (Fig. 6.23). It is from models of

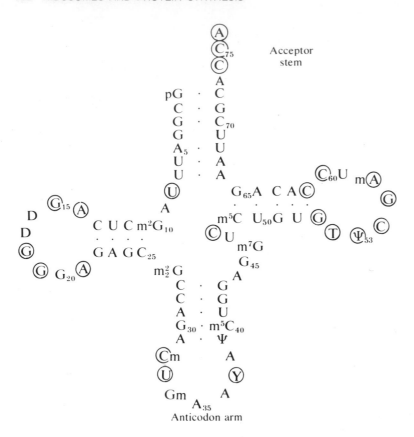

Figure 6.22
Generalized model of secondary structure in tRNA, showing bonding between some of the bases along the molecule, as well as the acceptor stem and anticodon loop. The amino acid binds to the acceptor stem. The site that recognizes the appropriate mRNA codon during translation is located in the anticodon arm of the molecule. (Reproduced with permission from Kim, S. H. *et al., Science* **185**:435–440, Fig. 1, 2 August 1974. Copyright © 1974 by the American Association for the Advancement of Science.)

this type that we hope to understand how tRNA interacts with enzymes and ribosomes during translation.

POLYPEPTIDE CHAIN ELONGATION. Until 1961 it was not known whether the polypeptide was formed by random insertions of correct amino acids until the whole chain was completed or by the addition of amino acids sequentially from the beginning to the terminus of the polymer molecule. Using reticulocytes, whose main protein was hemoglobin, H. M. Dintzis conducted a study which showed that amino acids were added sequentially and not through random insertions. In the test system, protein synthesis in reticulocytes was allowed to take place in media containing ^{14}C-leucine and other amino acid precursors. At various times during protein synthesis,

^{3}H-leucine was substituted for the original ^{14}C-leucine. Cells were collected and their unfinished proteins were isolated and digested before being "fingerprinted" by chromatography (Fig. 6.24). Since each digested fragment on the chromatograph could be identified as belonging to a known place in the hemoglobin chain, it was relatively simple to look for fragments with ^{14}C-label and those with ^{3}H-label. These fragments were then located within the hemoglobin sequence of amino acids to determine if the ^{3}H-leucine had been added randomly, with relation to ^{14}C-leucine, or added sequentially to pieces of the molecule that had already incorporated ^{14}C-leucine (Fig. 6.25). In each case, the first-formed parts of the molecule contained only ^{14}C-leucine, and ^{3}H-leucine was present only in the sequences added on later in

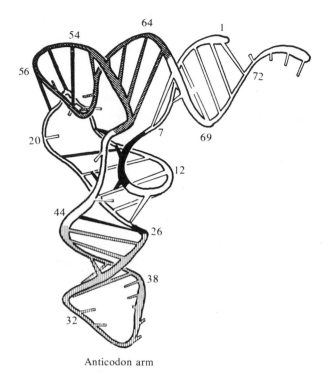

Anticodon arm

Figure 6.23
Model of the tertiary structure of yeast phenylalanine tRNA
(compare with Fig. 6.22). The numbers refer to nucleotides
in their primary structure locations in the sequenced tRNA
molecule. (Reproduced with permission from Kim, S. H.
et al., Science **185**:435–440, Fig. 3, 2 August 1974. Copyright © 1974 by the American Association for the Advancement of Science.)

1. Ribosomes.
2. tRNA.
3. Aminoacyl-tRNA synthetases.
4. Amino acids.
5. Protein factors for different stages of polypeptide synthesis.
6. Guanosine triphosphate (GTP), which cannot
   be replaced by other nucleotides.
7. ATP as energy source.
8. An ATP-producing system.
9. Mg^{2+} ions in appropriate concentration.
10. The monovalent cations K^+ and NH_4^+.

Many of these components can be isolated from the
sediments or nonsedimented fractions of lysed cells,
and they need not be in purified form in every case.

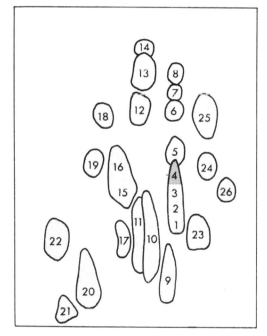

Figure 6.24
Distribution of peptide fragments from a digest of the β-
chain of hemoglobin showing the 26 pieces of the chain in
the locations they occupy after migration through solvents
applied to the paper sheet. The fragments are numbered to
correspond with their sequence in the intact β-chain.

sequence. There was no random insertion. Another
important observation was that the polypeptide always began to form at its *N*-terminus and not at its
C-end. This observation has since been verified for
other proteins.

Most evidence which supports the current ideas
about protein synthesis has come from *in vitro* studies.
Messenger RNAs may be synthetic, such as poly U,
or they may be natural mRNA from one or more
sources. In addition to adding mRNA to *in vitro* systems, other components are required:

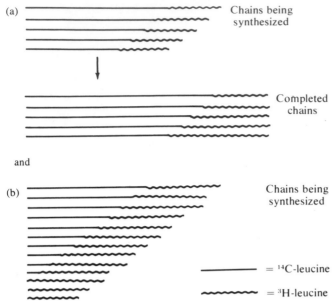

(a) ———————————— Chains being synthesized

and

(b) ———————————— Chains being synthesized

———— = ¹⁴C-leucine

~~~~ = ³H-leucine

Completed chains

Figure 6.25
Diagrammatic illustration of the Dintzis experiment show-ing that the distribution of ³H-leucine is always in the more recently synthesized region of the hemoglobin β-chain. The proportion of ¹⁴C-leucine to ³H-leucine in a chain depends on the time in which ³H-leucine replaced the ¹⁴C-leucine originally present in the medium.

For example, high-speed centrifugation causes sedi-mentation of ribosomes and subunits while cellular tRNAs and aminoacyl-tRNA synthetases remain in the fluid, or supernatant, phase of the preparation. Each component can be collected separately and used *in vitro* without purifying the specific ingredients be-forehand. At the same time ribosomes are harvested, protein factors for initiation of polypeptide synthesis are also isolated because they sediment with the ribo-some particles. Protein factors for elongation and termination remain in the supernatant and are col-lected along with tRNAs and synthetases in that fraction. Mixtures of components can be used, or individual ingredients can be purified for more exact-ing kinds of experiments.

Various lines of evidence show that the ribosome subunits have specific sites of activity, which contrib-ute to overall ribosome participation in protein syn-thesis. For example, if 70S ribosomes are separated into their constituent 30S and 50S subunits by lower-ing the concentration of Mg²⁺ ions, each type of sub-unit is recovered carrying a different load. The small subunits retain mRNA, while the large subunits sediment together with the growing polypeptide chains. The polypeptides are still attached to tRNA, which is also found in the sedimented 50S subunit fraction. These observations taken together with other kinds of information, lead to the belief that specific sites on the two subunits interact to (1) accept in-coming aminoacyl-tRNA, (2) add it on to the growing peptidyl chain of amino acids, and (3) prepare the ribosome for the next aminoacyl-tRNA in the se-quence until synthesis is completed.

In the current view, only two sites on a ribosome are thought to accommodate tRNAs at any one time and, therefore, only two molecules of tRNA can be attached simultaneously to an active ribosome. The **A site** is the place of entry, where incoming amino-acyl-tRNA is accepted at the ribosome. The **D site** is the place of release for free tRNA which is discharged from the ribosome after depositing its peptidyl chain onto the incoming aminoacyl-tRNA at the A site. The tRNA is released after a sequence of actions in which the peptidyl chain has been lengthened by one amino acid unit (Fig. 6.26).

According to this model, when a ribosome engaged in protein elongation carries the peptidyl-tRNA in its D site, the ribosome can accept an incoming amino-acyl-tRNA at its A site at the same time. The particu-lar aminoacyl-tRNA accepted depends on the exposed codon of the mRNA. Once in place, peptide bond formation takes place to link the incoming amino acid to the existing peptidyl chain. Bond formation occurs by transfer of the peptidyl chain from its tRNA at the D site to the aminoacyl-tRNA just brought to the A site. The reaction is catalyzed by **peptidyl trans-ferase.** Once the peptide bond has been formed, the

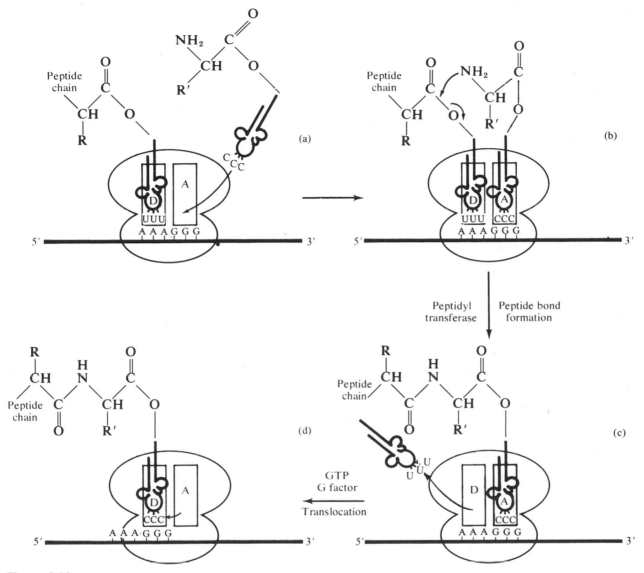

Figure 6.26

The sequence of events that takes place at the ribosome during peptide chain elongation: (a) Incoming aminoacyl-tRNA enters the A site on the ribosome, recognizing the mRNA codon *GGG*. Peptidyl-tRNA carrying the nascent polypeptide chain occupies the D site. (b) Peptide bond synthesis takes place by transfer of the polypeptide chain attached to the tRNA at the D site to the aminoacyl-tRNA at the A site. (c) Formation of the peptide bond is completed, catalyzed by peptidyl transferase. The A site is now occupied by a peptidyl-tRNA, while the freed tRNA at the D site is released to recycle. (d) The peptidyl-tRNA is translocated to the D site, a process that requires GTP and G factor. The A site is again free to accept the next aminoacyl-tRNA, the ribosome is one codon farther along the message (moving toward the 3' end of mRNA), and the polypeptide chain is one amino acid longer.

lengthened peptidyl-tRNA at the A site is translocated back into the D site. Translocation requires GTP and a **translocase** protein called the **G factor.** Before the peptidyl-tRNA can return to the D site, however, the now uncharged tRNA must be expelled from the place where it has remained after discharging its peptidyl chain. Once this is accomplished, the ribosome is restored to its previous state of readiness to accept an aminoacyl-tRNA at the A site, but the polypeptide is now one amino acid unit longer and is temporarily attached to a different tRNA than it was at the beginning of the episode.

The ribosome moves to the next triplet codon after every cycle of peptide bond formation, from the 5' toward the 3' end of the mRNA strand attached to the 30S subunit. If GTP is not available or not hydrolyzed, further synthesis stops because the A site remains blocked until peptidyl-tRNA is translocated back to the D site. There is no place for entry of another aminoacyl-tRNA until the A site is opened again. Similarly, G factor must also be present if translocation is to occur and the A site made accessible to incoming aminoacyl-tRNA.

In addition to G factor (translocase) and GTP, a protein **T factor** is also required for polypeptide chain elongation to proceed normally. T factors combine with GTP, and the T-GTP complex combines with incoming aminoacyl-tRNA. In the form of T-GTP—aminoacyl-tRNA, the incoming unit is brought to the A site of the ribosome where it can then participate in polypeptide chain elongation (Fig. 6.27).

The sequence of events in polypeptide chain elongation proceeds, therefore, as a result of close interaction among the various molecular and structural constituents described. Once it is complexed with T-GTP, an aminoacyl-tRNA will bind only to the A site of a ribosome that has its D site filled with a peptidyl-tRNA. Once bound to the ribosome, GTP of this complex is hydrolyzed so that T-GDP is released and the GTPase site on the ribosome becomes open to binding of G factor and a free second molecule of GTP. This second GTP is hydrolyzed during

translocation of the peptidyl-tRNA from the A to the D site. At the same time, GDP and G factor are released and the A site is opened again to entry of the next aminoacyl-tRNA. From these considerations, it is believed that the ribosome must have at least four active sites or centers (Fig. 6.28):

1. The **D site** situated largely on the 30S subunit.
2. The **A site** located largely on the 50S subunit.
3. A **peptidyl transferase center** exclusively on the 50S subunit.
4. A **GTPase site** situated on the 50S subunit.

It is not entirely clear how the two GTP hydrolysis events contribute to protein synthesis. It has been proposed that the first hydrolysis provides energy for a conformational change in the ribosome, and the second hydrolysis drives the translocation of peptidyl-tRNA from the A site to the D site. According to one suggestion, GTP hydrolysis would provide the energy to drive a pulsating ribosome contraction after formation of the peptide bond. This contraction would move the ribosome along the mRNA and change the status of the two bound tRNAs in one concerted movement.

POLYPEPTIDE CHAIN INITIATION. Until 1963 there was a general belief that any amino acid could begin the new polypeptide from its N-terminus to its C-terminus at the end of the translation. J. Waller noted, however, that methionine was the N-terminal amino acid in 45 percent of *E. coli* proteins studied, and only a few other amino acids occurred in this position in other *E. coli* proteins. This observation indicated that there might be some special conditions for initiation of protein synthesis which differed from those required to add on amino acids during elongation of a polypeptide. When *N*-formyl-methionyl-tRNA (**fmet-tRNA**) was discovered in *E. coli* extracts, studies were begun to determine its role in polypeptide chain initiation. This compound cannot participate directly in chain elongation because its amino group is blocked and it therefore cannot contribute to pep-

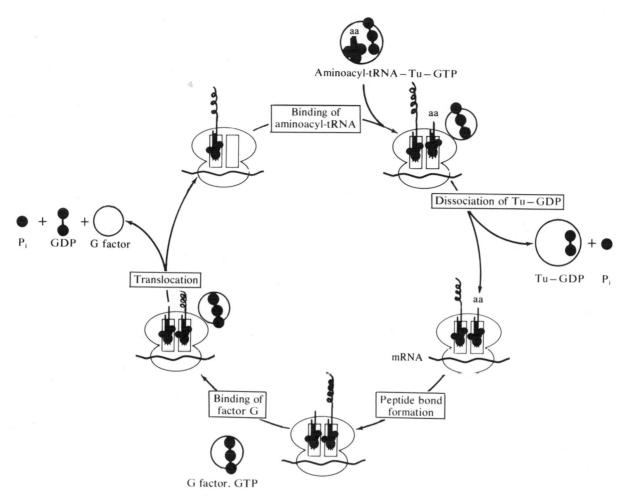

Aminoacyl-tRNA–Tu–GTP

Binding of aminoacyl-tRNA

Dissociation of Tu–GDP

Tu–GDP P_i

mRNA

P_i +   GDP   +   G factor

Translocation

Peptide bond formation

Binding of factor G

G factor, GTP

Figure 6.27
Summary diagram showing the elongation factors and other components involved in lengthening the peptide chain by one amino acid residue during an elongation episode.

tide bond formation, but *N*-formyl-methionine can act as the first amino acid at the *N*-terminus (Fig. 6.29). This modified amino acid can easily be converted to conventional methionine, the amino acid in the first position in almost half the proteins of *E. coli*. A *deformylase* can remove the formyl group and leave methionine, and a proteolytic enzyme known as an *aminopeptidase* can cleave methionine itself from a

chain and leave some other amino acid in the *N*-terminus position of a finished polypeptide.

There are two kinds of methionyl-tRNA, one of which can be formylated (met-tRNA$_f$) to produce fmet-tRNA$_f$ and another that cannot be formylated (met-tRNA$_m$), even in the presence of the formylase enzyme. Both types of aminoacyl-tRNA are stimulated to bind to mRNA at the AUG codon. However,

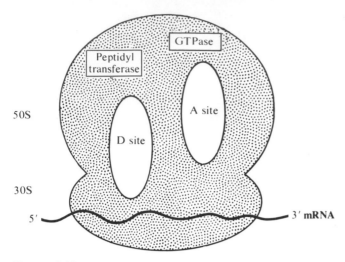

Figure 6.28
Probable locations of four active sites of the ribosome that are concerned with protein synthesis.

fmet-tRNA$_f$ differs from met-tRNA$_m$ in at least two more ways: (1) fmet-tRNA$_f$ is an initiator that is incorporated into the N-terminal position of a polypeptide, while met-tRNA$_m$ is inserted almost exclusively into internal positions of a chain; and (2) fmet-tRNA$_f$ will respond to either AUG or GUG initiator codons of a messenger, whereas met-tRNA$_m$ responds only to AUG codons.

When GUG occurs in any place other than the initiator codon of a messenger, valine is incorporated into the polypeptide since internal GUG codons are recognized by valyl-tRNA. Interaction between messenger codon and aminoacyl-tRNA anticodon, at least in these cases, depends partly on base-pairing recognition and partly on mRNA codon location. Most information about initiation in prokaryotes has come from studies of two bacterial species: *E. coli* and *Bacillus subtilis*. There is enough information from some other systems, however, to show that N-formyl-methionine is the initiating amino acid in polypeptide synthesis of other prokaryotes, too.

Protein synthesis in eukaryote cytoplasm is initiated much like prokaryote systems, except that the met-tRNA initiator is not formylated. There are two different met-tRNA entities, however: met-tRNA$_f$ which can be enzymatically formylated *in vitro* and met-tRNA$_m$ which cannot be formylated. Even though this compound has the capacity to be formylated, a formylase does not seem to be present in eukaryote cytoplasm. Because of the absence of a formylase, the usual initiator is met-tRNA$_f$. It is interesting to note that protein synthesis in mitochondria and chloroplasts of eukaryotes proceeds in the same way as it does in prokaryote systems. Both kinds of organelle utilize fmet-tRNA$_f$ as the initiating aminoacyl-tRNA. A formylase is present in these organelles, even when it is absent in the cytoplasm of the very same cells of eukaryotes.

One point deserves mention here since so many studies *in vitro* can be conducted successfully even when the synthetic messenger RNA lacks AUG or GUG initiator codons and fmet-tRNA$_f$ or met-tRNA$_f$ are not present in the mixture. For example, poly U can direct synthesis of polymers containing only phenylalanine residues. How can this be possible in view of requirements for initiation? *In vitro* protein-synthesizing systems can proceed even when messengers lack initiation codons only if the Mg^{2+} ion concentration of the medium is about 15 millimolar (mM) or higher. High Mg^{2+} ion concentrations apparently stabilize the complex between translation components. This fact went unobserved in earlier work because at least 20 mM Mg^{2+} ions are routinely included in *in vitro* systems. In 1967 H. G. Khorana showed that the only successful messengers at low Mg^{2+} concentrations were those with initiator codons, whereas various artificial RNAs could act as messengers when higher Mg^{2+} ion levels were provided. For example, only poly UG acted as a messenger at 4 mM Mg^{2+} (since GUG codons were present in the UGUGUG—molecules). But at higher levels of Mg^{2+}, poly UC, poly AG, and poly AC also could function as messengers as well as poly UG *in vitro*.

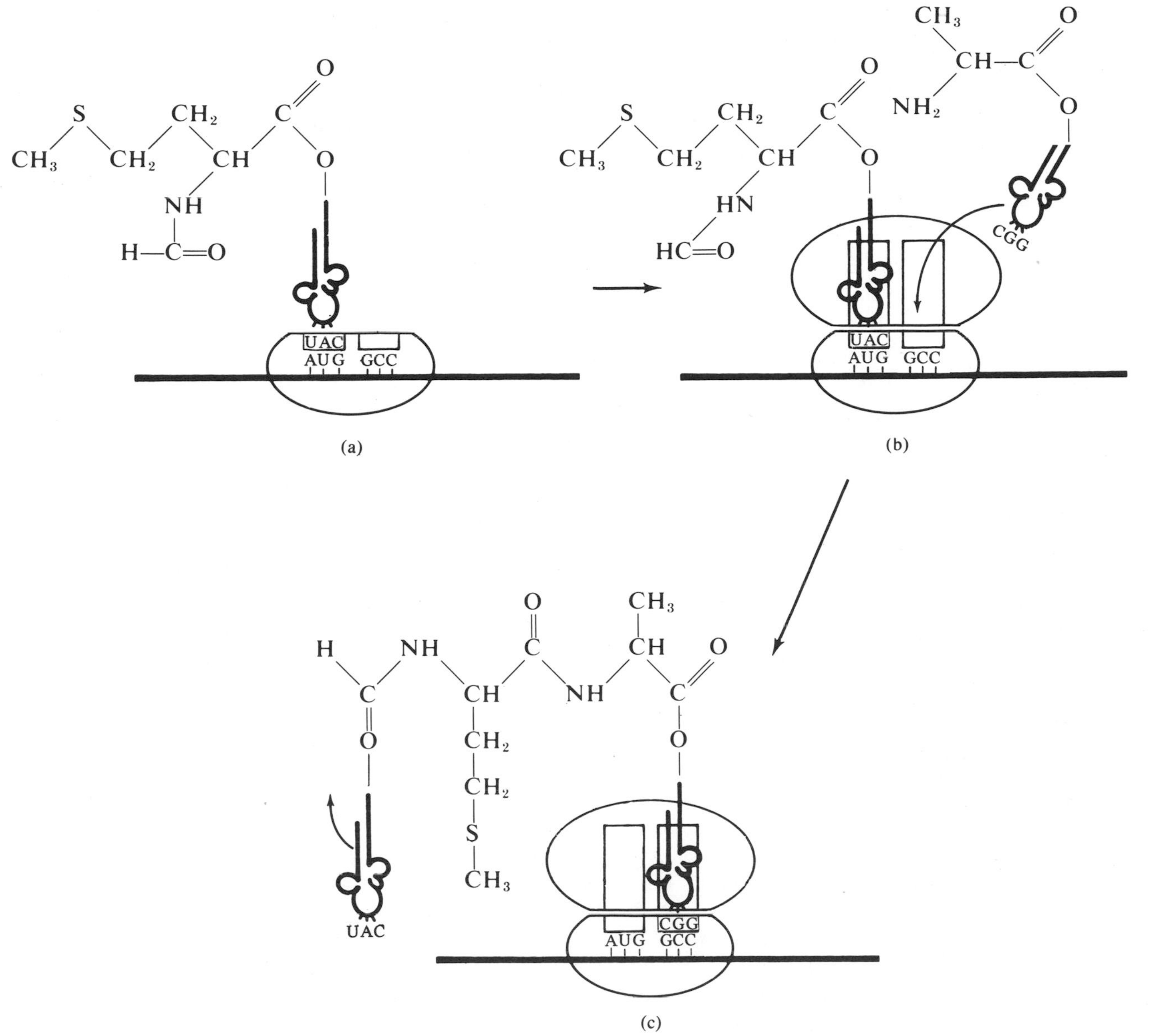

Figure 6.29

Polypeptide chain initiation at the ribosome: (a) fmet-tRNA binds at the D site of the 30S ribosome subunit, the anticodon UAC recognizing the AUG initiator codon in mRNA. (b) The aminoacyl-tRNA carrying alanine is entering the A site where the tRNA anticodon CGG recognizes the complementary mRNA codon GCC. (c) Peptide bond synthesis occurs producing a dipeptidyl-tRNA at the A site. Elongation continues according to the steps shown in Fig. 6.26.

Discrimination between initiating and internal AUG codons is twofold:

1. Ribosomes must bind to messenger RNA at or near the initiator AUG codon and not to AUG located elsewhere in the message.
2. The fmet-tRNA$_f$ must respond to a ribosome engaged in initiation, whereas met-tRNA$_m$ must respond to ribosomes engaged in elongation of a polypeptide chain.

The small subunit of the ribosome acts discriminately, not the larger subunit. In fact, the requirement that only small subunits need to participate in initiating polypeptide chains provided early clues to the notion that subunits rather than free ribosome monomers composed the pool of inactive ribosomal particles in the cytoplasm. Subunits come together into monomers after initiation of polypeptide synthesis, and subunits dissociate after the polypeptide is completed. Under normal conditions of active protein synthesis in prokaryotes and eukaryotes, whole ribosomes are found mainly as units in polysome groups. These ribosomes are actively engaged in synthesis. Once released from the polysome when the message has been translated, ribosome subunits separate from each other. They come together again at another round of polypeptide formation, but continue to recycle in successive syntheses (Fig. 6.30).

Initiation is mediated by the small subunit of the ribosome once it has accepted and bound a messenger RNA. At this stage the large subunit is not involved and is not attached to the small subunit. Initiation begins when fmet-tRNA$_f$ or its equivalent in eukaryotes is accepted by the 30S−mRNA at the AUG codon which begins the message. Only this aminoacyl-tRNA is recognized because it is accepted from **initiation factors** associated with the 30S subunit before the addition of the 50S subunit to form the ribosome monomer. After the 70S ribosome has been formed, chain elongation takes place. In this stage the ribosome will not accept fmet-tRNA$_f$ but will accept met-tRNA$_m$, in the presence of elongation factors.

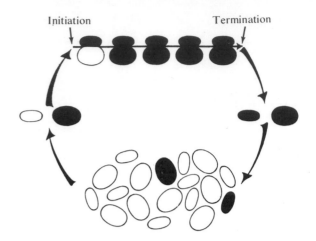

Figure 6.30
The subunit cycle in prokaryotic and eukaryotic cells.

Initiation depends only on the 30S subunit and can occur *in vitro* even if the system lacks 50S subunits. Elongation, on the other hand, takes place only if the entire 70S monomer is present. This event suggests that both subunits contribute to elongation events, since polypeptide chain extension cannot take place on 50S subunits alone.

The initiation complex formed from 30S subunit, mRNA, and initiator fmet-tRNA$_f$ requires GTP and three kinds of initiation factors (Fig. 6.31). The initiation factors are only found in association with 30S subunits and not with 70S monomers. This observation by itself indicates that the factors are utilized only during chain initiation. Further support comes from *in vitro* studies which show that the factors are not needed if synthetic mRNA is used (at Mg^{2+} ion levels of 15 mM or higher), but factors are essential for initiation if a natural mRNA is part of the *in vitro* protein-synthesizing system.

One of the initiation factors, called **IF3,** is believed to have two functions: (1) it stimulates binding of 30S subunits to appropriate recognition sites on mRNA; and (2) it assists in dissociating 70S monomers into 30S and 50S subunits. This second function also contributes to initiation because it leads to a pool

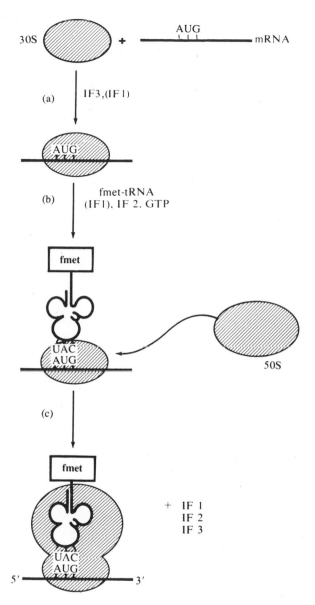

of available 30S subunits. Initiation factor **IF2** promotes binding of fmet-tRNA$_f$ to the 30S subunit—mRNA complex that formed first. The IF2 protein has GTPase activity, which suggests that the first step in binding the initiator aminoacyl-tRNA to 30S—mRNA may involve an association between IF2 and GTP. The actual role of GTP hydrolysis by GTPase (IF2) action is not known, but it would involve energy transfer. The role of initiation factor **IF1** is uncertain at present, except that the factor somehow assists both IF2 and IF3 in their activities during initiation. When initiation is complete, the 50S subunit binds to the 30S—mRNA—fmet-tRNA$_f$ initiation complex. Initiation factors are then released from the 70S ribosomes, which continue to participate in polypeptide chain elongation in conjunction with elongation factors described earlier.

POLYPEPTIDE CHAIN TERMINATION. Before the finished polypeptide can be released from the ribosome, the link must be broken between the last amino acid of the chain and the tRNA to which it is attached. Separation does not take place spontaneously. Separation does not take place efficiently even if terminator codons UAA, UAG, or UGA are present in the messenger. Proper release of tRNA and the completed polypeptide from the ribosome requires the terminator codon signal, and an active termination process in response to such a signal in the mRNA sequence. Protein termination factors have been found in eukaryotes and prokaryotes.

The most likely situation is that the termination factor enters the ribosome at the A site, otherwise used by an incoming aminoacyl-tRNA. The exact mechanism of the release action at the ribosome is not known, however.

Many other questions also remain to be answered,

Figure 6.31
Formation of the initiation complex in protein synthesis at the ribosome: (a) a 30S subunit binds to mRNA, stimulated by initiation factor IF3 and possibly IF1; (b) the binding of fmet-tRNA to the 30S—mRNA complex is promoted by IF2 and GTP and possibly IF1; and (c) the 50S subunit

binds to the 30S—mRNA—fmet-tRNA initiation complex, and all three initiation factors are released from the 70S ribosome.

but it is a tribute to molecular biologists that so much has been learned about the incredibly complex and intricate interactions in translation of proteins from coded information.

Protein Synthesis at the Rough Endoplasmic Reticulum

Ribosomes in many eukaryotic cells are attached to various membranes as well as occurring as free subunits and polysomes in the cytoplasm (Fig. 6.32). Most studies have concentrated on ribosomes at-

tached to the rough endoplasmic reticulum (rough *ER*). As early as 1960 it was suggested that free ribosomes were especially active in synthesizing proteins that were retained within the cell, while ribosomes of the rough *ER* synthesized proteins that were exported from the cell. There is little doubt that membrane-bound ribosomes do synthesize exported proteins, but it is not certain that free polysomes are responsible for synthesizing all other proteins. In fact, various eukaryotic cells have little *ER* and

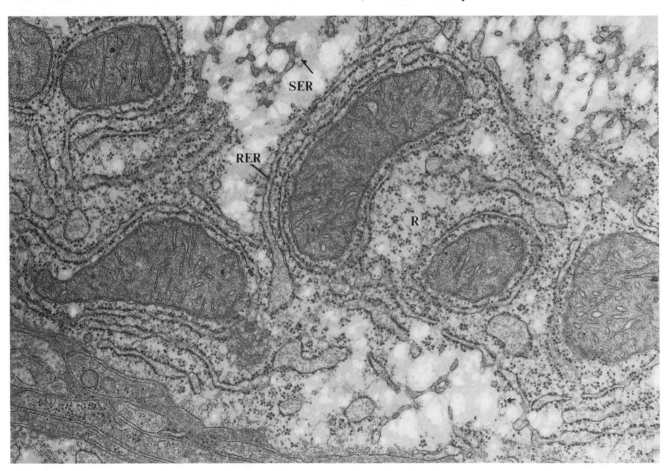

Figure 6.32
Portion of rat liver cell showing free ribosomes (R), rough endoplasmic reticulum with attached ribosomes (RER), and smooth endoplasmic reticulum (SER). × 25,000. (Courtesy of M. Federman)

probably synthesize almost all their proteins on free polysomes.

There is little agreement on the density and spacing between ribosomes of the rough *ER*, a disagreement more likely due to variations in technique used in different studies than to genetic distinctions between species or tissues. Typical electron micrographs show widely spaced ribosomes displayed along the outer faces of *ER* channels. Other electron microscopic studies, however, have shown rather closely spaced ribosomes or even continuous carpets of particles with no space between them. Whatever the spatial distribution and absolute shape of ribosomes and their subunits (another controversial topic), it is agreed that

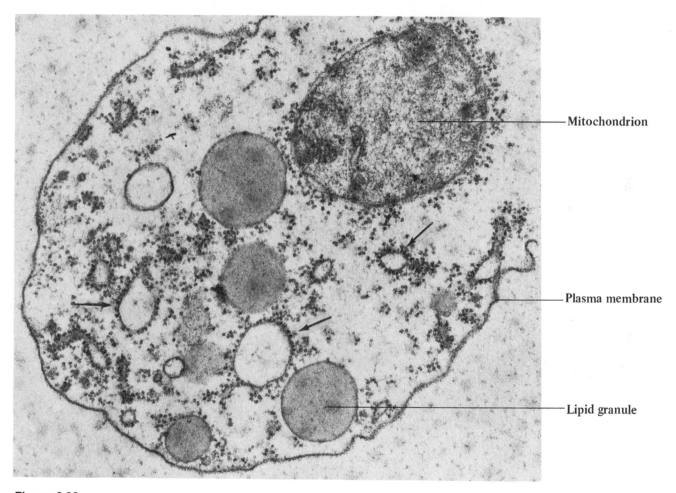

Figure 6.33
Part of a broken yeast cell showing membranous vesicles with ribosomes attached to their outer surface (at arrows). Such components constitute a substantial part of the microsome fraction of centrifuged cell-free lysates but are rarely observed in intact or undamaged yeast cells. × 43,300. (From Szabo, A., and C. J. Avers, 1969, *Ann. N. Y. Acad. Sci.* **168**:302–312, Fig. 2.)

the 60S subunit is tightly bound to the *ER* membranes.

One convenient way to study rough *ER* is to isolate the material by density gradient centrifugation and then to apply numerous biochemical tests to the materials. Although the endoplasmic reticulum appears to be a tortuously folded sheet of membrane(s) in the living cell, it is isolated in small vesicle fragments by centrifugation. These fragments were called **microsomes** by Albert Claude when he first examined such preparations in the 1940s. This convenient term is still used even though microsomes are an artifact of isolation and preparation and do not exist, as such, in the cell. Each microsome is a vesicle which may have ribosomes attached to its outer surface, just as ribosomes are attached to the outer cytoplasm-facing surface of the *ER in vivo* (Fig. 6.33).

Using isolated microsome materials, David Sabatini and associates studied ribosome-membrane interactions in protein synthesis. Microsomes are enzymatically and ultrastructurally representative of rough *ER*, and they can be made to function *in vitro* as miniature synthesizing units. In earlier reports, Sabatini showed that amino acids were incorporated into polypeptides at the ribosomes of rough *ER* vesicles (rough microsomes). If these active microsomes were treated with the antibiotic puromycin, their polypeptides were released prematurely. Puromycin mimics tRNA and is accepted at the ribosome in the same way as tRNA, but when puromycin is present, it picks up the growing peptidyl chain and is released from the ribosome along with the nascent polypeptide. The drug acts as an inhibitor of further chain elongation, therefore, but it also provides a very useful tool for studying the distribution of newly synthesized polypeptides.

When puromycin was added to active microsomes, the puromycin—polypeptide chains were released vectorially to the microsome interior (Fig. 6.34). Along with other information, these results fit nicely into a scheme that postulates growth of the polypeptide *into* the lumen of the *ER*. Once completed, the polypeptide is transported through *ER* channels to

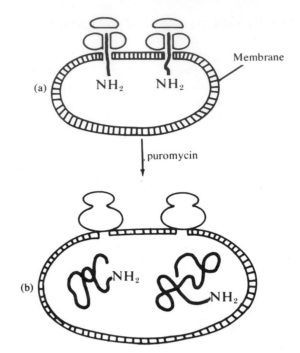

Figure 6.34
Diagrammatic illustration of polypeptide chain growth at 80S ribosomes attached to the outer surface of the *ER* or to a microsomal representative of the rough *ER:* (a) The growing polypeptide chain extends through a groove in the 60S subunit into the microsomal vesicle (or *ER* lumen). This process is demonstrated after puromycin treatment in b, showing released nascent polypeptide chains within the microsomal vesicle. The *N*-terminal end of the polypeptide chain is indicated by NH₂.

be packaged at the Golgi apparatus or elsewhere in the cell. The packaged proteins may then be exported from the cell, or retained and used in the cell. This topic is presented again in Chapter 10.

CHEMICAL INHIBITORS OF PROTEIN SYNTHESIS

A number of antibiotics and other chemicals are known to inhibit or modify protein synthesis, just as other specific agents influence manufacture or activity of DNA and RNA. Each kind of agent can be used as a specific probe to analyze some molecular or structural component or some biosynthetic step and, therefore, to aid in understanding cellular activities and interactions at the molecular level. We have already mentioned the usefulness of puromycin in cell studies in the preceding section.

Drugs that inhibit one or more steps in protein synthesis are shown in Table 6.4. Two examples will be given to illustrate the value of chemicals in experimental analysis. In one case, an antibiotic may be used to control protein synthesis by selectively poisoning one subcellular compartment and not another. Cycloheximide and chloramphenicol will serve as examples of such drugs. In the second case, we will consider the mode of action of streptomycin as an inhibitor of protein synthesis and as a probe used in studying the ribosome itself.

Selective Inhibition of Protein Synthesis

Cycloheximide is a selective inhibitor. It prevents chain elongation on eukaryote cytoplasmic ribosomes but has no detectable effects on prokaryotic systems. The mechanism of action has been shown to involve the step in which peptidyl-tRNA is translocated from the A to the D site on the ribosome. Since cycloheximide interferes with this step, the A site remains occupied and there is no place for incoming aminoacyl-tRNA. Ribosome movement along the messenger RNA stops and further synthesis of polypeptide chains is inhibited.

Another drug that inhibits translocation of peptidyl-tRNA is showdomycin. In this case the chemical prevents protein synthesis, in both eukaryotes and prokaryotes, by specifically interfering with the action of the translocase protein factor in the chain elongation system.

The antibiotic **chloramphenicol** is representative of a group of chemicals that inhibit peptidyl transferase activity. Protein synthesis stops because peptide bond formation is prevented. This inhibitor acts only on ribosomes from prokaryote cytoplasm and not on eukaryotic cytoplasmic ribosomes. Its effects

Table 6.4 Characteristics of inhibitor action on protein synthesis

STAGE INHIBITED	INHIBITOR	MODE OF ACTION OF INHIBITOR	EFFECTIVE IN	
			PROKARYOTES	EUKARYOTES
Initiation	Aurintricarboxylic acid	Prevents association of ribosome subunit with messenger RNA	+	+
	Streptomycin	Releases bound fmet-tRNA from initiation complex	+	−
	Cycloheximide	Prevents formation of initiation complex *in vitro*	−	+
Elongation	Streptomycin	Inhibits binding of aminoacyl-tRNA to ribosome; inhibits translocation on the ribosome	+	−
	Chloramphenicol	Stops amino acid incorporation into polypeptides by inhibiting peptidyl transferase	+	−
	Puromycin	Terminates nascent chain through action as amino acid analog	+	+
	Fusidic acid	Inhibits translocation GTPase	+	+
	Cycloheximide	Generally inhibits tRNA movement on ribosome	−	+
Termination	Various drugs	Inhibit releasing factors; inhibit ribosome release from messenger RNA	+	+

are therefore the opposite of cycloheximide action, in terms of general site of cellular effectiveness. In addition, chloramphenicol inhibits protein synthesis on mitochondrial and chloroplast ribosomes. Cycloheximide, on the other hand, does not prevent organelle ribosome activities.

Each of these inhibitors acts at the stage of polypeptide chain elongation, but each can be used discriminately in studying the patterns and contributions made by different cellular compartments in total protein synthesis. In a typical experiment using eukaryotic cells, the effects of each inhibitor can be judged by comparing the patterns of incorporation of radioactively labeled amino acids into polypeptides.

Cultures inhibited by cycloheximide contain little or no labeled proteins in the cytoplasmic compartment, whereas chloramphenicol-inhibited cells have labeled cytoplasmic proteins and little or no labeled organelle proteins. If both inhibitors are present, almost no label is incorporated into protein synthesized during the experimental period. All these values must be compared with the control cells which had no inhibitor included in the culture medium. Using experiments designed along these lines, it has been determined that each cell compartment can carry out its own program of protein synthesis. It has also been determined that some proteins are made on cytoplasmic ribosomes and some are made on organelle ribosomes. This division has been made by looking at the patterns of inhibition caused by cycloheximide and chloramphenicol. We will consider some of these studies further in Chapter 9.

Effects of Streptomycin on the Ribosome

Studies of streptomycin have been relatively broader than for many other kinds of inhibitors. In 1964 it was found that the antibiotic exerted its effects specifically on the 30S subunit of the bacterial ribosome. Strains of *E. coli* which were genetically streptomycin-resistant (*str^r*) and streptomycin-sensitive (*str^s*) were used to isolate ribosomes for study. The ribosome monomers were dissociated into their 30S and 50S subunits, so that they could be combined into monomers in various ways for *in vitro* protein synthesis (Fig. 6.35). Poly U was used as a messenger RNA and formation of phenylalanine polymers was followed in the presence or absence of streptomycin. When streptomycin was present, polypeptides were synthesized only in those systems with a 30S subunit from *str^r* strains. The source of the 50S subunit did not matter. When streptomycin was omitted from the system, all subunit combinations were equally effective in polyphenylalanine synthesis.

It was further shown in 1965 that low levels of streptomycin caused **misreading** of codons in streptomycin-sensitive *E. coli* strains. If poly U is provided as mRNA, we expect phenylalanine to be the only amino acid incorporated into the polypeptide because its codon is UUU. Sensitive strains, however, were found to make polypeptides that contained isoleucine and serine as well as phenylalanine. Resistant strains produced only polyphenylalanine. Using other artificial messengers, it was found that misreading usually

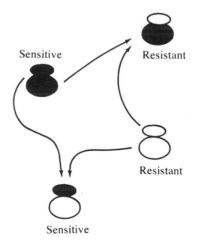

Figure 6.35
Recombining 30S and 50S subunits isolated from streptomycin-sensitive (black) and streptomycin-resistant (white) strains of *E. coli* showed that protein synthesis occurred normally in the presence of streptomycin only when the ribosomes contained the 30S subunit from resistant strains.

involved only the pyrimidines U and C in a codon (Fig. 6.36). For example, poly C promoted incorporation of the usual proline, and also serine, histidine, and threonine, when sensitive strains were presented with low concentrations of streptomycin. In every

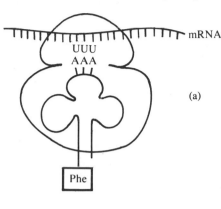

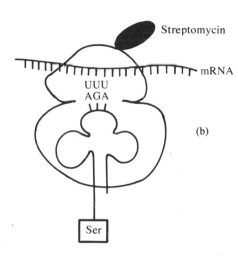

Figure 6.36
Misreading of codons: (a) Normal reading of UUU specifying phenylalanine occurs in the absence of streptomycin in sensitive wild-type or resistant mutant strains, but (b) in the presence of streptomycin, ribosomes from sensitive strains may misread codons. For example, mRNA codon UUU may be translated as UCU. A serine residue will be placed in the polypeptide chain instead of the phenylalanine that belongs there according to the coded message.

case it seemed that U in a codon was misread as C and that C was misread as U (occasionally either was misread as A). Misreading occurs only in prokaryotic streptomycin-sensitives and not in genetically similar eukaryotes. The reason for this difference is not known.

These results seem to indicate that streptomycin exerts its lethal effects on bacteria by flooding the cells with nonfunctional, misread proteins. This is not the case, however, as was shown by additional studies. For example, poly AC messengers are not misread, but protein synthesis is inhibited and cells that are streptomycin-sensitive are killed, even when misread proteins are not made. Streptomycin apparently inhibits protein synthesis and causes cell death because the drug interferes with initiation of the polypeptide chain. Streptomycin causes the release of fmet-tRNA$_f$ from the initiation complex after the 50S subunit has joined on. There is no inhibition of the formation of initiation complex itself.

Streptomycin leads to disturbances in the proportions of subunits, monomers, ribosomes, and polysomes in sensitive cells. Ribosome monomer levels increase while fewer polysomes and fewer free subunits are found. Apparently, ribosomes "fall off" the messenger when the drug prevents polypeptide initiation. These ribosomes do not readily dissociate into 30S and 50S subunits. The result of these events is a decline in polysomes and an increase in ribosome monomers. Streptomycin therefore acts at two levels to inhibit protein synthesis: (1) initiation is inhibited by release of fmet-tRNA$_f$, and (2) if the ribosome escapes this restriction, synthesis inhibition is assured because few subunits are generated for initiation from tightly associated particles of the 70S monomers.

The specific lesion in streptomycin-sensitive strains has been traced to a single protein of the 30S subunit. In 1969 Nomura showed that protein S12 was different in sensitive and resistant strains. If 30S subunits are assembled so that S12 comes from a sensitive strain and the other 20 proteins come from a resistant strain, the entire subunit behaves like a sensitive

str^s particle. Similarly, if S12 comes from a resistant strain then the 30S subunit behaves like a str^r particle, no matter what the source of the other subunit protein components. If the S12 protein comes from a resistant strain, the assembled 30S subunits are resistant to both misreading and to inhibition of protein synthesis when streptomycin is present in the system.

Studies of streptomycin effects on susceptible cells have thus proven useful in many ways. Not only do we have a clearer understanding of the mode of action of the drug at both the cellular and subcellular levels, but we have also been able to use the drug as a probe to analyze structure and function of the ribosome. We will see how streptomycin has been useful in dissecting organelle ribosome and organelle genetic properties in sections of Chapter 9.

SUGGESTED READING

Books, Monographs, and Symposia

Cold Spring Harbor Symposia on Quantitative Biology. 1969. *The Mechanism of Protein Synthesis*, vol. 34. New York: Cold Spring Harbor Laboratory.

Lewin, B. 1974. *Gene Expression*, vols. 1 and 2. New York: John Wiley.

Nomura, M., ed. 1973. *Ribosomes*. New York: Cold Spring Harbor.

Articles and Reviews

Brown, D. D. 1973. The isolation of genes. *Scientific American* 229(2):20–29.

Brown, D. D., and Gurdon, J. B. 1964. Absence of ribosomal RNA synthesis in the anucleolate mutant of *Xenopus laevis*. *Proceedings of the National Academy of Sciences* 51:138–146.

Clark, B. F. C., and Marcker, K. A. 1968. How proteins start. *Scientific American* 218(1):36–42.

Gall, J. G. 1968. Differential synthesis of the genes for ribosomal RNA during amphibian oogenesis. *Proceedings of the National Academy of Sciences* 60:553–560.

Gall, J. G., and Pardue, M. L. 1969. Formation and detection of RNA-DNA hybrid molecules in cytological preparations. *Proceedings of the National Academy of Sciences* 63:378–383.

Gorini, L. 1966. Antibiotics and the genetic code. *Scientific American* 214(4):102–109.

Held, W. A., Ballou, B., Mizushima, S., and Nomura, M. 1974. Assembly mapping of 30S ribosomal proteins from *Escherichia coli*. Further studies. *Journal of Biological Chemistry* 249:3103–3111.

Kim, S. H. *et al.* 1974. Three-dimensional tertiary structure of yeast phenylalanine transfer RNA. *Science* 185:435–440.

Lande, M. A., Adesnik, M., Sumida, M., Tashiro, Y., and Sabatini, D. D. 1975. Direct association of messenger RNA with microsomal membranes in human diploid fibroblasts. *Journal of Cell Biology* 65:513–528.

Lipmann, F. 1969. Polypeptide chain elongation in protein synthesis. *Science* 164:1024–1031.

Miller, O. L., Jr. 1973. The visualization of genes in action. *Scientific American* 228(3):34–42.

Miller, O. L., Jr., and Beatty, B. R. 1969. Visualization of nucleolar genes. *Science* 164:955–957.

Nanninga, N. 1973. Structural aspects of ribosomes. *International Review of Cytology* 35:135–188.

Nomura, M. 1973. Assembly of bacterial ribosomes. *Science* 179:864–873.

Pace, N. R. 1973. Structure and synthesis of the ribosomal ribonucleic acid of prokaryotes. *Bacteriological Reviews* 37:562–603.

Rich, A. 1963. Polyribosomes. *Scientific American* 209(6):44–53.

Ritossa, F. M., and Spiegelman, S. 1965. Localization of DNA complementary to ribosomal RNA in the nucleolus organizer region of *Drosophila melanogaster*. *Proceedings of the National Academy of Sciences* 53:737–745.

Sabatini, D. D., and Blobel, G. 1970. Controlled proteolysis of nascent polypeptides in rat liver cell fractions. *Journal of Cell Biology* 45:146–157.

Weissbach, H., and Brot, N. 1974. The role of protein factors in the biosynthesis of proteins. *Cell* 2:137–144.

Zamecnik, P. C. 1958. The microsome. *Scientific American* 198(3):118–124.

Chapter 7

Energy Flow Through the Mitochondrion

The mitochondrion has been called the "powerhouse" of the cell. This description refers to the fact that the metabolic reactions occurring within the mitochondrion yield the required energy that sustains aerobic life. Similar energy-yielding reactions in prokaryotes are catalyzed by enzymes located within the plasma membrane. The mitochondrial enzymes that catalyze aerobic respiration also are bound to a membrane, specifically to the inner membrane of the organelle itself. These membrane-bound catalysts interact with enzymes and reactions both outside the mitochondrion and within its matrix center.

Except for photosynthetic organisms that obtain energy from the sun's radiations, other life forms must derive energy for all their activities by breaking down fuel molecules. The simpler end-products of these reactions can then be used as building blocks for cellular syntheses, while the energy released during metabolic reactions is conserved in energy carriers such as ATP and NADH. Conserved energy is thus made available for synthesis in reactions closely coordinated with fuel breakdown, or wherever such needs must be met in cellular activities.

The primary fuel is carbohydrates, but lipids and

amino acids also can be catabolized to provide energy and carbon for growth and reproduction. Since carbohydrates are the principal organic nutrients, it is no surprise to find that a number of alternative pathways exist by which these fuel molecules can be dismantled to serve the active cell. Breakdown products from lipid and amino acid oxidations can also be channelled into pathways that lead to carbohydrate processing. The interactions in metabolism provide illustrations of the fundamental mechanisms by which many kinds of activities in different parts of the cell are coordinated. The end result is a harmonious set of complex events that characterize cellular orderliness amidst potential chaos.

Mitochondria have been a focus of interest for other reasons as well as for their metabolic contributions to cellular growth. The organelle membranes act as highly selective barriers to entry and exit of ions and molecules, and as model systems in which functional proteins as well as structural proteins contribute to a dynamic cell structure. The presence of DNA and an independent ribosomal machinery for protein synthesis has led to new studies of mitochondrial genetics and to important questions about the origin of the organelle during eukaryote evolution. Mitochondrial formation, or biogenesis, is another area of interest to cell biologists. Assembly of new membranes with uniqueness and specificity can now be studied with mitochondrial systems. New insights into eukaryote compartmentation and organization can thus be expected as many methods converge in the exploration of mitochondrial structure and function.

FORM AND STRUCTURE OF THE MITOCHONDRION

The discovery of mitochondria is usually attributed to R. Altmann, who first demonstrated the existence of these structures in 1886 through a method which showed mitochondria as magenta-colored granules when seen with the light microscope. Mitochondria can also be seen with the light microscope in living cells that have not been stained, or in living cells exposed to a so-called "vital" dye. They appear as highly active granules and threadlike components that move about rapidly and undergo dramatic changes in shape and size, when living cells are examined (Fig. 7.1). Nothing can be seen at these light microscope magnifications to characterize internal structure, or even to be certain about the form of the mitochondrion. The development of electron microscopy in the early 1950s, especially due to the efforts of G. Palade, K. Porter, and F. Sjöstrand, opened a new dimension of observation. Mitochondrial ultrastructure was an early subject of their studies.

Structure

The usual mitochondrial images seen in eukaryotic cell thin-sections are spherical, ovoid, or tubular outlines of structures bounded by two separate membranes (Fig. 7.2). The outer membrane is smooth and uninterrupted, but the inner membrane invaginates at various sites along its surface. These infoldings, called **cristae,** are a unique identifying feature for mitochondrial profiles as seen by electron microscopy. Cristae provide a substantial increase in the amount of inner membrane that can be accommodated within the mitochondrion boundary. The increased inner membrane is an obvious device to allow the insertion of many more of the respiratory enzymes that are localized within this membrane of the mitochondrion, and thus increase metabolic efficiency in the confined space. A. L. Lehninger has calculated that cristae in rat liver cell mitochondria provide ten times more membrane surface area than the plasma membrane around the cell. Since rat liver mitochondria are not particularly rich in cristae, even greater increases in mitochondrial membrane surface are realized in other cell types.

The **matrix** is a semisolid system surrounded by the inner membrane. Except for ribosomes and DNA fibers that can be seen in cells prepared with special care, no other regularly structured components occur in the mitochondrial matrix. Cells fixed in potassium

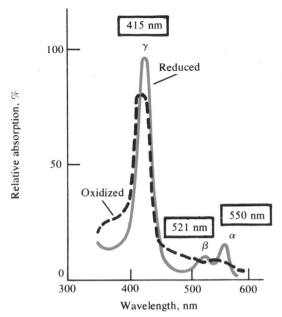

Figure 7.14
The absorption spectrum of cytochrome c in its oxidized (broken line) and reduced (solid line) states. The wavelengths of the α, β, and γ peak absorption are shown.

oxidase, an enzyme complex made up of cytochromes $a + a_3$. This enzyme can react with molecular oxygen, whereas reduced forms of other cytochromes cannot be reoxidized by oxygen. When cytochrome a is separated from cytochrome a_3, it can be shown that only cytochrome a_3 can react directly with oxygen and that cytochrome a cannot. The whole enzyme complex is formally called cytochrome c oxidase because the Fe^{3+} form of cytochrome $a + a_3$ accepts electrons from reduced cytochrome c. The Fe^{2+} form is then reoxidized in turn back to its Fe^{3+} state by molecular oxygen.

The sequence of electron transfer reactions is supported by different kinds of evidence, only some of which will be mentioned briefly in order to introduce other important features of the system. An examination of the standard reduction potentials of the different electron carriers shows a consistency

between the postulated sequence and the decreasing amount of free energy released during oxidation of the carrier by oxygen (Table 7.1). The order of the reactant in the respiratory chain should reflect the fact that the reactions proceed "downhill," from a higher to a lower energy potential. The greatest decrease in free energy occurs during the oxidation of NADH to NAD^+; the lowest free-energy decrease occurs during the oxidation of cytochrome a. These reactions would thus be placed at the beginning and end of the sequence, respectively.

More direct evidence has come from use of different and selective inhibitors of steps in the reaction sequence. It is expected that those components on the oxygen side of the inhibitor block will become more oxidized because they are unable to receive electrons from carriers earlier in the chain. The carriers on the substrate, or NAD^+, side of the inhibition, earlier in the sequence, will become more reduced since no electron acceptor is available past the blocked point. For example, when the antibiotic antimycin A is introduced into an active suspension of mitochondria, pyridine nucleotides, flavoprotein, and cytochrome b all remain reduced, while cytochromes c and a are fully oxidized. These results indicate that cytochromes c and a are located on the oxygen side of the transport block. When cyanide is added to active mitochondria, all the carriers remain reduced. Since cyanide is known to inhibit cytochrome a_3, its action shows that cytochrome a_3 is the terminal com-

Table 7.1 Decline in free energy during electron flow down the respiratory chain

ELECTRON CARRIER (red/ox)	E_0' (volt)	$\Delta G^{0'}$ (kcal/mole)
$NADH/NAD^+ + H^+$	−0.32	−52.6
Flavoprotein red/ox	−0.12	−43.4
CoQ red/ox	+0.10	−33.2
2 cytochrome b red/ox	+0.05	−35.6
2 cytochrome c_1 red/ox	+0.22	−27.8
2 cytochrome c red/ox	+0.25	−26.2
2 cytochrome a-a_3 red/ox	+0.28	−25.0
Water/oxygen	+0.82	0

Porphin

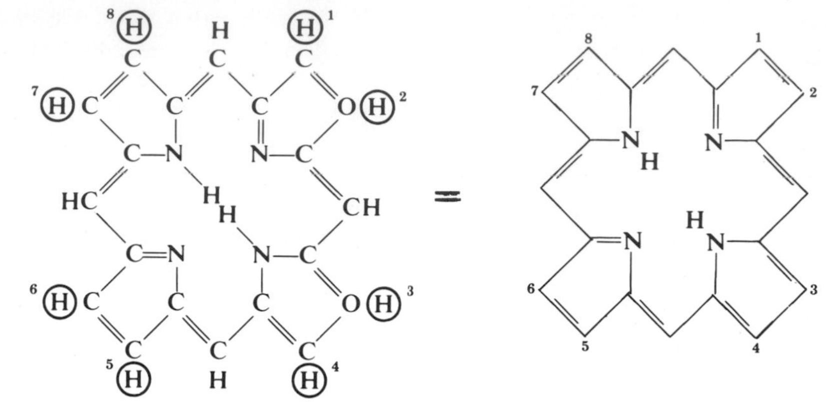

= (Shorthand notation)

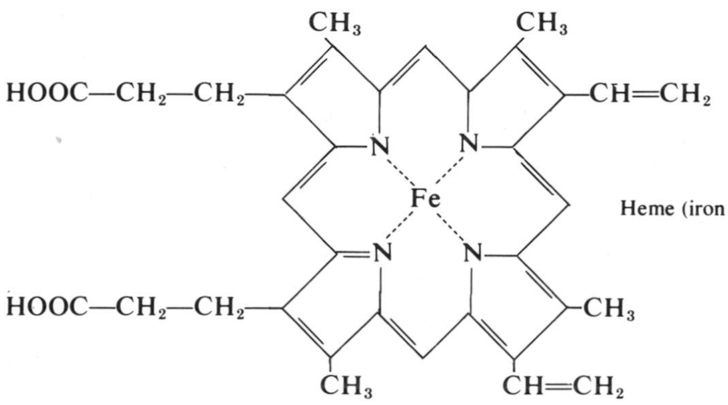

Heme (iron chelate of protoporphyrin IX)

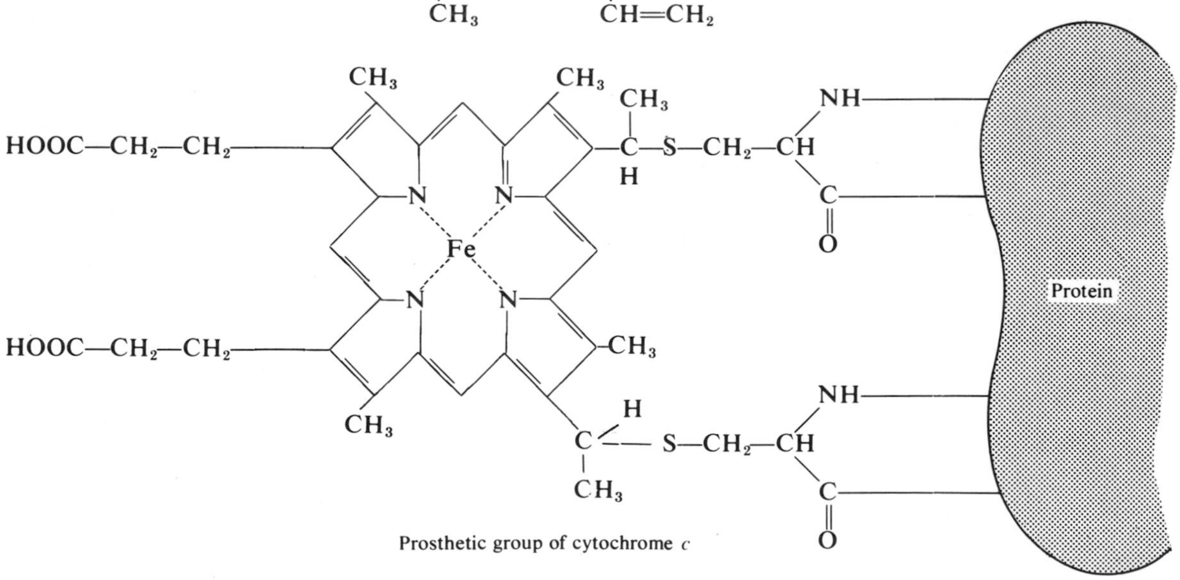

Prosthetic group of cytochrome *c*

Protein

ponent of the electron transport chain. Through these and other lines of experimental evidence, many of the components have been located specifically in the sequence. Other carriers, however, have not yet been located precisely.

The electron carriers occur in fixed numbers and relative proportions within the mitochondrial inner membrane. It is believed they exist in **respiratory assemblies** because of their strict proportionality and because of the nature of their interactions. Since there are no intermediates formed during electron transport there must be some physical arrangement or mechanism that brings adjacent reactive groups into apposition for interactions to be possible. This could be accomplished more readily if the enzyme carriers existed in a common complex of molecules, or a respiratory assembly.

Free-energy change calculations show that the passage of one pair of electrons from NADH to oxygen is accompanied by a sufficiently large decline in free energy so that several molecules of ATP could be formed from ADP, provided a mechanism exists that *couples* electron transport to phosphorylation of ADP. There are three sites along the respiratory chain where a large enough change in free energy occurs (at least 9–10 kcal) to make ATP formation possible: between NAD^+ and the flavin group of NADH-dehydrogenase, between cytochromes b and c, and between cytochrome a and molecular oxygen.

ATP SYNTHESIS DURING ELECTRON TRANSPORT. By the late 1930s there was enough evidence to postulate that the phosphorylation of ADP was coupled to respiration, and functioned in recovery of energy from aerobic oxidations. Between 1948 and 1951 important

Figure 7.15
Molecular formulae showing the nature of the porphin parent compound; the iron-chelate of a protoporphyrin, or heme; and the particular heme that functions as the tightly bound prosthetic group of cytochrome c.

new information was obtained by Lehninger and others who conducted biochemical studies using isolated mitochondria and tissue samples. They showed in several ways that:

1. Phosphorylation was an oxygen-dependent process.
2. Isolated mitochondria catalyzed this oxygen-dependent phosphorylation in association with oxidations of intermediates of the Krebs cycle.
3. When pure NADH was added to broken mitochondria as the only outside substrate, it was rapidly oxidized to NAD^+ at the expense of molecular oxygen by way of the respiratory chain.

These studies thus demonstrated the occurrence of the interconnected systems of aerobic respiration in the mitochondrial compartment of the cell.

Phosphorylation of ADP to form ATP occurs in other metabolic pathways, such as glycolysis. The particular process of ATP formation that is coupled to transport of electrons along the cytochrome chain toward molecular oxygen is generally referred to as **oxidative phosphorylation.** A somewhat similar process in which electrons pass from a large negative (reducing) potential to a positive (oxidizing) potential via cytochromes and other electron carriers, conserving the change in free energy in ATP formed by phosphorylation of ADP, also occurs in the chloroplast in a strictly light-dependent process called **photophosphorylation** (see Chapter 8).

A customary expression of the ATP yield from oxidative phosphorylations is the **P/O ratio,** expressed as the moles of inorganic phosphate recovered in organic form per oxygen atom consumed. Each oxygen atom taken up represents one pair of electrons transported along the carrier chain to its terminus. In assays utilizing various Krebs cycle intermediates as substrates, an average P/O value of 2.0 or 3.0 is obtained (Table 7.2). This average indicates that two or three molecules of ATP are formed

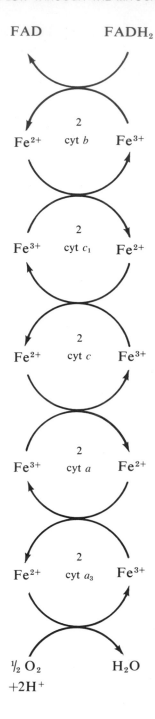

Table 7.2 P/O ratios of oxidations in the Krebs cycle

REACTION	P/O
Pyruvate → acetyl CoA	3
Isocitrate → α-ketoglutarate	3
α-ketoglutarate → succinate	4*
Succinate → fumarate	2
Malate → oxaloacetate	3

* One of these is a substrate-level phosphorylation.

for a particular substrate under study. A P/O ratio of 3.0 is obtained when electrons from NADH are transported to oxygen but the P/O ratio is only 2.0 when $FADH_2$ electron transport takes place.

Any one of the substrates involved in the five oxidative steps of the Krebs cycle can undergo oxidation that leads to reduction of NAD^+. The NADH formed in this reaction will be transported along the cytochrome chain and provide the electrons whose energy is conserved in the formation of ATP during oxidative phosphorylations (Fig. 7.17). If all the five oxidative steps are determined, the total P/O ratio is 15.0 for complete oxidation of pyruvate (the source of original substrate in the form of acetyl CoA that is processed in the Krebs cycle). Of these 15 molecules of ATP, 14 are formed during electron transport and 1 ATP by another kind of phosphorylation at the α-ketoglutarate step. This represents a conservation of about 40 percent of the free energy of pyruvate oxidation.

Oxidative phosphorylation of ADP can be dissociated from respiration by specific chemical **uncoupling agents,** such as 2,4-dinitrophenol. Respiration and electron transport continue normally, or may even be stimulated, but there is no coupled formation of ATP from phosphorylation of ADP in

Figure 7.16
Diagram showing the pathway of electron transfer to O_2 along the cytochromes of the electron transport chain, emphasizing the valency changes of the iron atom in cytochromes between oxidized Fe^{3+} and reduced Fe^{2+}.

these cases. The effect is specific since 2,4-dinitrophenol has no effect on other phosphorylation reactions, such as those in glycolysis or other pathways.

COUPLING MECHANISMS. Despite intensive investigation very little is known about the enzymic mechanisms of oxidative phosphorylation. Various kinds of inhibitors have been exploited in determining the nature of the coupling phenomenon and mechanisms by which the reactions of electron transport underwrite coupling. Since uncoupling agents like 2,4-

dinitrophenol only prevent phosphorylation and do not inhibit respiration, it is clear that the two sets of reactions are completely separable but coupled in some manner in the intact respiring cell. The uncoupled system continues to transport electrons along the respiratory chain but energy is lost in the form of heat instead of conserved in ATP. Some agents, such as the antibiotics oligomycin and rutamycin, inhibit both phosphorylation and electron transport. When 2,4-dinitrophenol is added to such an inhibited system, electron transport resumes, but oxidative phosphorylation remains inhibited. This observation

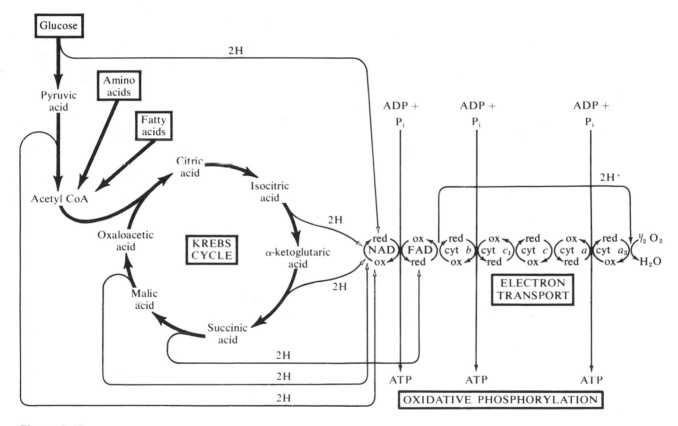

Figure 7.17
Summary of the steps by which glucose is oxidized to CO_2 and H_2O. A substantial part of the free energy content of glucose is conserved in ATP formed during oxidative phosphorylation coupled to electron transport toward molecular oxygen.

suggests that oligomycin and rutamycin act primarily on the energy-coupling mechanism and not on the electron carriers, and these agents can thus be useful probes in studies of the coupling components.

Three major hypotheses that attempt to explain the mechanism by which the reactions of electron transport are coupled to oxidative phosphorylation have been developed. Each one seeks the means by which energy conservation during electron transport will explain three kinds of observed events: (1) ADP phosphorylation, (2) accumulation of cations in the mitochondrial matrix against an osmotic gradient during ATP formation, and (3) structural changes in the inner membrane of the mitochondrion as phosphorylation proceeds. All these events depend upon a functional electron-carrier system and should therefore be understandable in terms of a single coupling mechanism.

The **chemical coupling mechanism** is associated with E. C. Slater, a Dutch biochemist, who presented the idea in 1953. This mechanism is based on the concept of energy transfer by a series of high-energy phosphorylated intermediates that are formed in addition to ATP itself. These intermediates, formed at phosphorylative sites along the respiratory chain, would be collected by an enzyme and transferred to ADP to form ATP (Fig. 7.18). The intermediates would thus be regenerated to continue phosphorylations in repeated rounds and might, therefore, be expected to occur only in trace amounts.

There are two principal difficulties that hinder wider acceptance of the chemical-coupling hypothesis: (1) The postulated high-energy intermediates have never been detected, much less isolated. (2) The mechanism ignores the observation that a reasonably intact membrane or segment of a membrane is required for all of the three kinds of events to proceed: oxidative phosphorylation, cation transport against a gradient during oxidative phosphorylation, and changes in membrane configuration during phosphorylation.

An attractive feature of the proposal is that it includes only well-characterized reactions known to occur in various metabolic pathways and can therefore be assimilated rather easily into the broad framework of conventional metabolism. Although chemical coupling can explain many observed activities, other mechanisms can explain these same observations just as easily. Such ambivalence leads some investigators to accept the hypothesis and others to reject it with equal enthusiasm.

The **chemiosmotic** or **electrochemical mechanism** has been developed most extensively since 1961 by Peter Mitchell, a British biochemist. The hypothesis requires an intact mitochondrial membrane but there is no common high-energy intermediate as proposed in the chemical-coupling mechanism. The chemiosmotic coupling hypothesis states that energy released during transport of electrons along the carrier chain is conserved in a hydrogen-ion and electrical gradient, which then drives the oxidative phosphorylation of ADP to form ATP. The development of such a H^+ ion gradient forms the basis for Mitchell's hypothesis and leads to predictions of particular observations that can verify the hypothesis.

A central postulate is that the mitochondrial inner membrane is selectively impermeable to H^+ and OH^- ions, so that conditions inside and outside the inner compartment of the mitochondrion lead to pH differences that constitute an energy-rich gradient across the membrane. Each pair of electrons carried along the respiratory chain produces three pairs of H^+ ions, which are ejected to the outside of the membrane via enzymatic transfers. The hypothesis states that the enzymes involved in uptake or production of H^+ ions are so precisely positioned in the plane of the membrane that H^+ ions can only be extracted from the lumen and lost only to the outside of this compartment of the mitochondrion (Fig. 7.19). The energy of electron transport is thus conserved in the energy-rich state of a H^+ ion gradient. The pH differences due to high OH^- concentration inside and high H^+ concentration outside the membrane maintain the electrical gradient in a steady

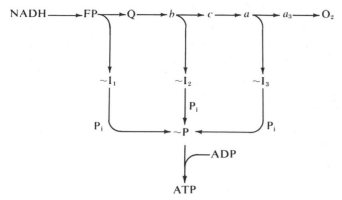

Figure 7.18
Summary diagram describing the postulated chemical coupling mechanism in which electrons transported from NADH to flavoprotein (FP) and cytochromes b, c, a, and a_3 via the intermediate Q are coupled to ATP formation through transfer of energy via intermediates (I). The squiggle mark ($\sim$) denotes a high free-energy level in the residue.

state dependent on selective impermeability of the membrane to H^+ and OH^- ions and the "pumping" action of electron transport that removes H^+ ions from the inner compartment.

The energy-rich H^+ gradient drives the synthesis of ATP from ADP and inorganic phosphate (P_i) by leading to removal of H^+ and OH^- ions of the H_2O (HOH) also formed during this dehydration reaction

$$ADP + P_i \rightleftharpoons ATP + H_2O \text{ [removed]} \qquad [7.5]$$

In effect, an ATP/ADP equilibrium exists. The chemiosmotic mechanism proposes that this equilibrium is shifted toward ATP formation because the hydrolysis of ATP leads to an accumulation of ADP, and the energy-rich H^+ gradient promotes the removal of HOH as H^+ and OH^- ions.

The chemiosmotic coupling hypothesis thus states that electron transport activities lead to the conservation of energy in the energy-rich pH and electrical gradients across membranes and that these gradients provide the energy required to generate

ATP by oxidative phosphorylation of ADP. The synthesis of ATP proceeds because the equilibrium is shifted in this direction as the H^+ and OH^- ions of water are removed as fast as they are formed in the accompanying dehydration of the ADP molecules. The reverse reaction of the equilibrium leads to an accumulation of ADP when ATP is hydrolyzed, thus continuing the cycle. A question has been raised about the reversal of the ATP/ADP equilibrium which energetically favors ADP. It is thought that there may not be an adequate potential to reverse the equilibrium toward ATP synthesis. Whether the underlying calculations of the equilibrium energetics are valid or not remains to be determined. The experimental evidence supporting the chemiosmotic coupling mechanism can also be explained in most instances on the basis of chemical-coupling as well.

D. Green has proposed the most recent coupling mechanism which is based on energy-linked changes in the conformation of the inner membrane of the mitochondrion, as measured optically by differences in light-scattering. The **conformational coupling mechanism** postulates that the macromolecular components of the mitochondrial inner membrane exist in at least two conformational states, with a sufficient free-energy change to account for the driving force of oxidative phosphorylation. The energy of electron transport is directly converted to an energy-rich conformational state of the inner membrane, and when this conformational energy is converted to chemical energy, oxidative phosphorylation occurs. In addition to optical measurements showing energy-dependent changes in mitochondrial volume which occur very rapidly, electron microscopical evidence shows mitochondria have two different ultrastructural states (Fig. 7.20). Hackenbrock has shown that mitochondria exist in the **orthodox state**, which is similar to the usual profiles seen in cell sections when the mitochondria respire in the absence of ADP. They assume the **condensed state** when actively phosphorylating ADP to produce ATP. A profound modification of the inner membrane arrangement and a highly

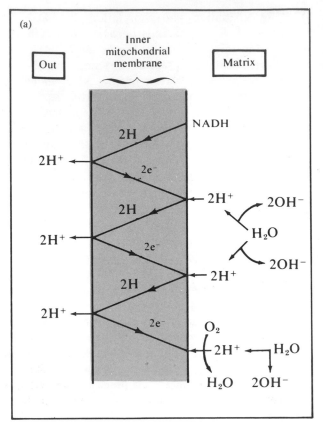

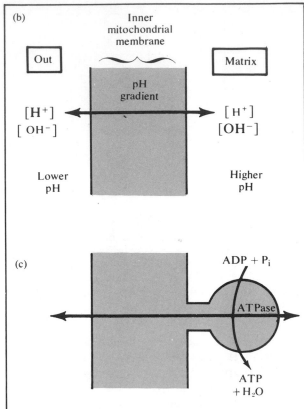

Figure 7.19
Summary diagram of the chemiosmotic coupling mechanism as proposed by Mitchell: (a) Transfer of energy during electron transport leads to a higher outside H^+ ion concentration; (b) an energy-rich pH gradient is thereby produced across the membrane; and (c) this gradient energy drives the oxidative phosphorylation of ADP to form ATP, the form in which free energy of the system is conserved.

condensed inner compartment are seen in mitochondria respiring at a maximal rate.

The three alternative coupling mechanisms are similar in many of their explanations of energy conservation during electron transport. They differ in their postulates about the primary form in which oxidation-reduction energy is conserved as electrons move from NADH to molecular oxygen (Fig. 7.21). The chemical-coupling hypothesis postulates that the transformation of energy by the coupling enzymes directs the **chemical work** of oxidative phosphoryla-tion. The chemiosmotic mechanism stipulates that the primary transformation guides the **osmotic work** needed to accumulate ions. The conformational-coupling mechanism is based on the primary transformation of conserved energy, which subsidizes the **mechanical work** of membrane conformational change. It is still uncertain which of these energy transformations actually is primary.

COUPLING FACTORS AND MEMBRANE ULTRASTRUC-TURE. The usual sectioned preparations reveal dif-

Orthodox conformation

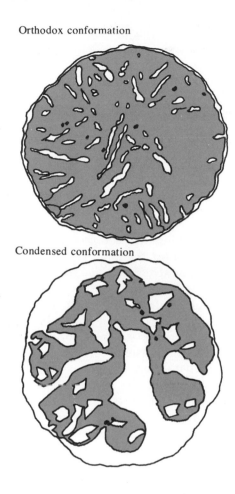

Condensed conformation

Figure 7.20
Drawings from electron micrographs of thin sections showing two conformational states of the mitochondrial inner membrane.

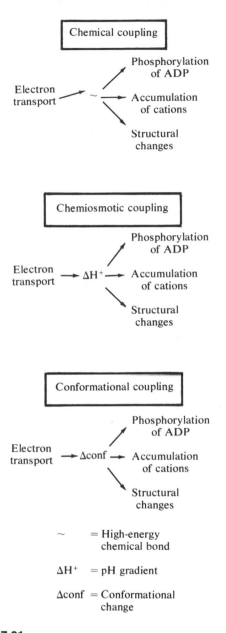

~ = High-energy chemical bond

ΔH^+ = pH gradient

Δconf = Conformational change

Figure 7.21
Comparison of the similarities and differences among the three postulated coupling mechanisms.

ferent patterns of cristae in different cell types. Regardless of the particular variation in crista form, the photographs usually show the mitochondrial inner compartment in its orthodox conformational state. The cristae may be arranged transversely which is the commonest pattern, parallel to the long axis of the mitochondrial profile, concentrically, or in the form of tubular, branching invaginations of the membrane (Fig. 7.22). Ciliated protozoa generally display tubular mitochondrial cristae.

A correlation is usually observed between cells with high respiratory activity and cells with the greatest amount of infolding of the inner membrane. Liver cells or yeast thus contain relatively few and irregular cristae, whereas heart mitochondria have so many invaginations they almost fill the visible profile area. These inward folds provide increased

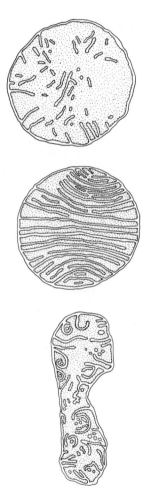

Figure 7.22
Drawings of mitochondrial profiles from thin sections showing various patterns of cristae arrangement.

surface area of inner membrane, where the enzymes of electron transport and oxidative phosphorylation are situated. In mutant yeast cells which lack some of the cytochromes and thus cannot respire aerobically, the cristae may be distorted or irregular or lacking altogether (Fig. 7.23). The inner membrane is retained, however, even in mutant strains completely devoid of mitochondrial DNA and respiratory activity. In these cases the inner and outer membrane envelopes run parallel to each other at all points along their surface area.

The outer mitochondrial membrane never infolds and is thus different from the inner membrane in this respect in all cells. In addition to differences in surface conformation, the outer and inner membranes contain different structural and functional subunit particles. In 1963–1964, D. F. Parsons and H. Fernández-Morán independently demonstrated the presence of previously undetected outer membrane subunits and inner membrane subunits in preparations of animal mitochondria. The mitochondria were isolated from cells and then broken into small fragments of membranes. Instead of staining the membranes directly, a method of negative staining was used. In this kind of preparation, the stain, usually a solution of phosphotungstate, is not adsorbed or absorbed by the biological materials so that they stand out in light contrast against the dark background of phosphotungstate stain. When viewed edge-on after negative staining, the mitochondrial membranes bristled with projecting subunits (Fig. 7.24). It is virtually certain that the staining treatment has some traumatic effect on the membranes and leads to the protrusion of particles otherwise usually embedded further within the thickness of the membrane.

Much more attention has been directed toward the **inner membrane subunits,** also called "elementary particles," because of great interest in determining the relationship between subunits and functional enzyme assemblies of electron transport and oxidative phosphorylation known to be located within the inner membrane. The inner membrane subunits occur

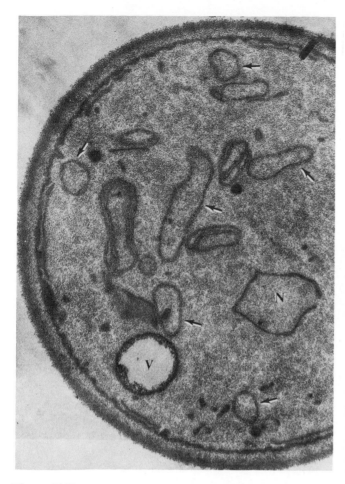

Figure 7.23
Thin section through a cell of respiration-deficient yeast. The inner membrane is infolded in several of the mitochondrial profiles but not in conventional cristae typical of normal yeast. Other profiles (at arrows) lack visible infolds of the mitochondrial inner membrane. A small section of the nucleus (N) and a vacuole (V) are also present. × 38,650. (Photograph by M. Federman)

remains within the membrane even as the rest of the particle is extruded. According to detailed measurements reported by Fernández-Morán, there were 2000–4000 subunits per μm^2 of inner membrane surface in beef heart mitochondria.

The first studies converged on the possible presence of electron transport components in these subunits. Lively controversies existed as different investigators produced evidence for and against the presence of electron transport cytochromes in these

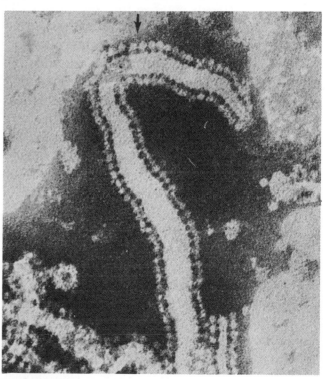

Figure 7.24
Negatively stained portion of cristae of the adult bee isolated mitochondria. The inner membrane subunits (arrow) line the surface of the crista bordering the mitochondrial matrix. Subunits are not found on the opposite surface of the membrane bordering the intracristal space. × 264,500. (Reproduced with permission from Chance, B., and D. F. Parsons, *Science* **142**:1176–1180, Fig. 5, 29 November 1963. Copyright © 1963 by the American Association for the Advancement of Science.)

only along the inner surface of the membrane, facing the matrix of the mitochondrion. The particle consists of a spherical headpiece attached by a stalk to the membrane inner surface. It is uncertain whether or not the stalk, in turn, is attached to a basepiece that particles. The problem was resolved by careful and convincing studies conducted by Efraim Racker and colleagues beginning in the mid-1960s. They showed that the inner-membrane-subunit headpiece was equivalent to a **coupling factor** (F_1), one of the proteins active in promoting oxidative phosphorylation. The subunit had no electron transporting activity.

Racker's experiments took advantage of the fact that isolated mitochondria could be broken into membrane fragments on exposure to high-frequency sound waves. The fragments obtained from these physically disrupted organelles were fully capable of carrying out electron transport coupled to oxidative phosphorylation of ADP to form ATP. The submitochondrial "particles" are actually closed vesicles formed from membrane fragments whose edges have resealed following mitochondrial disruption. These vesicles provided a convenient system for study, and they also were useful in ultimately isolating and purifying the coupling factor proteins. The experiments were designed so that actively functioning inner membrane fragments could be modified and reconstructed, using biochemical and electron microscopical methods to monitor the system.

When submitochondrial particles are treated with agents that strip the protein coupling factors, the membranous vesicles retain electron transport capacity but can no longer carry out ATP synthesis. When purified coupling factors are added to such stripped membranes, oxidative phosphorylation is restored and coupled to electron transport, although with reduced efficiency. When these modified and reconstructed membranes were examined by electron microscopy, it was clear that stripped membranes lacked inner membrane subunits, whereas restored membranes were peppered with these subunits (Fig. 7.25).

In addition to the evidence demonstrating that inner membrane subunits were equivalents of oxidative phosphorylation and not of electron transport, electron micrographs of purified F_1 protein revealed that this coupling factor existed as a spherical particle

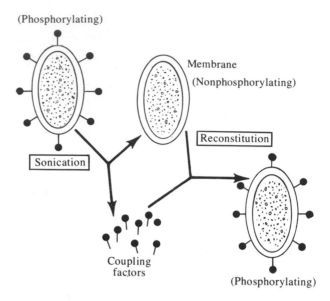

Figure 7.25
Diagrammatic representation of Racker's reconstruction experiment that showed oxidative phosphorylation was a function of the inner membrane subunit of the mitochondrial inner membrane. See text for details.

measuring about 80 Å in diameter. This size was exactly the one predicted if the F_1 factor and the headpiece of the inner membrane subunit were one and the same entity.

The F_1 coupling factor has a molecular weight of about 280,000 daltons and functionally is an ATPase. The purified factor catalyzes the hydrolysis of ATP. When the factor is incorporated into the mitochondrial membrane along with other factors, the enzyme catalyzes synthesis of ATP coupled to electron transport. The purified F_1 preparation is not inhibited when oligomycin is added to the suspension, but the ATPase activity of the factor is inhibited by the drug when intact membrane fragments are assayed. Another coupling factor, F_0, appears to be the component that is oligomycin-sensitive, since F_1 and F_0 together are sensitive to oligomycin in purified preparations. It has been suggested that the stalk portion of the inner membrane subunit may harbor, or even be, this F_0 oligomycin-sensitivity-conferring component.

The physical association of oxidative phosphorylation and electron transport components within the inner mitochondrial membrane structure must contribute to their functional interactions. We have not yet determined the nature of their spatial relationships, but their functional coordination is clearly responsible for the coupled reactions in which energy of electrons derived from substrate oxidations is conserved in phosphate-bond energy in ATP molecules.

Overall Energetics of Glucose Oxidation

If glucose is burned completely to CO_2 and H_2O in air under standard conditions, then the standard free-energy change between starting materials and end-products can be shown as

$$C_6H_{12}O_6 + 6\ O_2 \rightarrow 6\ CO_2 + 6\ H_2O$$
$$\Delta G^{0\prime} = -686\ \text{kcal/mole} \qquad [7.6]$$

The significant difference in terms of energy change in the cell is that some of the energy of glucose is conserved in ATP and only a part of the potential of 686 kilocalories is dissipated as heat. There is an input of about 7 kcal of energy per mole of ATP that is formed

$$ADP + HPO_4^{2-} \rightarrow ATP + H_2O$$
$$\Delta G^{0\prime} = -7\ \text{kcal/mole} \qquad [7.7]$$

Since 38 ATP are formed during oxidation of glucose in glycolysis, pyruvate oxidation to acetyl CoA, and transfer of electrons from NADH and $FADH_2$ to oxygen, we can calculate the efficiency of energy conservation (Fig. 7.26). Energy conserved in the form of ATP is 266 kcal/mole of glucose oxidized to completion (38 ATP × 7 kcal/mole of ATP formed = 266 kcal/mole of glucose oxidized). On this basis, the efficiency of conservation is 266/686 × 100 = 39 percent. The remaining 420 kcal

is dissipated as heat. The overall equation for oxidation of glucose in which 38 molecules of ATP are accumulated is shown as

$$C_6H_{12}O_6 + 6\ O_2 + 38\ ADP + 38\ HPO_4^{2-} \rightarrow$$
$$6\ CO_2 + 44\ H_2O + 38\ ATP \qquad [7.8]$$
$$\Delta G^{0\prime} = -420\ \text{kcal/mole}$$

The difference in free-energy change between the starting and end products of reactions is lower in equation [7.8] than in [7.6] because of ATP formation. The free-energy change is still a large negative value, which indicates that the reactions proceed "downhill," but 39 percent of the energy potential of glucose has been retained for cellular work in the form of ATP. The free-energy change of −420 kcal/mole in equation [7.8] is obviously derived by arithmetic (686 − 266 = 420).

Other Cellular Oxidations

Although glucose is the major fuel in most cells, energy can also be derived from other metabolites. Glucose can be degraded initially through glycolysis or other pathways. We will briefly consider some of the alternative pathways that are important to cells under certain growth conditions and some that enhance the flexibility of cells in deriving energy and building block molecules for metabolism.

PHOSPHOGLUCONATE PATHWAY. The phosphogluconate sequence provides an alternative for glucose degradation, but it is not a mainline pathway for glucose oxidation. Like glycolytic reactions, the phosphogluconate pathway enzymes also are situated in the cytosol and are not compartmented within the mitochondrion or other organelles. This pathway has been described by several synonyms: **pentose phosphate pathway, hexose monophosphate shunt,** and **pentose shunt.** As the term "shunt" implies, the glycolytic route and the Krebs cycle are bypassed when glucose oxidation is shunted into alternative paths. The primary function of the phosphogluconate

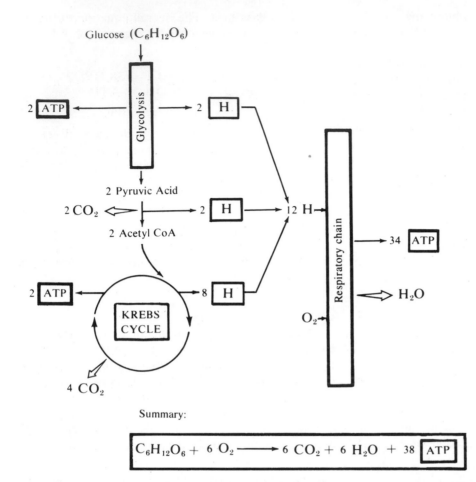

Glucose ($C_6H_{12}O_6$)

2 ATP

Glycolysis

2 H

2 Pyruvic Acid

$2 CO_2$

2 Acetyl CoA

2 ATP

KREBS CYCLE

8 H

$4 CO_2$

2 H

12 H

Respiratory chain

34 ATP

H_2O

O_2

Summary:

$$C_6H_{12}O_6 + 6 O_2 \longrightarrow 6 CO_2 + 6 H_2O + 38 \;\text{ATP}$$

Figure 7.26
Summary of the major events during the oxidation of glucose to CO_2 and H_2O.

pathway is the generation of reducing power in the cytosol, in the form of reduced $NADP^+$. This electron acceptor was discovered by O. Warburg after the pathway had been explored. A secondary function of the pathway is pentose sugar production, especially D-ribose for synthesis of nucleotides and nucleic acids.

In the first reaction, glucose 6-phosphate is oxidized to 6-phosphogluconate by **glucose 6-phosphate dehydrogenase** action (Fig. 7.27). Next, 6-phosphogluconate is oxidatively decarboxylated to form D-ribulose 5-phosphate in the presence of **6-phos-**

phogluconate dehydrogenase. Each of these two reactions also generates one molecule of NADPH, which acts in electron-transferring oxidation-reduction reactions in the cytosol in conjunction with $NADP^+$-linked enzymes. Further reactions lead to D-ribose 5-phosphate, a precursor of RNA and nucleotide synthesis. The phosphogluconate pathway may rejoin the glycolytic sequence at various points in the oxidation steps, or it can lead to more complex oxidation of glucose 6-phosphate. In photosynthetic cells, a reverse phosphogluconate pathway functions during the formation of glucose from carbon dioxide

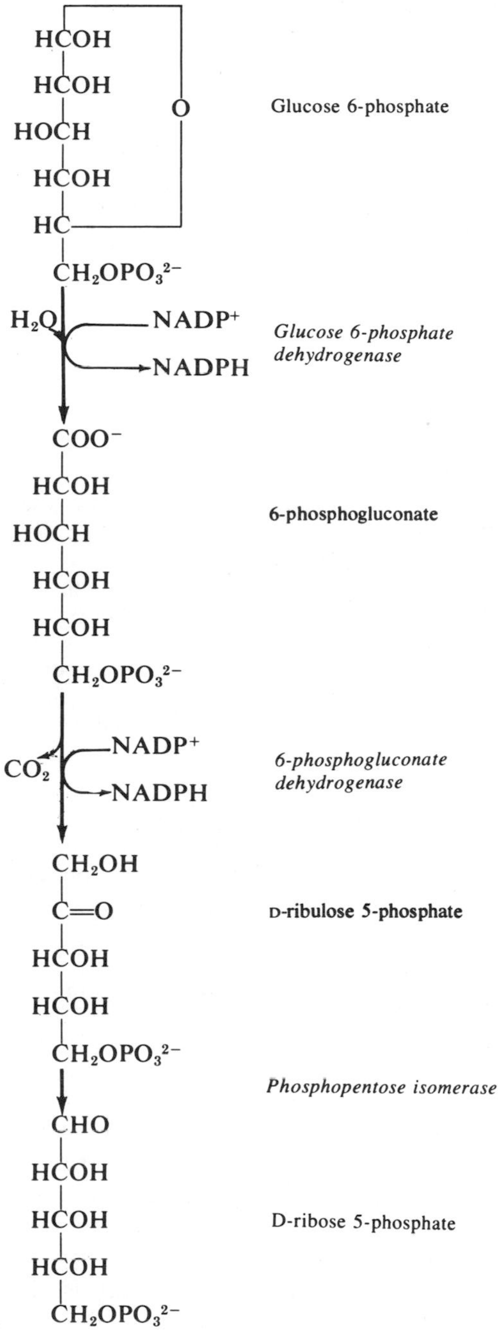

HCOH
HCOH
HOCH ⟩O   Glucose 6-phosphate
HCOH
HC
$CH_2OPO_3^{2-}$

H_2O ⟍ $NADP^+$ *Glucose 6-phosphate*
⟋ $NADPH$ *dehydrogenase*

COO^-
HCOH
HOCH   **6-phosphogluconate**
HCOH
HCOH
$CH_2OPO_3^{2-}$

CO_2 ⟍ $NADP^+$ *6-phosphogluconate*
⟋ $NADPH$ *dehydrogenase*

CH_2OH
$C=O$ D-**ribulose 5-phosphate**
HCOH
HCOH
$CH_2OPO_3^{2-}$

    Phosphopentose isomerase

CHO
HCOH
HCOH   D-ribose 5-phosphate
HCOH
$CH_2OPO_3^{2-}$

in the so-called "dark" reactions associated with the photosynthetic process (see Chapter 8).

OXIDATION OF FATTY ACIDS. Fatty acids provide an important source of energy in organisms that are capable of storing neutral fat, but glucose is the major fuel in most organisms, whether or not fat is stored. Storage fat droplets in the cytoplasm and phospholipids from membranes are sources of fatty acids after the large lipid molecules have been degraded by **lipases** to their constituent free fatty acids during metabolic turnover in many kinds of cells. Fatty acid utilization apparently governs the pathway and rate of intracellular lipid metabolism.

Earlier studies showed that fatty acids were synthesized and degraded by addition and subtraction, respectively, of two-carbon fragments. Oxidation of fatty acids was shown to require molecular oxygen and ATP, and in 1948–1949 Lehninger showed that these oxidative reactions occurred exclusively in the mitochondria. We now know that the fatty acid oxidation system also occurs in a microbodylike organelle (glyoxysome) during the brief time of germination of fat-storing seeds such as those of castor bean, peanut, and others (see Chapter 10).

Later studies by F. Lynen, D. Green, S. Ochoa, and others led to clarification of the overall pathway of fatty acid oxidation. First, long-chain fatty acids are activated by an energy-requiring step in which the fatty acid ester of coenzyme A is formed at the expense of ATP. Second, the fatty acids enter the mitochondrion in ester form, where they undergo a series of successive dismantlings within the matrix.

At each passage through a specified set of reactions, the fatty acid chain loses a two-carbon fragment in the form of acetyl CoA and two pairs of hydrogen atoms to specified acceptors. The shortened

Figure 7.27
The phosphogluconate pathway, a set of reactions that bypasses the glycolytic and Krebs cycle routes for glucose oxidation in the cell.

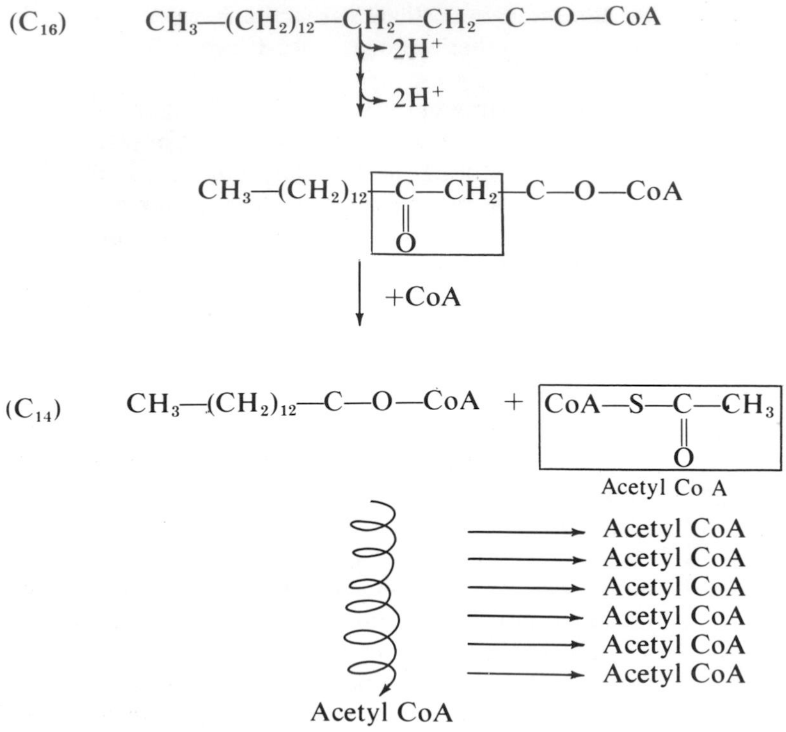

(C_{16}) $CH_3—(CH_2)_{12}—CH_2—CH_2—C—O—CoA$

$$\downarrow 2H^+$$
$$\downarrow 2H^+$$

$CH_3—(CH_2)_{12}\boxed{—C—CH_2—}C—O—CoA$
            $\|$
            O

$$\downarrow +CoA$$

(C_{14}) $CH_3—(CH_2)_{12}—C—O—CoA + \boxed{CoA—S—C—CH_3}$
                                              $\|$
                                              O

Acetyl Co A

⟶ Acetyl CoA
⟶ Acetyl CoA
⟶ Acetyl CoA
⟶ Acetyl CoA
⟶ Acetyl CoA
⟶ Acetyl CoA

Acetyl CoA

Figure 7.28
Schematic summary of the processing of palmitic acid in β-oxidation through fatty acyl CoA to the final products of 8 molecules of acetyl CoA. The first steps occur once in each β-oxidation cycle, but removal of each of the 8 acetyl CoA fragments occurs in repeated steps (indicated by the spiral) exactly like the one that renders the C_{16} molecule to a C_{14} intermediate.

chain continues to be processed in this way until it has been completely rendered. In the case of palmitic acid with 16 carbon atoms, the molecule is processed during seven such cycles to yield 8 molecules of acetyl CoA and 14 pairs of hydrogen atoms (Fig. 7.28).

The acetyl CoA of this system may then enter the Krebs cycle, while the hydrogen atoms enter the respiratory chain of electron transport in the form of NADH and FADH$_2$. As the electrons are transported from these carriers to molecular oxygen, the expected number of phosphorylations of ADP to form ATP occur. The overall processing of a molecule such as palmityl CoA within the mitochondrion produces 35 molecules of ATP and 8 acetyl CoA. These 8 molecules of acetyl CoA enter the Krebs cycle and lead to the formation of 96 molecules of ATP during oxidation and coupled oxidative phosphorylations. The overall equation would be

$$\text{palmityl CoA} + 23\ O_2 + 131\ HPO_4^{2-} + 131\ ADP$$
$$\downarrow \qquad\qquad [7.9]$$
$$CoA + 16\ CO_2 + 146\ H_2O + 131\ ATP$$

About 40 percent of the free energy of palmitic acid is recovered in the form of phosphate-bond energy.

OXIDATION OF AMINO ACIDS. Although amino acids are most familiar as precursors of proteins and some other biologically important molecules, they are also common sources of energy in cells. Intact proteins and most peptides cannot pass through the cell membrane, so their complete hydrolysis to their constituent amino acids must take place before the amino acids can be absorbed into the cell.

In mammals, the action of the enzyme *pepsin* initiates hydrolysis of proteins and peptides, or **proteolysis.** The enzymes in these species are usually synthesized in the form of inactive precursors called **zymogens,** and they are converted to their active forms afterward. The inactive form of pepsin is *pepsinogen,* which is synthesized in the stomach. The polypeptides formed in the stomach by pepsin action are then exposed to proteolytic enzyme action in the small intestine. Some of these enzymes are secreted in the pancreas as the inactive zymogens which are enclosed within a membrane during passage through the Golgi apparatus. These zymogen granules with their inactive enzyme contents leave the pancreas through the pancreatic duct and enter the small intestine where zymogens such as *trypsinogen* and *chymotrypsinogen* are converted to their active forms, *trypsin* and *chymotrypsin,* respectively.

By combined action of the proteolytic enzymes secreted in the stomach, pancreas, and small intestine, amino acids are formed by complete hydrolysis of ingested proteins. These amino acids are absorbed by epithelial cells lining the small intestine through an active transport process and are circulated to all tissues through the bloodstream. They are absorbed by individual cells by the same active transport processes, and then enter metabolic pathways within the cells. The active transport of amino acids and ions across membranes will be discussed shortly (see also Chapter 4).

Each of the twenty commonly-occurring amino acids is oxidized by a different multienzyme sequence, but all the sequences ultimately converge into a few terminal pathways leading to the Krebs cycle within the mitochondrion. Depending on the particular amino acid, the molecule is converted to acetyl CoA, α-ketoglutarate, succinyl CoA, or oxaloacetate, and undergoes the usual oxidations in Krebs cycle reactions.

MITOCHONDRIAL COMPARTMENTATION

We have been discussing metabolic reactions and energy production which take place in different regions of the mitochondrion and in different parts of the cell. Reaction pathways can proceed most efficiently if they are separated from competing activities occurring elsewhere, but the overall efficiency of cellular activity must involve mechanisms by which the separate pathways intermesh. Regulation and coordination of potential and actual metabolism is enormously complex, but considerable progress has been made toward defining particular mechanisms that contribute to a harmonious moment-to-moment flux of activities in the dynamic open system of the cell.

A substantial component underlying orderly interaction among compartments within an organelle, or between it and the rest of the cell, or between the cell and its environment is the selective permeability of membranes. Compartmentation within the cell provides a means for localizing high concentrations of reactants, but movement of reactants, end-products, and energy-carrying agents is dependent on the properties of membranes and their structural and functional constituents. Some of these qualities were discussed earlier in Chapter 4.

Communication By Carrier Systems

The two membranes of the mitochondrion have very different permeability properties. The outer membrane essentially is freely permeable to most solutes of low molecular weight. The inner membrane is permeable only to water, short-chain fatty acids, and a few uncharged molecules such as glycerol. It is not permeable to cations such as Na^+, K^+, or Mg^{2+}, anions such as Cl^- and NO_3^-, sugars, most amino acids, reduced and oxidized NAD^+ and $NADP^+$, nucleoside $5'$-phosphates (mono-, di-, and tri-), or coenzyme A and its esters. It is obvious from this listing that the inner compartment of the mitochondrion contains an internal pool of coenzymes and nucleosides which are functionally distinct and physically separated from the extramitochondrial, or cytosolic, pools of constituents. It is equally clear that mechanisms must exist by which transfers can be effected between compartments within the cell. Carrier proteins provide one such mechanism.

Transfers of specific metabolites across the inner membrane of the mitochondrion are achieved in conjunction with certain inner membrane protein components called **permeases,** one of the kinds of **carriers.** Permeases are not enzymes although they have many enzymelike qualities, hence the suffix -ase. Only a few carrier proteins have been identified specifically, and many others are assumed to exist because there is some experimental evidence for selective transfers of other substances. Carriers are responsible for transport of certain sugars and amino acids across the membrane, as well as various intermediates of the Krebs cycle. An important carrier system exists in aiding transport of ADP and ATP across the mitochondrial inner membrane; it sponsors entry of one ADP molecule for each ATP that exits.

Taking all this evidence together, it has been generally concluded that ADP, phosphates, fuel molecules, and metabolites must enter the inner compartment through the mitochondrial inner membrane before oxidations can take place within. Krebs cycle oxidations, electron transport, and oxidative phosphorylations take place in the matrix and along the inner surface of the inner membrane, within a common compartment. The ATP formed during these activities must leave this inner compartment to reach the cytoplasm where it is used to perform work. The mitochondrion must, therefore, serve as a distribution center for ATP, as well as the major site of its synthesis. Communication between the pools of ADP and ATP in the mitochondrion and cytosol is effected via specific carrier systems. These lines of communication participate in extremely complex interchanges of metabolic intermediates and phosphates between the two regions of the cell. The control and integration of various pathways depend both on compartmentation and on communication between compartments, particularly in relation to energy-yielding systems. Biosyntheses also require these coordinations, particularly in reactions involving Krebs cycle intermediates.

One advantage of selective permeability is that molecules can occur locally in high concentrations; another advantage is that molecules may be admitted to a compartment in direct relation to particular needs at particular times in metabolism. In assisted passage, or **transport,** across a membrane, a substance may move in the expected direction of its lower concentration by **facilitated diffusion** or by **active transport** against a gradient in energy-expending processes. Since very few molecules cross the membrane by **free diffusion,** mitochondrial products and reactants generally are involved in one or the other transport process. Carriers are usually highly specific and can thus move particular molecules through a membrane, ignoring the many other molecules found in the same environment. The directionality of movement into or out of a compartment thus serves as a feature of regulation and control over coordinated activities. A unifying concept has been evolved in recent years to explain active-transport system organization in the cell. The action of an active pumping of one substance out of a compartment or cell is thought to provide the driving force for active transport of various other substances into the compartment or cell.

The solutes most actively pumped into cells are sugars, amino acids, and certain ions. The driving force for this inward transport usually is inferred to be an ion gradient across the membrane, created by active transport of the ion to the outside of the cell or organelle. Na^+ ion gradients in animal cells and H^+ ion gradients in bacterial cells have been studied the most. In the case of a Na^+ pump, the ion gradient is increased as Na^+ is pumped against the Na^+

gradient to the outside. The internal Na$^+$ concentration is left low and the external Na$^+$ concentration remains high. The energy required to pump Na$^+$ against the gradient by active transport is provided by ATP which is hydrolyzed by ATPase that is firmly bound to the inside surface of the membrane. More than one ATPase is known to operate, including the well-characterized enzyme of the inner mitochondrial membrane.

While sugars and amino acids are driven into the cell or organelle by existing ion gradients, their entry must be assisted by carrier proteins that form a part of the membrane itself. The active transport of metabolites against their own concentration gradients requires energy, usually provided by hydrolysis of ATP mediated by some ATPase. As metabolites are added to internal pools, they undergo continuing and vigorous oxidations. These reactions lead to ATP formation, and some portion of this ATP is used to derive energy for further active transport and the maintenance of an ion pump.

Metabolites can enter the cell by alternative active transport systems that do not require ion pumps. In bacterial cells, for example, sugars are phosphorylated as they pass through the membrane. Phosphorylation is accomplished by the sugar-carrier system that is part of the plasmalemma structure. Once inside the cell, the charged phosphorylated sugars are unable to escape in this particular form. More sugars continue to be transported into the cell against a concentration gradient, but no ion pump is involved in these events.

Communication by Shuttle Systems

NADH, NADPH, NAD$^+$, and NADP$^+$ cannot penetrate the mitochondrial membrane. The pools of NADH in the cytosol and mitochondrion are therefore segregated, and it might seem that any NADH formed in the cytosol would not be available for oxidation through the mitochondrial electron transport chain. Furthermore, it might seem that reoxidized NAD$^+$ could not be exchanged between compart-

ments and that metabolic efficiency in the whole cell might be affected. While these oxidized and reduced electron-carriers cannot penetrate the membranes directly, they can be indirectly routed among compartments by **shuttles.** One of the better understood systems is the **glycerol phosphate shuttle,** which can serve as an example of the general operation of this type of mechanism for intracellular metabolic interactions. (Fig. 7.29).

NADH in the cytosol first reacts with dihydroxyacetone phosphate, an intermediate of glycolysis. In this reaction, glycerol 3-phosphate is

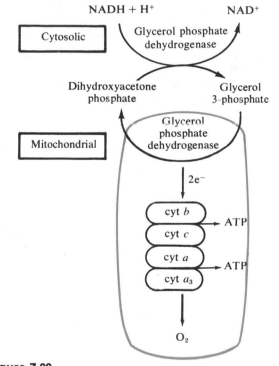

Figure 7.29
Diagram summarizing the transfer of reducing power from NADH in the cytosol to the mitochondrial electron transport chain via reactions of the glycerol phosphate shuttle system.

formed through the mediation of the cytosolic enzyme **glycerol phosphate dehydrogenase,** whose coenzyme NADH is oxidized to NAD^+. The electrons from NADH have been passed on to glycerol 3-phosphate, which can readily penetrate the mitochondrial membrane. NADH reducing power, in the form of its transferred electrons, has been provided to the mitochondrion by this indirect route. Further reactions take place so that glycerol 3-phosphate is oxidized back to dihydroxyacetone phosphate, and the electrons originally derived from cytosolic NADH are finally given to the mitochondrial electron transport chain. They pass along the chain until they are accepted by oxygen, but ATP is formed along the way through coupled oxidative phosphorylations. The dihydroxyacetone phosphate can diffuse out of the mitochondrion back to the cytosol, and accept electrons from another molecule of cytosolic NADH. Electrons are thus passed from one compartment to another via shuttles, in either direction. NADH produced within one compartment can move into another only indirectly. By shuttling metabolites from one place to another, however, the reducing power of NADH can be used in enzyme-catalyzed oxidations on both sides of the membrane barrier.

The Pasteur Effect

When cells use glucose anaerobically by glycolysis or alcoholic fermentation, the rate of glucose breakdown is much faster than under aerobic conditions in the very same cells. Since fermentations yield only 2 ATP from glucose to ethyl alcohol or lactate end products, an additional 36 ATP result when glucose is oxidized completely in Krebs cycle and the respiratory chain reactions. Eighteen times more glucose is required to get the same yield of ATP in glycolysis as in aerobic respiration. Relatively huge amounts of glucose are metabolized anaerobically. As Louis Pasteur first noted in his studies on wine-making, when a suspension of anaerobically-metabolizing cells is aerated, immediate changes in glucose degradation patterns are noted. Oxygen consumption begins immediately, the rate of glucose consumption declines rapidly to a fraction of the anaerobic activity, and formation of lactate or ethyl alcohol stops. The inhibition of glucose uptake and the cessation of the accumulation of end-products of fermentations once respiration begins is called the **Pasteur effect.**

In the case of cells carrying on glycolysis, for example, the introduction of oxygen leads to metabolic changes. In air there is a competition for cytosolic NADH between lactate dehydrogenase of the glycolysis pathway and a shuttle system which funnels electrons from NADH into the mitochondrial respiratory chain. Lactate dehydrogenase loses out because of its lower affinity for NADH. As a result, pyruvate is not reduced in the cytosol to lactate by lactate dehydrogenase action. Instead, pyruvate is "pulled" into the mitochondrion where it is oxidized to acetyl CoA and carbon dioxide. If oxygen is absent, then pyruvate is reduced in the cytosol and lactate accumulates there.

Although there are many levels of primary and secondary control over the integration of glycolysis and respiration, this example shows one way in which cell metabolism is modulated in response to environmental conditions. Another way in which respiration can inhibit glycolysis is through competition for ADP by particular enzymes of the two pathways. Phosphorylations can occur in mitochondria even when ADP is in low concentration because of the relatively high affinity for ADP in that system. If ADP becomes unavailable in the cytosol because it is preferentially channelled to mitochondria, glycolysis would be inhibited because little ADP would remain in the cytosol for phosphorylations.

Interactions involving $NAD^+/NADH$ and ADP/ATP are the most crucial in regulating compartmented metabolism (Fig. 7.30). Together with many other factors, including membrane permeability, carrier and shuttle systems, and the regulating effects of various pacemaker enzymes, the ebb and flow of cellular metabolism continues with little or no trauma.

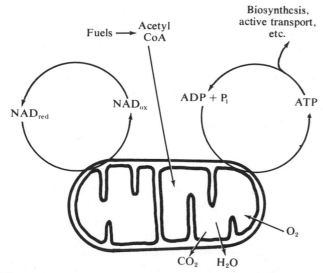

Figure 7.30
Compartmented metabolism involving the mitochondrion is regulated through interactions involving NADH/NAD⁺ and ADP/ATP. Reducing equivalents from NADH power the mitochondrial processing of fuel molecules entering in the form of acetyl CoA, and free energy is conserved in ATP, which then is available for many energy-requiring cellular activities.

The utilization of substrates and the production of essential materials for growth, repair, and reproduction are carried out with relatively high efficiency.

Enzyme Localization

Using suitable methods, Carl Schnaitman and John Greenawalt succeeded in separating the outer membrane of rat liver mitochondria from the remainder of the structure and concentrating this material by centrifugation. The inner compartment was also isolated in a relatively uncontaminated form. With such preparations, each compartment could be analyzed individually to locate mitochondrial enzymes. In addition, the membranes and intermembrane spaces of mitochondria could be assayed for protein and lipid content.

The outer mitochondrial membrane from rat liver cells contains the enzyme *monoamine oxidase*. Once it was established that this enzyme was located exclusively in this membrane, all other preparations could be monitored for relative purity using monoamine oxidase as a marker for outer membrane. Some cell types do not have mitochondrial monoamine oxidase so it is not a universal indicator. Other enzymes have also been found in outer membrane of rat liver mitochondria (Table 7.3).

Other parts of the mitochondrion were also assayed and mapped for enzymes after removal of the outer membrane. When the inner compartment is intact so that undamaged inner membrane surrounds the matrix, various enzymes can be found there. Among these are enzymes of the Krebs cycle and the respiratory chain, ATP synthesizing enzymes, various dehydrogenases, and others as shown in Table 7.3. More precise localizations were made when the inner membrane and matrix were separated and analyzed individually. Many of these enzymes had previously been localized in a more general way. For example, enzymes that "leaked" from broken mitochondria were assumed to be in the unstructured matrix. Enzymes that sedimented with membranes from broken

Table 7.3 Compartmentation of some enzymes in rat liver mitochondria

LOCATION	ENZYMES
Outer membrane	Monoamine oxidase Rotenone-insensitive cytochrome c reductase
Space between membranes	Adenylate kinase Nucleoside diphosphokinase
Inner membrane	ATP-synthesizing enzymes Respiratory chain enzymes Succinate dehydrogenase
Matrix	Krebs cycle enzymes (citrate synthase, isocitrate dehydrogenase, fumarase, malate dehydrogenase, aconitase) Glutamate dehydrogenase Fatty acid oxidation enzymes

mitochondria were assumed to be present in or on the membrane, which explained why they did not "leak" out. In addition to confirming these earlier assumptions, new and specific localizations were made in the 1968 study and in ones reported later.

The distribution is not random. We can assume from this fact that there are correlations between enzyme location and the regulation of enzyme activity, interactions with the nucleus, and interactions with surrounding extramitochondrial compartments in the cytoplasm. Not only have we rejected the naive notion that the cell is a "bag of enzymes," but we must now reject the similar view that an organelle contains a randomized assemblage of structures and catalysts. Cell order is maintained by a regulatory framework which extends out in both space and time and permits maximum interaction, coordination, and efficiency. Characteristics of the mitochondrial genetic apparatus and its interactions with the nuclear genetic system must also be understood in relation to the intricacy and smooth operations of the organelle in the cell. These features will be discussed in some detail in Chapter 9.

BIOGENESIS OF MITOCHONDRIA

Three general hypotheses have been proposed to explain how new mitochondria are formed:

1. *De novo* (formed anew) from submicroscopic precursors in the extramitochondrial compartment.
2. From nonmitochondrial membranes, such as the nuclear or plasma membrane.
3. By growth and division of pre-existing mitochondria.

The first two hypotheses have enjoyed less favor, but the weaker of the two is the *de novo* mode of mitochondrial biogenesis.

Two sets of experiments conducted by David Luck provided almost the only evidence for a growth-and-division process; nevertheless, the idea has gained wide acceptance. Using a mutant of *Neurospora* that required choline for growth, it was possible to follow mitochondrial membrane changes by providing choline labeled with heavy or radioactive isotope markers. Each set of experiments was designed to show that one of the three alternative hypotheses was correct and simultaneously show that the other two were incorrect. In some ways the experiments were designed along the lines of the classic Meselson–Stahl studies of the mode of DNA replication. There were some important differences, however, between Luck's and the other experiments.

The second set of studies that was reported by Luck in 1965 was more easily interpreted than the earlier work. Cultures of choline-requiring *Neurospora* were grown in media containing a particular concentration of choline. At this particular concentration, *dense* mitochondria are formed and they can be identified easily according to their pattern of sedimentation in sucrose density gradients. If *Neurospora* is grown in media with higher choline concentrations, then *light* mitochondria are formed. The light mitochondria can be distinguished from dense mitochondria on the basis of sedimentation in a density gradient. To make sure that choline had been incorporated into the mitochondrial membranes, the heavy isotope

Figure 7.31
Summary diagrams describing the mitochondrial biogenesis experiments of D. Luck: (a) preliminary experimental information required to distinguish mitochondria grown in media containing different concentrations of choline, and (b) the density transfer experimental procedure leading to one of three possible results. Each outcome is unique and indicative of one of the three possible mechanisms of biogenesis. The pattern predicted for mitochondrial biogenesis by growth and division was actually observed.

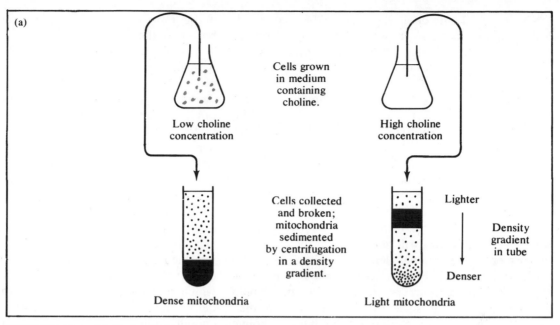

(a)

Cells grown in medium containing choline.

Low choline concentration

High choline concentration

Cells collected and broken; mitochondria sedimented by centrifugation in a density gradient.

Lighter

Density gradient in tube

Denser

Dense mitochondria

Light mitochondria

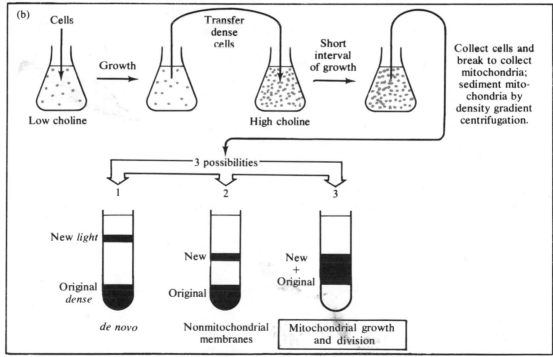

(b)

Cells

Transfer dense cells

Short interval of growth

Collect cells and break to collect mitochondria; sediment mitochondria by density gradient centrifugation.

Growth

Low choline

High choline

3 possibilities

1   2   3

New *light*

Original *dense*

de novo

New

Original

Nonmitochondrial membranes

New + Original

Mitochondrial growth and division

^{15}N was tagged onto choline so it could be recognized. Sedimentation patterns, however, were based on the relative lightness or density of mitochondria according to the concentration of choline in the culture medium.

The cultures were first grown in choline concentrations that led to formation of dense mitochondria. Afterward, these cultures were shifted to a higher choline concentration in which new light mitochondria would form. The distribution of the original population of dense mitochondria in relation to newly formed light mitochondria would be different in all three alternative hypotheses (Fig. 7.31):

(1) If mitochondria form *de novo,* two populations of mitochondria will be present at the end of the experiment. The original dense mitochondria would remain unchanged. New mitochondria formed from submicroscopic precursors, including choline, would be exclusively light since they would have formed after the culture was shifted to the higher choline concentration. The original and new mitochondria would be unrelated to each other but related directly to the amount of choline present at the time of organelle formation. The two distinct populations of mitochondria would band in the region for dense and the region for light mitochondria; there would be no intermediate densities.

(2) If mitochondria form from nonmitochondrial membranes, then the original dense mitochondria should remain unchanged since they do not contribute to formation of new organelles. Since choline is also incorporated into nonmitochondrial membranes in the first part of the experiment, newly formed mitochondria would be partially derived from these and partially from membranes unaffected by the choline concentration. The newly formed mitochondria would therefore be intermediate in density at first but would later be light mitochondria as shown by sedimentation in sucrose density gradients. The nature of the newly formed mitochondria would determine whether one or the other of these two hypotheses was correct, even though the original

dense mitochondria would be present according to both hypotheses.

(3) If mitochondria grow and divide, then the original dense mitochondria would gradually disappear. As new mitochondria form in the transfer medium, light membrane added on to dense membrane during mitochondrial growth would lead to intermediate densities. Division of such intermediate-density mitochondria would also maintain this density level while the experiment lasted.

The evidence showed that the original dense population disappeared after cultures were transferred to media with higher choline concentrations. A single region of mitochondria with intermediate density was found instead of one dense and one light or intermediate mitochondrial population. This result effectively ruled out the two alternative hypotheses which had each predicted retention of the original dense mitochondria. Growth of pre-existing mitochondria and then division to form more mitochondria seemed to have been demonstrated by Luck's experiments.

Growth and subsequent division of the mitochondrion have been observed directly in yeast cells. The number and size of mitochondria in growing cells were analyzed by electron microscopy and by three-dimensional reconstructions from a consecutive series of sections through individual cells in their entirety (Fig. 7.32). The large mitochondrion increases in size during cell budding, until it is twice the size of the original mitochondrion in the mother cell. When the wall pinches off the bud from the mother cell, each cell has a whole mitochondrion. The process is repeated at each cycle of budding. On the basis of these observations, division of mitochondria is thought to be a passive, rather than an active process. The two parts of the growing mitochondrion are separated when the cell wall forms at the end of the budding cycle and not by any process initiated by the mitochondrion itself. In this sense, therefore, division of mitochondria is quite different from division of a bacterial or eukaryotic cell. Further studies of cells

Figure 7.32
Three-dimensional reconstruction of a single, giant mitochondrion from a yeast cell with a large bud. The mitochondrion is continuous in the mother cell and bud (at the left). The model was reconstructed from a complete series of consecutive thin sections through the entire budding cell complex. (Courtesy of H.-P. Hoffmann)

from series of sections should clarify many of the current problems and questions about mitochondrial increase in number.

SUGGESTED READING

Books, Monographs, and Symposia

Lehninger, A. L. 1970. *Biochemistry*. New York: Worth.

Lloyd, D. 1974. *The Mitochondria of Microorganisms*. New York: Academic Press.

Munn, E. A. 1974. *The Structure of Mitochondria*. New York: Academic Press.

Tandler, B., and Hoppel, C. L. 1972. *Mitochondria*. New York: Academic Press.

Articles and Reviews

Atkinson, A. W., Jr., John, P. C. L., and Gunning, B. E. S. 1974. The growth and division of the single mitochondrion and other organelles during the cell cycle of *Chlorella*, studied by quantitative stereology and three dimensional reconstruction. *Protoplasma* 81:77–109.

Fernández-Morán, H., Oda, T., Blair, P. V., and Green, D. E. 1964. A macromolecular repeating unit of mitochondrial structure and function. *Journal of Cell Biology* 22:63–100.

Hackenbrock, C. R. 1966. Ultrastructural bases for metabolically linked mechanical activity in mitochondria. I. Reversible ultrastructural changes with change in metabolic steady state in isolated liver mitochondria. *Journal of Cell Biology* 30:269–297.

Hoffmann, H.-P., and Avers, C. J. 1973. Mitochondrion of yeast: Ultrastructural evidence for one giant, branched organelle per cell. *Science* 181:749–751.

Jagendorf, A. T., and Uribe, E. 1967. ATP formation caused by acid-base transition of spinach chloroplasts. *Proceedings of the National Academy of Sciences* 55:170–177.

Luck, D. J. L. 1965. Formation of mitochondria in *Neurospora crassa*. A study based on mitochondrial density changes. *Journal of Cell Biology* 24:461–470.

Mitchell, P. 1972. Chemiosmotic coupling in energy transduction: A logical development of biochemical knowledge. *Journal of Bioenergetics* 3:5.

Plattner, H., and Schatz, G. 1969. Promitochondria of anaerobically grown yeast. III. Morphology. *Biochemistry* 8:339–343.

Racker, E. 1968. The membrane of the mitochondrion. *Scientific American* 218(2):32–39.

Schnaitman, C., and Greenawalt, J. W. 1968. Enzymatic properties of the inner and outer membranes of rat liver mitochondria. *Journal of Cell Biology* 38:158–175.

Chapter 8

The Chloroplast and Photosynthesis

It is no exaggeration to say that the course of evolution on Earth owes some of its major features to the processes of photosynthesis. These processes provide the principal sources of foods that nourish all nongreen forms of life, as well as green cells and plants themselves. Photosynthetic cells are the beginning of the food chain, serving as food for animals, fungi, and at least some bacteria and protists. These organisms, in turn, provide food for other life forms and a cycle of carbon and nitrogen use and re-use occurs. In addition to providing foods made in reactions that employ energy originally derived from the sun's radiations, photosynthetic species constantly replenish the oxygen of our atmosphere. Life originated and evolved for at least 1 billion years before the appearance of photosynthesis in prokaryotes. About 1 billion years later, eukaryotic photosynthesizers appeared. Once these organisms multiplied and released molecular oxygen as a by-product of photosynthesis, the atmosphere changed profoundly and surviving life became predominantly aerobic about 1 billion years later. Some anaerobic forms and facultatively anaerobic species still exist, but these are a very small fraction of the total number of living organisms.

With the appearance of photosynthetic forms, a constant source of organic fuels was assured as long as the sun kept shining and green cells captured light energy to make sugars. The maintenance of an atmosphere with 21 percent oxygen content is entirely due to release of this gas by green cells during daylight hours when they manufacture foods from carbon dioxide, water, a few minerals, and light energy.

The photosynthesizing system appears deceptively simple because of the simple raw materials used by green cells and organisms to grow, reproduce, repair, and develop. These systems are far from simple. Photosynthesis and its related pathways are extremely complex and intricate. The membranes in photosynthetic prokaryotes and in the chloroplasts of eukaryotic cells are remarkable devices by which light energy is captured, converted to chemical energy, and applied toward the performance of all the work in the living organism. The major forms in which energy is conserved are ATP and reducing power of $NADP^+$ after it has accepted electrons from reactants and carrier systems in an intricate set of pathways. The relationship between membrane structure and the organization of reactants and functional molecules within the photosynthetic membranes is an active area of study today, as are the physical and chemical processes in chloroplasts as a whole.

GENERAL STRUCTURE OF PHOTOSYNTHETIC SYSTEMS

Chloroplasts in all cells are readily visible even at low magnification and resolution of light microscopes. They were described as early as the seventeenth century by Nehemiah Grew and Antonie van Leeuwenhoek. Experimental studies during the eighteenth and nineteenth centuries, sometimes of considerable ingenuity, established the relationship between absorption of light by chlorophyll in chloroplasts and the evolution of molecular oxygen, as well as conversion of carbon dioxide to stored starch. After the development of electron microscopy in the 1940s, more specific localizations were made.

Chloroplasts of land plants (mosses and other bryophytes, ferns, and seed-bearing plants) are usually discoid or lens-shaped and may number 50 or more per cell. Algae, on the other hand, are more variable in chloroplast number and shape. Various chloroplast shapes are well known among the algae, including the familiar ribbonlike, helical plastid in *Spirogyra*, the cup-shaped organelle in *Chlamydomonas*, and others (Fig. 8.1). The average chloroplast is about 1–10 μm long, but some algal plastids may be much larger. Chloroplasts are not present in prokaryotic photosynthesizing species.

Structure of Eukaryote Chloroplasts

Each eukaryotic chloroplast is separated from its cytoplasmic surroundings by two membranous envelopes; a space is usually seen between the envelopes but it is sometimes obscured. Contained within the boundary of the inner membrane is a third system of folded membranes bathed in an unstructured

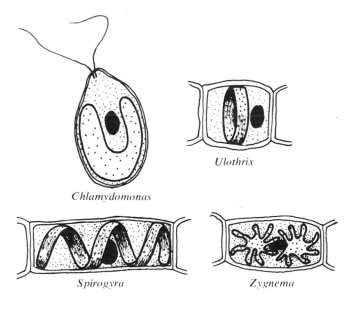

Figure 8.1
Various shapes of chloroplasts: cup-shaped in *Chlamydomonas*, band-shaped in *Ulothrix*, spiral in *Spirogyra*, and stellate in *Zygnema*.

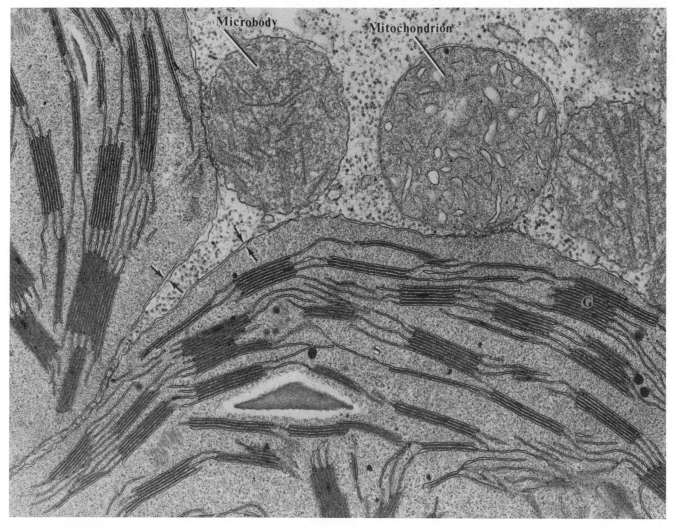

Figure 8.2
Thin section of part of a mesophyll cell of oat (*Avena sativa*) leaf showing thylakoids in parts of two chloroplasts. Stroma thylakoids (S) traverse the matrix and also connect the stacks of grana thylakoids (G). The appressed two membranes of the chloroplast envelope are visible (at arrows). Note that the mitochondrion also has two enclosing membranes while there is only a single limiting membrane for the microbody. × 42,000. (Courtesy of E. H. Newcomb and P. J. Gruber)

matrix called the **stroma** (Fig. 8.2). During the 1930s and 1940s increasing experimental evidence indicated the existence of membranous aggregates called **grana,** but their existence was not fully verified until later. In 1947 Keith Porter and Samuel Granick provided substantiating evidence from electron microscopy. Grana are recognized as stacks of flattened discoid components, arranged like neat piles of coins (Fig. 8.3). In-

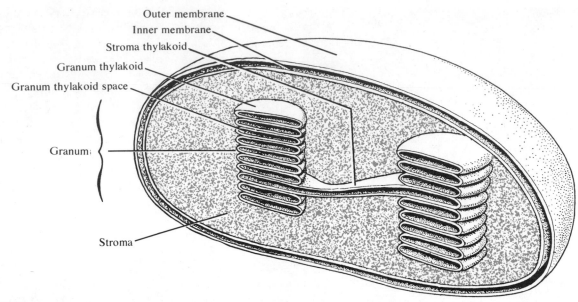

Outer membrane
Inner membrane
Stroma thylakoid
Granum thylakoid
Granum thylakoid space

Granum

Stroma

Figure 8.3
Diagram of the membranes and nonmembranous regions of
the chloroplast.

dividual grana elements are separate, flattened, membrane-bound sacs enclosing an inner space.

The term **thylakoid,** proposed by W. Menke in 1962, is generally used to refer to any flattened membrane sac in chloroplasts and related systems. The coin-shaped component in a granum is called a **grana thylakoid.** Some of these structures are continuous with counterparts in neighboring grana by means of connecting **stroma thylakoids.** One of several synonyms for thylakoid is **lamella,** a term often used in describing photosynthetic membranes. The channels within stroma thylakoids or lamellae are clearly continuous with the **thylakoid spaces** in some individual grana thylakoids. More complex and interacting channels and connections may also occur, and another terminology to describe these structures has been used by T. E. Weier and A. A. Benson. While the actual three-dimensional internal structure of the chloroplast has not been completely determined as yet, some observations are generally agreed upon. Each individual grana thylakoid is separated by its

membrane from neighbors above and below it in the same stack. An opaque fusion line can be seen in the region of membrane contact. The line may be related to the organization of particles contained within the membrane thickness.

One hundred or more stacked thylakoids may be found in a granum, but many fewer occur in most cell types. Some differentiated cells in flowering plants seem to have long, single stroma thylakoids that are not organized into grana. Most green cells of land plants, however, and many of the green algae contain grana or granalike assemblies of thylakoids. The stroma in which the thylakoids are suspended is granular and essentially unstructured, as far as can be seen by electron microscopy. Spherical darkly-staining granules with some unknown function are seen, along with numerous ribosomes about 150–200 Å in diameter and aggregations of 25 Å-wide fibrils of DNA (Fig. 8.4).

Thylakoid patterns vary among eukaryote groups. Grana or granalike arrangements characterize many of the green algae as well as land plants. Other

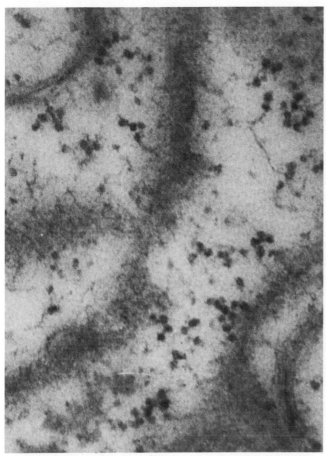

Figure 8.4
Thin section of *Euglena gracilis* chloroplast with fine DNA fibrils and ribosomes associated with these fibrils. Some polysomes are evident. It appears that the DNA and polysomes are distributed along the thylakoid membranes rather than occurring at random in the stroma. × 148,200. (Courtesy of H.-P. Hoffmann)

velope with properties of selective permeability to ions and solutes; (2) genetic capacities reflected in unique DNA and ribosomal machineries within the organelle; and (3) lipoprotein membranes containing tightly-bound components, including electron transport and ADP phosphorylating systems. There are many differences as well which are related to their different functions in aerobic respiration and photosynthesis. One of these differences bears mentioning even though its significance is unclear at present. When mitochondria are fragmented, aerobic respiration can only take place if intact inner-membrane segments are present. When chloroplasts are fragmented into functional photosynthetic particles, only the thylakoids need be present; the inner (or outer) chloroplast membrane does not participate in light-requiring reactions of photosynthesis. Thylakoid membranes are physically separate from the chloroplast inner membrane. Cristae are infolded regions of the inner membrane in mitochondria.

Thylakoid Structure in Prokaryotes

Three very different groups of prokaryotes are also photosynthetic: blue-green algae, purple bacteria, and green bacteria. Blue-green algae, also known as cyanophytes, are aerobic photosynthesizing species. Both groups of bacteria, however, exist anaerobically or at least carry out photosynthesis under anaerobic conditions even if otherwise capable of some aerobic activities. Blue-green algae regularly release molecular oxygen as a byproduct of photosynthesis, but purple and green bacteria never evolve oxygen. In fact, oxygen is poisonous to the strictly anaerobic green bacteria and those species of purple bacteria that live anaerobically. These differences reflect chemical rather than structural differences in their photosynthesizing systems.

Thylakoids are never enclosed within organelles in blue-green algae. They are distributed instead as single elements in the cytoplasm with a fairly regular spacing (Fig. 8.7). The thylakoids are disposed peripherally and are rarely seen to be attached to the plasma membrane, although it is believed they are de-

groups of algae and most protist groups have long thylakoids parallel to the long axis of the organelle or near the periphery. Thylakoids occur singly in red algae, in pairs in diverse groups such as diatoms, dinoflagellates, and euglenoids, and in threes in the brown algae (Fig. 8.5 and 8.6.).

Mitochondria and chloroplasts share many features. Both have (1) more than one membrane en-

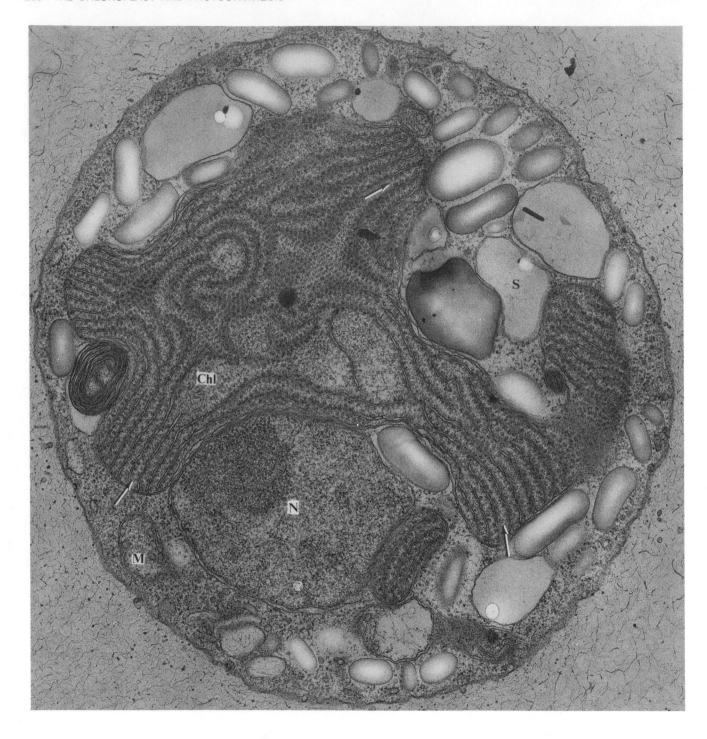

rived from it. The photosynthetic membranes are about 150–160 Å thick and show the usual dark-light-dark staining pattern. Together with aggregations of DNA fibrils, densely packed ribosomes, and photosynthetic thylakoids, blue-green algae bear an uncanny resemblance to chloroplasts of some eukaryotes.

Thylakoids develop as deeply infolded regions of the plasma membrane in all purple bacteria and are probably continuous with the plasmalemma at all times. A great deal of variation is found among species, but three general patterns of photosynthetic membranes have been recognized in purple bacteria. The commonest pattern is a vesicular type in which plasmalemma invaginations take the form of numerous small vesicles 400–500 Å wide, or elongated tubules of about this width (Fig. 8.8). This pattern occurs in most species of sulfur purple bacteria, including *Chromatium*, and in some species of nonsulfur purple bacteria, such as *Rhodopseudomonas*. A second internal arrangement is seen as parallel arrays of long, branched tubules with a regular diameter of about 450 Å. The third major pattern is found in certain purple bacteria which reproduce by budding, such as most of the species of *Rhodospirillum*. In these cells the photosynthetic membranes occur in disc-shaped stacks superficially similar to grana. These membranes are connected with the plasmalemma at only one or two places, rather than being continuous at every point as in the first two patterns described. These variations may reflect differences in nonphotosynthetic components within the membranes, since

no correlations have been found for membrane pattern and either pigments or mechanisms of photosynthesis in the various species.

Green bacteria, such as *Chlorobium*, differ from all other photosynthesizing prokaryotes in that their photosynthetic pigments are contained in special **chlorobium vesicles.** These vesicles are bounded by an unusual thin membrane 20–30 Å in width. The structures are completely separated from the plasmalemma, although they are distributed in a region under and parallel to the cell boundary. These cigar-shaped vesicles are about 500 Å wide and 1200–1500 Å long. Infolded regions of plasmalemma occur as common features of green bacteria and some other groups. These regions are called **mesosomes** (Figs. 8.9 and 8.10). It has been suggested that they function in respiration because they contain a high concentration of electron transport enzymes. They may also be involved in cell division because they usually are found where the new wall is forming. They apparently have no role in photosynthesis because negligible amounts of photosynthetic pigments have been found in isolated mesosomes. Differentiation of chlorobium vesicles may be the equivalent of compartmentation in eukaryotes, except that the very thin membrane is quite different from usual cellular membrane types.

Photosynthetic Pigments

There are three classes of photosynthetic pigments in photosynthetic cells: **chlorophylls, carotenoids,** and **phycobilins.** One or more kinds of chlorophyll and carotenoid occur in all photosynthetic species, but phycobilins are found in red algae and blue-green algae. Except for the green bacteria whose pigments are located in special chlorobium vesicles, all other photosynthetic species contain chlorophylls and carotenoids in tightly bound association with thylakoid membranes. Phycobilins are loosely associated with these membranes and can be separated very easily from other photosensitive pigments. Phycobilins are concentrated in granules called **phycobilisomes** which look like amorphous particles

Figure 8.5
Thin section of the unicellular red alga *Porphyridium cruentum.* Granular phycobilisomes (at arrows) containing accessory pigment are displayed along both surfaces of the extensive photosynthetic thylakoid membranes, in which chlorophyll is located. In addition to the nucleus (N) near the chloroplast (Chl) there are numerous starch granules (S) and occasional mitochondrial (M) profiles visible in the cytoplasm. × 33,000. (Courtesy of E. Gantt, from Gantt, E., and S. F. Conti, 1965, *J. Cell Biol.* **26**:365–381, Fig. 7.)

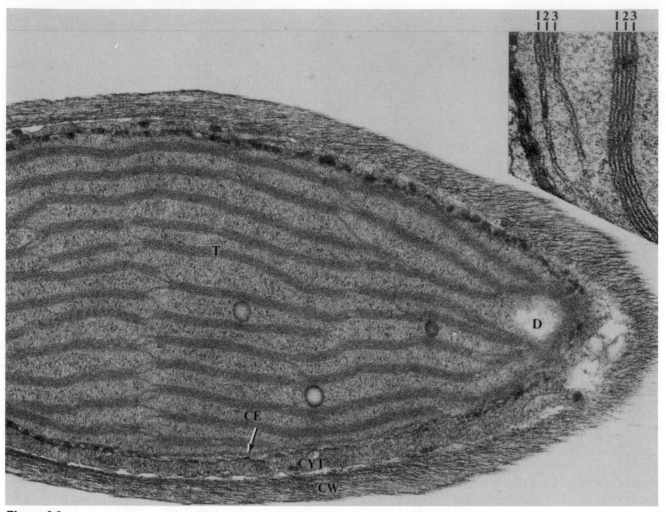

Figure 8.6

Tangential thin section of part of a chloroplast in a zoosporangial mother cell from the brown alga *Nereocystis.* The photosynthetic thylakoids (T) are in threes, as seen even more clearly in the insert photograph (123). The chloroplast envelope (CE) is bounded by a layer of cytoplasm (Cyt) and the cell wall (CW). A region of DNA (D) is present in the chloroplast. × 48,000; insert, × 114,000. (Courtesy of T. Bisalputra and D. C. Walker)

in electron micrographs of these algal cells (see Figs. 8.5 and 8.7). Photosynthetic pigments are thus concentrated and localized in some fashion in all cells and are not dispersed at random in either stroma or cytosol. Taken in some combination, these pigments provide the cell with the capacity to absorb solar energy across the entire span of the visible spectrum from 400 to 700 nm and into the far-red and infra-red range in some cases (Fig. 8.11).

CHLOROPHYLLS. This predominating pigment type absorbs energy in the blue and red wavelengths. Since

Table 8.1 Distribution of chlorophylls and other photosynthetic pigments

ORGANISM	CHLOROPHYLL				BACTERIOCHLOROPHYLL				CAROTENOIDS	PHYCOBILIPROTEINS
	a	b	c	d	a	b	c	d		
Eukaryotes:										
Mosses, ferns, seed plants	+	+	−	−					+	−
Green algae	+	+	−	−					+	−
Euglenoids	+	+	−	−					+	−
Diatoms	+	−	+	−					+	−
Dinoflagellates	+	−	+	−					+	−
Brown algae	+	−	+	−					+	−
Red algae	+	−	−	+					+	+
Prokaryotes:										
Blue-green algae	+	−	−	−	−	−	−	−	+	+
Sulfur purple bacteria					+ or +	−	−		+	−
Nonsulfur purple bacteria					+ or +	−	−		+	−
Green bacteria					+	−	+ or +		+	−

its pattern of light absorption coincides with the wavelengths which are most efficient in inducing oxygen evolution or some other photochemical activity, it is clear that chlorophyll is the principal light-capturing molecule in the cell (Fig. 8.12). An **absorption spectrum** is a plot of degree of absorbance of different wavelengths of a spectrum by a test substance, and an **action spectrum** is a plot of the efficiency of different wavelengths of light in supporting some activity, such as oxygen evolution. Comparisons of absorption and action spectra for whole chloroplasts or individual pigments have thus established the range of light-capturing capacity in different photosynthetic systems. Since chlorophylls transmit green wavelengths but absorb others, we see them as green substances in a test tube or in landscapes over the vegetated regions of the world.

Chlorophylls are lipid-soluble and easily extracted and purified of contaminating proteins. The pigment is constructed of a **porphyrin** whose central nitrogen atoms coordinate with a Mg^{2+} ion and a long **phytol** side-chain which is added during chlorophyll biosynthesis (Fig. 8.13). Different kinds of chlorophylls have different substitutions in the porphyrin ring. The porphyrin part of the chlorohyll molecule

is very similar to porphyrin prosthetic groups of hemoglobin and cytochromes, but these latter compounds have a central iron atom rather than magnesium (see Fig. 7.15).

Higher plants and algae, including prokaryotic blue-greens, all contain chlorophyll *a*. Eukaryotes contain a second chlorophyll, which varies according to the species group. The second chlorophyll may be *b* (higher plants and most green algae), *c* (diatoms, dinoflagellates, and brown algae), or *d* (red algae). Bacteria, which do not evolve molecular oxygen, have no chlorophyll *a*. Instead, they contain one or more of the four known **bacteriochlorophylls** (Table 8.1). All the chlorophylls absorb and hold the energy of visible light efficiently because of their numerous conjugated double bonds.

CAROTENOIDS AND PHYCOBILINS. The yellow carotenoids and the red or blue phycobilins also absorb light energy. Carotenoids absorb maximally in the violet to green range (400–550 nm), and phycobilins show maximum absorption of the green to orange region (550–630 nm) of the visible spectrum. Their colors, like chorophylls, are due to transmission of particular wavelengths of the visible spectrum.

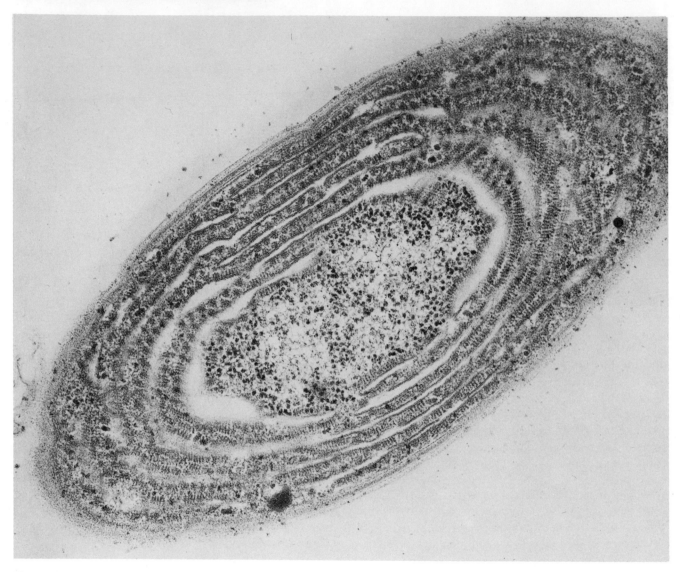

Figure 8.7
Thin section through the blue-green alga *Synechococcus lividus*. The photosynthetic lamellae are concentrically folded and are lined on both surfaces by granular phyco-bilisomes. × 62,400. (Courtesy of E. Gantt, from Edwards, M. E., and E. Gantt, 1971, *J. Cell Biol.* **50**:896–900, Fig. 1.)

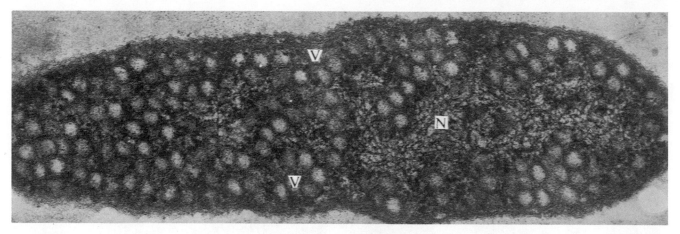

Figure 8.8

Longitudinal section through the purple bacterium *Rhodopseudomonas spheroides*. The cytoplasm contains numerous electron-transparent vesicular photosynthetic membranes (V). × 102,000. (Reproduced with permission from Cohen-Bazire, G., 1971, *Biological Ultrastructure: The Origin of Cell Organelles*, Oregon State Univ. Press, pp. 65–90. Copyright © 1971 by Oregon State University Press.)

Various carotenoids have different long-chain carbon regions with many conjugated double bonds and unique termini on either end of the long chain. A common carotenoid responsible for color in carrots and tomatoes is β-carotene (see Fig. 8.13). These lipid-soluble pigments may transfer absorbed light energy to chlorophyll *a* or its equivalent, which in turn is directly involved in conversion of light energy to chemical energy during the light reactions of photosynthesis. Carotenoids also protect chlorophyll in plants and algae from the degradative effects induced by molecular oxygen in the light.

Phycobilins are accessory pigments found in eukaryotic red algae and prokaryotic blue-green algae. These pigments are different in that they are conjugated to specific proteins. The protein conjugate is a **phycobiliprotein,** the pigment itself is a **phycobilin.** The protein conjugate of the pigment phycoerythrobilin in most red algae is called **phycoerythrin,** and its analog in almost all blue-green algae is **phycocyanin.** A few red algae have the blue pigment and a few blue-green algae have the red pigment, which sometimes leads to incorrect identification and assignment of one of these algal species.

In contrast with the cyclic porphyrin component in chlorophyll, phycobilins are open-chain porphyrin-types (see Fig. 8.13) firmly bound to protein. Like carotenoids, phycobilins transfer their absorbed light energy to chlorophyll *a*, which is the primary photoreactive pigment in all plants and algae. Energy is transferred more efficiently by phycobilins than by carotenoids to chlorophyll *a*.

OVERALL REACTIONS OF PHOTOSYNTHESIS

In all photosynthesizing cells, the energy of solar radiation is absorbed by light-sensitive pigments, transferred to electron acceptors, and conserved as chemical energy for subsidizing cellular work. Except for the purple and green bacteria, water serves as the donor of electrons and hydrogens to reduce carbon

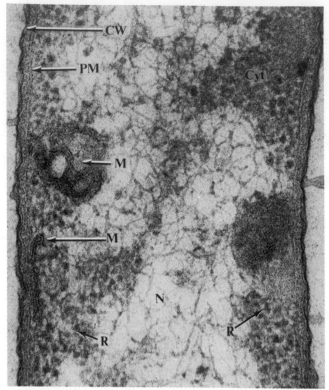

Figure 8.9
Longitudinal section through part of the bacterium *Pseudomonas aeruginosa*. Two mesosomes (**M**) are closely associated with the plasma membrane (**PM**), which is bounded by the cell wall (**CW**). Nucleoid (**N**) and cytoplasm (**Cyt**) are present, with ribosomes (**R**) within the cytoplasmic region. × 100,000. (Courtesy of H.-P. Hoffmann, from Hoffmann, H.-P. *et al.*, 1973, *J. Bacteriol.* **114**:434–438, Fig. 1.)

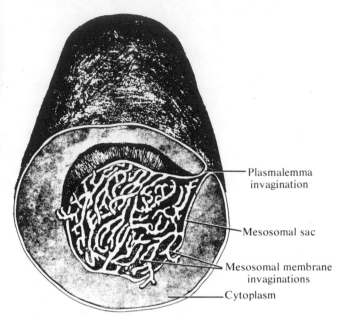

Figure 8.10
Three-dimensional reconstruction of a cutaway view of a rod-shaped bacterial cell showing the mesosome as multiple branching tubular invaginations of a sac, which is itself an invagination of the plasma membrane. (Reproduced with permission from Rucinsky, T. E., and E. H. Cota-Robles, 1974, *J. Bacteriol.* **118**:721, Fig. 4.)

dioxide or other electron-accepting substrates. As a result, molecular oxygen is evolved when water is oxidized. In the overall view, therefore, nonbacterial photosynthesis is a process in which light energy is used to reduce ("fix") CO_2 into organic compounds, with H_2O as the electron and hydrogen donor for this reduction:

$$n\ H_2O + n\ CO_2 \xrightarrow{\text{light}} [CH_2O]_n + n\ O_2 \qquad [8.1]$$

As first proposed by C. B. van Niel, bacterial and plant photosynthesis could be viewed as essentially similar processes if the reaction reflected a more general hydrogen donor than water. For example, the reaction might also include H_2S, or some organic compound such as lactic acid used by nonsulfur purple bacteria.

$$2\ H_2D + CO_2 \xrightarrow{\text{light}} [CH_2O] + H_2O + 2\ D \qquad [8.2]$$

According to this formulation, various hydrogen donors (H_2D) may be involved. H_2D could be H_2O, H_2S, or another hydrogen donor, depending on the species. The donor would be oxidized to D, as its hydrogens were transferred for fixation of carbon dioxide.

As van Niel noted, the molecular oxygen evolved during aerobic photosynthesis must originate from H_2O and not from CO_2 molecules. Later studies showed this indeed to be the case. Using the heavy ^{18}O isotope of ordinary ^{16}O, the evolved molecular oxygen contained atoms of oxygen originally present in H_2O, but none from CO_2:

$$n\ H_2^{18}O + n\ C^{16}O_2 \xrightarrow{\text{light}} [CH_2O]_n + n\ ^{18}O_2 \qquad [8.3]$$

In the reciprocal experiment

$$n\ H_2^{16}O + n\ C^{18}O_2 \xrightarrow{\text{light}} [CH_2O]_n + n\ ^{16}O_2 \qquad [8.4]$$

Furthermore, it is known that CO_2 is the major acceptor of hydrogens and electrons from a donor, but

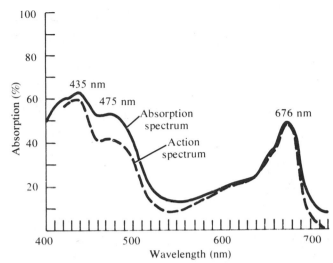

Figure 8.12
The absorption spectrum for a cell suspension of *Chlorella* (solid line) compared with the photosynthetic action spectrum (broken line) of these cells.

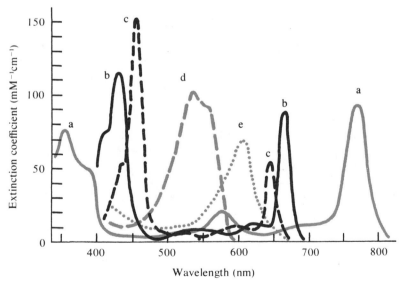

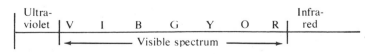

Figure 8.11
Absorption spectra of (a) bacteriochlorophyll, (b) chlorophyll *a*, and (c) chlorophyll *b*, all in ether; and (d) phycoerythrin and (e) phycocyanin, in aqueous solution.

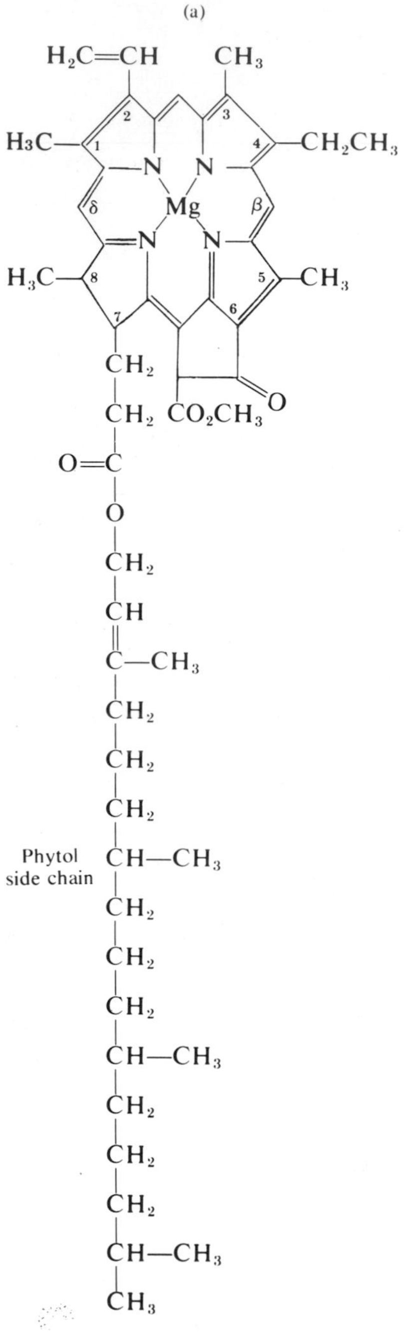

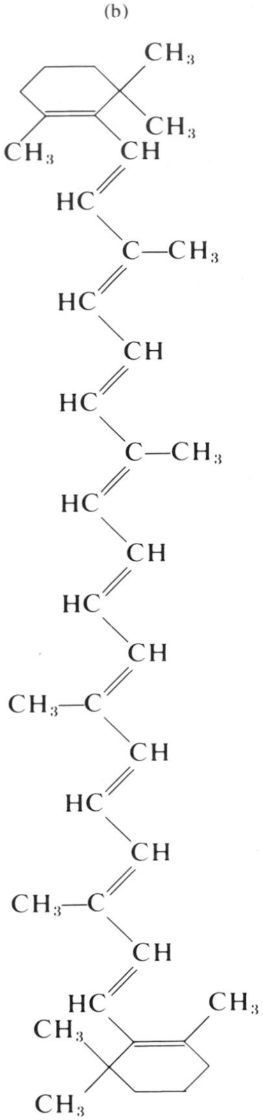

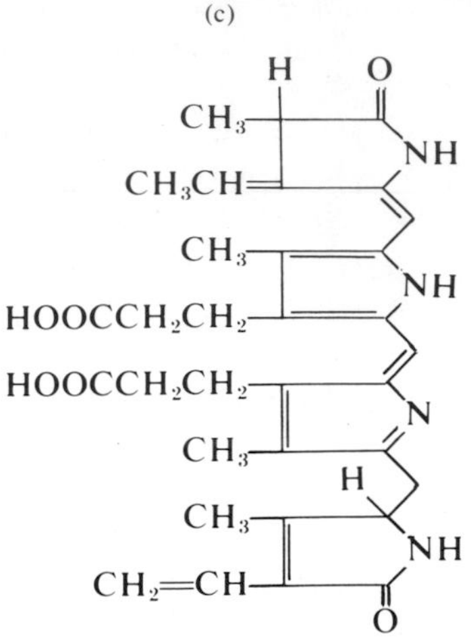

(a)

(b)

(c)

$H_2C=CH$ CH_3

H_3C CH_2CH_3

H_3C CH_3

H_3C CH_3

Mg

CH_2
CH_2 CO_2CH_3
$O=C$
O
CH_2
CH
$C-CH_3$
CH_2
CH_2
CH_2
Phytol side chain $CH-CH_3$
CH_2
CH_2
CH_2
$CH-CH_3$
CH_2
CH_2
CH_2
$CH-CH_3$
CH_3

CH_3
CH_3
CH_3
CH
HC
$C-CH_3$
HC
CH
HC
$C-CH_3$
HC
CH
HC
CH
CH_3-C
CH
HC
CH
CH_3-C
CH
HC CH_3
CH_3
CH_3

CH_3 H O
$CH_3CH=$ NH
CH_3
$HOOCCH_2CH_2$ NH
$HOOCCH_2CH_2$ N
CH_3
H
CH_3
$CH_2=CH$ NH
O

Figure 8.13
(a) Structural formula for chlorophyll *a*. (b) Structural formula for β-carotene. (c) Structural formula for phycoerythrobilin.

not the only one. Most plants can use nitrate or even H^+ ions, and some nitrogen-fixing plants can use molecular nitrogen as an electron acceptor during photosynthesis. Since various electron donors and acceptors may be used by different species, the more general photosynthetic reaction for any organism would be

$$H_2D + A \xrightarrow{\text{light}} H_2A + D \qquad [8.5]$$

in which H_2D is the donor and A is the acceptor of hydrogens and electrons.

Regardless of the specific electron donor or acceptor, the light-induced flow of electrons is toward the system of greater electronegative potential, that is, *against* the usual gradient of potentials of acceptor-donor systems. In contrast with respiration processes that flow "downhill" from an electronegative (energy-rich) state to an electropositive (energy-poor) state, the force of light energy absorbed by chlorophyll in photosynthesis causes an "uphill" flow of electrons and energy (see Table 2.6).

It may be noted here that water is a very weak reductant, and photosynthesis is the only metabolic process known to use it as an electron donor. Water has a strongly positive standard electrode potential of + 0.8 volt and, therefore, a very high affinity for electrons. It has very little tendency, theoretically, to lose electrons and form O_2.

Photoexcitation of Molecules

We are only able to see the small portion of the electromagnetic radiation spectrum that falls between wavelengths of 400 to 700 nm; this is the "visible" spectrum to us. These energetic wavelengths reach us from the sun or artificial light sources in discrete packets called **photons.** The only difference between visible light and other electromagnetic radiations (x-rays, radio waves, ultraviolet light, and others) is in the frequency of vibrations of the radiation source. The energy of a photon is called a **quantum,** and the amount of energy in one quantum for a particular photon is given by the equation

$$E = h\nu \qquad [8.6]$$

where h is Planck's constant (1.585×10^{-34} cal-sec) and ν is vibrations/sec. In 1900 Max Planck developed an equation by which the energy in one quantum is related to the wavelength of the particular photon.

$$E = \frac{hc}{\lambda} \qquad [8.7]$$

where h is Planck's constant; c is the speed of light, or 3×10^{10} cm/sec; and λ is the wavelength of emitted radiation. When an atom or molecule absorbs a photon of light, it is absorbing a quantum of energy. Since photon energy is inversely proportional to wavelength, photons of the shorter wavelengths have the higher energy content. This relationship is implied in equation 8.7.

The energy equivalent is most usefully expressed in some form which permits comparisons and relationships with numbers of molecules. One way to express photon energy equivalents is in calories or kilocalories per **einstein.** An einstein can be considered as one "mole" of light, or 6.023×10^{23} quanta, expressed in calories. Since the number of molecules in a mole is given by Avogadro's number (6.023×10^{23}), it is necessary to multiply the energy per molecule by this number to give the energy associated with one mole. Thus, 1 mole of pigment containing 6.023×10^{23} molecules and absorbing 1 einstein of light at 600 nm will absorb the energy equivalent of 47.67 kilocalories. Or, 1 einstein of photons at 600 nm wavelength has an energy equivalent of 47.67 kilocalories. The simplified equation is:

$$N_{Av}E = \frac{N_{Av}hc}{\lambda} \text{ (expressed in kcal/einstein)} \qquad [8.8]$$

where N_{Av} is 6.023×10^{23}, or Avogadro's number, h is Planck's constant in cal-sec, c is the speed of light in cm/sec, and λ is the wavelength in centimeters. Values for other wavelengths are given in Table 8.2. Another expression used quite often is the **quantum yield** of photosynthesis, or amount of molecular oxygen evolved per einstein of light.

An atom or molecule may gain energy by **excitation,** an event which converts the substance to an energy-rich state. When the electrons occupy orbitals of lowest accessible energy level around the nucleus of an atom, the atom is in its **ground state.** When a sufficient packet of energy is absorbed by the atom so that an electron may move from one orbital to another of higher accessible energy level, the atom enters an **excited state.** In its energy-rich excited state, the atom can transfer an electron to a different atom or molecule with lower energy content.

The electron transfer proceeds "downhill" because the excited atom temporarily is at a higher energy level than the atom receiving the electron. When photons of particular wavelength (energy content) succeed in exciting a given atom in a molecule, events occur very rapidly. It takes 10^{-15} second or less for excitation to occur, and the excited state lasts only about 10^{-9} to 10^{-8} second. Excited atoms are thus extremely unstable. This short period is usually sufficient time, however, to allow trapping of some or most of the light energy. On return to the ground state, the exact amount of energy is released from the atom or molecule as was originally absorbed. Released energy may be dissipated as heat, emitted as radiation (called fluorescence), or converted to chemical energy, depending on the lifetime of the excited state of each atom or molecule involved. The significant

Table 8.2 Energy equivalent of the einstein at different wavelengths

WAVELENGTH (nm)	SPECTRUM COLOR	kcal per einstein
400	Violet	71.5
500	Blue	57.2
600	Yellow	47.7
675	Red	42.3
700	Near-red	40.9

energy of photoexcitation in biological systems is the proportion of the total that is converted to chemical energy. A greater efficiency of absorption is possible because different pigments receive photons of different wavelengths, making more light energy available to be trapped in the chloroplast or photosynthetic cell.

Separability of Light and Dark Reactions

Information suggesting that photosynthesis could be separated into two sets of reactions began to accumulate early in this century and Robert Emerson demonstrated this division conclusively in the early 1930s. One set of reactions was strictly dependent on light; another set was independent of the presence of light. The first experimental demonstration that a separate light reaction was localized in chloroplasts was reported by Robert Hill in 1937. When a preparation of chloroplasts from cells was illuminated in the presence of artificial hydrogen acceptors such as reducible dyes, molecular oxygen was evolved and the acceptors were reduced simultaneously. Water served as the exclusive hydrogen donor in this system and most importantly, CO_2 was not required for the reaction. If CO_2 reduction did occur, its reduction products did not accumulate. This observation further supported the separability of light and dark reactions, with the first leading to oxygen evolution and the second to carbon dioxide fixation (reduction). The **Hill reaction** follows the general sequence

$$2\ H_2O + 2\ A \xrightarrow{\text{light}} 2\ AH_2 + O_2 \qquad [8.9]$$

where A is the hydrogen acceptor and AH_2 is its reduced form. This formulation indicates the activity of the photosynthetic light reaction in functioning preparations of chloroplasts, chloroplast fragments, or cells.

S. Ochoa and R. Vishniac showed in 1950 that $NADP^+$ could substitute as the hydrogen acceptor in the Hill reaction. This observation supported the prediction that an end product of the light reaction was a reducing agent which could reduce carbon dioxide in subsequent dark reactions. NADPH was already known at that time to act as an electron donor in cellular biosyntheses. A very significant discovery which clearly showed that light and dark reactions of photosynthesis were separable was made by Daniel Arnon in 1954. He demonstrated that illuminated spinach chloroplasts could form ATP from ADP and inorganic phosphate. CO_2 was neither required nor consumed in this light-dependent reaction called **photophosphorylation.** The view thus gradually developed that $NADP^+$ reduction and ADP phosphorylation during the initial light reactions produced NADPH and ATP, which were then utilized in dark reaction pathways to reduce carbon dioxide and other electron acceptors.

During the 1950s it was shown that oxygen evolution and photophosphorylation reactions were localized within chloroplast grana, the same membrane system that contained chlorophyll. The dark reactions, on the other hand, took place in the stroma region of the chloroplast. One of the pivotal enzymes of the CO_2 fixation pathways is **ribulose 1,5-diphosphate carboxylase.** This enzyme links dark and light reactions of photosynthesis. It is usually present in the stroma right along thylakoid membrane surfaces, rather than being randomly distributed. Spatial organization includes individual molecules as well as membrane compartmentation.

THE LIGHT REACTIONS OF PHOTOSYNTHESIS

Before the 1950s it was generally believed that absorption of light energy by chlorophyll was linked to oxygen evolution, reduction of $NADP^+$, and phosphorylation of ADP to form ATP, all in one common pathway. Evidence against this notion began to accumulate in the late 1950s, and in support of the existence of two separate photosystems that interacted in plant photosynthesis. Using the alga *Chlorella*, Robert Emerson noted that efficient photosynthesis did not take place at wavelengths longer than 680 nm, even though these wavelengths were within the range of absorption of chlorophyll *a*. This decrease in photosynthetic efficiency has been called the *red drop effect*.

Photosynthesis efficiency could be increased if shorter-wave red light was supplied along with longer wavelengths, even if the two wavelengths were provided in alternate flashes in intervals lasting several seconds. These results indicated that two different photosystems operated coordinately through the mediation of chemical products that linked the two systems. If they did not, excited molecules would be expected to return quickly to their ground state, with the release of fluorescence and heat. The increased efficiency of the system demonstrated that energy had been conserved in chemical form. This *enhancement effect* with two different wavelengths and the single-wavelength red drop effect were explained later on the basis of a common set of photochemical activities. With the discovery of cytochromes in chloroplasts, it seemed likely that these electron carriers could act as links between two light-dependent systems acting in series. The red drop effect was suggested to be the result of only one photosystem being in operation, and the interactions of both photosystems were the possible basis for the enhancement effect. L. Duysens, R. Hill, and others contributed to the formulation of the series scheme for photosynthesis (Fig. 8.14), which has since been substantiated by a large body of evidence.

Photosystems I and II

The significant component of **photosystem I (PS I)** in plants and algae is absorption of light by chlorophyll *a,* with an absorption maximum of 683 nm. Oxygen evolution is not associated with photosystem I events. All oxygen-evolving photosynthetic species possess a second photosystem, **photosystem II (PS II)** which includes chlorophyll *a* with an absorption maximum of 672 nm, and either chlorophyll *b* or some other chlorophyll or phycobilin as an accessory pigment. The significant light-absorbing pigment in PS II is chlorophyll *a*. All light energy absorbed by the ac-

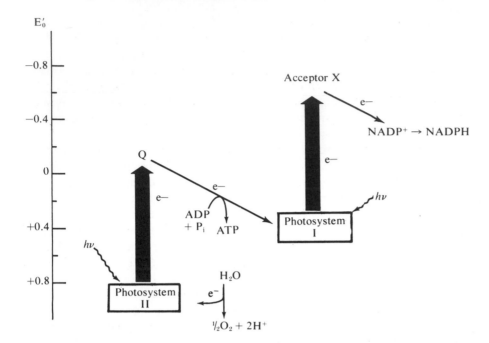

Figure 8.14
Summary diagram showing photosystem I and photosystem II activities coupled in series, in the light reactions of photosynthesis.

cessory PS II pigments is channeled to chlorophyll *a* which is the single kind of pigment that participates directly in conversion of light energy to chemical energy in plants and algae.

Photosynthetic bacteria, which do not evolve oxygen, possess only a PS I equivalent. The various bacteriochlorophylls present in different species are the light-absorbing pigments involved in trapping radiant energy in chemical form (Table 8.3). The chlorobium chlorophylls are very similar to chlorophyll *a* in absorption, while peak absorption of other bacteriochlorophylls is in the near-red part of the spectrum.

Emerson and others showed that 1 molecule of oxygen was evolved, or 1 molecule of carbon dioxide was fixed, by 8 quanta of light energy. A brief flash of light sufficient to saturate the mechanism led to release of one molecule of oxygen for every 3000 chlorophyll molecules present. These data, along with many observations which showed that light absorbed by one pigment could be utilized by another, led to a

central postulate. This postulate stated that chlorophyll and associated pigments exist in light-harvesting assemblies which collaborate in bringing together the necessary 8 quanta and delivering this energy to an active center for conversion to chemical energy. The

Table 8.3 Peak absorption wavelengths of photosynthetic pigments in organic solution

PIGMENT	WAVELENGTH (nm)
Chlorophyll a	660
Chlorophyll b	640
Bacteriochlorophyll a	770
Bacteriochlorophyll b	795
Bacteriochlorophyll c	660
Bacteriochlorophyll d	650
Carotenoids	
β-carotene	450–470
Xanthophyll	480–540
Phycobiliproteins	
Phycoerythrin	540–560
Phycocyanin	610–630

pigment light-harvesting assemblies and the **reaction center** constitute the **photosynthetic unit.** According to some, the photosynthetic unit includes 2500–3000 molecules of chlorophyll and other pigments as well as all the components of photosystems I and II. Others consider the photosynthetic unit to include only 250–300 molecules of chlorophyll and 1 reaction center on a per quantum basis. Eight of these smaller units would be needed to evolve 1 molecule of oxygen.

The reactive center of PS I is believed to be a special form of chlorophyll *a,* called **P700** in recognition of its maximum absorption band at 700 nm. There is approximately 1 molecule of P700 for every 300 to 400 molecules of chlorophyll *a.* P700 appears to act as a trap for collecting quanta of excitation energy from chlorophyll in PS I, subsequently passing on electrons. In this way P700 acts as an electron lead from PS I pigment to electron acceptors. It remains to be determined whether the PS II chlorophyll *a* (672 nm maximum absorption) acts as a reaction center in that photosystem.

Interrelationships between PS I and II posed important questions concerning their coordination, the phosphorylation process in ATP formation, and the identity of electron carriers in the electron transport sequence connecting the two photosystems. The series formulation (see Fig. 8.14) provides an increasingly acceptable framework, but many details remain to be filled into this broad outline. The sequence of PS II → PS I was determined after numbers had been assigned to the photosystems, and this numbering has been retained even though it does not accurately depict the light reaction sequence. It does, however, reflect the evolutionary sequence since PS II arose after PS I was in existence.

Electron Transport in Chloroplasts

The primary event in photosynthesis is the raising of an electron to a higher energy level in chlorophyll. After that electron is transferred to an acceptor, there is an electron "hole" in chlorophyll which invites filling. The "hole" is filled by electrons from water, thus converting chlorophyll to its original ground state. In the initial PS II reaction, electrons from water are raised from a potential of + 0.8 volt to 0.0 volt by transfer of energy (Fig. 8.15). The details are under intensive study, but it is believed that electrons from water are received by excited chlorophyll, transferred to reaction center chlorophyll, which then contributes these electrons to some electron acceptor. One suggested acceptor compound has been termed Q, because chemically it is a quinone and also because it quenches fluorescence. There are many uncertainties about the pathway of oxidation of water forming electrons, protons, and molecular oxygen.

The electrons at 0.0 volt potential are then moved along a carrier chain, losing energy in a "downhill" sequence. Some of this energy is conserved during **noncyclic photophosphorylation,** in which ATP is formed from ADP in a coupled reaction. Recent evidence makes it uncertain whether the electrons move from **plastoquinone** to **cytochrome-559,** as was originally believed, or from cytochrome-559 to plastoquinone, as recent studies have suggested. In either case the electrons move to **cytochrome-553** (also known as cytochrome *f*), and then to **plastocyanin.** At approximately + 0.4 volt, these electrons are accepted by P700, the reactive center of PS I in its excited state. In the PS I pathway, P700 ejects electrons that are accepted by a high-potential substance and then transferred to **ferredoxin-reducing substance** at − 0.6 volt. From here, electrons are transferred to **ferredoxin** and thence to the enzyme **ferredoxin-NADP reductase.** The final step is reduction of $NADP^+$ to NADPH in which chemical energy is conserved for subsequent biosyntheses within the chloroplast and cell.

During transfers of electrons and protons from water, initial weak oxidants and reductants are led through steps ultimately leading to the formation of NADPH and ATP. Light energy absorbed by the two photosystem pigments provides the differential between the initial low-energy materials and the high-energy products of photosynthesis. NADPH and ATP are utilized in reduction of carbon dioxide to carbohydrates in the subsequent "dark" reactions of

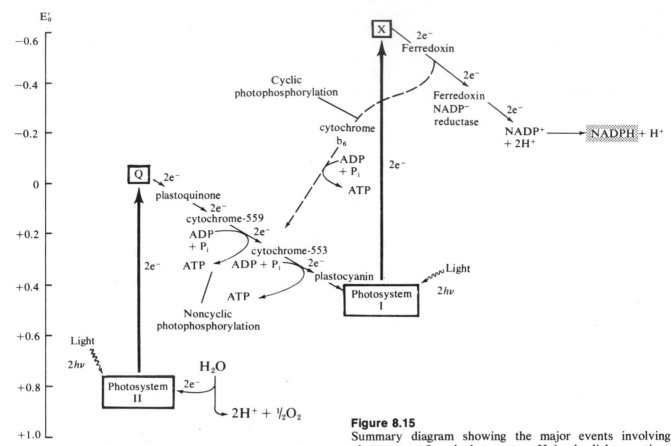

Figure 8.15
Summary diagram showing the major events involving photosystem I and photosystem II in the light reactions of photosynthesis. See text for details.

photosynthesis. The efficiency of photosynthetic conversion of light energy to chemical energy in NADPH is calculated to be 25–35 percent, which is a reasonable figure when compared with exceptionally low values achieved by man-made devices for converting light energy, such as solar batteries or heaters.

Photophosphorylation

Suspensions of isolated chloroplasts or bacterial photosynthesizing particles can form ATP by phosphorylation of ADP in the light. The process is absolutely light-dependent and may occur in the absence of oxygen evolution or CO_2 fixation. Photophosphorylation provides stored chemical energy in the form of ATP, and together with reducing agents like NADPH, biosyntheses can occur at the expense of the chemical energy stored in the light reactions of photosynthesis.

In plants and algae, a process of **noncyclic photophosphorylation** is coupled to the transport of electrons along the carrier chain linking the two photosystems (see Fig. 8.15). The process is noncyclic since new electrons from donor molecules are required for each phosphorylation sequence. The electrons continue to P700 of photosystem I and are incorporated into NADPH at the end of the sequence. Energy from these electrons is thus conserved as ATP and NADPH and is utilized in subsequent fixation of carbon dioxide in the dark reactions.

If uncoupling agents are introduced into illuminated chloroplast suspensions, noncyclic phosphorylation of ADP stops, thus demonstrating that it is coupled to electron transport. If the herbicide dichlorophenyldimethylurea (DCMU) is added to active chloroplasts, electron flow and ATP formation are both blocked. When a reducing agent is added as a source of electrons in a DCMU-inhibited system, the flow of electrons to $NADP^+$ resumes through photosystem I, but ATP formation remains blocked because DCMU continues to inhibit the carrier chain link between PS I and PS II (Fig. 8.16). This demonstrates that NADPH formation is a photosystem I function, whereas ATP formation in plants requires PS I and II linked by a chain of electron carriers.

In **cyclic photophosphorylation,** electrons recycle without replenishment by external donors (Fig. 8.17). There is some uncertainty about the association between cyclic formation of ATP in the light and electron flow during photosynthesis. It is difficult to see how these events can be linked when some kinds of uncoupling agents do not prevent cyclic photophosphorylation. It is known, however, that ATP formed in cyclic light-driven electron flow in bacteria is not associated with NADPH formation. The ATP produced in cyclic photophosphorylation is vital for bacterial metabolism, regardless of its associations with photosynthesis.

Coupling Between Electron Transport and Phosphorylation

Photosynthetic electron transport and phosphorylation of ADP to form ATP have been postulated to occur in a coupled system driven by a mech-

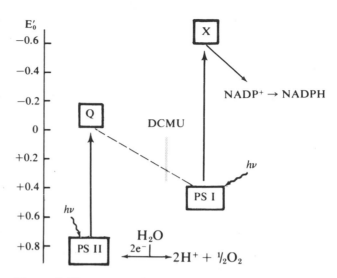

Figure 8.16
The inhibitor DCMU blocks electron flow and ATP formation, but electron flow to $NADP^+$ in photosystem I can be resumed if a reducing agent is added to such a DCMU-inhibited chloroplast preparation. ATP formation remains inhibited, however, because of the DCMU block to electron flow between PS I and PS II.

anism similar to the one in mitochondria. The same chemical-coupling, chemiosmotic-coupling, or conformational-coupling alternatives have been suggested to operate in the chloroplast. During photosynthetic electron transport, ions and protons are translocated across the thylakoid membrane into the thylakoid spaces. The flow of ions and protons induces changes in the electrical potential across the thylakoid membrane. If the driving force of ion and proton flux is inhibited, photophosphorylation comes to a stop.

While there is no clear evidence to rule out any of the three possible alternative mechanisms in either chloroplasts or mitochondria, one striking set of experiments by André Jagendorf has been cited very often as providing support for Mitchell's chemiosmotic coupling hypothesis. In brief, it was shown that a suspension of chloroplasts could synthesize

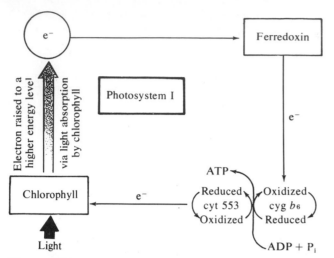

Figure 8.17
Cyclic photophosphorylation. When chlorophyll absorbs a photon of light of sufficient energy, it sends an electron into a high energy state. The electron reduces ferredoxin and the cytochrome system in turn. The reduction of cytochrome is coupled to ATP formation. No "outside" electron donor is required for the process since the electron returns to chlorophyll, at a lower energy level.

ATP in response to an artificial pH gradient (Fig. 8.18). A temporary gradient of H^+ ion concentration was created by first exposing chloroplasts to a medium of pH 4.0 and then transferring them rapidly to a medium of pH 8.0. The H^+ ion concentration was thus higher inside and lower outside the thylakoid membranes. The effect was the same as a physiological pumping of H^+ ions into the thylakoid space in illuminated systems. When ADP and inorganic phosphate were added at the time the H^+ gradient was created, ATP was formed.

The particularly significant aspect of the data was that the amount of ATP formed by this treatment was approximately equal to the amount of ATP generated during about 100 cycles of photochemical reactions in chloroplasts. These results are therefore physiologically significant. These data have been interpreted as supporting the coupling between electron transport and ATP formation according to a difference

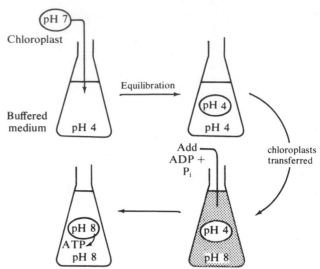

Figure 8.18
Sequence of events in the Jagendorf experiment. Spinach chloroplasts at an initial pH of 7 were transferred to an acid medium in which they acquired a higher concentration of H^+ ions after equilibrating in the acid medium. When pH 4 chloroplasts were transferred to a medium of pH 8 and ADP + P_i were then added, ATP synthesis was observed to occur at the expense of the temporary H^+ ion concentration (pH) gradient.

in H^+ ion concentration on either side of the membrane and the electrical potential thus generated across this membrane.

THE DARK REACTIONS OF PHOTOSYNTHESIS

ATP and NADPH formed in the light-dependent reactions are sources of energy for the formation of carbohydrates from atmospheric carbon dioxide in green cells. These sequences of CO_2 reduction to carbohydrates are independent of light, and have therefore been called the **dark reactions.** Many of the later individual chemical reactions also take place in nongreen cells, such as the events of glycolysis or synthesis of large molecules, but there are two major pathways by which the *initial* reactions of CO_2

fixation occur exclusively in green cells. One of these leads to the formation of an initial three-carbon product and the other is a set of reactions in which a four-carbon product is formed when CO_2 is reduced in chloroplasts. These pathways are sometimes referred to as the C_3 and C_4 cycles, respectively.

Carbon Dioxide Fixation by the C_3 Cycle

During the 1940s and 1950s, Melvin Calvin with Andrew Benson and other associates unravelled the outlines of CO_2 fixation into carbohydrates in photosynthetic plants and algae. It had been believed earlier that hexose sugars such as glucose were the initial products of dark reactions. Using the newly available radioactive isotopes of carbon, Calvin and colleagues applied $^{14}CO_2$ in intervals lasting only seconds, and identified the earliest labelled products. Most of the scheme we now accept was provided by Calvin's studies. The enzymes of the cycle were identified later by E. Racker and B. Horecker (Fig. 8.19).

In the initial step of the cycle, CO_2 from the air is joined with the five-carbon sugar **ribulose 1,5-diphosphate (RuDP)** to form a transient six-carbon compound. The reaction is catalyzed by the abundant and important enzyme **ribulose 1,5-diphosphate carboxylase (RuDP carboxylase)** that is found in the stroma of the chloroplast next to thylakoid membranes. The transient C_6 compound breaks down to form two molecules of the three-carbon sugar **3-phosphoglycerate.** This triose sugar is next phosphorylated at the expense of ATP to form an activated molecule able to accept hydrogens and electrons from NADPH. The activated three-carbon sugar is **3-phosphoglyceraldehyde.** This molecule is reduced in further reactions to form hexose sugars and more complex carbohydrates.

This set of reactions cannot account for the fact that all six carbons of the hexose sugar are ultimately derived from CO_2 during photosynthesis. Calvin and Benson thus proposed that there was a cyclic pathway in which one molecule of RuDP was regenerated for

each molecule of CO_2 fixed in photosynthesis. The regeneration sequence is quite complex, involving at least twelve different enzymes that catalyze steps leading to various 5-, 6-, and 7-carbon intermediates (Fig. 8.20). By continuous regeneration of RuDP, continuous CO_2 fixation can occur.

The link from the Calvin cycle, which is also known as the C_3 **cycle** because the initial important product of CO_2 fixation is a triose sugar, to the formation of hexose sugar is believed to involve six turns of the cycle. In this way, six molecules of CO_2 are fixed for each hexose and all six carbons of the hexose sugar are thus derived from CO_2 fixation within the chloroplast stroma.

From the complex sequence of RuDP regeneration, through a variety of sugar-phosphate intermediates, a molecule of fructose 6-phosphate is formed. Since all its six carbon atoms are derived from CO_2, we may state that 6 CO_2 produce $C_6H_{12}O_6$. The biosynthesis of sucrose and polysaccharides (such as starch), as well as other large molecules, proceeds from fructose 6-phosphate and its hexose derivatives. One of the vital products of these biosyntheses is a sugar-containing lipid which is a constituent of chloroplast membranes.

Each mole of hexose formed from CO_2 and H_2O in the classic equation of photosynthesis requires an input of 686 kilocalories. This amount of energy is based on the known standard free-energy change (ΔG^0) of the reverse reaction in which 686 kcal/mole is released when glucose is completely oxidized to CO_2 and H_2O. The classic equation for photosynthesis is

$$6 \ CO_2 + 12 \ H_2O \xrightarrow{\text{light}} C_6H_{12}O_6 + 6 \ O_2 + 6 \ H_2O$$
$$\Delta G^0 = 686 \text{ kcal/mole of glucose} \quad [8.10]$$

(*Note:* ^{18}O tracer studies showed that all O_2 was derived from H_2O, so 12 H_2O must be assimilated for each glucose formed in photosynthesis, to balance the equation.)

To determine the relative efficiency of photosyn-

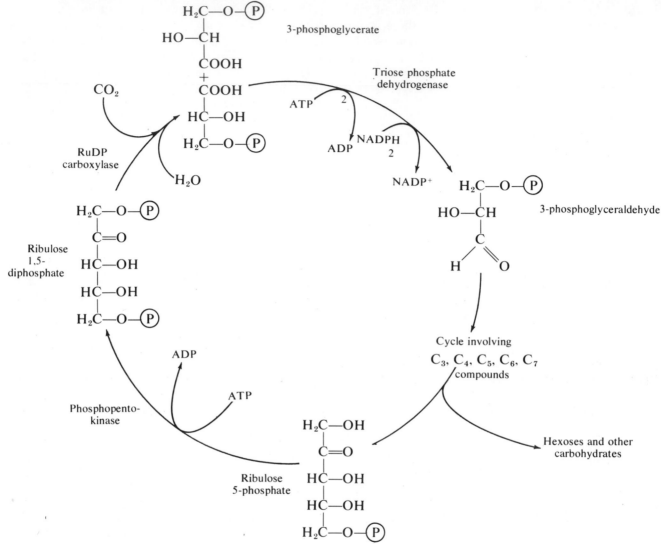

Figure 8.19
Reaction intermediates and enzymes of the C_3 cycle in which CO_2 is reduced to carbohydrate in the dark reactions of photosynthesis.

thetic dark reactions involving the C_3 pathway, we must consider the events at each turn of the C_3 cycle

$$CO_2 + 2\ NADPH + 3\ ATP \rightarrow$$
$$[CH_2O] + 2\ NADP^+ + 3\ ADP + 3\ P_i \qquad [8.11]$$

Since it requires six turns of the cycle to form one molecule of glucose, a total of 12 NADPH ($6 \times$ 2 NADPH) and 18 ATP (6×3 ATP) are utilized to form one molecule of glucose. If we accept the standard free-energy change for ATP hydrolysis to

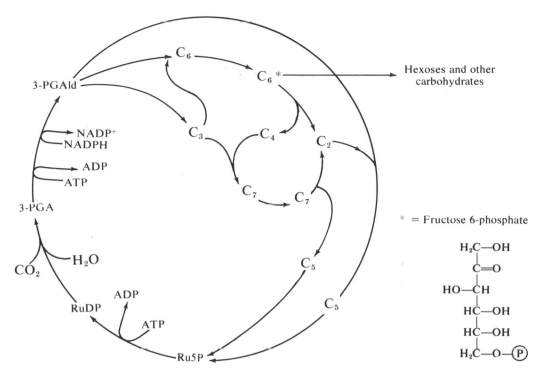

Figure 8.20
Regeneration of RuDP by a cyclic reaction series that includes various intermediates, one of which (*) is fructose 6-phosphate, a compound that can be diverted to other pathways of carbohydrate synthesis.

be 7 kcal/mole, and 62.6 kcal/mole for the oxidation of NADPH, then the total free-energy increment to form one mole of glucose is 877 kcal. The overall efficiency of photosynthetic dark reactions would then be $686/877 \times 100 = 78$ percent. This remarkably high efficiency is rarely achieved in biological activities.

It may be noted here that the overall equation for photosynthesis, in which glucose and oxygen are formed from carbon dioxide and water, bears a superficial resemblance to the overall equation for mitochondrial processing of glucose in aerobic respiration (see Chapter 7). The two processes are *not* a forward and reverse sequence of a common set of reactions. Completely different enzymes are involved in synthesis of glucose during photosynthesis and its aerobic degradation during respiration. Only 18 ATP are required to form a molecule of glucose in photosynthesis, whereas 44 ATP are produced in the combined reactions of glycolysis and aerobic respiration. The electron carriers in photosynthesis are $NADP^+$ groups, reduced to 12 NADPH for subsequent use in cellular biosyntheses. NADH and NAD^+ are involved in respiration reactions. Although it may seem that glucose formation from CO_2 and H_2O is a reverse sequence from respiratory degradation of glucose to CO_2 and H_2O, this is far from true, as the above-mentioned information should clearly indicate.

Carbon Dioxide Fixation by the C_4 Cycle

In 1966 M. D. Hatch and C. R. Slack reported on an alternative pathway for CO_2 fixation in photosynthesis. Their study of sugarcane, a tropical grass of agricultural importance, was based in turn on earlier studies by other investigators. The occurrence of an alternative mechanism for CO_2 fixation has led to interesting insights into evolutionary adaptations that ensure greater success for plants that grow in relatively harsh environments. These studies also hold some theoretical promise for improvements which may be incorporated into crop plants that are a major food source for countries in hot, sunny, arid climates.

When radioactively labeled carbon dioxide was presented to green cells in Calvin's studies, the earliest labeled product was 3-phosphoglycerate rather than hexose sugar. This result led to the concept of a C_3 cycle which ultimately produced the hexose. When similar labeling studies were conducted by Hatch and Slack, they found the earliest products to be an equilibrium mixture of three particular four-carbon dicarboxylic acids: oxaloacetate, malate, and aspartate. If they extended the time allowed for $^{14}CO_2$ labeling before examining the products of CO_2 fixation, a small amount of 3-phosphoglycerate was labeled as well as the dicarboxylic acids. These results with sugarcane were, therefore, contrary to expectations based on the activity of a C_3 cycle to produce the initial reduction of CO_2 in photosynthetic dark reactions.

Hatch and Slack proposed a scheme, which has since been expanded and confirmed, in which the carboxylation reaction fixing CO_2 into an organic product occurred via a different enzymatic pathway initially (Fig. 8.21). Instead of RuDP carboxylase action as in the C_3 cycle, the new cycle was postulated to begin with CO_2 fixation by action of **phosphoenolpyruvate carboxylase (PEP carboxylase)**. When three-carbon **phosphoenolpyruvate** is carboxylated by addition of CO_2, the four-carbon product is **oxaloacetate**. This breaks down readily to **malate** and **aspartate,** thus explaining the presence of the group of dicarboxylic acids that were labeled first in the $^{14}CO_2$ experiments. Neither malate nor aspartate can continue in reactions leading to carbohydrate synthesis, however, so other steps must intervene between this

stage and the formation of hexose. It has been proposed that the C_4 acids are degraded enzymatically to yield free CO_2 and pyruvate, a three-carbon intermediate. When pyruvate is phosphorylated at the expense of ATP to form phosphoenolpyruvate, the C_4 cycle can undergo another turn. The cycle continues to repeat as new CO_2 enters the system and is fixed into a C_4 product.

The CO_2 which is released during the C_4 cycle is then picked up by the RuDP carboxylase system in the Calvin C_3 cycle. There CO_2 undergoes reduction to carbohydrates as it does in the C_3 plants discussed earlier. In essence, C_4 plants have an additional CO_2-fixing sequence that is coordinated with the usual C_3 cycle and RuDP carboxylase. The CO_2 originally fixed by PEP carboxylase is ultimately handed over to RuDP carboxylase within the leaf, rather than being absorbed directly into the C_3 cycle as in C_3 plants.

This alternative CO_2-fixing pattern has since been shown to occur in about 100 species of plants belonging to many unrelated families (including some of the algae). All the higher plants with the C_4 **cycle** ability also have two other features in common: (1) anatomically similar pattern of cells in the leaf, and (2) ability to grow successfully in situations where low levels of CO_2 in the air are limiting to the rate of photosynthesis. In fact, these plants share the ability to survive under conditions that incapacitate or even kill C_3 plants. Examination of leaf anatomy is very revealing, as is the compartmentation of the enzymes of the C_3 and C_4 cycles. The two features are correlated in their occurrence. Plants with both a C_3 and C_4 cycle have one pattern of leaf cells and tissues, while plants with only the C_3 cycle enzymes have another anatomical pattern (Fig. 8.22).

The Leaf System in C_3 and C_4 Plants

C_4 plants can thrive at concentrations of CO_2 as low as 1–2 parts per million of air, whereas C_3 plants stop photosynthesizing near 50 parts per million of CO_2. The usual concentration of CO_2 in our atmosphere is about 300 parts per million or 0.03 percent.

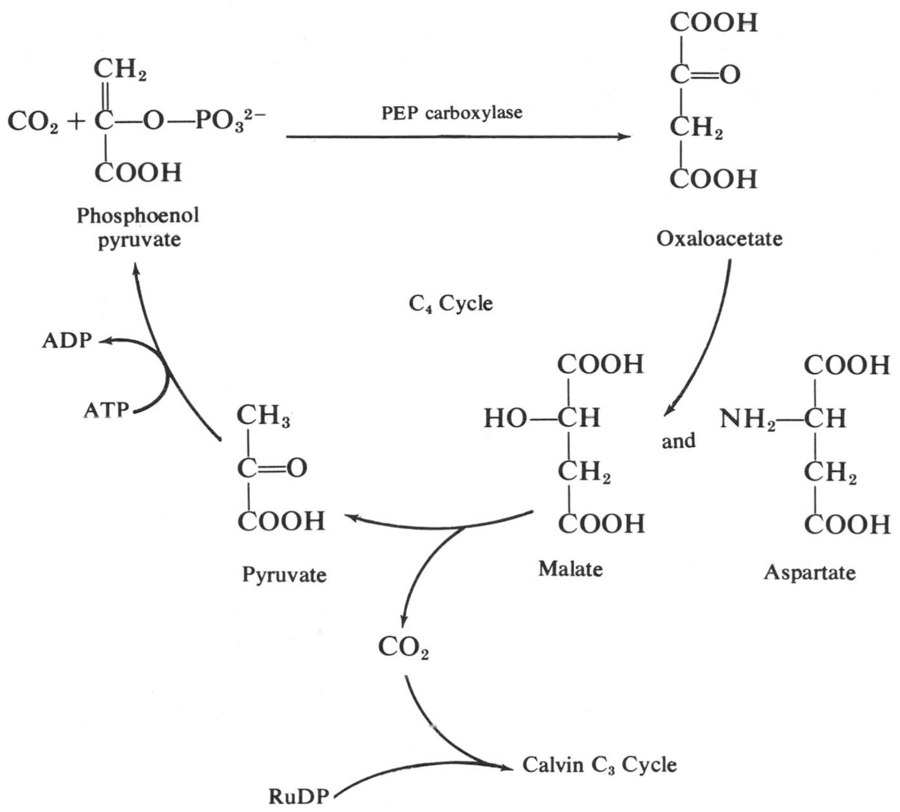

Figure 8.21
Reactions of the accessory C_4 cycle of CO_2 reduction as proposed by Hatch and Slack for some plants. The site of interaction of the C_4 and C_3 cycles that operate in the same plants are also shown.

Gas exchanges between leaves and the outside air take place mainly through pores in the outer epidermal tissue of the leaf and not through the thick, waxy, nonporous areas. These pores (known as stomates) open and close, and gas exchange requires open or partially open pores.

In regions where there is high light intensity, high temperature, and very little water, leaf pores are almost closed. This response permits the leaf tissues to lose the least amount of water by diffusion toward the drier air, but when these pores are partially closed very little CO_2 enters the leaf. In C_4 plants, PEP carboxylase is very efficient in absorbing CO_2 and incorporating it into C_4 acids. The action of this carboxylase keeps the concentration of CO_2 inside the leaf very low and helps to maintain a greater concentration difference between CO_2 outside and inside the leaf. Because of this, CO_2 continues to enter C_4 leaves even though outside concentrations are very low and pores are barely open. In C_3 plants, on the other hand, RuDP carboxylase is much less efficient in absorbing the CO_2 that enters the leaf, so the inside concentra-

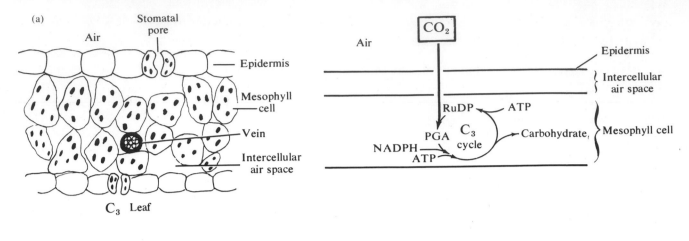

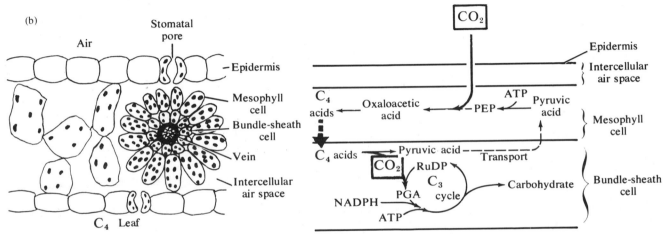

Figure 8.22
Comparison of the leaf anatomy and CO_2 reduction pathways in C_3 and C_4 plants: (a) leaves having only the C_3 pathway contain large mesophyll cells with chloroplasts and air spaces saturated with water vapor. CO_2 enters through stomatal pores and moves into the chloroplasts of mesophyll cells, where reduction occurs via the C_3 cycle. (b) Leaves with the accessory C_4 cycle have most of their chloroplast-containing cells in a unique arrangement of mesophyll cells in a layer around a layer of bundle-sheath cells, which surrounds a leaf vein. CO_2 enters the meso-phyll cell chloroplasts first and undergoes a reaction with the 3-carbon compound phosphoenolpyruvate (PEP) to form C_4 oxaloacetic acid. Further processing of the C_4 acids occurs in the bundle-sheath chloroplasts, where C_4 acids are decarboxylated to yield free CO_2. CO_2 enters the major C_3 reduction cycle. The remaining 3-carbon pyruvate is returned to the mesophyll cells where it is phosphorylated to PEP, and the C_4 cycle is completed and ready for another CO_2 fixation event.

tion remains higher. When the CO_2 in the air immediately around the leaf drops below 50 parts per million, the difference in concentration is not enough to maintain a diffusion gradient. CO_2 no longer enters the C_3 leaf, and photosynthesis stops. In laboratory experiments, C_4 leaves continue to carry out photosynthesis even when the air contains only 1–2 parts per million CO_2, because a diffusion gradient is maintained as CO_2 is swept away by PEP carboxylase activity in the C_4 cycle. CO_2 thus continues to enter C_4 leaves. Some plants can grow very well at noon in Death Valley summers when the temperature around the plant is near 50°C. These plants are of the C_4 type which grow well and photosynthesize efficiently under relatively hostile conditions.

Two kinds of programs directed toward improving food supplies in hot, dry climates are currently under study. If favorable characteristics of C_4 plants can be bred into crops that have only the C_3 pathway, yields might increase since these plants could continue to grow well even in poor seasons. It is also possible to look for potentially higher-yielding strains in C_4 plants that provide staple foods in hot, dry regions of the world. By selecting higher-yielding individuals in a long-range program, more productive harvests could be obtained. This type of program would be particularly important in increasing crop yields in underdeveloped regions in which crops grow in scrubby or semi-desert zones in parts of Asia, Africa, and other continents of the world. Some high-yield crop plants already form important foundations for world agriculture, including C_4 species such as sugarcane, sorghum, and corn. Some undesirable but successful C_4 plants also exist, such as the crabgrass of otherwise well-tended lawns.

PHOTORESPIRATION

The process of **photorespiration** involves consumption of oxygen and release of carbon dioxide in the light; that is, it is considered to be a light-enhanced respiration. The process is energetically wasteful since it diverts the flow of energy from CO_2 reduction to O_2 reduction. As much as 50 percent of the reducing power generated in photosynthesis may be diverted in this process. Based on measurements of the amount of CO_2 released in the light, C_4 plants do not appear to carry out photorespiration whereas C_3 plants do photorespire. The following discussion will therefore be restricted to studies of C_3 species.

The elegant studies of N. E. Tolbert and his associates have provided a large part of the current understanding about processes affiliated with leaf photorespiration. Compartmentalized reactions within chloroplasts, mitochondria, microbodies, and cytosol of leaf cells are closely coordinated to recoup some of the reducing power which is lost in photorespiration.

It was mentioned earlier that C_3 plants stop photosynthesizing when the CO_2 air level reaches about 50 parts per million. At this concentration, photorespiration takes place, resulting in the release of CO_2 in the light rather than its incorporation into carbohydrates in photosynthesis. At the same time that CO_2 concentration is low, a high concentration of O_2 is found in the leaf cells and in the surrounding air. Under these twin conditions of low CO_2 and high O_2, the enzyme RuDP carboxylase acts catalytically as an oxygenase more than as a carboxylase. Instead of forming 3-phosphoglycerate exclusively (see Fig. 8.19), after fixing CO_2 by its carboxylase ability, the enzyme sponsors oxygen consumption and incorporation of an oxygen atom into RuDP by virtue of its oxygenase activity. When this happens, RuDP remains a five-carbon compound and does not become converted to a six-carbon intermediate as in the C_3 cycle. As RuDP is oxygenated, therefore, it breaks down to form some 3-phosphoglycerate and the two-carbon compound **phosphoglycolate** (Fig. 8.23). The 3-phosphoglycerate is fed back into the Calvin C_3 cycle where RuDP is regenerated. Phosphoglycolate formed in the chloroplast by RuDP carboxylase-

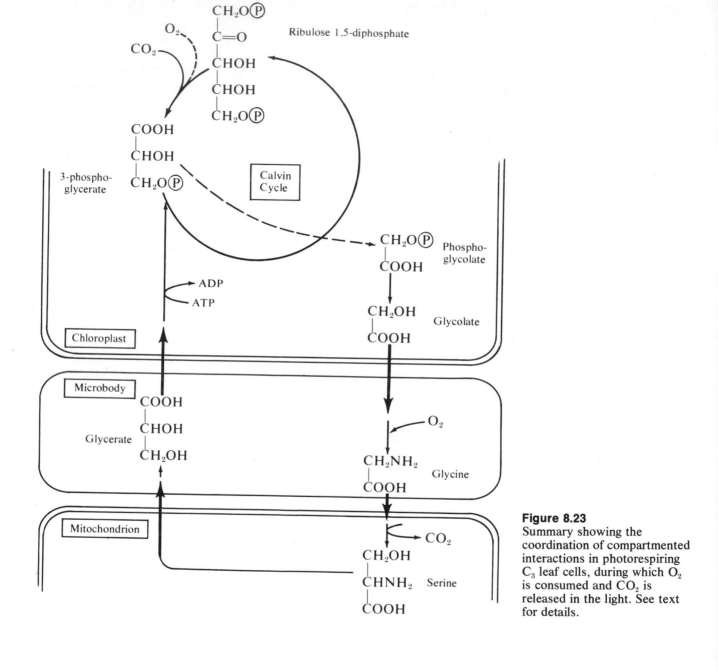

Figure 8.23
Summary showing the coordination of compartmented interactions in photorespiring C_3 leaf cells, during which O_2 is consumed and CO_2 is released in the light. See text for details.

oxygenase, however, undergoes a complex set of reactions which involves three other cellular compartments.

Phosphoglycolate is converted to **glycolate** within the chloroplast, after which glycolate is transferred to microbodies. Within the microbodies, glycolate is processed in sequential reactions and is converted to the amino acid **glycine.** This amino acid is then transferred to mitochondria where reactions take place leading to formation of another amino acid, **serine.** Serine next leaves the mitochondrion and enters the microbody where it is converted to **glycerate.** Glycerate is converted to **3-phosphoglycerate.** At this point, as occurs normally in the C_3 cycle, 3-phosphoglycerate is converted to sugar-phosphates and more complex carbohydrates. The latter reactions probably occur in chloroplasts and the cytosol since enzymes of the pathway are found in both compartments.

These reactions explain the basis for the characteristic pattern of gas exchange in photorespiration. Oxygen is consumed during the step catalyzed by RuDP oxygenase activity, producing phosphoglycolate and 3-phosphoglycerate from RuDP in the chloroplast. Carbon dioxide is released from mitochondria during the conversion of glycine to serine which takes place there. Some oxygen also is taken up by microbodies in the initial step when glycolate is reduced.

Under unfavorable daylight conditions of high O_2 and low CO_2, chloroplast RuDP is converted to phosphoglycolate and 3-glycerophosphate molecules. As phosphoglycolate is converted to glycolate in this organelle, photorespiration reactions are triggered. If glycolate production in the light comes to a halt, photorespiration also stops. Although there are energetically wasteful steps in the compartmented activities described, there is continued flow of carbon from RuDP and the Calvin cycle in chloroplasts through the glycolate pathway in microbodies and mitochondria and back to 3-phosphoglycerate in the chloroplast and cytosol. Some additional carbohydrate synthesis can take place, therefore, as more of this vital three-carbon intermediate is produced.

The photorespiration sequence is remarkably coordinated through a circuitous route and may be aided physically by the fact that all three kinds of interacting organelles exist in closely apposed associations (Fig. 8.24). Microbodies and mitochondria appear to be literally squeezed up against chloroplasts in leaf cells. Even though photosynthetic dark reactions take place at reduced rates, the RuDP molecules continue to serve as a link between light and dark reactions, and some carbohydrates continue to be synthesized despite little or no CO_2 fixation. Since C_4 plants are almost unaffected by low CO_2 concentration in the air, they maintain high rates of photosynthesis. They are therefore very efficient in producing carbohydrates for growth in identical environments which sharply inhibit C_3 plant photosynthesis.

FINE-STRUCTURAL FEATURES OF PHOTOSYNTHESIS

A rough sketch of chloroplast structure was given earlier in this chapter. This sketch showed that thylakoids exist within the stroma and that two parallel enveloping membranes enclose these systems. Considerable evidence shows that photosensitive pigments and other components of the light-dependent reactions are housed within the thylakoid membranes. Enzymes of the dark reactions and related biosyntheses are located within the stroma of the chloroplast. In addition to these general localizations, the nature and location of membrane-bound photosynthetic assemblies have been studied using electron microscopy.

In 1963 Roderick Park reported on the first of a series of fine-structure studies on thylakoid membranes prepared by freeze-fracture methods (Fig. 8.25). Daniel Branton has shown that thylakoid and other membranes are actually split during freeze-fracturing because the hydrophobic bonds primarily responsible for holding the membrane together are

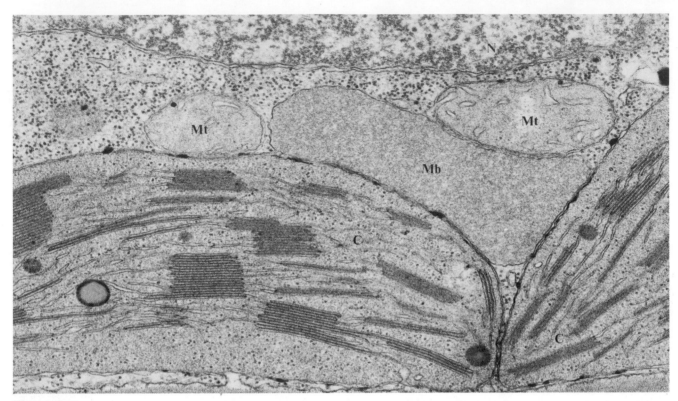

Figure 8.24
Thin section of leaf mesophyll cell of tobacco (*Nicotiana tabacum*). The microbody (Mb) is wedged against the two chloroplasts (C) and one of the mitochondria (Mt). A part of the nucleus (N) is at the top of the photograph. Free flow of metabolites among the organelles is facilitated by their close physical proximity to one another. $\times$ 45,000. (Courtesy of E. H. Newcomb and S. E. Frederick)

Figure 8.25
Steps in freeze-fracture or freeze-etch preparations of biological materials: (a) freezing. Specimen is in ice at $-100\,°C$ or under liquid nitrogen at $-196\,°C$, on a specimen support. (b) Fracturing. Specimen is cut with a cold knife, exposing the inner surfaces of the cleaved membrane. If the temperature is lower than $-120\,°C$, ice will not sublime and etching cannot and does not take place. Fresh fracture faces are replicated as shown in d. (c) Etching. If the temperature is above $-120\,°C$ then etching occurs as ice sublimes. The longer the time allowed between fracturing and replication, the deeper the etching. (d) Replica formation by coating. The surfaces (fresh or etched) are coated with platinum and carbon. These replicas of the freeze-fractured or freeze-etched surfaces of split membranes and membrane surfaces are then observed and photographed using the electron microscope. (e) Freeze-etch preparation showing the nucleus (N) with numerous pores, part of the endoplasmic reticulum (ER). and an assortment of vesicles and fracture faces. $\times$ 60,000. (Courtesy of D. Branton)

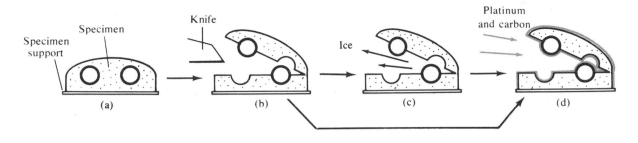

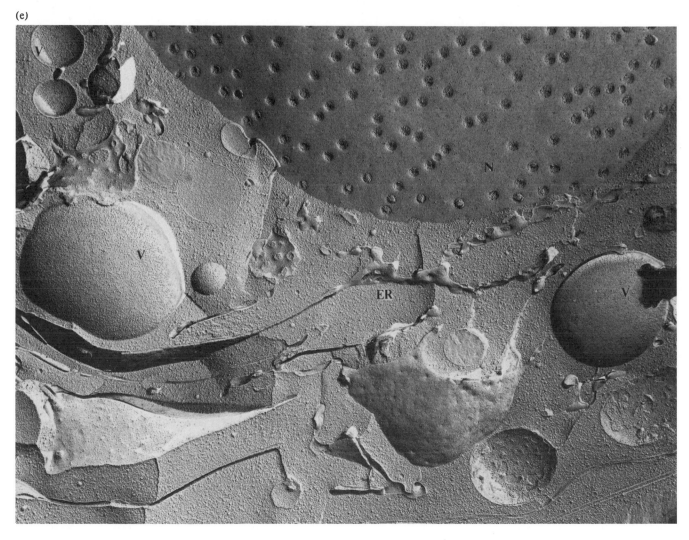

(e)

weakened during the freezing step. The membrane splits along this weakened plane in the fracture step and opens so that the peeled-back halves reveal their *intra*-membrane features (Fig. 8.26).

Park originally described a particle embedded within the thylakoid membrane wherever a break had occurred during freeze-fracturing to reveal the inner features of the membrane (Fig. 8.27). He termed the particle a *quantasome,* and he suggested it was the morphological equivalent of the assembly of photosynthetic molecules in a photosynthetic unit. The quantasome measured about 175–200 Å in its broadest dimension, and Park calculated that a particle of just this size would have a molecular weight of about 2 million daltons. This particle would adequately accommodate about 230–250 chlorophylls, carotenoids, electron transport components, and other essentials of photosystems I and II.

The original quantasome particle is now considered to be an artifact. There is no question, however, that particles do exist within these membranes and can be visualized by freeze-fracture methods. The current controversies revolve around the numbers of different kinds of particles and their chemical makeup. It now seems less likely that any one of these particle types is the embodiment of the photosynthetic unit, or quantasome, but little else is known at the present time. In a recent study, Peter Satir has described four sizes of particles in thylakoid membranes. These particles are distributed homogeneously within a membrane from a single thylakoid. When individual thylakoids come together in pairs, some particles redistribute within the membrane wherever a zone of contact occurs. Stacking of thylakoids into organized arrays appears to be correlated with particle distributions and pattern. The relationship between these events, however, remains undetermined. We will discuss some of these features in a later section of this chapter.

The two principal kinds of particles within the stroma are 70S ribosomes of the chloroplast protein-synthesizing machinery, and molecules of RuDP carboxylase. The carboxylase molecules tend to be found along the outer face of the thylakoid and may sediment along with the membranes in unwashed preparations. On washing, carboxylase is easily removed

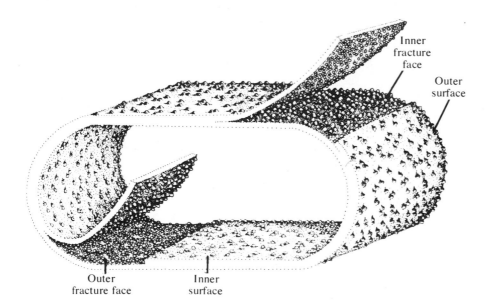

Figure 8.26
Drawing representative of a single chloroplast thylakoid that has been fractured and etched to various degrees. The inner and outer *fracture faces* (exposed interiors of the membrane) are exposed by fracturing with minimal etching, while the outer and inner *surfaces* are exposed when a fractured thylakoid is deep-etched. (Reproduced with permission from Garber, M. P., and P. L. Steponkus, 1974, *J. Cell Biol.* **63**: 28, Fig. 5.)

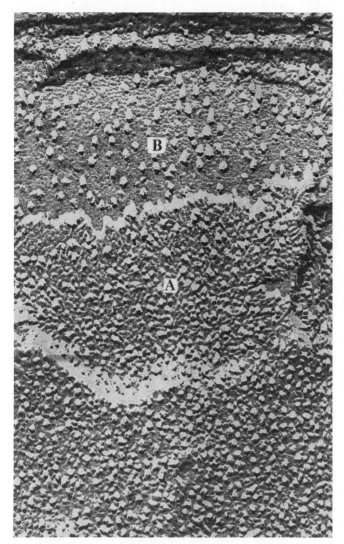

Figure 8.27
Freeze-cleave preparation of *Euglena* chloroplast thylakoids. The outer fracture face (A) closer to the stroma has many more particles than occur at the inner fracture face (B) closer to the space within the thylakoid. × 125,000. (Courtesy of U. W. Goodenough and K. Miller)

from the remaining thylakoid structure and is not, therefore, considered to be a component of this membrane system. Aggregates of DNA fibrils are also present in the stroma. They usually are found close to the inner membrane of the chloroplast near the periphery of the organelle and may even be attached to this membrane according to some investigators.

CHLOROPLAST BIOGENESIS

There is no substantiated case for *de novo* origin of chloroplasts. Occasional reports of spontaneous appearance of plastids have repeatedly been disproven by later observations. In higher plants and occasional lower plants or unicells like *Euglena,* chloroplasts differentiate from a primordium known as a **proplastid.** The sequence of development from proplastid to mature chloroplast has been determined from many ultrastructural studies of normal cells and from some useful mutants that are defective in particular steps of the maturation process.

Synthesis of chlorophyll and formation of thylakoids are independent events since mutants incapable of thylakoid development will still produce normal chlorophyll, and normal thylakoids may develop in mutants having very low levels of chlorophyll. This pigment is rapidly destroyed in some kinds of mutants, presumably because it fails to become integrated into protective membranes or it is degraded by chemical reactions in the light.

Multiplication and Differentiation of Proplastids

Proplastids usually are about 0.5–1.0 × 1.0–1.5 μm in size. Their two bounding membranes surround an unstructured matrix in which DNA fibrils, ribosomes, and some form of starch or other carbohydrate may be present. Some tubular membrane elements or recognizable thylakoid primoridia may exist in the undeveloped proplastid (Fig. 8.28). On exposure to light, vesicular and tubular invaginations of the inner membrane pinch off and gradually become dispersed

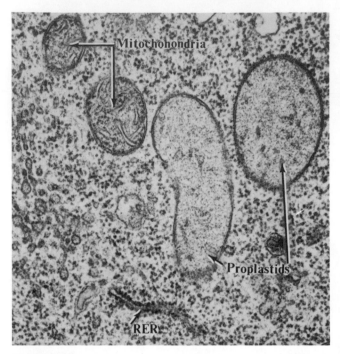

Figure 8.28
Undifferentiated proplastids in a thin section of *Arabidopsis thaliana* root cell. × 37,000. (Courtesy of M. C. Ledbetter)

within the stroma. These membrane elements ultimately become disposed into the long, flattened stroma thylakoids and where they are folded into multiple layers, grana thylakoid stacks may also be evident (Fig. 8.29). Synthesis of photosynthetic pigments occurs simultaneously. At maturity, typical chloroplast ultrastructure and chemistry have evolved from a primordial proplastid initial. There seems to be no fusion of proplastids to produce the larger chloroplast; instead it is believed that a general growth of new membranes leads to enlargement of the mature organelle.

Increase in proplastid number involves a pinching-in similar to cleavage and later separation of the two products. There is no particular evidence showing that new proplastids bud from a mother structure. In some species a cleavagelike process also occurs in mature or

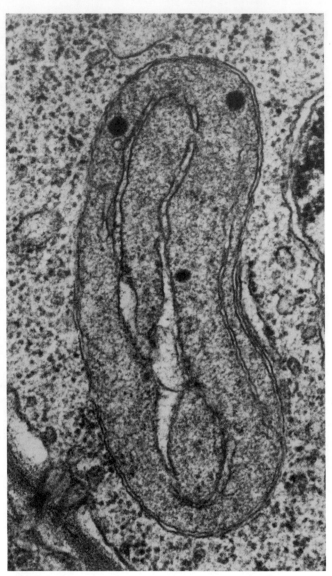

Figure 8.29
The beginnings of thylakoid development in a proplastid from a cell of plantain (*Plantago*). × 62,000. (Courtesy of M. C. Ledbetter)

nearly mature chloroplasts that leads to numerical increase. A constriction process therefore underlies formation of new proplastids or of new chloroplasts in most green eukaryotic cells.

When green plant cells are returned to the dark for a protracted period, chloroplasts are transformed into **etioplasts** that lack typical thylakoid structures. The aggregation of membranes in etioplasts may be somewhat disorganized or may show a lattice-work arrangement. In either case the membrane aggregation has been called a **prolamellar body.** Similar membrane organization develops in plant cells that have been grown in the dark; that is, proplastids will differentiate into etioplasts in the dark or chloroplasts in the light. Upon exposure to light, etioplasts are converted or reconverted into functional green chloroplasts which have typical photosynthetic ultrastructure and chemistry (Fig. 8.30).

Etioplasts contain *protochlorophyll,* a precursor of chlorophyll *a.* Protochlorophyll is localized within the prolamellar body. In the light, this precursor is reduced to chlorophyll *a.* Although it may take only 1 minute for protochlorophyll conversion to chlorophyll *a* in the light, several hours elapse before detectable amounts of new chlorophyll *a* and chlorophyll *b* appear after illumination. The conversion of prolamellar-body tubules into flat sheets of thylakoids begins within minutes after etioplasts have been illuminated.

Biogenesis of Thylakoid Membranes

The mode of formation of photosynthetic membranes, their organization into functional arrays, and the origin of membrane proteins have been studied intensively. Particularly useful mutant strains of unicells such as *Chlamydomonas* and flowering plants such as *Hordeum* (barley) have been exploited profitably in a number of investigations. In addition to information concerning membrane biogenesis and organization, some of the underlying qualities of chloroplast genetics are now understood as a result of analysis of mutant forms.

At the outset it should be clearly stated that virtu-ally all inherited traits related to chloroplast structure, function, and biochemistry have been traced to nuclear genes. A few inherited traits follow a non-Mendelian pattern typical of genes outside the chromosomes of the nucleus. Some of these mutations have been associated with changes in chloroplast DNA, while others have not been localized specifically in one region or another of the cytoplasm even though the affected trait is expressed in the chloroplast or its activity.

Nuclear Gene Control of Chloroplast Development

Barley plants have been under study by Diter von Wettstein, a Danish biologist, for many years. He has shown that many different nuclear genes cooperate in directing normal membrane development as proplastids mature to functional chloroplasts. In a series of mutants described as *xantha,* there may be almost no thylakoid development, aberrant organization of stroma and grana thylakoids, no grana formation, and other developmental variations. Sometimes no chlorophyll is synthesized in these mutants or some disturbance in normal synthesis is exhibited, but other mutants of the *xantha* series continue to produce normal amounts of functional chlorophylls. It would thus appear that thylakoid formation and chlorophyll synthesis are under independent genetic control but usually proceed in coordinated events in nonmutant cells. Modification of one set of processes need not influence normal events in the accompanying set of processes, as shown by *xantha* mutants.

Genetic analysis has shown that each of the specific mutant defects was due to a different nuclear gene. In spite of this genetic simplicity, the chloroplast malformations are relatively gross so it is difficult to make fine distinctions in the association of structural and chemical development of the organelle. Some observed effects could well be due to secondary effects arising as a result of some primary gene malfunction, but the relationships are too obscure to be observed directly.

A recent report by von Wettstein concerned mu-

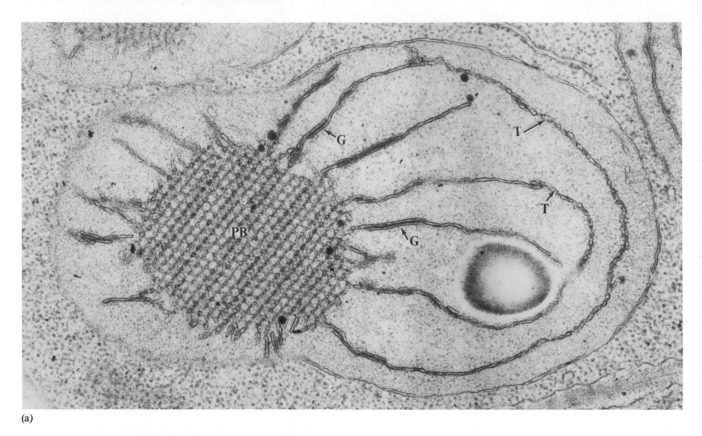

(a)

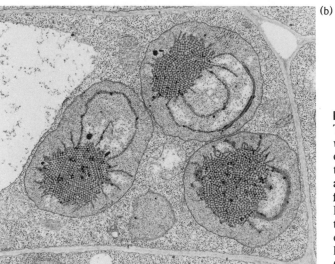

(b)

Figure 8.30
Thylakoid differentiation in etioplasts of bean (*Phaseolus vulgaris*) leaf grown in the light. (a) Note the crystalline display of the prolamellar body (PB) from which the thylakoids develop, and that grana thylakoids (G) have already begun to stack (at arrows). Thylakoids form from fused vesicles (T, at arrows). × 39,000. (Courtesy of M. C. Ledbetter) (b) Etioplast differentiation is at approximately the same stage throughout the leaf cell population in 11-day-old seedlings. Three organelles in part of one cell are shown here. × 12,500. (Courtesy of E. H. Newcomb and P. J. Gruber)

tants that were defective in synthesis of precursors for chlorophyll and β-carotene; characteristic thylakoid abnormalities were also present. Genetic analysis showed that different structural genes guided chemical formation of the pigments in pathways that were very closely coordinated. The preliminary information also pointed to action of regulatory genes that coordinated the separate pathways specifying chlorophyll and carotenoid biosynthesis. In chloroplast development, therefore, genes act not only in the determination of specific molecules for structure and function, but also in regulating complex sequences of development. Regulatory gene action provides a coarse control over the enzymatic reactions leading to functional and structural molecules that are unique to the chloroplast.

Chlamydomonas reinhardi is a favorite unicellular organism for genetic studies of chloroplast structure, function, and inheritance patterns. The organism is easily manipulated under various conditions for growth and development, and its single large chloroplast can be studied by biochemical and electron microscopical methods. R. P. Levine and colleagues have reconstructed part of the probable sequence of electron transport between photosystems I and II by noting defects in electron transport in mutant strains with defective or missing components.

Levine and Ursula Goodenough also have attempted to relate thylakoid formation and organization to photosynthetic deficiencies in a number of mutants (Table 8.4). Membrane and thylakoid development appeared to be relatively normal in these mutants, even in one strain whose chlorophyll levels were about 25 percent of wild-type amounts. Two principal correlations were found: (1) fusion of thylakoids into stacks (or "stacking") differed between wild-type and 6 of the 7 electron-transport mutants, and (2) mutants with similar disabilities showed similar "stacking" patterns even when the causative nuclear genes were on different chromosomes (unlinked).

Mutants defective in cytochrome-559 all had long, single, unstacked thylakoids. Mutants lacking one or the other of the adjacent electron-transport components, cytochrome-553 and plastocyanin, had long pairs of thylakoids; P700 mutants showed "hyperstacked" thylakoids; and so forth. Only the transport mutant *ac-21* resembled wild-type in having a confluent system of short and long thylakoids in stacks averaging 3 units (range: 2–5). Two other mutants whose photosystems were intact also resembled the wild-type in this regard. One of these two, strain *ac-1*, had low levels of chlorophyll *a;* the second of these, strain *F-60,* lacked an enzyme involved in CO_2 fixa-

Table 8.4 Comparisons of chlorophyll content and thylakoids in wild-type and mutant strains of *Chlamydomonas*

STRAIN	DEFICIENCY	TOTAL CHLOROPHYLL (μg/10^6 CELLS)	RATIO CHL a:b	THYLAKOID "STACKING" PATTERN	
				AVERAGE NO./STACK	LENGTH AND ORGANIZATION
Wild-type	none	3.0–4.6	2.4	3	Short, long; confluent
ac-115	Cytochrome-559	3.7	2.4	1	Long; singles mostly
ac-141	Cytochrome-559	2.6	2.6	1	Long; singles mostly
F-34	Cytochrome-559	—	—	1	Long; singles mostly
ac-206	Cytochrome-553	2.3	2.7	2	Long; paired
ac-208	Plastocyanin	3.3	2.6	2	Long; paired
ac-21	"M" protein	2.8	2.6	2–3	Resembles wild-type
ac-80a	P700	3.9	1.7	5 or more	Long; wide "stacks"
F-1	P700	3.3	1.6	5 or more	Long; wide "stacks"
*ac-1**	Chlorophyll	0.3–1.3	—	3	Resembles wild-type
*F-60**	CO_2 fixation	2.9	2.3	3	Resembles wild-type

* Strains are not deficient in electron carriers or photosystem components.

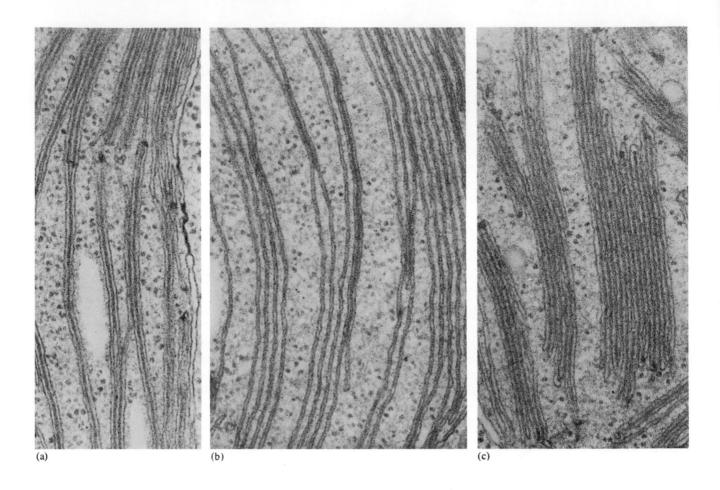

(a)   (b)   (c)

Figure 8.31
Variations in thylakoid "stacking" pattern in *Chlamydomonas reinhardi:* (a) wild type; (b) mutant strain *ac-115,* deficient in cytochrome-559, with single, unstacked thylakoids predominating; (c) mutant strain *F-1,* deficient in P700 chlorophyll, with "superstacked" thylakoids. × 65,000. (Courtesy of U. W. Goodenough, from Goodenough, U. W., and R. P. Levine, 1969, *Plant Physiol.* **44**:990–1000.)

tion. Both *ac-1* and *F-60* showed typical wild-type chloroplast ultrastructure (Fig. 8.31).

These observations underscore the importance of another dimension of chloroplast development, namely, *organization* of thylakoids into particular arrays. It can be reasonably inferred that particular photosynthetic capacities are related to particular organizational patterns of participating membrane systems. There are other hints that such relationships are significant. Each of the major groups of algae displays an invariant pattern of thylakoid organization (see Figs. 8.5 and 8.6). This observation indicates that thylakoid grouping is at least correlated with major patterns of photosynthetic chemistry and action.

The two simplest evolutionary groups of algae are blue-green and red species, and each of these has single, unstacked thylakoids. Peter Satir has shown that intramembrane particles are distributed differently in single thylakoids and in fused thylakoids at the zone of contact. It will be interesting to see whether particle distribution is related to thylakoid fusion alone or to normal photosynthetic activity in stacked thylakoids and deficient activity in unstacked units. Since these organizational variations have been shown to be under gene control, it is clear that formation of functional chloroplasts depends on many levels of gene-directed activities. Genes act in specifying structure, regulation, and organization of chloroplast components.

SUGGESTED READING

Books, Monographs, and Symposia

Clayton, R. K. 1971. *Light and Living Matter,* vol. 2. New York: McGraw-Hill.

Kirk, J. T. O., and Tilney-Bassett, R. A. E. 1967. *The Plastids.* San Francisco: W. H. Freeman.

Whittingham, C. P. 1974. *The Mechanism of Photosynthesis.* London: Edward Arnold.

Articles and Reviews

Arnon, D. I. 1960. The role of light in photosynthesis. *Scientific American* 203(5):104–118.

Bassham, J. A. 1962. The path of carbon in photosynthesis. *Scientific American* 206(6):88–100.

Björkman, O., and Berry, J. 1973. High-efficiency photosynthesis. *Scientific American* 229(4):80–93.

Cohen-Bazire, G. 1971. The photosynthetic apparatus of procaryotic organisms. In *Biological Ultrastructure: The Origin of Cell Organelles,* pp. 65–90. Corvallis: Oregon State Univ. Press.

Goodenough, U. W., and Levine, R. P. 1969. Chloroplast ultrastructure in mutant strains of *Chlamydomonas reinhardi* lacking components of the photosynthetic apparatus. *Plant Physiology* 44:990–1000.

Goodenough, U. W., and Levine, R. P. 1970. Chloroplast structure and function in *ac-20,* a mutant strain of *Chlamydomonas reinhardi.* III. Chloroplast ribosomes and membrane organization. *Journal of Cell Biology* 44:547–562.

Govindjee, and Govindjee, R. 1974. The absorption of light in photosynthesis. *Scientific American* 231(6):68–82.

Gunning, B. E. S. 1965. The greening process in plastids. I. The structure of the prolamellar body. *Protoplasma* 60:111–130.

Hatch, M. D., and Slack, C. R. 1966. Photosynthesis by sugar-cane leaves: A new carboxylation reaction and the pathway of sugar formation. *Biochemical Journal* 101:103–111.

Hill, R. 1937. Oxygen evolved by isolated chloroplasts. *Nature* 139:881–882.

Jagendorf, A. 1967. Acid-base transitions and phosphorylation by chloroplasts. *Federation Proceedings* 26:1361–1369.

Levine, R. P. 1967. Genetic dissection of photosynthesis. *Science* 162:768–771.

Levine, R. P. 1969. The mechanism of photosynthesis. *Scientific American* 221(6):58–70.

Ojakian, G. K., and Satir, P. 1974. Particle movements in chloroplast membranes: Quantitative measurements of membrane fluidity by the freeze-fracture technique. *Proceedings of the National Academy of Sciences* 71:2052–2056.

Shumway, L. K., and Weier, T. E. 1967. The chloroplast structure of *iojap* maize. *American Journal of Botany* 54:773–780.

Weier, T. E., and Benson, A. A. 1967. The molecular organization of chloroplast membranes. *American Journal of Botany* 54:389–402.

Whatley, F. R., Togawa, K., and Arnon, D. I. 1963. Separation of the light and dark reactions in electron transfer during photosynthesis. *Proceedings of the National Academy of Sciences* 49:266–270.

Chapter 9

Genetic Activities of Organelles

Our understanding of organelle genetics and molecular constructs which underlie it has reached its present status by a very erratic course of events. Some difficulties which delayed this understanding were caused by technical problems in tackling organelle systems genetically. Others were caused by very strong reluctance to accept the existence of a molecular genetic apparatus in an organelle. Biologists were comfortable with the idea of genes in the nucleus but not with the presence of additional genes outside the nucleus. While much remains to be learned about the genetic activities of organelles, considerable progress has been made in describing and defining the molecular components that underwrite the extranuclear genetic system. Evidence showing that DNA, RNA, and protein are the basic molecules involved is comforting proof of the universality of genetic material. Since the organelle genetic system can be accommodated into the acceptable framework of replication, transcription, and translation there is additional evidence of the legitimacy of such a system.

Mitochondria and chloroplasts are the only two eukaryotic organelle types that have been shown to contain their own components for genetic activities.

They not only have unique DNA, which is replicated within the organelle itself, but mitochondria and chloroplasts also maintain their own ribosomal protein-synthesizing machinery on which translation takes place. Reports of the presence of DNA or RNA or ribosomes in other organelles appear periodically. To date, none of these studies has shown convincingly that these materials represent components of a genetic apparatus in these organelles. Centrioles appear to lack DNA. Although earlier studies produced some experimental support for its existence in these structures, more stringent evidence from recent investigations apparently rules out its occurrence. Some studies have reported DNA in microbodies, but these observations have been rejected because the DNA has been shown to be contamination from other cellular compartments. Further discussion of genetic systems in organelles other than mitochondria and chloroplasts appears in chapters devoted to the different compartmented systems (Chapters 10 and 11). In this chapter we will be concerned only with features of the established genetic systems in the mitochondrion and chloroplast of eukaryotes.

ORGANELLE DNA

Almost all known genetic systems are based on informational DNA duplex molecules. Exceptions occur among some viruses, such as those containing RNA and others with single-stranded DNA. If genetic DNA does exist in an organelle, we would expect the molecule to be double-stranded and to behave like any other DNA in chemical and physical tests. One way to identify DNA, therefore, is to see whether or not it stains like genetic material. Two particular staining tests are widely used: (1) the *Feulgen test* which reveals chromosomal DNA as magenta-stained material, and (2) *acridine orange* staining which yields a greenish-yellow fluorescence with DNA and an orange fluorescent color when combined with RNA.

Various studies of the staining reactions of organelle genetic material with Feulgen and acridine orange procedures have been reported over the years. Mitochondria of certain protozoa were shown to contain Feulgen-positive material as early as 1926, but little attention was paid to the results at that time or for many years later. A positive reaction after acridine orange staining was reported in 1962 for chloroplasts of *Chlamydomonas*, but, again, there was limited acceptance of these results as evidence supporting the presence of DNA in the organelle. The greatest difficulty in accepting such evidence was that similar staining tests almost always failed to yield positive results with other species or, sometimes, with the same species.

The first important breakthroughs came from electron microscopy studies in the early 1960s. Using newly developed methods originally designed to visualize bacterial DNA, Hans Ris showed that fibrils in mitochondria and chloroplasts resembled fibrils of bacterial DNA exposed to the same test conditions (Fig. 9.1). When Margit Nass and colleagues verified these results using the same methods for both plant and animal cells, the existence of organelle DNA was finally accepted. Once the idea was accepted, numerous reports began to appear almost immediately for various species studied in various laboratories. These initial observations led to a landslide of reports and to further analysis of the nature of organelle DNA molecules.

Double Helix Structure

The stained DNA fibrils were about 25 Å in diameter, a size which implied that the molecules were duplex. Biochemical analysis revealed that organelle DNA contained the usual four bases and that adenine and thymine were present in equal amounts, as were guanine and cytosine. Equimolar 1 : 1 ratios of these bases immediately imply that the molecule is double-stranded and that the duplex is held together by conventional base-pairing. This conclusion was confirmed by biophysical studies of isolated organelle DNA.

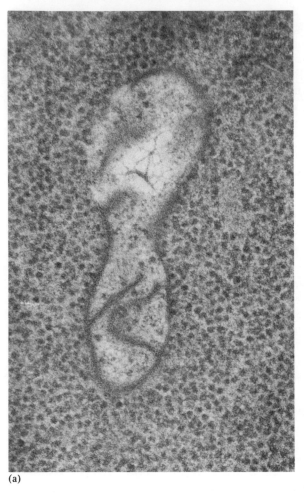

(a)

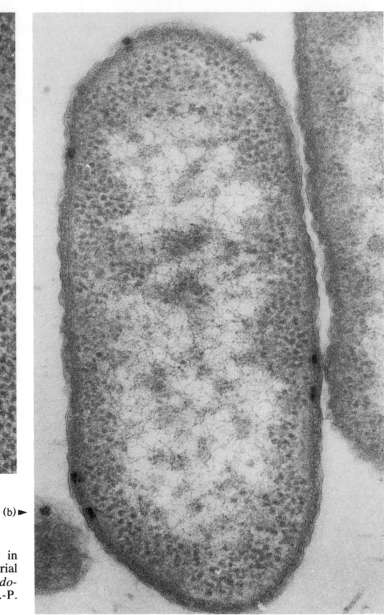

(b) ▶

Figure 9.1
Naked fibrils of DNA are distributed similarly in organelles and prokaryotic cells. (a) Mitochondrial profile of yeast. × 121,700. (b) The bacterium *Pseudomonas aeruginosa*. × 106,000. (Courtesy of H.-P. Hoffmann)

DNA from organelles and chromosomes are chemically similar since the same four kinds of bases are attached to the conventional sugar—phosphate "backbone." If DNA is centrifuged to the point of equilibrium in cesium chloride (CsCl) density gradients, different DNAs come to rest at different places in the gradient according to their buoyant density in CsCl (Fig. 9.2). When total DNA is purified from cells and examined after centrifugation in CsCl, more than one population of molecules usually can be seen. If each of these fractions is isolated and then re-centrifuged, each comes to rest in the same place it had occupied in the original test. This result demonstrates the differences are real since each group of molecules behaves consistently. Much better evidence, however, comes from examination of DNA isolated separately from purified nuclei, mitochondria, and chloroplasts. Each of these sources contains *one* of the DNAs found in the total cell DNA seen earlier. From results such as this, it was possible to identify and characterize each kind of organelle DNA from many different species.

If organelle DNA is duplex, as seemed most probable from base composition analysis and electron microscopy, then the two strands should be separable. If organelle DNA is heated to the point of strand separation (melting), the expected **hyperchromic shift** is seen (see Chapter 2). The increase in absorbance of melted DNA takes place without an increase in the concentration of DNA in solution, as is expected for duplex molecules (Fig. 9.3). When the melted strands are cooled, the complementary partners reanneal to restore the duplex molecule conformation. Biophysical evidence therefore confirms the duplex nature of mitochondrial and chloroplast DNA molecules.

Molecule Circularity

In 1966 several important articles showing the size and shape of mitochondrial DNA molecules in electron micrographs were published independently (Fig. 9.4). Using either purified DNA or purified

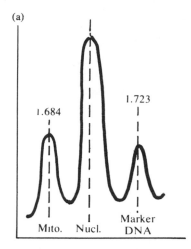

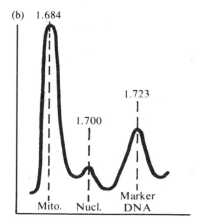

Figure 9.2
Separation of yeast mitochondrial and nuclear DNAs in CsCl density gradients from (a) whole-cell preparations and (b) purified mitochondrial preparations. The marker DNA provides a reference point for ρ values of the experimental DNAs. Nuclear DNA may contaminate a mitochondrial preparation, or there may be more than one kind of DNA isolated in yeast mitochondrial samples as an explanation for the small amount of 1.700 DNA in experiment b.

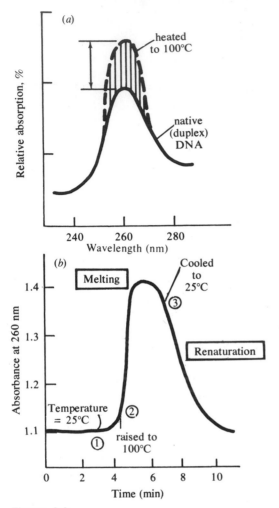

Figure 9.3
Experimental evidence for the duplex nature of organelle DNA: (a) a hyperchromic shift may be observed when melted DNA is scanned at various wavelengths of light, with a peak absorbance increase around 260 nm; or (b) increase in absorbance can be measured at 260 nm over a period of time during DNA melting and renaturation. The T_m of the DNA can be determined from data obtained in b.

mitochondria, the preparations were spread on an air-liquid phase system. Pure DNA molecules were spread out on the liquid surface and were collected for microscopy, using a method originally developed by A. Kleinschmidt to observe viral and bacterial DNA (Fig. 9.5). The same "Kleinschmidt" monolayering technique was also useful for preparations of mitochondria. In this case, the mitochondria burst when they are layered on the liquid surface. DNA molecules are released from such osmotically shocked organelles and can be collected for microscopy in the same way as purified DNA preparations. Analysis of DNAs from mitochondria of many eukaryotic species and of chloroplast DNA from a few species has provided some very important information that would otherwise be difficult to obtain in conventional biochemical or biophysical tests.

MITOCHONDRIAL DNA. Mammalian and chicken mitochondrial DNAs were the first to be studied in 1966. Each species was found to have circular DNA measuring about 5 μm in contour length. Since then it has been shown that every species of vertebrate and invertebrate animal also has such molecules. The range of lengths is between 4.5 and 5.9 μm, but the average for all metazoan animals still comes out to be about 5 μm (Table 9.1). The molecules are twisted in particular ways in the natural state, but they may open out into simple circles if one of the strands of duplex DNA is nicked or broken. Breaking a DNA strand relaxes the twisted molecule and leads to an open circle (Fig. 9.6). If both strands of the duplex DNA are broken by mechanical damage or by enzyme action, the twisted DNA is converted to a linear molecule with two free ends.

In contrast with the essential uniformity of length and conformation in metazoan animals, other eukaryote groups may have greater variability in mitochondrial DNA. For example, three different protozoan species not only differ from the metazoans but also differ from one another in their mitochondrial DNA. The ciliate *Tetrahymena pyriformis* possesses linear molecules measuring 17.6 μm in length. *Paramecium,* a closely related ciliate, also has such molecules in its mitochondria. The soil ameba *Acanthamoeba* belongs to a different group of protozoa. *Acanthamoeba* mitochondrial DNA is circular but varies in length, with an average length of about 13 μm. A third group of protozoa that has been studied are trypanosomes, which include parasitic hemoflagellates. These unicells have a meshwork of mitochondrial DNA. The complex is made up of tiny circles less than 1 μm in contour length, plus a tangle of longer molecules (Fig. 9.7).

Table 9.1 Size and conformation of mitochondrial DNAs

SPECIES	CONTOUR LENGTH* (μm)	CONFORMATION
Animals		
Vertebrates		
Mammals	4.7–5.6	Circular
Birds	5.1–5.4	Circular
Amphibians	4.7–5.9	Circular
Fish	5.4	Circular
Invertebrates		
Insects	4.6–5.2	Circular
Echinoderms	4.5–4.9	Circular
Various worm groups	4.8–5.9	Circular
Plants		
Flowering plants	30	Circular
Fungi		
Yeast (*Saccharomyces*)	<1–26	Circular*
Neurospora	<1–26**	Circular
Water mold (*Saprolegnia*)	14	Circular
Protists		
Protozoa		
Ciliates	17.6	Linear
Amoebae (*Acanthamoeba*)	12.8	Circular
Trypanosomes	0.3–0.8***	Circular
Euglenoids		
Chlamydomonas	4.6	Circular

* Both circular and linear molecules have been reported.
** Differences have been found in various strains.
*** These "minicircles" are connected in a meshwork of DNA.

Several species of fungi have been analyzed, and an unusual degree of variation is observed, even in two strains of the same species. Different investigators have reported different DNA molecular lengths for yeast and *Neurospora* mitochondria (see Table 9.1). In some cases the molecules have been described as linear and in others as circular. The extreme fragility of mitochondrial DNA from fungi almost certainly leads to breakage of circular duplex molecules. Whether this is the cause of all observed linear molecules or of size variability has not yet been determined. Since both yeast and *Neurospora* are favorite species for mitochondrial genetic studies, the issue is of some importance in interpreting genetic data.

Only a few higher plants have been examined for mitochondrial DNA. In each case it has been shown that the molecules are circular and measure about 30 μm in contour length. *Chlamydomonas* is classified as a protist by some but as an algal member of the plant kingdom by others. This distinction may become important since *Chlamydomonas* mitochondrial DNA consists of circles only 4.6 μm in average contour length.

It will be interesting to see what mitochondrial DNA will be found in other protists, algae, and green land plants. This one feature alone should not be given a disproportionate weight in assigning evolutionary relationships, but clear-cut differences between mitochondrial DNAs of various green cell types could indicate some degree of divergence during evolution. In a similar vein, the greater variability seen among protozoa and fungi, as compared to that among metazoan animals, could be a reflection of divergences related to the duration of each group in the history of life on Earth. Animals and land plants arose most recently, perhaps hundreds of millions of years after eukaryotic unicells were already present and diverging during their evolution.

CHLOROPLAST DNA. The few species of organisms studied have been found to contain circular DNA about 40–45 μm in contour length (Fig. 9.8). Greater

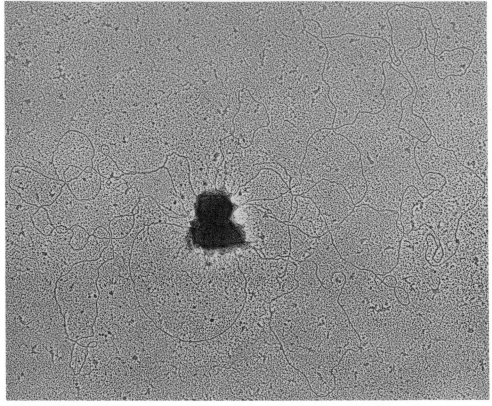

(a)

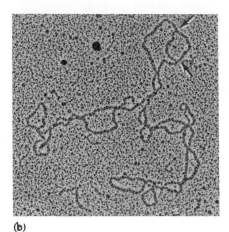

(b)

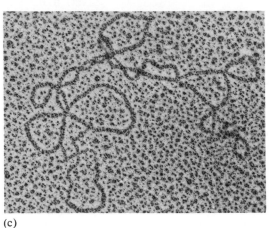

(c)

Figure 9.4
Mitochondrial DNA molecules: (a) loops of DNA measuring a total of 48.5 μm in contour length remain associated with a fragment of membrane from osmotically-ruptured mitochondria of yeast. × 29,000. (From Avers, C. J. *et al.,* 1968, *Proc. Natl. Acad. Sci.* **61**:90–97, Fig. 3.) (b) and (c) Purified rat liver mitochondrial DNA molecules measuring approximately 5 μm in contour length. Note the duplex replicated region defined by arrows in b. × 65,900. (Courtesy of D. R. Wolstenholme, from Wolstenholme, D. R. *et al.,* 1973, *Cold Spring Harbor Sympos. Quant. Biol.* **38**:267–280, Figs. 1b and 2b.)

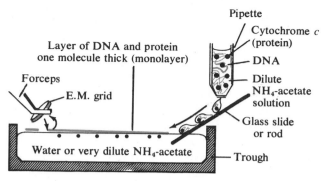

Figure 9.5
Diagram illustrating the monolayering method developed by Kleinschmidt. A suspension of purified DNA, cells, organelles, or viruses is carefully added to the surface of an aqueous medium in a trough. After the DNA and protein binder have spread out in a monomolecular film, the material can be picked up on metal grids and prepared for electron microscopy.

technical problems were encountered in isolating unbroken DNA from chloroplasts, so the first information did not become available until 1972, six years later than the first reports on mitochondrial DNA.

Duplex DNA molecules isolated from chloroplasts of peas, spinach, and other plants and those from the protist *Euglena* are approximately the same size. It is too soon to tell whether this unexpected consistency will continue to be found for chloroplast DNAs of other species.

Replication and Transcription

In Chapter 5 we discussed evidence showing that mitochondrial and chloroplast DNA replicate by a semiconservative mechanism. Chloroplast DNA was monitored in an experiment designed according to the classic plan of Meselson and Stahl (see Fig. 5.18). Each generation of duplex DNA consisted of molecules with one parental and one newly-synthesized complementary partner strand. Parental DNA is semiconserved since each new duplex has only one of the two parental strands, but all parental strands are retained and transferred from one generation to

the next. Genetic material is not routinely metabolized.

Mitochondrial DNA also replicates semiconservatively, but the particular way in which complementary DNA is synthesized is somewhat different than usual. Copying of one new complement begins along a parental template strand and continues for some time. After a while, synthesis of a complementary strand begins along the second parental strand of the duplex DNA (Fig. 9.9). Staggered copying and a unique pattern of **D-loop synthesis** has so far been found for several species, but it is possible that all mitochondrial DNA replicates in this same way (see Fig. 5.25).

If organelle DNA functions as genetic material then we expect it to store coded information. One of the simpler ways to determine whether DNA is informational is to hybridize single strands of melted DNA with single strands of organelle RNA. If RNA is transcribed, it will have a complementary base sequence to the strand of DNA from which it was produced.

Ribosomes from mitochondria or chloroplasts can be isolated and purified, and their RNA and proteins recovered separately. When ribosomal RNA from an

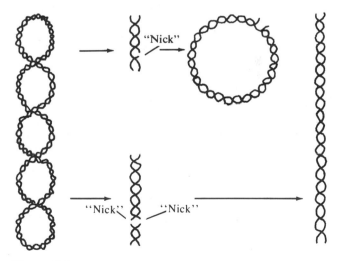

Figure 9.6
Diagram showing the effects of breaks in one or both strands of a twisted circular duplex DNA molecule.

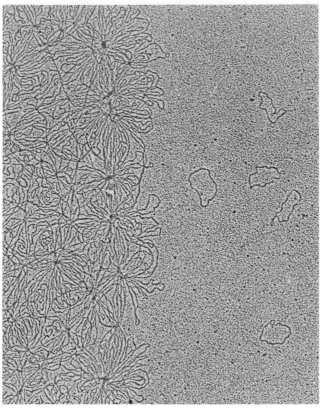

Figure 9.7
Kinetoplast DNA association from the trypanosome *Crithidia acanthocephali.* Free, circular molecules, 0.8 μm in contour length, can be seen next to the meshwork made of longer DNA molecules. × 85,700. (Courtesy of D. R. Wolstenholme and J. E. Manning)

organelle type is *hybridized competitively* with DNA from the same kind of organelle, its relationship to the DNA can be determined. In every case so far, all the ribosomal RNA from mitochondria hybridizes only with DNA from the same mitochondria; the same situation is true for chloroplast rRNA and DNA. Ribosomal RNAs from an organelle do not hybridize with nuclear DNA of the same cells. These results clearly show that each kind of organelle DNA contains information which specifies the organelle ribo-

somal RNA. Using hybridization-saturation assays, it has been found that there is one gene for each major type of rRNA in each kind of organelle DNA (Fig. 9.10).

Using DNA—RNA hybridization-saturation assays, one gene copy has been estimated to occur for each of as many as 20 different transfer RNA species in mitochondria. The situation is not as clear as in the case of ribosomal RNA, but many, if not all, organelle tRNAs are believed to be transcribed from organelle DNA information. Most tRNA hybridizes with the "heavy" strand of duplex DNA, but a few tRNAs hybridize specifically with the "light" strand. The separated strands of duplex DNA are identified after density gradient centrifugation, and each strand can be tested separately for its ability to hybridize with RNA. Most information in DNA molecules in general is found in the "heavy" strand.

RIBOSOMES IN ORGANELLES

Little thought was given to organelle ribosomes before it was accepted that DNA was present in mitochondria and chloroplasts. It had been reported some time earlier that protein synthesis did take place in mitochondria and chloroplasts, but these observations were not linked with a ribosomal machinery until the latter part of the 1960s. An extensive body of information has been accumulated within the past decade, especially after organelle ribosomes were purified and characterized.

Ribosome Size and Composition

Organelle ribosomes can be identified by sedimentation behavior in the same way as any other kind of particle. Chloroplast ribosomes from many sources seem to be relatively similar 70S monomers, whereas mitochondrial ribosomes vary from 55S to about 80S (Table 9.2). Both kinds of ribosome monomers can be dissociated into two subunits. Chloroplast ribosome subunits are essentially 30S and 50S, like those

Table 9.2 Sedimentation values for mitochondrial ribosomes*

SPECIES	MONOMER	SMALL SUBUNIT	LARGE SUBUNIT
Metazoan animals	55–60S	30–35S	40–45S
Flowering plants	78S	44S	60S
Fungi	80S	40S	52S
Protists			
Protozoan (*Tetrahymena*)	80S	55S	55S
Euglenoid (*Euglena*)	71S	32S	50S

* The reference standards are *E. coli* 70S monomers (30S and 50S subunits) and rat liver cytoplasmic-ribosome monomers (80S) and subunits (40S and 60S).

from bacterial cytoplasm. Ribosomal RNAs are also very similar in chloroplast and bacterial systems. Mitochondrial ribosomes, however, are too varied to be characterized as "bacteria-like." They are not necessarily like cytoplasmic ribosomes either.

There has been some confusion concerning absolute S values for mitochondrial ribosomes from some species. Different results using the same materials may be due to variations in technique, conditions for storage of the particles used in experiments, different degrees of folding or unfolding of the structures, and so forth. Despite some disagreements, generalizations can be made.

The smallest known mitochondrial ribosomes have been isolated from metazoan animals. These particles from vertebrates and invertebrates sediment at 55–60S, and they are sometimes referred to as "mini-ribosomes." At first it was assumed that the particle was only a large subunit or even an unusual ribosome made up of a single 55–60S structure. In 1970 it was clearly shown that the mitochondrial ribosomes from the toad *Xenopus laevis* consisted of two unequal-sized subunits. Both subunits of these ribosomes were required for protein synthesis in an *in vitro* system. After this important model study, it was also shown that mitochondrial ribosomes from other animals were similar to those described in *Xenopus*. Animal mitochondrial ribosomes therefore are smaller than usual, but they are otherwise of conventional function and construction.

There is some resistance to the acceptance of mito-

chondrial ribosomes as large as 80S, even though these sizes have been reported for some protozoa and some fungi. Since mitochondrial ribosomes behave in some ways like prokaryotic 70S particles and not like 80S cytoplasmic ribosomes of eukaryotes, it is assumed by many that the organelle ribosome should be as small as those in prokaryotes. Nevertheless, two particular mitochondrial systems may well have 80S ribosomes.

The size of ribosomes from *Tetrahymena* mitochondria has been reported as 80S. The ribosome dissociates into two equal-sized 55S subunits. At present there has been no further subdivision of the 55S particles. This system is unusual if the ribosomes are indeed as large as 80S, and even more unusual if the subunits are the same size. All other systems have unequal-sized subunits. It has also been suggested that the 55S particle is a ribosome monomer, and that two such monomers aggregate to form the 80S dimer. Even if this were the case, it would still be a unique ribosomal system when compared with all the others we know.

Mitochondrial ribosomes from yeast and *Neurospora* were reported that range in size from 72S to 80S. A recent study of *Neurospora* has provided some useful guidelines for resolving the size controversy. Depending on experimental conditions during ribosome isolation, the particles were 73S, 80S, or both sizes in the same fractions. The 73S and 80S ribosomes were identical on the basis of several criteria: (1) both had the same responses to inhibitors; (2)

Figure 9.8
A circular molecule 44 μm in contour
length of chloroplast DNA from
spinach (*Spinacia oleracea*). × 25,700.
(Courtesy of D. R. Wolstenholme,
from Manning, J. E. *et al.,* 1972,
J. Cell Biol. **53**:594–601, Fig. 3.)

Table 9.3 Sedimentation values of organelle rRNAs*

ORGANELLE	SPECIES	rRNA FROM	
		SMALL SUBUNIT	LARGE SUBUNIT
Chloroplast	Various species	16S	23S + 5S
Mitochondrion	Metazoan animals	12–13S	16–17S
	Fungi	14–17S	21–24S
	Protists		
	Euglena	16S	23S
	Tetrahymena	14S + 21S**	

* The reference standards are 5S, 16S, and 23S rRNA from *E. coli* and 5S, 18S, and 25S or 28S from eukaryotic cytoplasmic ribosomes.
** There is no way of knowing which rRNAs come from the two 55S ribosome subunits (see Table 9.2).

both types dissociated into 39S and 52S subunits; and (3) these subunits contained 17S and 24S ribosomal RNA in all cases. It was possible to convert 80S ribosomes to 73S, but the reverse was not possible. The tentative conclusion was that 80S monomers were the native ribosomes in *Neurospora* mitochondria. These ribosomes could be conformationally or chemically modified under some conditions of preparation and changed to the 73S artifact particle.

In *Neurospora* as in yeast, *Tetrahymena,* and other species with 80S mitochondrial ribosomes, there are other features that distinguish cytoplasmic 80S ribosomes from mitochondrial particles with the same sedimentation behavior. Although each is 80S, they differ in responses to inhibitors of protein synthesis, size of subunits, size of rRNAs, base composition of rRNAs, and other features. We will discuss some of these traits in the following sections.

Ribosomal RNA and Protein

All chloroplast and mitochondrial ribosomes contain a smaller RNA in the smaller subunit and a larger RNA molecule in the larger subunit; they resemble cytoplasmic ribosomes of all cells in this regard. Chloroplast ribosomes further resemble cytoplasmic particles in having one molecule of 5S RNA in the large subunit. There has been no convincing evidence

for the existence of such a molecule or one like it in mitochondrial ribosomes. Since almost nothing is known about the function of 5S rRNA, little can be said about the significance of the observed difference in relation to ribosome activities.

Evidence has been obtained from studies of rRNA synthesis showing that a larger precursor RNA is cleaved to form the two mature RNAs of the subunits. Organelle ribosomes are, therefore, like other ribosomes in their mode of rRNA formation. Chloroplast rRNAs are about 16S and 23S molecules, like those in bacteria. Mitochondrial rRNAs, however, are as variable as their ribosome and subunit sizes (Table 9.3). Animal rRNAs sediment near 12–13S and 16–17S, compared with *E. coli* 16S and 23S molecules. Like many other unicells, rRNA from protozoa, fungi, and some protists are 14–17S and 21–24S. Plant mitochondrial rRNA molecules are larger than equivalent rRNAs in chloroplasts of the same species. The size of the rRNA correlates generally with the size of the subunit from which it comes.

Studies of organelle ribosomal proteins have only recently begun. More attention is being directed toward these components because they may be products of organelle gene action. If individual proteins can be identified, as they have been in *E. coli* ribosomes, it would be a useful tool for studying the effects of extranuclear mutations in antibiotic-resistant

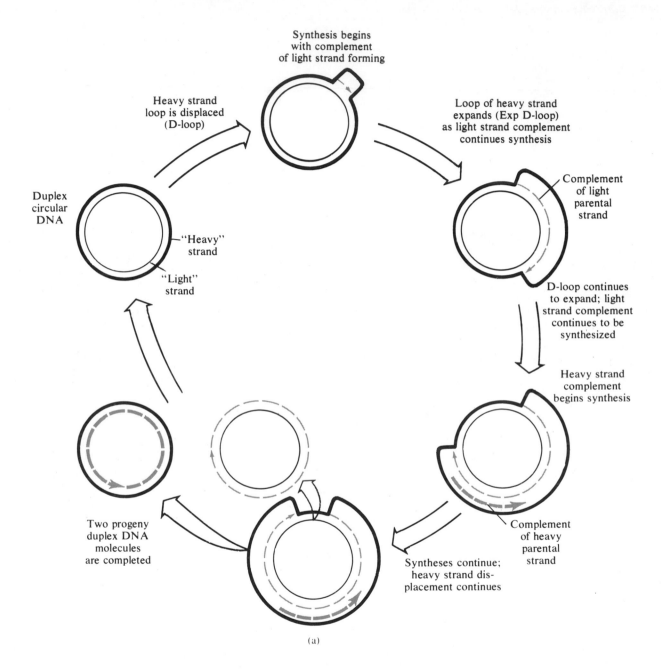

Synthesis begins
with complement
of light strand forming

Heavy strand
loop is displaced
(D-loop)

Loop of heavy strand
expands (Exp D-loop)
as light strand complement
continues synthesis

Duplex
circular
DNA

Complement
of light
parental
strand

"Heavy"
strand

"Light"
strand

D-loop continues
to expand; light
strand complement
continues to be
synthesized

Heavy strand
complement
begins synthesis

Two progeny
duplex DNA
molecules
are completed

Complement
of heavy
parental
strand

Syntheses continue;
heavy strand dis-
placement continues

(a)

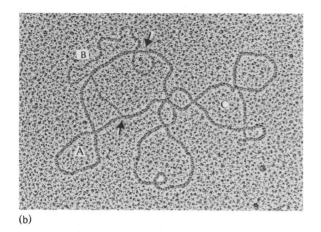

(b)

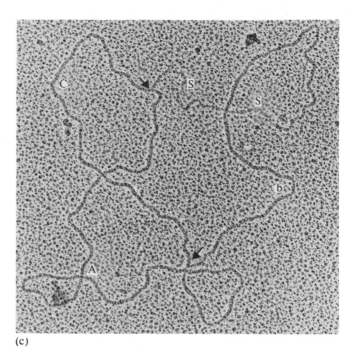

(c)

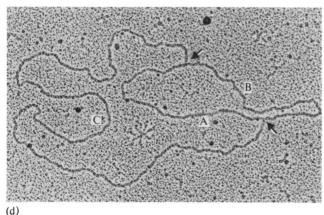

(d)

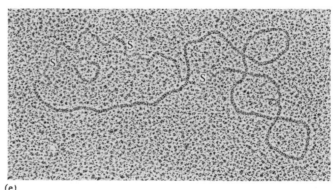

(e)

Figure 9.9

D-loop synthesis of mitochondrial DNA: (a) diagrammatic summary of sequence. Replicating mitochondrial DNA molecules from rat liver (b–e): (b) Double-forked, circular molecule in which all of one daughter segment (B), delimited by the forks (arrows), is single-stranded. A = 1.37 μm, B = 1.29 μm, C = 3.79 μm, A + C = 5.16 μm. (c) Double-forked, circular molecule in which a region of one daughter segment (B), in association with a fork, appears to be single-stranded (S), whereas the region of the same segment, in association with the other fork, appears to be double-stranded. The arrows indicate the forks. A = 3.47 μm, B = 3.41 μm, C = 1.67 μm, A + C = 5.14 μm. (d) Double-forked, circular molecule in which all three segments delimited by the forks (at arrows) appear to be entirely double-stranded. A = 1.06 μm, B = 1.06 μm, C =

3.79 μm, A + C = 4.85 μm. (e) Simple, circular molecule with a single-stranded region (S) measuring 1.4 μm (approximately 27 percent of the total contour length of 5.01 μm). The variety of replicative intermediate forms leads to the suggestion that D-loop synthesis of mitochondrial DNA may not be the only mechanism by which these molecules replicate. All micrographs × 56,000. (Courtesy of D. R. Wolstenholme, from Wolstenholme, D. R. *et al.*, 1973, *Cold Spring Harbor Sympos. Quant. Biol.* **38**:267–280, Figs. 1g, 2d, 1a, 2f, respectively.)

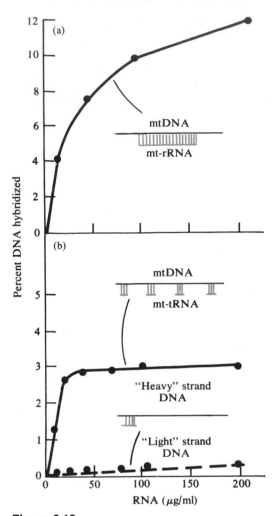

Figure 9.10
Analysis of complementarity between mitochondrial DNA and mitochondrial rRNA and tRNA: (a) Hybridization-competition assays show that rRNA from mitochondria hybridizes only with mitochondrial (mt) DNA. (b) Mitochondrial tRNA genes can be localized specifically on the "heavy" and/or "light" strands of mtDNA by hybridization-saturation assays, which also indicate the number of tRNA genes present in each strand of mtDNA. Similar data have been obtained for various species.

strains. We will discuss these systems later in the chapter.

Rather careful studies of *Neurospora* mitochondrial ribosome proteins showed that these were completely different from the proteins of cytoplasmic ribosomes in the same cells. Early results indicate at least 30 different proteins are found in the larger subunit and about 23 different proteins are found in the small subunit of the mitochondrial ribosome. With two major RNAs and more than 50 different proteins, mitochondrial (and chloroplast) ribosomes are just as complex as their equivalents in the cytoplasm of the cell.

Protein Synthesis at the Ribosome

Labeled amino acids are incorporated into polypeptides in mitochondria and chloroplasts studied *in vitro* or in the living cell. Stringent controls in experimental systems have ruled out the possibility that these polypeptides are contaminants made in the cytoplasm or produced by bacteria which may be present in some cases. Ribosomes can be isolated from organelles and used to guide protein synthesis in the presence of artificial or natural messenger RNAs and other essential ingredients for biosynthesis.

Ample evidence has been presented since 1968 to show that mitochondria and chloroplasts contain unique populations of transfer RNAs, aminoacyl-tRNA synthetases, enzymes, and protein factors required for initiation, elongation, and termination of polypeptides. The translation apparatus of organelles closely resembles that of prokaryotic systems rather than that of eukaryotic ribosomal systems. Resemblances are apparent from their similar responses to inhibitors of nucleic acid and protein synthesis, and the fact that enzymes, tRNAs, and synthetases often are interchangeable between organelle and prokaryotic systems.

Like the prokaryotic systems, mitochondria and chloroplasts possess an active formylase that converts its methionine to *N*-formyl-methionine. The initiator aminoacyl-tRNA is fmet-tRNA$_f$ which binds to the *AUG* codon of the messenger RNA. There are some variations from one species to another, but the overall features of organelle translation are very similar to prokaryotic equivalents.

Active polysomes have been demonstrated in

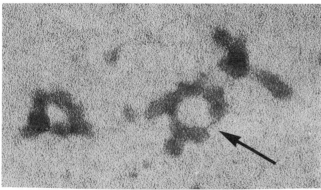

(a)

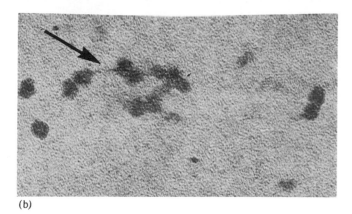

(b)

Figure 9.11
Polysomes isolated from purified yeast mitochondria after centrifugation. The presumptive messenger RNA strand is shown at the arrows. (a) × 255,000. (b) × 160,000. (From Cooper, C. S., and C. J. Avers, 1974, *The Biogenesis of Mitochondria* [ed. Kroon, A. M., and C. Saccone], Academic Press, pp. 289–303, Fig. 6.)

mitochondria and chloroplasts by electron microscopy and biochemical tests. Polysomes have been isolated and shown to be held together by a thin strand of RNA, which is digestible with ribonuclease (Fig. 9.11). Nascent polypeptides are found predominantly in polysome fractions isolated from mitochondria, showing that they form there. The growing polypeptides can be stripped from polysomes if puromycin is added to the active mixture (Fig. 9.12). When active polysomes are carefully isolated from *Euglena* cells, polypeptides continue to be synthesized for as long as two hours on these isolated aggregates. This is another significant line of evidence demonstrating the reality of polysomes in organelles.

Polysomes containing different numbers of ribosome monomers have been isolated from yeast cells. The lengths of polysomes coincided closely with the lengths expected for some polypeptides known to be synthesized within the organelle. Since protein length is reflected in messenger RNA length, it is possible that the observed polysomes were engaged in manufacturing mitochondrial proteins that would later be incorporated into the organelle. A recent study using *Drosophila* mitochondria provided striking photographs of transcription-translation complexes similar to those obtained from *E. coli.* When the mitochondria were burst open by osmotic shock, the 5 μm-long circular DNA was released. Attached to this molecule were 70–80 polysome groups (Fig. 9.13). This observation is another example of the similarities between organelle and prokaryotic translation systems.

Messenger RNA is assumed to be transcribed from organelle DNA, but little direct evidence is available. Transcription-translation complexes, such as those isolated from *Drosophila* mitochondria, suggest that mRNA is made along organelle DNA and that translation begins before the messenger is released from its DNA template. Studies using labeled RNA precursors have also provided some evidence showing that a certain amount of the newly made RNA is not ribosomal or transfer RNA and is, there-

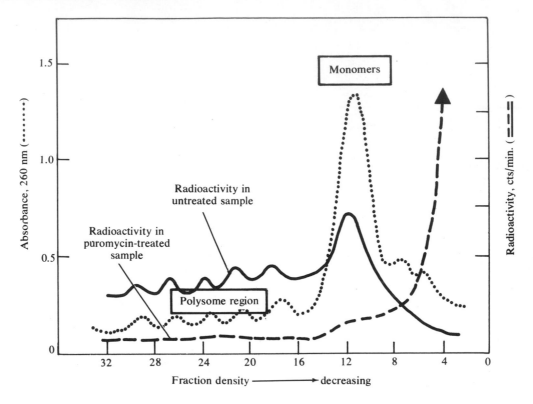

Figure 9.12

Protein synthesis at yeast mitochondrial polysomes is indicated by the concentration of radioactively-labeled polypeptide associated with these fractions in the density gradient. Verification that these polypeptides are nascent chains is obtained if they are released from the polysomes after puromycin treatment because this treatment terminates polypeptide synthesis prematurely.

fore, most likely to be messenger RNA still associated with polysomes to which this material is attached.

A very strong line of supporting evidence for the existence of organelle mRNA was reported in 1974 by A. H. Scragg. He isolated RNA which had been transcribed from mitochondrial DNA and added this RNA to a translation system isolated from *E. coli* cells. The proteins that were synthesized in this system were then isolated and compared with mitochondrial membrane proteins which were known to be made in the organelle itself. Three major proteins from the known mitochondrial group proved to be identical with three of the proteins translated from organelle RNA in the *in vitro E. coli* system (Fig. 9.14). The simplest interpretation is that the mitochondrial RNA was messenger RNA that had been synthesized along mitochondrial DNA templates. This mRNA was then translated into mitochondrial membrane proteins in the experimental test system, just as they are manufactured normally in the mitochondrion itself.

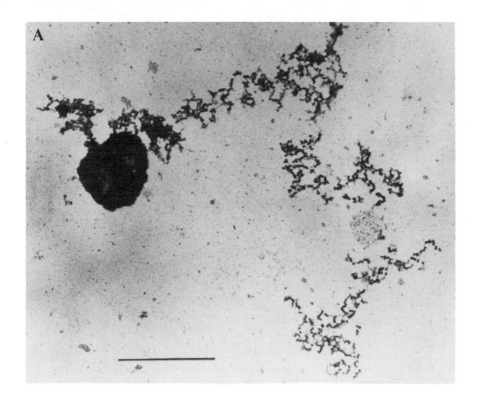

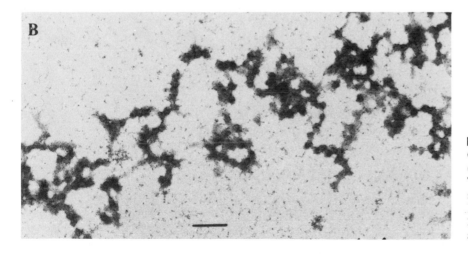

Figure 9.13
Osmotically ruptured mitochondrion (M) from *Drosophila melanogaster*. The strings of polysomes are well spread to both sides of the central DNA filament. (a) × 25,000. (b) × 90,000. (Courtesy of C. D. Laird and W. Y. Chooi)

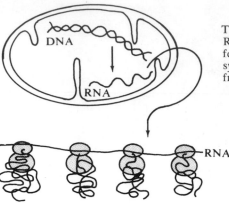

Transcribed mitochondrial RNA is isolated and transferred to a cell-free protein synthesizing system derived from *E. coli.*

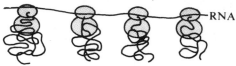

RNA

Translation occurs in the *E. coli* system.

Proteins are isolated and characterized by electrophoresis.

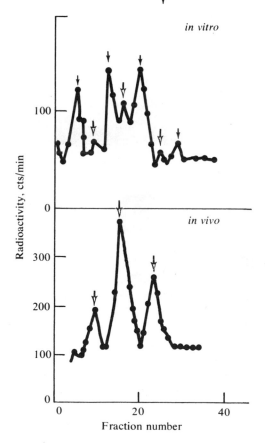

Seven proteins are recognized in the sample taken from the cell-free *E. coli* system, three of which are the same as the three major proteins synthesized in mitochondria within intact yeast cells (*in vivo*).

GENES IN ORGANELLES

We have seen so far that mitochondria and chloroplasts contain all the required components for a functional genetic system. Organelle DNA is stable, being transferred in continuity from one generation to the next. This stable material replicates by a semiconservative mechanism like other known genetic DNAs in viruses, prokaryotes, and the eukaryotic nucleus. DNA information is transcribed into all the known classes of RNA: ribosomal, transfer, and messenger. DNA information also appears to be translated into certain proteins that form part of the organelle structure and function.

These molecular themes combine into a powerful circumstantial case for the existence of organelle genes, that is, for informational DNA which is transcribed and translated. If genes do exist then we would also expect to find mutations. One of the unique qualities of a gene is its property of mutation. Organelle genetics has been technically difficult to approach, but considerable progress has been made in recent years.

Three of the four properties of genetic material were discussed in previous sections: (1) semiconservative replication, (2) information storage, and (3) information transfer. We will now examine the evidence showing that the fourth property also exists, namely, mutation. Once the occurrence of mutations in organelle systems is established, a few related questions of great importance will be considered: Are the mutations localized within organelle DNA? What proteins are encoded in organelle DNA? What portion of organelle structure and function is specified by organelle genes and what portion by nuclear genes? These and other questions are under intensive study today.

Detection of Organelle Gene Mutations

One fundamental approach in searching for mutations in organelles is to conduct a breeding analysis and determine the *pattern of inheritance* of the wild-type and mutant alternatives of the trait under study. In conventional Mendelian inheritance we expect the distribution of genes to coincide with the distribution of chromosomes to the progeny. In a case of single-gene inheritance we find a ratio of 3 : 1 in F_2 progeny of diploids and 1 : 1 in haploid progeny generations (Fig. 9.15). So-called *non-Mendelian inheritance* patterns do not conform to patterns of inheritance established for genes on chromosomes. Once the mutation is localized outside the chromosome, the case in favor of gene location in some other DNA system begins to accumulate. If the trait being studied affects the mitochondrion or the chloroplast, we may tentatively assign the gene to the organelle. A non-Mendelian inheritance pattern is only one of a number of features we would expect to find if organelle genes are responsible for the characteristics under study.

Location and identification of mitochondrial genes in yeast has progressed very rapidly in recent years. Because of these advances and the general points which can be illustrated, it will be instructive to explore this particular system as a model for understanding other organelle gene systems.

Mitochondrial Genes in Yeast

Beginning in 1949, B. Ephrussi, P. Slonimski, and others have provided some remarkable insights into the operations of yeast mitochondrial genetics. Ephrussi first described a mutant that was unable to carry out aerobic respiration. The mutant survived by fermentation, but produced very small colonies when grown on solid media. This slow-growing, respiration-deficient form was called *petite,* and its

Figure 9.14
Flow diagram and results of the Scragg experiments showing that yeast mitochondrial proteins are synthesized *in vitro* on *E. coli* ribosomes, presumably by translation of messenger RNAs transcribed in the mitochondria from mtDNA.

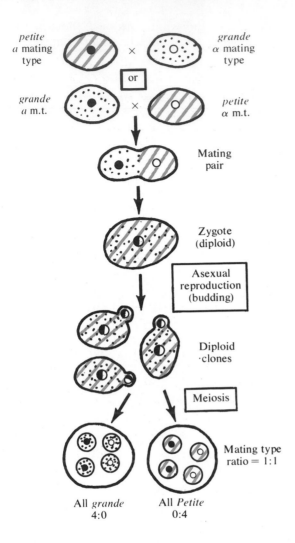

Figure 9.15
Diagram illustrating the different inheritance patterns for chromosomal and nonchromosomal traits in yeast. Alleles of the chromosomal mating-type gene (*a* and *α*) segregate 1:1 after meiosis, whereas the nonchromosomal alternatives (*petite* and *grande*) segregate 0:4 or 4:0 after meiosis. *Petite* mutants lack aerobic respiratory capacity, while *grandes* are wild-type and can carry out aerobic respiration.

wild-type alternative was named *grande*. In crosses between *grandes* and *petites* a non-Mendelian inheritance pattern was found. Instead of a ratio of 1 : 1 there were usually only *petites* or only *grandes* in the progeny. *Petites* were true-breeding. Since they maintained their altered characteristics for countless generations, they were clearly mutants rather than wild types whose phenotypes varied in response to environmental conditions.

Another feature which indicated that *petites* were **nonchromosomal** mutants was found in mutation-induction studies. Using chemicals that had no effect on chromosomal genes, it was possible to convert almost 100 percent of a population of *grandes* to *petites*. This behavior does not typify all nonchromosomal mutants, but it does show the inherited factors are outside the chromosomes. Because *petites* have defective mitochondrial function and structure, it was assumed that the mitochondrion itself had undergone mutational change (Fig. 9.16). Once it was established, in the 1960s, that mitochondria contained DNA, attention was focused on finding altered mitochondrial DNA in *petites*. If altered DNA could be found, it would show that the inherited *petite* traits were the result of mutated genetically informational organelle DNA. Using CsCl density gradient centrifugation, Slonimski found that *petite* mitochondrial DNA was physically different from *grande* DNA (Fig. 9.17). Later he showed that progeny from crosses between *petite* and *grande* could have *recombined* DNA. These results, along with others, showed that mitochondrial DNA has a coding function and can undergo genetic exchanges, just like DNA in chromosomes. Mitochondria had genes.

As it turned out, *petites* were not the result of a mutational change in one gene. Their mitochondrial DNA was extensively modified, having lost much material. In some cases, a *petite* may lose all its mitochondrial DNA. These mutants have proven to be very valuable in recent mitochondrial genetic studies. Using "neutral" *petites* with no mitochondrial DNA at all, it has been possible to see which organelle traits are coded by nuclear genes and which are directed by mitochondrial genes. Those characteristics which are retained must be coded by chromosomal genes in the nucleus, since mitochondrial DNA is lacking altogether.

Other nonchromosomal genes can be located directly in mitochondrial DNA by a relatively straightforward test. If mitochondrial genes are present, then they should be lost when cells are converted to *petites* by a mutagenic chemical. Since DNA can be eliminated entirely by the treatment which induces *petites,* parallel loss of a gene suspected to be of mitochondrial origin would show that it was located within mitochondrial DNA. Tests showing parallel gene loss and loss of mitochondrial DNA during induction of *petites* have been used successfully to identify a number of mitochondrial genes in yeast.

In addition to identifying mitochondrial genes in yeast, mutants with altered mitochondrial traits have also been found in *Neurospora* and *Paramecium*. These traits are also believed to be expressions of mitochondrial gene activity. Further evidence is needed to demonstrate that these traits are directly related to mitochondrial DNA.

Chloroplast Genes

Studies of nonchromosomal inheritance involving chloroplast traits have been in progress for many years, using species such as corn, tobacco, *Euglena,* and *Chlamydomonas*. Ruth Sager has provided a substantial body of evidence showing the existence of many nonchromosomal genes that affect chloroplast activities in *Chlamydomonas*. These genes have been placed on a linkage map that Sager believes is representative of chloroplast DNA (Fig. 9.18).

In recent studies, Sager was able to show that certain antibiotic-resistant nonchromosomal mutants had undergone a change in chloroplast ribosome activities. In these experiments, ribosomes were isolated from chloroplasts of five different nonchromosomal mutant strains. These ribosomes were tested in an *in vitro* protein-synthesizing system using poly U as messenger. By examining the ability of these ribosomes to guide synthesis of polyphenylalanine, the defects of the mutant could be compared to wild-type chloroplast ribosome activities in the same system.

When 30S and 50S subunits from wild-type and mutant chloroplasts were recombined to make hybrid ribosome monomers, the defects could be assigned more specifically to one or the other of these subunits (Fig. 9.19). Three of the mutants were shown to have altered 30S subunits, and the other two mutants had modified 50S chloroplast ribosome subunits. Knowing that similar ribosomal defects in mutant bacteria were due to their genetically modified ribosomal proteins, Sager reasoned that the *Chlamydomonas* mutations must have been located in chloroplast DNA. Alteration of chloroplast genes produced altered protein products which were presumably coded by these altered genes, just as gene mutations in similar bacteria led to inherited defects in their ribosomes. We will return to this topic later in the chapter when we discuss interactions between nuclear and organelle genes and their relative inputs to the development of mitochondrial and chloroplast structure and function.

One of the abundant proteins of chloroplasts is the enzyme *RuDP carboxylase* which functions in the dark reactions of photosynthesis. The whole enzyme is composed of a number of polypeptide subunits. According to studies of mutant inheritance patterns, some of these polypeptides are coded by nuclear genes and some by genes outside the nucleus. The most likely location for the nonchromosomal genes is believed to be chloroplast DNA. There is very little direct evidence showing that chloroplast DNA codes for the polypeptides inherited nonchromosomally, but as in most cases of organelle inheritance, it is assumed that the DNA involved is located in the same structure whose proteins or activities have been genetically altered.

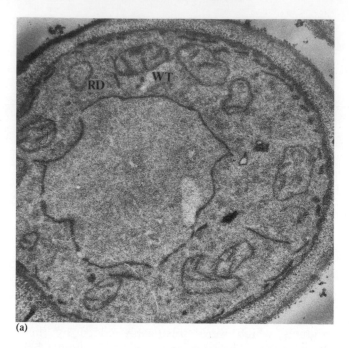

(a)

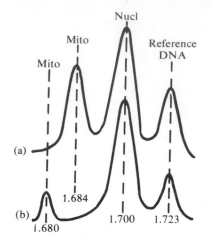

Figure 9.17
Identification of different yeast mitochondrial DNAs in (a) a wild-type *grande* strain and (b) a mutant *petite* strain. The different ρ values of 1.684 and 1.680 reflect physical and chemical variations in the two DNA populations.

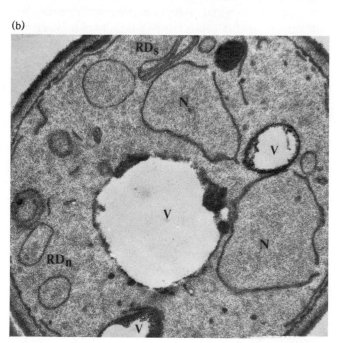

(b)

Localizing genes that code for proteins requires appropriate mutants, as just discussed. The existence of genes that specify ribosomal or some other RNA can be accomplished more easily and directly using DNA-RNA hybridization tests. Since we know that ribosomal and transfer RNAs are synthesized from chloroplast as well as from mitochondrial DNA templates, each kind of organelle DNA has at least the genes that code for these RNAs. We suspect that other genes will be found since we already know of

Figure 9.16
Thin sections of yeast cells fixed in permanganate: (a) Wild-type cells have mitochondria showing the usual cristae infolds (WT) of the inner mitochondrial membrane in all profiles except one (RD). (b) Mutant *petite* cells have mitochondrial profiles that either show the absence of crista development (RD_n) or have unusually elaborated infolded inner mitochondrial membrane (RD_s). The nucleus in b appears in two separate portions due to the sectioning plane. Vacuoles (V) are characteristic of yeast cells. $\times$ 31,000. (From Federman, M., and C. J. Avers, 1967, *J. Bacteriol.* **94**:1236–1243, Figs. 1 and 2.)

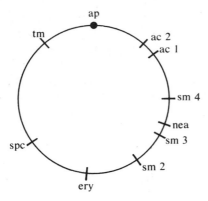

Figure 9.18
Circular linkage map of nonchromosomal genes in *Chlamydomonas* as proposed by R. Sager. The map is assumed to describe the genetic properties of chloroplast DNA based on inheritance studies of antibiotic response and other traits.

some that have been identified. Furthermore, only 2 or 3 μm at the most would be coded for transfer and ribosomal RNAs. If a chloroplast DNA molecule is 40 45 μm long, what is the function of the majority of this DNA other than to specify structural and regulatory proteins that form parts of the organelle? In the next section we will be concerned with some of these proteins and with the relative contribution made by nuclear genes to organelle structure and function.

PROTEINS SYNTHESIZED IN ORGANELLES

It is very clear that many organelle components are specified by nuclear genes. There are many chro-

Figure 9.19
Diagram summarizing results of experiments by Schlanger and Sager. Recombinations of chloroplast ribosome subunits from nonchromosomal mutants and wild-type strains of *Chlamydomonas* helped to indicate the specific subunit responsible for response to five different antibiotics.

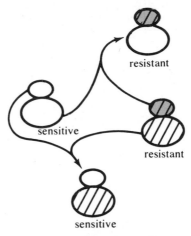

Antibiotic resistance/sensitivity to:

> streptomycin
> spectinomycin
> neamine

depends on the characteristics of the small subunit of the ribosome (30S)

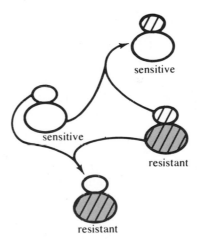

Antibiotic resistance/sensitivity to:

> carbomycin
> cleocin

depends on the characteristics of the large subunit of the ribosome (50S)

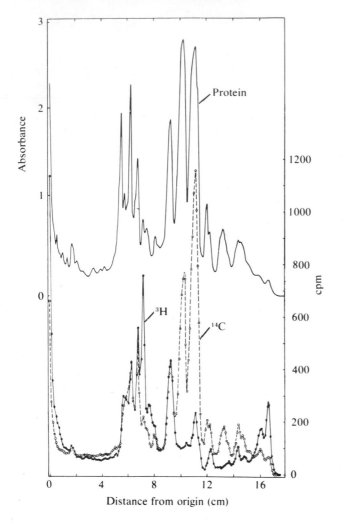

Figure 9.20
Results of experiments to identify the cellular compartment in which chloroplast thylakoid membrane proteins are synthesized in *Chlamydomonas reinhardi*. The total protein complement of thylakoid membranes is shown by the solid line measuring absorbance of the electrophoretically separated proteins. The proteins made in the cytoplasm are labeled with ^{14}C-arginine, which was present during the time the cells were growing in chloramphenicol. The proteins made in the chloroplast are labeled with ^{3}H-arginine, which was present in the medium containing cycloheximide later in the synthesis experiment. Cycloheximide inhibits

mosomal gene mutants whose inherited modifications are expressed in mitochondrial or chloroplast structure and function. In addition, a molecule of organelle DNA is not large enough to code for hundreds of different proteins that occur in a mitochondrion or chloroplast. At least 50 different proteins are found in the ribosome alone, and many more enzymes and structural proteins exist as well. These simple facts, by themselves, tell us that nuclear information must specify the production of many proteins of an organelle. In this section we will examine the kinds of proteins known to be synthesized within the organelle and explore some of the interactions between organelle and nucleocytoplasmic systems which lead to assemblies of enzymes, membranes, and ribosomes in mitochondria and chloroplasts. Most of this information has come from two general approaches: (1) analysis of mutant proteins and requirements for protein synthesis; and (2) study of synthesis in the presence of inhibitors that selectively shut down translation on cytoplasmic or organelle ribosomes.

Mitochondrial Protein Synthesis
When cells are incubated in the presence of labeled amino acids, newly synthesized proteins can be recognized by their radioactivity. These proteins appear in all parts of the cell, including the mitochondria. Some mitochondrial proteins are synthesized in the organelle itself, while other proteins are made in the cytoplasm and then transported across the membrane to be assembled inside the mitochondrion. The origin of these two groups of proteins can be determined if protein synthesis is inhibited in one compartment and allowed to continue in another part of the cell.

cytoplasmic ribosome-mediated protein synthesis, while chloramphenicol inhibits protein synthesis at the chloroplast ribosomes. Comparable studies have been reported for mitochondrial protein synthesis of membrane components. (Reproduced with permission from Hoober, J. K., 1970, *J. Biol. Chem.* **245**:4327, Fig. 6.)

One of the widely used inhibitors of protein synthesis at cytoplasmic ribosomes is **cycloheximide.** It does not affect synthesis along mitochondrial ribosomes.

If cells are allowed to synthesize proteins from labeled amino acids in the presence of cycloheximide, only mitochondrial systems are active. Any labeled polypeptides that are found would therefore be molecules that were made on mitochondrial ribosomes since the cytoplasmic ribosomes are poisoned. In a parallel set of experiments, mitochondrial synthesis should be shut off while cytoplasmic ribosomes remain unaffected. Labeled polypeptides found within mitochondria must have been made in the cytoplasm from labeled amino acids and then entered the mitochondrion (Fig. 9.20). Two inhibitors of mitochondrial protein synthesis often used are **chloramphenicol** and **erthryomycin.** Both of these drugs, as well as cycloheximide, act at the ribosome in some step required for polypeptide chain elongation (see Table 6.4).

Two systems that have been examined after selective inhibition of protein synthesis are proteins of the mitochondrial ribosome and the ATPase enzyme which is associated with the mitochondrial inner membrane. When yeast cells are allowed to synthesize proteins in the presence of cycloheximide, no labeled amino acids are found in the mitochondrial ribosomal proteins. This observation indicates that these proteins are synthesized along cytoplasmic ribosomes.

The interpretation of this evidence is confirmed when yeast cells are presented with chloramphenicol instead of cycloheximide. In these cells, labeled amino acid is present in all proteins of the mitochondrial ribosomes. Since mitochondrial protein synthesis was shut off in this experiment, these molecules must have been made on cytoplasmic ribosomes and transferred to the mitochondria afterward. Once inside, these proteins assembled with ribosomal RNAs (made in mitochondria) to form functional organelle ribosomes. It is possible that a few of these proteins are synthesized within the mitochondrion but that they escaped detection in the particular test system that was followed. Almost all, if not all, mitochondrial ribosomal proteins are made outside the organelle and shipped in for assembly.

A. Tzagoloff has studied formation of inner membrane ATPase using yeast and *Neurospora*. On the basis of selective inhibition of cytoplasmic and mitochondrial protein synthesis, he has shown that parts of the enzyme are made in both compartments. Of the ten polypeptide subunits of the ATPase, four components are synthesized at mitochondrial ribosomes and six are synthesized in the cytoplasm. The latter are transferred to the mitochondrion from the cytoplasm, where they assemble with the other four polypeptides. All four of the mitochondrially-synthesized subunits form part of the complex that lies within the inner membrane as a lipoprotein constituent. Of the six cytoplasmically-synthesized polypeptides, five are joined to form the catalytic component of the enzyme which is the ATPase coupling factor F_1. The sixth polypeptide corresponds to the F_0 coupling factor that probably binds F_1 to the membrane (see Chapter 7). The mitochondrial ATPase is formed therefore by cooperative action of two translation systems, each of which presumably translates genetic information from its own compartmented DNA. This ATPase is the membrane-bound catalyst that acts in oxidative phosphorylation coupled to electron transport.

The cytochrome oxidase protein complex has been studied by Tzagoloff and by G. Schatz and colleagues. This enzyme is also synthesized as subunits in both the mitochondrial and cytoplasmic compartments, and it is then assembled within the mitochondrion itself. Mutant strains have been used to determine the sites of polypeptide synthesis, in conjunction with selective modulation of protein synthesis. Of the seven polypeptides making up cytochrome oxidase, the four smallest are made in the cytoplasm and the three largest in the mitochondrion.

In yeast mutants with altered mitochondrial DNA but an intact set of nuclear genes, Schatz found that all three mitochondrially-synthesized subunits were missing. This study shows that (1) mitochondrial DNA probably codes for these three polypeptides; (2) the polypeptides are made in the mitochondrion, probably from messenger RNA transcripts of mitochondrial genes which specify these polypeptides; and (3) the whole functional enzyme is under cooperative control of mitochondrial and nucleocytoplasmic genetic systems. Nuclear genes specify four polypeptides which are translated on cytoplasmic ribosomes. Mitochondrial genes specify three polypeptides which are translated on mitochondrial ribosomes. All seven polypeptides assemble within the mitochondrion to form the functional enzyme that is a structural component of the inner membrane.

In all three of these cases, there is cooperation between organelle and nucleus systems. Ribosomes assemble from RNAs coded and transcribed within the mitochondrion and proteins made exclusively or predominantly by the nucleocytoplasmic system. The two enzymes, as just described, follow a similar biosynthetic pathway involving different cellular compartments and DNAs. Why are some components made in one compartment and some components in the other? One suggested answer to this question is based on the hydrophobic nature of mitochondrially-synthesized polypeptides.

The mitochondrially-synthesized subunits of ATPase and cytochrome oxidase have a high percentage of nonpolar amino acids, giving a highly hydrophobic character to these polypeptides. Polypeptides such as these would be very unlikely to cross the mitochondrial membranes if they were made in the cytoplasm. If they are synthesized on mitochondrial ribosomes, the hydrophobic molecules can become situated in the interior of the mitochondrial inner membrane. The polypeptides made in the cytoplasm, however, can move through the outer membrane to the surface of the mitochondrial inner membrane. They may become incorporated there or move completely through to the inner surface of the membrane facing the organelle matrix (Fig. 9.21).

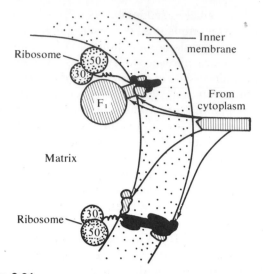

Figure 9.21
Hypothetical scheme illustrating cooperation between mitochondrial and cytoplasmic protein synthesis in the assembly of cytochrome oxidase and ATPase of the inner mitochondrial membrane in yeast. Mitochondrially-synthesized enzyme subunits (solid black) are formed on mitochondrial ribosomes (dotted) close to their site of deposition within the membrane. Cytoplasmically-synthesized enzyme subunits (cross-hatched) are imported from the cytoplasm outside the mitochondrion and are situated close to the membrane surface. The two curved lines signify the thickness of the mitochondrial inner membrane. (From Ebner, E., T. L. Mason, and G. Schatz, 1973, *J. Biol. Chem.* **248**:5369–5378, Fig. 12.)

The cytoplasmically-synthesized components are relatively hydrophilic. They are also the parts of the enzyme with catalytic and binding properties. Just how these sets of polypeptides assemble in the proper order, in both time and space, remains an important question for current research. Regulation of synthesis and assembly has hardly been studied to date. Both ATPase and cytochrome oxidase end-products eventually are bound to the membrane, held there tenaciously by their embedded polypeptide subunits.

Chloroplast Protein Synthesis

There are many parallels between methods used to study chloroplasts and those used to study mitochondria and between the cooperative synthesis and assembly of membranes, ribosomes, and enzymes in these organelles.

BIOGENESIS OF CHLOROPLAST MEMBRANES. J. K. Hoober and others have shown that one or more of the 15–20 membrane proteins of *Chlamydomonas* thylakoids are translated on 70S chloroplast ribosomes. The remainder are made at 80S cytoplasmic ribosomes. Wild-type strains can synthesize photosynthetic membranes and pigments in either the light or the dark (flowering plants can only make these components in the light). The nonchromosomal mutant *y-1* cannot manufacture functional chloroplasts in the dark. When *y-1* cells are transferred into the light, new chloroplast membranes are formed and then organized into thylakoid stacks. This mutant provides a convenient system, since biosynthesis can be controlled simply by transfers between darkness and light conditions. Selective inhibitors of protein synthesis can then be added during growth in the light to determine where the membrane proteins are made. Cycloheximide and chloramphenicol have been used.

When *y-1* cells are transferred into the light in the presence of cycloheximide, very little chlorophyll or thylakoid membrane is formed. If thylakoids were present, however, they would fuse into stacks even though cycloheximide was also present. If cells are placed in the light and chloramphenicol is provided, only a slight reduction in synthesis of new membranes occurs, but these thylakoids do not fuse into the usual grana stacks.

Together with other experiments, it is clear that both chloroplast and nucleocytoplasmic systems are required for biogenesis of functional organelles. If there are defects in either genetic component or in their ribosomal machineries, the total complement of photosynthetic and membrane proteins will not be manufactured. Proper fusion of thylakoids into functional grana displays also requires both functional synthesizing systems.

In a study of the nuclear gene mutant strain *ac-20*, R. Levine and U. Goodenough found they could increase the amount of photosynthetic protein manufactured if they grew *Chlamydomonas* mutant cells in particular media. Before any increase in protein manufacture occurred, however, they found a prior increase in levels of chloroplast ribosomes in these cells. When wild-type cells were inhibited from forming new chloroplast ribosomes, their activities mimicked the *ac-20* mutants. Both observations indicate that certain chloroplast functions require proteins that are manufactured on chloroplast ribosomes. Most organelle proteins, however, are synthesized in the cytoplasm from nuclear information. These protein products of nuclear gene action are then imported by chloroplasts, where they engage in various activities. Both sets of proteins must be present to achieve full expression of chloroplast structure and function.

RUDP CARBOXYLASE FORMATION. RuDP carboxylase is a critical enzyme of the photosynthetic dark reactions. It accounts for about 25 percent of the protein in chloroplast stroma. Using wild-type and mutant strains of tobacco plants, S. Wildman has shown that some polypeptides of the whole enzyme are made in the chloroplast on 70S ribosomes and others are made at 80S cytoplasmic ribosomes. The portion of the enzyme with catalytic properties is made in the chloroplast.

Presumably, chloroplast genes and nuclear genes specify different parts of the enzyme, and its constituent polypeptides are made in two different compartments. This situation is similar to the formation of mitochondrial ATPase and cytochrome oxidase.

COOPERATIVE FORMATION OF RIBOSOMES. As discussed earlier in this chapter, most ribosomal proteins are translated from nuclear gene information, but one or a few probably are specified by chloroplast genes. Ribosomal RNAs are transcripts of chloroplast DNA, made in the organelle. Assembly of ribosomes, therefore, also requires molecules made in different parts of the cell and their organization into functional units within the chloroplast.

At least two proteins of the 50S ribosome subunit have been implicated in altered response to the antibiotic erythromycin. This drug binds to the 50S subunit in sensitive strains and inhibits polypeptide chain elongation. Resistant strains have been isolated in several species, including *Chlamydomonas*. One report by L. Bogorad demonstrated that chloroplast ribosomes were altered in a nuclear mutant and in a nonchromosomal mutant. In each case one ribosomal protein was different in the 50S subunits of mutants as compared with wild-type, but the particular protein which had been modified was different in the two kinds of mutants. This difference indicated that 50S subunit proteins are coded separately in chloroplast and chromosomal DNA.

As with membrane and enzyme systems, ribosome formation requires cooperative action of organelle and nucleocytoplasmic systems. In all these examples, the greatest input is made by the nucleocytoplasmic compartment. Most proteins are coded by nuclear DNA and translated at cytoplasmic ribosomes, but normal structure and function cannot develop unless the chloroplast itself contributes translations and transcripts of its own genetic information. Development of mitochondrial structure and function also requires the integrated activities and capacities of compartmented information and translation systems in eukaryotic cells.

SUGGESTED READING

Books, Monographs, and Symposia

Kroon, A. M., and Saccone, C., eds. 1974. *The Biogenesis of Mitochondria.* New York: Academic Press.

Sager, R. 1972. *Cytoplasmic Genes and Organelles.* New York: Academic Press.

Articles and Reviews

Aloni, Y., and Attardi, G. 1972. Expression of the mitochondrial genome in HeLa cells. XI. Isolation and characterization of transcription complexes of mitochondrial DNA. *Journal of Molecular Biology* 70:363–373.

Clark-Walker, G. D., and Linnane, A. W. 1966. *In vivo* differentiation of yeast cytoplasmic and mitochondrial protein synthesis with antibiotics. *Biochemical and Biophysical Research Communications* 25:8–13.

Cooper, C. S., and Avers, C. J. 1974. Evidence of involvement of mitochondrial polysomes and messenger RNA in synthesis of organelle proteins. In *The Biogenesis of Mitochondria,* A. M. Kroon and C. Saccone, eds., pp. 289–303. New York: Academic Press.

Datema, R., Agsteribbe, E., and Kroon, A. M. 1974. The mitochondrial ribosome of *Neurospora crassa.* I. On the occurrence of 80S ribosomes. *Biochimica Biophysica Acta* 335:386–395.

Goodenough, U. W., and Levine, R. P. 1970. The genetic activity of mitochondria and chloroplasts. *Scientific American* 223(5):22–29.

Hoober, J. K. 1970. Sites of synthesis of chloroplast membrane polypeptides in *Chlamydomonas reinhardi y-1. Journal of Biological Chemistry* 245:4327–4334.

Kirschner, R. H., Wolstenholme, D. R., and Gross, N. J. 1968. Replicating molecules of circular mitochondrial DNA. *Proceedings of the National Academy of Sciences* 60:1466–1472.

Küntzel, H., and Blossey, H.-C. 1974. Translation products *in vitro* of mitochondrial messenger RNA from *Neurospora crassa. European Journal of Biochemistry* 47:165–171.

Manning, J. E., Wolstenholme, D. R., and Richards, O. C. 1972. Circular DNA molecules associated with chloroplasts of spinach, *Spinacia oleracea. Journal of Cell Biology* 53:594–601.

Mounolou, J. C., Jakob, H., and Slonimski, P. P. 1966. Mitochondrial DNA from yeast 'petite' mutants: Specific changes in buoyant density corresponding to different cytoplasmic mutations. *Biochemical and Biophysical Research Communications* 24:218–224.

Nagley, P., and Linnane, A. W. 1970. Mitochondrial DNA deficient petite mutants of yeast. *Biochemical and Biophysical Research Communications* 39:989–996.

Nass, M. M. K., and Nass, S. 1963. Intramitochondrial

fibers with DNA characteristics. I. Fixation and electron staining reactions. *Journal of Cell Biology* 19:593–611.

Robberson, D. L., Kasamatsu, H., and Vinograd, J. 1972. Replication of mitochondrial DNA. Circular replicative intermediates in mouse L cells. *Proceedings of the National Academy of Sciences* 69:737–741.

Sager, R. 1965. Genes outside the chromosomes. *Scientific American* 212(1):70–79.

Schlanger, G., and Sager, R. 1974. Localization of five antibiotic resistances at the subunit level in chloroplast ribosomes of *Chlamydomonas*. *Proceedings of the National Academy of Sciences* 71:1715–1719.

Scragg, A. H. 1974. A mitochondrial DNA-directed RNA polymerase from yeast mitochondria. In *The Biogenesis of Mitochondria*, A. M. Kroon and C. Saccone, eds., pp. 47–57. New York: Academic Press.

Swanson, R. F., and Dawid, I. B. 1970. The mitochondrial ribosome of *Xenopus laevis*. *Proceedings of the National Academy of Sciences* 66:117–124.

Chapter 10

Golgi Apparatus, Lysosomes, and Microbodies

Understanding of each of the three compartments to be discussed in this chapter was brought into sharper focus through modern electron microscopy and biochemical studies. Lysosomes and microbodies were unknown until the 1950s, and the Golgi apparatus was believed to be an artifact until it was shown to exist as a unique compartment by electron microscopy. These organelles share some basic characteristics: (1) each is bounded by a single membrane; (2) each is formed from parts of the cell membrane system and is incapable of self-perpetuation; and (3) each is found in most, but not all, kinds of eukaryotic cells.

The **Golgi apparatus** is a way station in the cell through which some proteins, destined for export, are modified chemically and wrapped in a membrane before exiting the cell. One of the packaged structures produced at the Golgi membranes is the **lysosome,** a container filled with digestive enzymes that acts as an intracellular disposal unit for many organic compounds and foreign or native materials that are either harmful to the cell or no longer required for its activities. Lysosomes look very much like microbodies but the two organelles are dramatically different in many ways. **Microbodies** contain different ·enzyme repertories, whereas lysosomes always have digestive enzyme contents. The functions of microbodies vary from one cell type to another, but the function of lysosomes is essentially the same in the many kinds of cells in which they are found. All three organelle systems are under active study in many laboratories today.

THE GOLGI APPARATUS

The first reported observations of the Golgi apparatus were made in 1898 by Camillo Golgi, an Italian cytologist. Using a specially developed silver stain he was able to see platelike and threadlike structures close to the nucleus in nerve cell cytoplasm. There was much controversy about the reality of this elusive structure because the staining method was difficult to apply to a variety of tissues and results were not readily repeatable. Another problem arose because the subcellular component was not visible in the brightfield microscope, either for unstained preparations or those stained in conventional ways. The reality and ubiquity of the Golgi apparatus finally became evident with the introduction of electron microscopy. The first ultrastructural evidence was presented independently in the 1950s by A. J. Dalton and M. D. Felix, and by F. Sjöstrand. Phase contrast microscopy can also be used to visualize Golgi membrane systems in unstained cells, but this method is unsuitable where the elements of the system are widely dispersed throughout the cytoplasm, as they are in many invertebrate, plant, protist, and fungal cells.

Ultrastructural Organization

The Golgi apparatus is composed of smooth membranes and is considered by some investigators to be a smooth endoplasmic reticulum (smooth *ER*) differentiation of the endomembrane system of the cell. In its usual distinctive form, it consists of stacked **cisternae** which are membrane sacs or cavities filled with fluid contents (Fig. 10.1). The sacs are bounded by smooth-surfaced membranes in the Golgi apparatus and usually appear as flattened regions that are continuous with a peripheral system of tubules and vesicles.

Although they vary in form, the cisternae of the Golgi regions have some features in common. The flattened platelike region is generally 0.5 to 1.0 μm in diameter and often perforated (fenestrated). The tubules are 0.03 to 0.05 μm in diameter and are con-

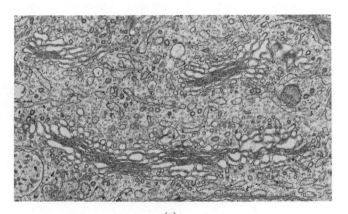

(a)

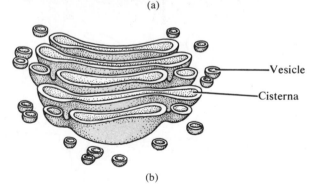

Vesicle

Cisterna

(b)

Figure 10.1
Part of a Golgi apparatus from epididymis of rat testis: (a) Electron micrograph showing the pattern of parallel cisternae, some of which have terminal dilations. Numerous vesicles are usually associated with the flattened membranous sacs, or cisternae. $\times$ 12,600. (Courtesy of H. H. Mollenhauer) (b) Three-dimensional representation of a stack of cisternae and associated vesicles.

tinuous with the peripheral region of the platelike area, extending for several micrometers from the edge of the plate (Fig. 10.2). The peripheral tubules often form an anastomosing network, most clearly seen in negatively-stained preparations of isolated Golgi fractions sedimented from cell-free lysates during centrifugation (Fig. 10.3).

When cisternae are organized into stacks, the stacks are called **dictyosomes** (Fig. 10.4). There usually are 5 to 8 cisternae per stack, but 30 or more are not uncommon in dictyosomes of lower organisms. A single Golgi apparatus may be composed of one or more dictyosomes, depending on whether or not membrane continuities link together the disjunct stacks of cisternae. Dictyosomes are polarized, a property demonstrated by the differences in form and composition of successive cisternae and by their associations with other membrane components (Fig. 10.5). The proximal pole, or forming face, in each dictyosome is associated either with the nuclear

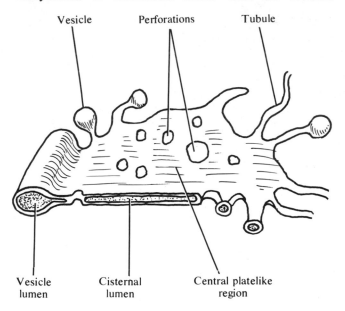

Figure 10.2
A three-dimensional representation of a region of a Golgi apparatus showing its various differentiated parts.

envelope or the *ER*. The cisternal membranes near the forming face are similar to *ER* in morphology and in staining properties. Toward the distal pole, or maturing face, the cisternal membranes undergo morphological and cytochemical transitions so that they resemble the plasma membrane. The cisternal membranes here are thicker, denser, and have a clearer tripartite staining pattern when seen in cross-section. Secretory vesicles generally form from the maturing face of a dictyosome.

Because the extent and form of the Golgi apparatus varies in relation to the type of cell and its metabolic state, controversies have arisen in interpreting the observed differences and similarities in the form and organization of the Golgi apparatus in eukaryotic species. The mammalian cell generally displays a single region of Golgi smooth membranes adjacent to the nucleus (Fig. 10.6). In these cells there is a compact Golgi region which appears as a unified compartment in electron micrographs. Similarly prepared thin-sections through cells of protists, fungi, invertebrate or higher plant tissues generally reveal single dictyosomes dispersed through the cytoplasm rather than a single localization (see Fig. 10.4). There may be 1–25,000 dictyosomes in a single cell, although certain cell types or developmental stages of cells may lack dictyosomes altogether. A single dictyosome may be considered a Golgi apparatus if it occurs alone in a cell or if tubular connections are lacking between cisternal stacks. In some kinds of cells, such as certain fungi, there may be single cisternae rather than stacks, or systems of vesicle-producing tubules that function as the Golgi apparatus.

Coated vesicles generally form from tubular elements, unlike the noncoated secretory vesicles produced from the flattened region of a cisterna in a typical dictyosome. The cytoplasmic area surrounding dictyosomes usually lacks such structured components as ribosomes, mitochondria, and chloroplasts. The coated vesicles of the Golgi apparatus are restricted to this zone of exclusion. Similar zones of

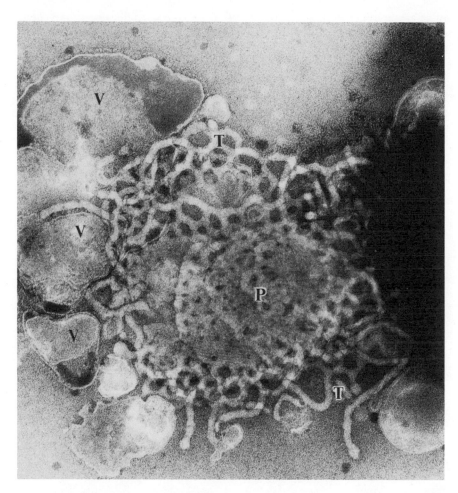

Figure 10.3
Negatively stained, isolated dictyosome from radish (*Raphanus sativus*) root. The dictyosome is partially unstacked and shows the central platelike region (P), peripheral tubules (T), and secretion vesicles (V). × 75,000. (Courtesy of H. H. Mollenhauer)

exclusion have been observed around microtubules, centrioles, and regions of centriole formation.

Functions of the Golgi Apparatus

For some years it was believed that chemical syntheses did not take place in the Golgi regions but that compounds produced elsewhere in the cell were packaged during their passage through the Golgi apparatus. After being wrapped in a membrane, the enzymes or other organic materials would then be discharged through the plasma membrane by a process called **exocytosis** (see Fig. 10.11). It was believed that

chemicals destined for transport could be sent in this manner to other parts of the cell or of the organism.

PROTEIN MODIFICATIONS. With the development of autoradiographic tracer methods, the sequence of protein synthesis and packaging and the participation at the Golgi apparatus in these processes could be studied. Lucien Caro and George Palade reported on experiments designed to follow the migration of radioactively labeled amino acids injected into pancreatic cells of the guinea pig. After 3 minutes of exposure to radioactive amino acids, cells were col-

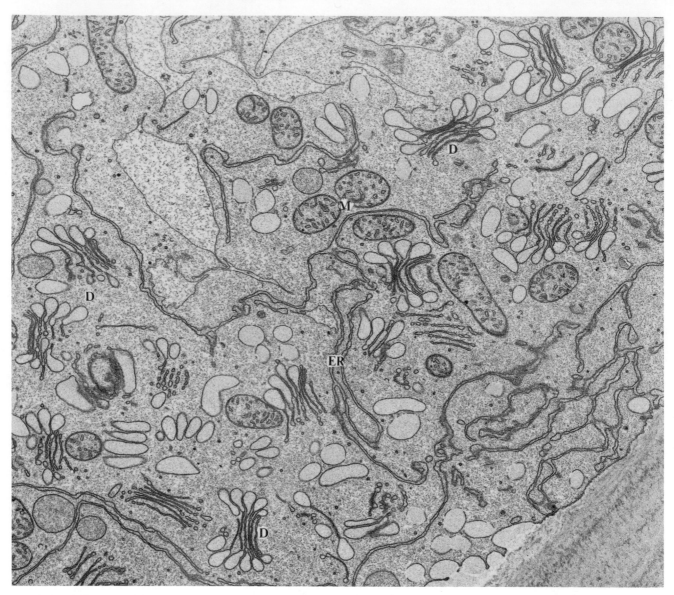

Figure 10.4
Dictyosomes (D) in permanganate-fixed thin section of root cap cell of maize (*Zea mays*). Mitochondria (M) and endoplasmic reticulum (ER) are also present. × 19,500. (Courtesy of H. H. Mollenhauer)

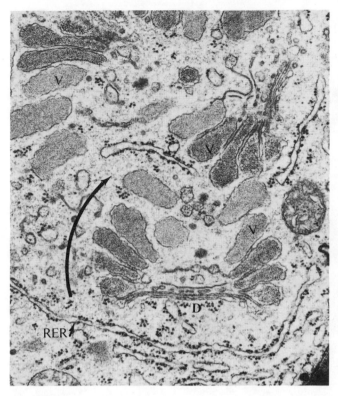

Figure 10.5
Dictyosomes (D) in osmium-fixed thin section of root cap cell of maize. The secretion vesicles (V) formed from dictyosomal components of the Golgi apparatus are filled with a dense material secreted elsewhere in the cell and packaged at the Golgi region. × 35,000. (Courtesy of H. H. Mollenhauer)

lected and prepared for electron microscope autoradiography. These cells contained radioactivity only in the rough *ER* regions, indicating that amino acids had become incorporated into proteins in these areas. Any unincorporated labeled amino acid would have been washed out during standard procedures, leaving insoluble polypeptides which could be distinguished by the presence of silver grains in the emulsion over these sites.

The 3-minute labeling, or "pulse," was also fol-lowed by a "chase" period begun by injecting un-labeled amino acids. The duration of the "chase" was varied in the experiment so that the sequence of events following protein synthesis could be studied and the fate of the original labeled amino acids introduced during the original "pulse" could be traced (Fig. 10.7). Because unlabeled amino acids were provided, syntheses would continue but only the original products manufactured in the first 3 minutes would contain the radioactive label. After a 3-minute pulse and 17-minute chase, the labeled amino acids were found to be distributed over rough *ER* near the Golgi region and over cisternae and secretion vesicles of the Golgi complex itself. If the chase period was 117 minutes in duration, the labeled amino acids were found exclusively in the **zymogen granules,** near the apex of the cell, and in some of the discharged contents in the lumen outside the cell. Zymogen granules contain pancreatic enzyme precursors that are distributed to other parts of the animal where digestive activities occur.

From these data and other observations made in experiments with various intervals of chase time, the sequence of events could be reconstructed as follows:

1. Proteins are synthesized along ribosomes of the rough *ER*.
2. The newly synthesized proteins move from the *ER* to the Golgi apparatus.
3. These proteins become surrounded by a membrane while within the Golgi region and can then be considered as contents of the zymogen granules.
4. The newly formed zymogen granules move to the apex of the cell and discharge their contents into the lumen bordering the pancreatic cell.

These experiments provided evidence that proteins were packaged at the Golgi sites, but they did not demonstrate that protein synthesis occurred in the Golgi elements. These studies did not provide direct evidence concerning the origin of the membrane of the newly formed zymogen granules, but it can be in-

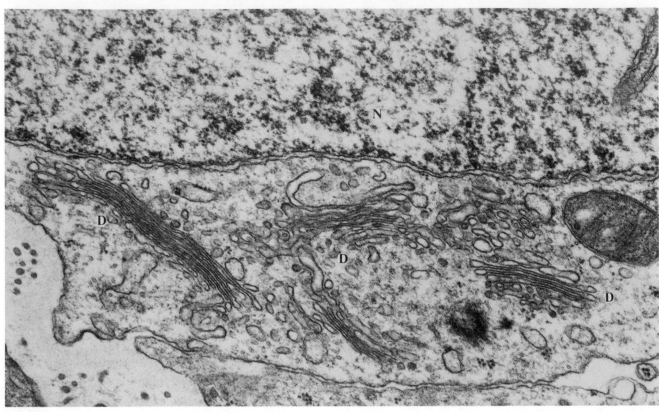

Figure 10.6
Portion of the Golgi apparatus of spermatogonium of rat testis. The closely spaced dictyosomal (D) components are localized adjacent to the nucleus (N). × 50,000. (Courtesy of H. H. Mollenhauer)

ferred that the membranes are derived from Golgi components, since no other reasonable site of manufacture was involved in the sequence of events traced during the autoradiographic studies.

A more informative series of experiments was reported by M. Neutra and C. P. Leblond for mucus granule formation in goblet cells of rat intestine. Microscopical studies by the Spanish histologist S. Ramón y Cajal in 1914 had provided early evidence of possible Golgi involvement in mucus droplet formation in intestinal goblet cells. He observed that mucus droplets were present in the Golgi region of these cells and suggested that the Golgi apparatus might be involved in production of these secretion droplets. The mucus produced in goblet cells is released to the intestinal lumen and covers the cells as a lining that provides resistance to foreign materials, such as bacteria, and is also protective in other ways.

As background for the Neutra–Leblond experiments, we need to describe the structure of goblet cells. Goblet cells are squeezed between other cells of the intestinal lining and are thus long and narrow in form. The Golgi region of goblet cells is cup-shaped and includes dictyosomes which are greatly compacted into a region of smooth membranes. Each dictyosome is made up of eight to ten cisternae, with

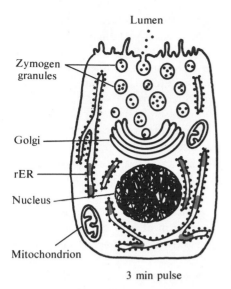
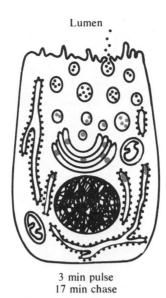
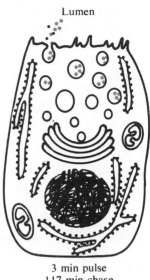

Figure 10.7
Diagram summarizing the autoradiographic results obtained in a "pulse-chase" experiment using mammalian pancreatic cells. The radioactive amino acids (gray) appear first in the rough *ER* and pass through the Golgi region before exiting the cell when the zymogen granule fuses with the plasma membrane in an exocytosis event. See text for details.

the forming face near the nucleus and the maturing face of each stack directed toward the apex of the cell, pointing toward the lumen of the intestinal lining. The cisternae at the forming face appear empty and greatly flattened; those at the maturing face appear swollen due to the presence of mucus-filled globules. The globules detach from the topmost cisterna and migrate up to the cell membrane where the contents are discharged from the cell into the lumen.

The experimental design utilized by Neutra and Leblond was similar to the one described for pancreatic cells. Labeled amino acids were injected into rat intestine and the time course of migration of incorporated labeled amino acids was followed autoradiographically. Labeled amino acids were first seen at the rough *ER*, then in the Golgi cisternae, and finally in the mucus droplet secretions that were ultimately discharged at the cell apex. These events were similar to migrations noted for labeled amino acids in the pancreas. But these studies went further in attempts to answer important questions raised by the preliminary observations. One major question was whether proteins pass through the Golgi apparatus only to be packaged, or whether proteins are processed *before* being packaged within a membrane. Mucus is typical of secretory proteins in having a sugar covalently linked to protein; that is, it is a **glycoprotein**. Experiments were designed to find out if sugar was linked to protein to form the mucus glycoprotein within the Golgi apparatus or elsewhere in the cell. These experiments were conducted, therefore, to determine if the Golgi apparatus processed the glycoprotein or merely packaged it.

Previous experiments with goblet cells and pancreas, as well as other types, had clearly revealed that protein was synthesized in rough *ER* regions and not within the Golgi apparatus. Radioactively labeled glucose was injected into rats, since it was known that glucose is processed into simpler carbohydrates such as those incorporated into the mucus glycoprotein molecule (Fig. 10.8). Within 15 minutes after injection of tritiated glucose, all the radioactivity was concentrated in Golgi cisternae and vesicles; after 20 minutes radioactivity began to appear in mucus globules; after 4 hours the labeled mucus was discharged into the intestinal lumen. From these observations it was found that a stack of eight to ten cisternae was converted to mucus droplets in about 40 minutes, and a new group of 8 to 10 cisternae had formed as replacements during this time.

The experiments thus demonstrated that proteins, which had been synthesized in the rough *ER*, pass on to the Golgi apparatus where carbohydrates, which had been synthesized from simple sugars, were then added to the protein (Fig. 10.9). In similar tracer experiments, other laboratories have discovered that sulfate groups are added on to glycoproteins within the Golgi cisternae. Various secretory cell types apparently produce conjugated proteins by processes in which the sugar residues are linked covalently to the protein molecules, during the time the forming molecules sojourn in the Golgi apparatus. Cartilage cells, for example, produce protein secretions which have sugars added to them while they are in the Golgi region of the cell. The synthesis of plant cell wall material containing cellulose and pectin has been shown to occur in the Golgi apparatus by autoradiographic studies conducted by D. Northcote and J. D. Pickett-Heaps in Great Britain. The pectins are synthesized and packaged in the Golgi apparatus and are used in construction of new cell walls during cell division events following mitosis (Fig. 10.10).

MEMBRANE TRANSFORMATIONS. In addition to their participation in processing cell secretions, Golgi complexes also appear to be involved in membrane

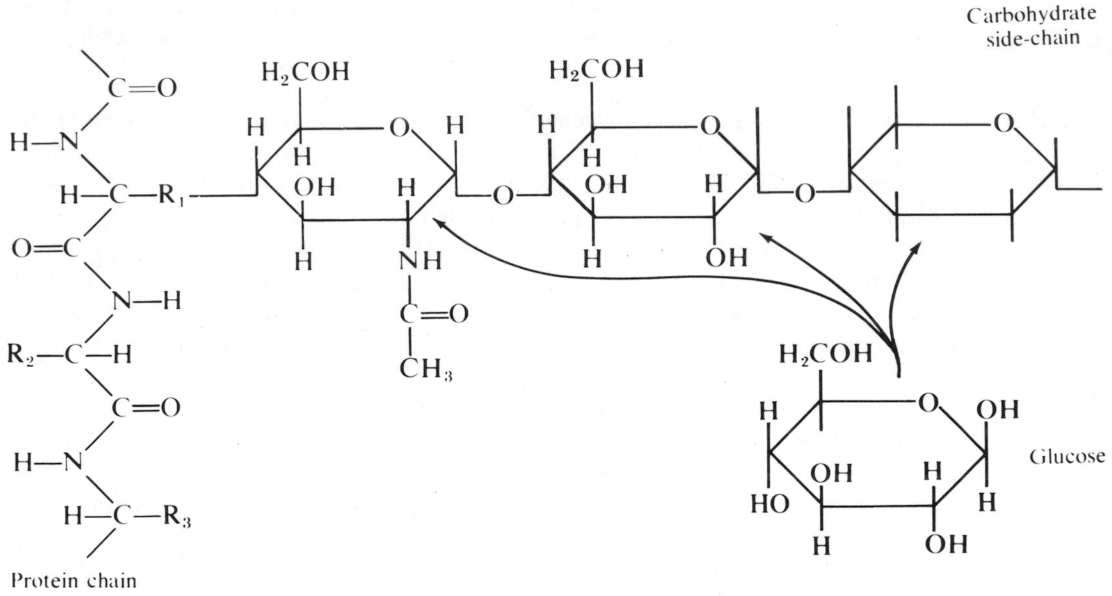

Figure 10.8
General formula for part of a mucus glycoprotein molecule which contains residues of glucose and glucose derivatives in the polysaccharide side chain.

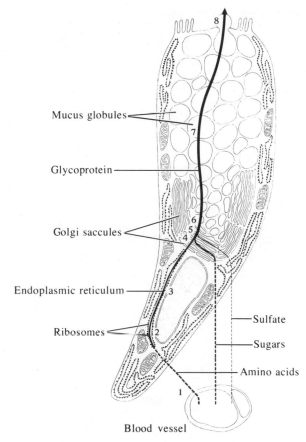

Mucus globules

Glycoprotein

Golgi saccules

Endoplasmic reticulum

Ribosomes

Sulfate

Sugars

Amino acids

Blood vessel

Figure 10.9
Summary diagram showing the flow of precursors and glycoprotein products through the intestinal goblet cell. This process has been deduced from autoradiographic experiments described in the text. (From "The Golgi Apparatus" by M. Neutra and C. P. Leblond. Copyright © 1969 by Scientific American, Inc. All rights reserved.)

transformation. The evidence for this major function is not as complete or convincing, but provocative lines of information suggest membrane transformation may be an even more important function than protein modification. Using a series of static images, not unlike those obtained from autoradiographic experiments, some investigators have followed the events

leading to formation of new plasma membrane components. Membranes of the *ER* are thinner, less dense, and lack contrasted dark-light-dark staining patterns when compared with the plasmalemma. Chemical analysis shows that phospholipids and sterols of *ER* and plasma membrane differ, and protein differences have been discovered in studies employing gel electrophoresis separation methods. The chemistry and morphology of the Golgi membranes are intermediate between *ER* and plasmalemma. These descriptive observations suggest that new membrane is synthesized at the *ER*, transferred to the Golgi apparatus where modifications occur, and fully modified membrane is then added to the plasmalemma during fusions of Golgi vesicles in episodes of exocytosis (Fig. 10.11). A gradation of membrane type from *ER*-like to plasmalemma-like is seen as one observes a stack of cisternae proceeding from the forming face near the *ER* to the maturing face nearer the plasmalemma. This gradation occurs in chemical, as well as morphological, characteristics. Studies of incorporation and turnover using ^{3}H-glycerol and ^{14}C-labeled precursors of membranes have also indicated transfer of phospholipids from *ER* membranes to Golgi membranes to the plasmalemma.

Even before these lines of experimental evidence became available, other evidence had suggested that new Golgi cisternae were formed at *ER*. Grimstone observed in 1959 that starved protozoa (*Trichonympha*) gradually stopped making new dictyosomes and showed reduced amounts or almost total absence of rough *ER*. A reduction in the number of cisternae per dictyosome stack was observed since secretory vesicles continued to form in these cells during the experimental period. When the protozoans were fed again new rough *ER* formed and the capacity to produce new dictyosome cisternae was restored. *ER* thus appeared to be involved in the formation of new Golgi cisternal membranes.

These observations indicate that there is a direction of membrane flow accompanied by modifications

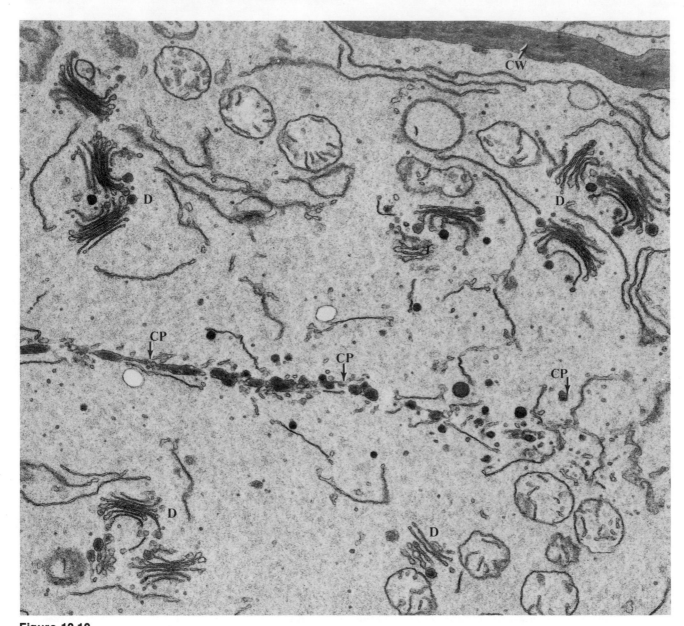

Figure 10.10
Thin section of permanganate-fixed maize epidermal cell. The dense vesicles within the region of the developing cell plate (CP) resemble vesicles formed at the dictyosomes (D) of the Golgi apparatus. This resemblance leads to the suggestion that some of the cell plate material may originate in the Golgi apparatus. Part of the cell wall (CW) is at the upper right of the photograph. × 14,000. (Courtesy of H. H. Mollenhauer)

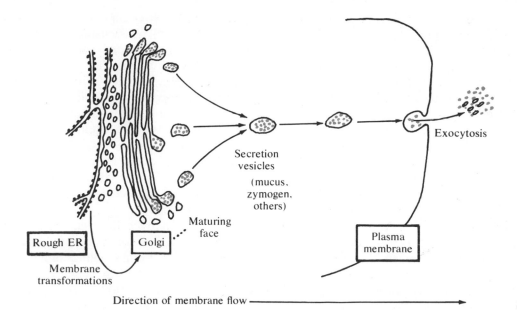

Exocytosis

Secretion
vesicles

(mucus,
zymogen,
others)

Maturing
face

Rough ER

Golgi

Plasma
membrane

Membrane
transformations

Direction of membrane flow

Figure 10.11
Summary diagram showing a postulated sequence of cellular membrane transformation and travel from sites of synthesis at the rough *ER* through the Golgi apparatus and the final fusion with the plasma membrane during exocytosis.

at each step in the progress of changes. While the direction is probably from the Golgi apparatus to the cell surface in most cases, some evidence exists for a reverse flow of membrane. A great deal remains to be learned about these fundamental systems and events in the living and dynamic cell.

LYSOSOMES

Unlike other organelles, lysosomes were first studied by biochemical methods and were not seen under the electron microscope until about six years after the biochemical studies. Christian de Duve and colleagues were studying the distribution of enzymes involved in carbohydrate metabolism in rat liver homogenates, and they included an assay for *acid phosphatase* activity as a control, since this enzyme was known to have no role in carbohydrate metabolism. In their report in 1949, de Duve and associates stated that acid phosphatase activity was higher in tissues that had been isolated in distilled water than others suspended in sucrose solution. Enzyme activity also increased in aged preparations as compared with acid phosphatase in freshly isolated materials, and the activity was not associated with sedimented particles from the aged preparations. These characteristics led de Duve to search for a particulate entity containing acid phosphatase activity, enclosed within a membrane that was sensitive to damage or could be disrupted and thus release its contents.

Subsequent studies by several laboratories showed that a number of hydrolytic enzymes were located in the same membrane-bound particles, all with an optimum pH of 5.0 for activity. About 40 acid hydrolases have been localized in these organelles, called **lysosomes,** which provide cells with the capability of digesting all the biologically significant groups of macromolecules by action of lipases, proteases, nucleases, phosphatases, glycosidases, and sulfatases (Fig. 10.12). The general hydrolysis reaction catalyzed by these enzymes is:

$$R_1—R_2 + H_2O \rightarrow R_1—H + R_2—OH \qquad [10.1]$$

Other kinds of evidence from biochemical studies indicated other probable characteristics for the hypothetical organelle. For example, the centrifugation pattern indicated a probable size of about 0.5 μm when compared with centrifuged mitochondria and other subcellular components of known size. The enzyme activities were latent and only became evident after structural damage had occurred, thus indicating that substrates probably entered the lysosome where they were digested. This sequence is a logical inference since release of hydrolytic enzymes into the cell would lead to digestion of the cell itself, which usually does not happen.

The lysosome was finally visualized in 1955 by de Duve's group. The lysosome had a single membrane surrounding electron-dense but unstructured matrix, and its average size was close to the predicted value of 0.5 μm, although there was an obvious range of sizes. Confirming microscopical evidence of lysosome structure was provided shortly after-

ward by A. B. Novikoff in the United States and S. J. Holt in England, each using cytochemical methods in which a precipitated enzyme reaction product could be seen by either light or electron microscopy of the cell (Fig. 10.13). Using cytochemistry it was possible to survey many kinds of tissue, including those that were difficult to analyze by centrifugation because of low yield or extensive cross-contamination of membranous materials. The cytochemical test for acid phosphatase activity still serves as an excellent marker for lysosomes, and it has contributed significantly to our understanding of the diversity of lysosome morphology and development.

Except for a few cell types, such as the mammalian red blood cell, lysosomes probably occur in cells of all protozoa and multicellular animals. Lysosomes have also been found in some kinds of plant cells, in yeasts and other fungi, and in green unicellular species such as *Euglena*. Most evidence has come from electron microscope cytochemical study, using acid phosphatase activity as the ubiquitous marker enzyme of the lysosome. The organelles range in size from vesicles of about 0.05 μm to droplets averaging several μm in diameter, but the usual dimension has been found to be about 0.5 μm. Most size variability can be attributed to changes in lysosome development during intracellular digestion.

Formation and Function

The lysosomal enzyme proteins are synthesized at the ribosomes just like any other translations of genetic information. While some evidence indicates

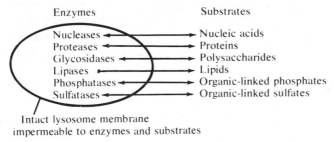

Figure 10.12
Summary of the general groups of digestive enzymes contained within the lysosome and the types of substrates upon which they act under appropriate cellular conditions.

Figure 10.13
Rat liver tissue fixed 4 hours after partial hepatectomy and then incubated for acid phosphatase activity in a cytochemical test system. (a) Light microscope photograph showing dark precipitates of the enzyme activity product in structures presumed to be developmental stages in the lysosome cycle. × 600. (b) Electron micrograph reveals precipitated product of acid phosphatase activity in a secondary lysosome (L) and two autophagic vacuoles (A). Note the single limiting membrane of these structures. × 31,000. (Courtesy of A. B. Novikoff and M. Mori)

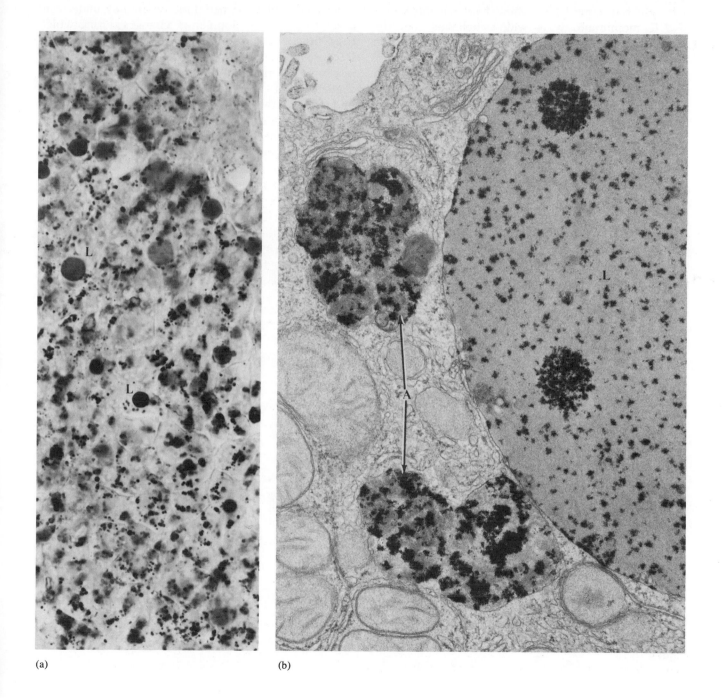

(a)

(b)

that lysosomes may be formed between the sheets of *ER* membrane, it is widely believed that the principal site of lysosome formation is the Golgi apparatus. Enzymes synthesized along the rough *ER* are transported through these channels to the Golgi region where they are packaged into lysosomes. The processes are not very well understood, but a large body of microscopical evidence shows that lysosomes develop within terminal dilations of the Golgi cisternae. These swellings enlarge and eventually separate to become independent organelles containing acid hydrolases (Fig. 10.14).

Lysosomal enzymes are inactive in recently formed structures, and they remain latent until the organelle membrane is damaged or some appropriate substrate enters the lysosome itself. Using the cytochemical staining test to locate acid phosphatase activity and thus identify the lysosomes in a cell,

it was gradually learned that lysosomes underwent a sequence of developmental changes during which they functioned as intracellular digestive systems. Because of their changes in size, shape, and contents, lysosome morphology was highly variable during sequences of digestion. Once these events were understood, the varied morphology of lysosomes were soon considered to be modifications of the original organelles seen at different stages of the digestive processes or undergoing different kinds of digestive activities.

Substances may enter the cell by a general process of **endocytosis,** in which the plasma membrane infolds around the entering material, a vesicle is formed, and the endocytotic vacuole or vesicle then moves within the cell. Endocytosis is a general term that includes the more specific processes of *pinocytosis* and *phagocytosis*. The former refers to inges-

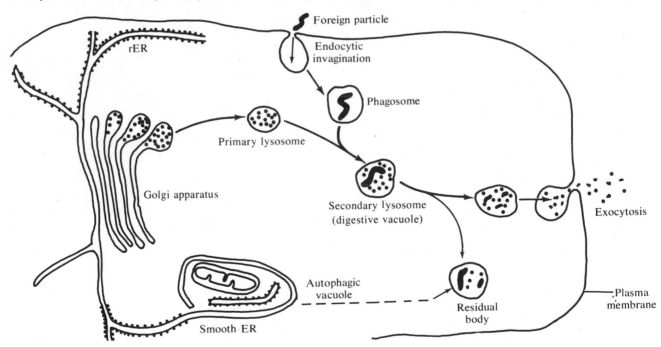

Figure 10.14
Summary diagram of the lysosome cycle. See text for details.

tion of materials in solution or of very small particles in suspension, while the latter is used to describe intake of large particles and structures, like bacteria, within a membrane enclosure. These membrane-bound phagocytized materials may be referred to as **phagosomes.** Since there is a gradation of entering particle sizes but a common process of plasma membrane infolding to aid particle entrance into the cell, the term endocytosis would seem to be a suitable compromise description of the process.

When an endocytotic vacuole enters the cell, it fuses with a **primary lysosome** that contains latent hydrolytic enzymes. The fusion product is called a **secondary lysosome** or, sometimes, a **digestive vacuole.** Once the substances are within a common membrane enclosure with the hydrolases, digestion proceeds within the secondary lysosome. If digestion is complete or nearly so, the remaining materials may be expelled from the cell by a reverse endocytosis, or **exocytosis,** which also involves plasmalemma folding. If undigested residue remains within the secondary lysosome for any length of time, the structure is called a **residual body.** Since all phases of the lysosome cycle may be found in a cell population, the controversy and confusion over lysosome structure until the sequence of events was unravelled is understandable.

Although most intracellular digestions produce the structures described above, there are occasional incidents in which some part of the cell itself may become associated with lysosomes and undergo at least partial digestion. It is not entirely clear how these **autophagic vacuoles** form. It is quite probable that membranes of the *ER* or mitochondria become enclosed by other cell membranes which subsequently fuse with primary lysosomes; digestion then takes place. Starved protozoan and animal cells often develop autophagic vacuoles but usually survive these incidents. In some cases during normal development or because of pathological conditions, the membrane of the primary lysosome may dissolve or leak its enzymic contents into the cell. Self-dissolution and

death of the cell then occurs. In each mammalian ovarian cycle in which fertilization does not occur, lysosomal enzymes cause the corpus luteum to degenerate. Lysosomes also are responsible in large measure for degeneration of larval tissues during metamorphosis. Studies by Rudolf Weber in Switzerland showed that the concentration of lysosomal enzymes increases just before tadpole metamorphosis and continues to increase as the tail is resorbed (Fig. 10.15).

The head of a spermatozoan is covered by an **acrosome,** a structure derived from the Golgi apparatus of the spermatocyte cell (Fig. 10.16). The details of its formation from coalescence of a collection of Golgi vesicles and the presence of hydrolytic

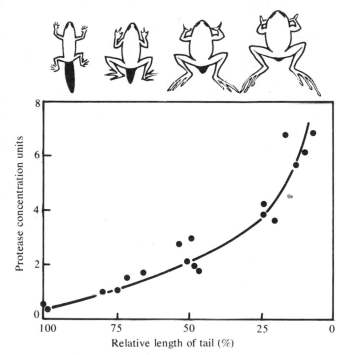

Figure 10.15
Increase in proteolytic enzyme activity during tadpole morphogenesis is correlated with resorption and dissolution of the tail as the tadpole develops to its adult form. The digestive enzymes are located in lysosomes in these cells.

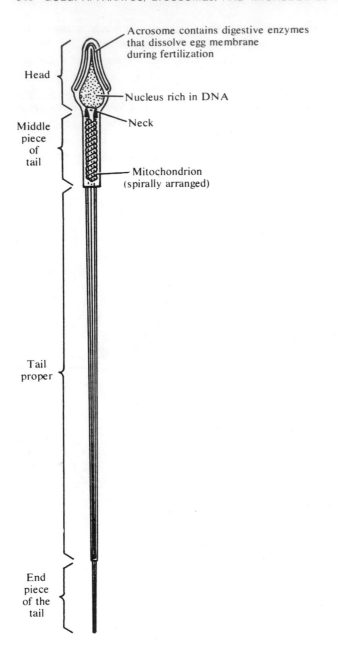

Acrosome contains digestive enzymes that dissolve egg membrane during fertilization

Head

Nucleus rich in DNA

Neck

Middle piece of tail

Mitochondrion (spirally arranged)

Tail proper

End piece of the tail

Figure 10.16
Drawing of mature human spermatozoan based on electron micrographs.

enzymes such as hyaluronidase suggest that the acrosome is a giant specialized lysosome uniquely associated with most animal spermatozoa. Within seconds after a sperm becomes attached to the outer coat of the egg, the acrosome membrane fuses with the plasma membrane of the sperm thus creating perforations through which acrosomal contents are dispersed into the surrounding medium. The outer coat and follicular cells are digested rather quickly and the sperm reaches the plasma membrane of the egg cell; this process takes about 30 minutes in some mammals. The egg and sperm plasma membranes fuse as the sperm nucleus becomes surrounded by the cytoplasm of the egg.

There are a number of almost instantaneous reactions by the egg to sperm penetration, one of which is disruption of cortical granules which themselves have enzymes and staining reactions common to lysosomes. The outer layers of the egg are broken down after dispersal of the contents of the cortical granules, and a new membrane that is resistant to enzyme degradation is formed. Cleavages are initiated after this and may also be induced by mechanical pricking of the egg with a needle. Mechanical stimulation may lead to parthenogenetic development of the egg, that is, development without prior fertilization. The numerous and varied events of fertilization depend on many factors, but lysosomal action in digestion of the egg coatings by the acrosome enzymes and the subsequent modifications of the egg by cortical granule enzymes are an integral part of the complex of processes.

Lysosomes and Disease

Except for protozoa and mammalian white blood cells, very few other kinds of cells regularly engulf materials by phagocytosis; many cell types do so on occasion. White blood cells, or leucocytes, constitute an important element of the body's defenses against infection and disease. Lysosomes develop quite extensively in leucocytes, even as membranous systems such as mitochondria and *ER* diminish, and these intracellular digestive structures appear to be stored against some future contingency. When a leucocyte engulfs some foreign material, such as bacteria, the lysosomes rapidly fuse with the material and digest it. The white blood cells usually die shortly afterward, having expended their stores of lysosomes. Phagocytes in tissues such as liver, lung, and spleen also contain large lysosomes that are important in the digestion of foreign materials.

Some kinds of bacteria, such as the tuberculosis bacterium, and nonliving materials, such as silica particles or asbestos fibers, are not destroyed by lysosomal enzymes. Tuberculosis bacteria have a thick, waxy coat which makes them immune to attack by lysosomal hydrolases. These organisms thus survive leucocyte and other phagocyte incursions and may then be able to initiate infections in the organism.

Studies by Anthony Allison in London and by other investigators have provided important insights into the relationship of lysosomes to disease processes and other aspects of cellular development. When silicon dioxide particles are inhaled into the lungs, they are taken up by phagocytes in the tissue. These cells die, releasing the silica which can be ingested by other phagocytes with the same result. Repeated phagocytic deaths ultimately lead to stimulation of fibroblast deposition of nodules of collagen fibers which decrease lung elasticity and thus impair lung function.

In experiments using tissue culture procedures, Allison showed that ingestion of nontoxic particles such as diamond dust resulted in sequestration of the material in secondary lysosomes which persisted for a long time and did not damage the cells. When silica was phagocytized by the cells in culture, secondary lysosome formation also occurred, but these lysosomes ruptured rapidly, releasing hydrolases and causing cell death. When red blood cells are suspended with different size, shape, and kinds of particles, they lyse only when silica is present and not in the presence of any nontoxic substances, even if the size and surface area of the nontoxic particles are

identical to silicon dioxide. Apparently the reactivity of silica is caused by the formation of silicic acid on the particle surface and hydrogen-bonding between hydroxyl groups of the acid and acceptor molecules of the lysosome membrane. Membrane disruption can be circumvented if some kinds of protective materials which react with the silicic acid hydroxyl groups and prevent attachment to membrane molecules are added to the system along with silica.

The chemical reactions between silica particles and the lysosome membrane lead to lysosome membrane damage and eventual breakage of the organelle. Formation of fibrous nodules of connective tissue apparently results from some factor which is released from the dead phagocytic cells, since a culture of fibroblasts will produce connective tissue fibers after addition of material from a culture of phagocytes that had been exposed to silica particles. The development of fibrous tissues in the lungs of people who have been exposed to asbestos fibers is one symptom of asbestosis, a disease similar in many ways to silicosis, a disease related to silica exposure.

Recent medical genetics studies have provided molecular and cytological information about certain kinds of "storage" diseases, in which a macromolecule remains incompletely processed and ultimately accumulates in the body leading to development of a pathological condition. Many of these inherited disorders affect the central nervous system and cause nervous system deterioration and premature death.

One well studied example of a metabolic defect which leads to accumulations of fatty materials in neurons and some other cells is *Tay-Sachs disease,* a condition inherited as an autosomal recessive trait. The disease is found predominantly among Ashkenazic Jews but some 5–10 percent of babies with this trait come from non-Jewish or Sephardic Jewish families. Afflicted children begin to show symptoms by about 6 months of age and undergo rapid deterioration afterward, until death occurs at some time between the ages of 2 and 4 years.

In Tay-Sachs disease, a 100 to 300-fold excess of

a particular glycolipid, known as a ganglioside, accumulates in the brain tissue of afflicted individuals. The accumulation is believed to be caused by greatly diminished degradation of the glycolipid and not by increased synthesis of the compound. One particular hexosaminidase is missing in all tissues of Tay-Sachs victims, and the lack of this enzyme accounts for failure to degrade the glycolipid molecules (Fig. 10.17). This hydrolytic enzyme normally is found in lysosomes, and lysosomes from children with the disease are filled with numerous concentrically arranged membranes containing the glycolipid.

In other kinds of storage disease, an accumulation of mucopolysaccharides is seen in lysosomes of affected individuals. (Fig. 10.18). Inherited conditions such as Hurler disease or Hunter disease are examples of afflictions caused by accumulation of mucopolysaccharides as a result of defective or missing lysosomal enzymes. Mucopolysaccharides are long protein chains to which polysaccharide groups are attached at intervals. The protein chains are apparently substantially degraded in affected individuals, but the carbohydrate portions and some attached protein are only partially degraded.

Experiments using labeled precursors provided to cells in culture have shown that normal and mutant cells both synthesize mucopolysaccharide at the same rate. Degradation of the macromolecules occurs within 8 hours in normal cells, but macromolecules may remain undegraded in mutant cells for a number of days. These studies, together with the microscopic evidence showing bloated lysosomes filled with partially degraded mucopolysaccharide, suggest that a defect in lysosomal hydrolase activity is the primary cause of the disease process.

MICROBODIES

The microbody was first described in 1954 in electron micrographs of mouse kidney tissue, and it was described in rat liver cells two years later. The emphasis of studies in the next ten years continued to focus on mammalian liver and kidney, using biochemical and biophysical methods along with electron microscopy. Since the mid-1960s a great variety of species and cell types has been studied, contributing to the view that microbodies are a ubiquitous eukaryotic organelle type.

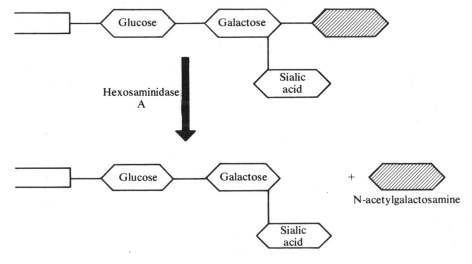

Figure 10.17
The enzyme hexosaminidase A cleaves the terminal N-acetylgalactosamine residue from the carbohydrate chain of the ganglioside type of glycolipid molecule.

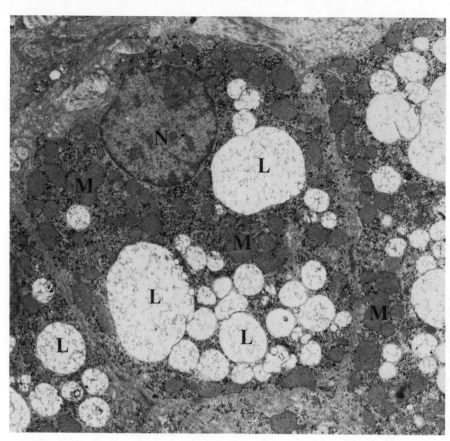

Figure 10.18
Thin section of liver biopsy from a patient with Hurler disease. The largest of the bloated lysosomes (L) are about the same size as the nucleus (N) and considerably larger than the mitochondrial (M) profiles in the group of cells included in the section. × 5,500. (Courtesy of F. Van Hoof)

Microbodies have an astonishing variety of enzyme repertories and functions depending on cell type, species, developmental stage, and a spectrum of physiological conditions. Such variability is totally unknown for any other kind of organelle. This observation lends considerable interest to the study of microbodies but has also produced little agreement on terminology or a unified conceptual framework into which these variations can be placed. Microbody functions in peroxide disposal, biosynthesis and degradation of cellular metabolites, replenishment of metabolic intermediates for respiration and other pathways, and other processes, have been discovered and evaluated with amazing speed within the past ten years. These advances have been made possible by the emphasis on biochemical and physiological approaches taken by most investigators.

Occurrence and Identification

The term **microbody** was first applied to a relatively nondescript ovoid or spherical structure whose granular matrix was enclosed by a single limiting membrane. The kidney microbodies described in 1954 had a homogeneous and unstructured matrix, but the microbodies in rat liver cells contained a substantial crystalline core. Cores or other kinds of inclusions have since been described in microbodies of various cell types and species (Fig. 10.19). These inclusions are sometimes fibrous or amorphous as well as crystalline.

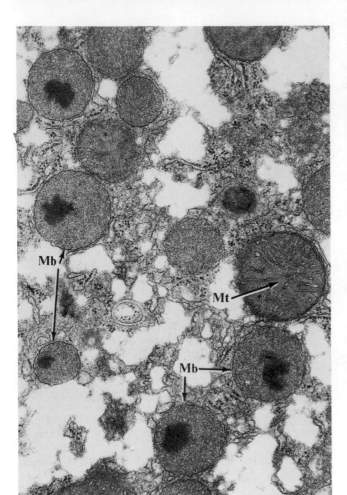

(a)

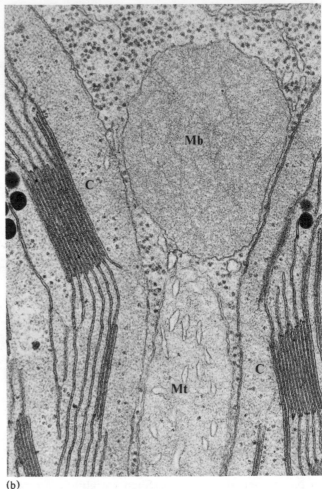

(b)

Figure 10.19
Microbodies. (a) A core inclusion occurs typically in microbodies (Mb) from rat liver. × 25,000. (Courtesy of M. Federman) (b) Cores do not occur regularly in microbodies (Mb) from leaf mesophyll cells, although fine fibrils are often present. The cells are from timothy grass (*Phleum pratense*). Note the membrane organization in the chloroplasts (C) and mitochondrion (Mt). × 66,000. (Courtesy of E. H. Newcomb and S. E. Frederick)

Considerable effort was devoted to defining the crystalline core in animal microbodies, especially during the mid-1960s. Chemical analysis of isolated core material indicated that it was composed of **urate oxidase,** an enzyme that catalyzes the oxidation of uric acid in the pathway of purine degradation. A survey of liver microbodies from various animal species showed that crystalline cores occurred only if the species possessed liver urate oxidase activity, an observation which further supported the notion that the core was made up of this enzyme material in many animal groups. Interest in these inclusions declined very quickly afterward but has since enjoyed a minor revival because many kinds of inclusions occur in plant cell microbodies. The significance of these inclusions has not been determined as yet, but substances other than urate oxidase have been found.

Microbodies range in length from about 0.2 to 1.7 μm, and may appear spherical, ovoid, or dumbbell-shaped in cell sections. The organelles usually occur in clustered groups rather than being randomly dispersed throughout a cell and usually are found adjacent to membranes of the endoplasmic reticulum. In some kinds of cells, microbodies are commonly found in close apposition to chloroplasts or mitochondria or both. Some three-dimensional reconstructions from serial sections have shown that the microbodies are actually larger, irregularly shaped structures. The separate organelles of a cluster are cross-sectional profile views of parts of a single microbody, although more than one of these may occur in a cell.

Microbodies or microbodylike structures have been seen in species belonging to all kingdoms of eukaryotes. They have been described in protists, fungi, plants, and animals, but not necessarily in all cell types or in all developmental stages. The major difficulty in identifying microbodies lies in their utter simplicity of ultrastructure, as well as the fact that they fluctuate in relation to the physiological or growth state of many cells. When yeast or *Euglena* cells are grown in liquid media containing glucose as a source of carbon and energy, microbodies can be found in only a few cell sections. If these same cells are transferred to media in which two-carbon compounds such as ethanol or acetate are the carbon sources, then microbodies appear in relative abundance in the cell cytoplasm. These responses to growth conditions provide flexibility in cellular metabolism, which we will discuss later, but they also made it difficult to find microbodies in some kinds of cells during the earlier years of investigation.

In seeking biochemical and functional criteria by which to identify microbodies, it became evident that a bewildering variety of microbody activities or types existed. As new enzyme activities were found to be associated with microbodies from different kinds of cells, it was comforting to note that at least one reaction was common to them all: removal of hydrogen peroxide by the enzyme **catalase.** It was even more comforting when a simple and reliable method for staining catalase-active structures using cytochemistry was developed (Fig. 10.20). Just as acid phosphatase cytochemical staining identified lysosomes of varied size, shape, and location, catalase cytochemical assays could be employed to identify microbodies regardless of their otherwise diverse enzymatic activities. It was especially helpful because most cells could not be broken gently enough to isolate undamaged microbodies for biochemical tests, and the staining reaction could be used in any case, even if there were very few microbodies or if isolation of unbroken organelles could not be achieved.

As a wider variety of species came under investigation, however, even the test for catalase activity could not be utilized universally. Some cell types apparently lacked catalase activity or lacked this activity in the microbody compartment specifically. Biochemical analysis remains the most reliable means of characterizing and identifying these organelles, but we also come back to the difficulties of defining a common basic pattern or unified concept which describes microbody contributions to cell metabolism and development. Some studies, which we will discuss, have been made possible by the fact that microbodies can be separated from other organelles after centrifugation to equilibrium in density gradients containing sucrose or some other appropriate carbohydrate (Fig. 10.21). The separated organelle populations can be collected from these density gradients and analyzed biochemically to determine their enzymatic capacities.

Each kind of organelle can be recognized by some test for a unique property. For example, mitochondria show cytochrome oxidase activity, chloroplasts or immature plastids contain chlorophyll or a precursor pigment, and lysosomes are identified by their

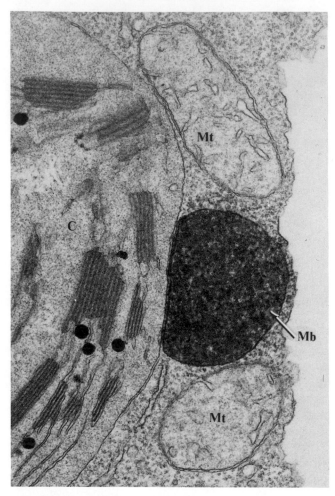

Figure 10.20
Thin section of leaf mesophyll cell from tobacco (*Nicotiana tabacum*), incubated for catalase activity in a cytochemical test. Enzyme activity product is localized in the microbody (Mb) but is absent from the closely appressed mitochondrion (Mt) and chloroplast (C) in the same cell. × 31,000. (Courtesy of E. H. Newcomb, from Frederick, S. E. *et al.*, 1975, *Protoplasma* **84**:1–29, Fig. 5.)

acid phosphatase activity. Microbodies from these same density gradients will thus possess a different group of biochemical properties from these neighboring organelle species, as well as occupying a

particular region within the density gradient in relation to the size and density of the organelle itself. The biochemistry of microbodies from a number of cell types has thus been established through enzyme activity assays of microbodies isolated after centrifugation.

Microbody Biochemistry and Functions

The pioneering studies of C. de Duve led to the first biochemical formulation for microbody activity and function. In his studies of mammalian liver and kidney, de Duve found that **catalase** disposed of harmful hydrogen peroxide that had been generated

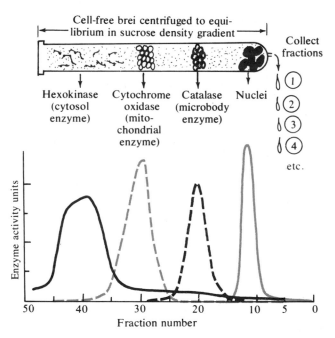

Figure 10.21
Diagram illustrating the methods for identifying different organelles after density gradient centrifugation. Samples can be collected sequentially beginning at the top of the tube as well as through a hole punctured in the bottom of the tube as shown here. Enzyme activity assays are performed after organelle regions have been located by light absorption at 540 nm (peak absorbance region for cell proteins).

in a prior reaction catalyzed by one or more oxidizing enzymes. Since these oxidizing enzymes were of the same general type but varied according to the specific substrate on which they acted, the general term **flavin oxidase** was used to identify the catalysts that guided reactions in which some substrate molecule was oxidized and its hydrogens transferred to oxygen to form hydrogen peroxide. The enzymes all possessed a flavin cofactor bonded to the protein part of the enzyme molecule and contributed to the general reaction

$$RH_2 + O_2 \xrightarrow{\text{oxidase}} R + H_2O_2$$

$$H_2O_2 \xrightarrow{\text{catalase}} H_2O + \tfrac{1}{2}O_2$$

[10.2]

REACTIONS INVOLVING PEROXIDE. Different flavin-linked oxidases handle different substrates, some of which are found only in some kinds of cells. Urate oxidase, which is found in liver cells of some animal species, was the first of these enzymes to be identified, but a more commonly occurring enzyme was found to be an α-hydroxy acid oxidase known as **glycolate (α-hydroxyacetate) oxidase.** Because of its widespread occurrence, it was believed that all microbodies might contain glycolate oxidase which catalyzes the transfer of hydrogens to molecular oxygen, forming glyoxylate and hydrogen peroxide (H_2O_2) in the process. Catalase then degrades the H_2O_2 while glyoxylate can be further processed in various ways, depending on the other enzymes that are present in the particular cell type. This particular oxidation reaction exemplifies the general pathway first proposed by de Duve

$$\begin{array}{c} \text{COOH} \\ | \\ \text{CH}_2\text{OH} \\ \text{Glycolate} \end{array} + O_2 \xrightarrow{\text{glycolate oxidase}} \begin{array}{c} \text{COOH} \\ | \\ \text{CHO} \\ \text{Glyoxylate} \end{array} + H_2O_2 \quad [10.3]$$

Hydrogen peroxide is a powerful reducing agent that is potentially harmful to living cells. Its disposal by catalase would therefore be advantageous to aerobic life, and catalase is found in all oxygen-using species. The reaction shown above is wasteful, however, because energy is not conserved in this hydrogen transfer but is discarded, instead, in the form of hydrogen peroxide. The term **peroxisome** was proposed by de Duve to identify any microbody that was shown to possess a flavin oxidase—catalase reaction pathway. The emphasis was placed on catalase disposal of hydrogen peroxide according to this identification, implying its importance in microbody function.

Several problems were soon encountered. The significance of microbody enzyme activities was shown to lie in glyoxylate metabolism rather than peroxide disposal, and some unicellular species produce glyoxylate in microbodies that lack catalase activity. While these features distinguish microbodies in *Euglena* and some other green unicells, other microbody reaction pathways were similar in *Euglena* and species with glycolate oxidase—catalase reaction systems. What had seemed at first to be a unifying basis for microbody function now appeared to be another variation in microbody enzymatic activities.

Although the name peroxisome remains in wide use today, there have been disagreements about its general acceptance as a synonym for all microbodies. Part of this terminology debate will become more evident as we explore the additional metabolic pathways that have been localized to microbodies in different species groups (Table 10.1).

REACTIONS INVOLVING GLYOXYLATE. The observation that certain bacteria could grow using acetate as the only source of carbon and energy for growth led to the discovery of the **glyoxylate cycle** by H. L. Kornberg and H. A. Krebs in 1957 (Fig. 10.22). Three of the five enzymes (**citrate synthase, aconitase, and malate dehydrogenase;** see Fig. 7.13) are the same ones involved in the Krebs cycle of the mitochondrion in which energy is conserved in aerobic reactions leading to formation of ATP, NADH, and $FADH_2$. The other two are novel enzymes of the

Table 10.1 Enzyme activities found in microbodies of selected cell types and species*

ENZYME	RAT LIVER	RAT KIDNEY	FROG LIVER	FROG KIDNEY	*Tetra-hymena*	YEAST	SPINACH LEAVES	CASTOR BEAN ENDOSPERM
Glyoxylate cycle:								
Isocitrate lyase	−	−	−	−	+	+	−	+
Malate synthase	−	−	−	−	+	+	−	+
Malate dehydrogenase	−	−	−	−	−	−	+	+
Citrate synthase	−	−	−	−	−	−	−	+
Aconitase	−	−	−	−	−	−	−	+
Catalase	+	+	+	+	+	+	+	+
Glycolate oxidase	+	+			+	+	+	+
Urate oxidase	+	−	+	+	−	−	−	+
Allantoinase			+		+			+
Transaminases (various)							+	+
Fatty acid oxidation system							−	+

* Enzyme present, +; enzyme absent, −; enzyme not tested, left blank.

glyoxylate cycle: **malate synthase** and **isocitrate lyase.** Kornberg and Krebs showed that the two-carbon acetate molecule was activated as *acetyl CoA* and entered the glyoxylate cycle at two different places to produce a four-carbon metabolite. The four-carbon compounds could then be channeled to various metabolic pathways involved in carbohydrate synthesis, aerobic respiration, and others. The glyoxylate cycle is an important subsidiary pathway in metabolism that allows cells to utilize a greater variety of fuels for growth and reproduction.

Bacterial cells have no compartmentation, so it was of some interest to see where the enzymes of the glyoxylate cycle were localized in eukaryotic cells. The first studies showed that glyoxylate cycle enzymes sedimented in an organelle fraction, but separations of particles in the density gradients were incomplete and glyoxylate enzymes appeared to be associated with the mitochondrial fraction. Careful studies of castor bean seedlings by H. Beevers and his associates in 1967, however, revealed that glyoxylate cycle enzymes sedimented with a novel particle which could be cleanly separated from mitochondria and proplastids in sucrose density gradients (Fig. 10.23). This same organelle fraction with glyoxylate cycle activity was also shown to have microbody enzymes,

including glycolate oxidase and catalase. In view of the greater significance of the glyoxylate cycle in organelle function, Beevers termed the organelle a **glyoxysome** and not a peroxisome, even though catalase activity was clearly present.

With purified glyoxysome preparations available, Beevers and his associates proceeded to analyze reactions of the glyoxysomes obtained from germinating fatty seeds such as those of castor bean. Stored foods used in germination and growth exist in the form of fats, which must be converted to sugars and other carbohydrates for synthesis and development in the dark. It was shown that stored fats were quantitatively converted to sugar. The fats are hydrolyzed to fatty acids which are dismantled to two-carbon acetyl CoA molecules. Acetyl CoA is retained and is then channeled into the glyoxylate cycle in these glyoxysomes, rather than transferred to the Krebs cycle in the mitochondrion. In this way glyoxysome metabolism provides metabolic intermediates that can be assimilated into cell structures and which do not enter into oxidations in the mitochondrial respiratory pathway (Fig. 10.24).

Glyoxysomes have been found in germinating fatty seeds from plants other than castor bean, including watermelon, cucumber, peanut, and others.

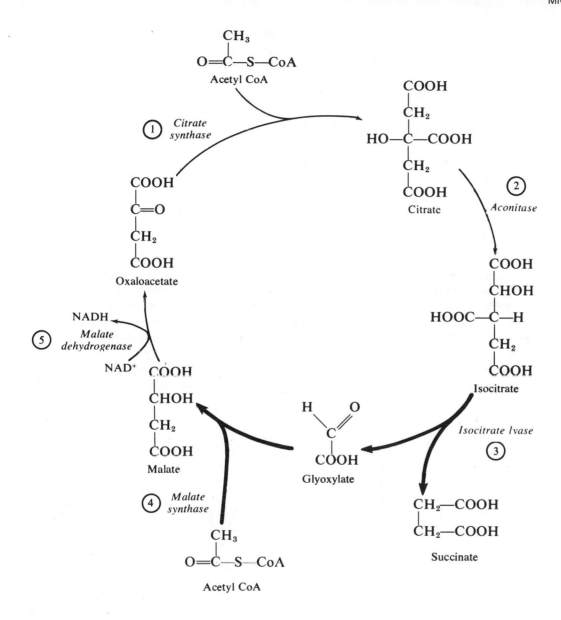

Figure 10.22
The glyoxylate cycle. The unique enzyme-catalyzed reactions in the glyoxylate bypass sequence are shown by heavy arrows.

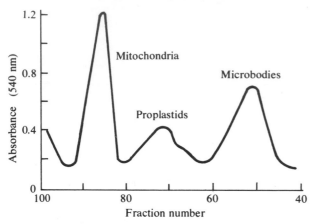

Figure 10.23
Gradient profile of light absorption by proteins of three classes of organelles separated according to physical and chemical differences. The materials were obtained from castor bean endosperm tissue.

The organelles first appear shortly after seed germination begins, reach a peak of activity within a few days, and decline rapidly afterward until no trace of these organelles or their enzymes can be found by about the tenth or eleventh day. The stored fat provides energy for the growing seedling and also contributes the carbon atoms and skeletons for **gluconeogenesis** reactions leading to sugar production required for biosynthesis of cellular materials. The processes of gluconeogenesis represent one of the two major pathways for synthesis of sugars from noncarbohydrate precursors. This pathway proceeds from fats and fatty acids via acetyl CoA and the glyoxylate cycle in seedlings of fat-storing plants and certain microorganisms and through pyruvate or Krebs cycle intermediates in other kinds of cells (Fig. 10.25). The second major pathway is nongluconeogenic which involves formation of sugars from carbon dioxide in photosynthetic species and some other autotrophic organisms (see Chapter 8).

Although glyoxysomes are rich in enzymes and metabolic potential, their major significance would certainly appear to be the synthesis of carbohydrates

and other cellular constituents from fats, via acetyl CoA and the glyoxylate cycle. For these reasons Beevers has continued to insist on the term "glyoxysome" in preference to "peroxisome," even though catalase activity is present. Glyoxysomes of fatty seeds are thus far the only microbodylike organelle shown to contain the entire glyoxylate cycle.

Many other microbodylike structures in protists, fungi, and lower plant and animal cells contain only the two unique enzymes of the cycle, malate synthase and isocitrate lyase. In such cases these two enzymes participate in a **glyoxylate bypass** sequence that fulfills most of the same advantageous functions as the intact glyoxylate cycle. Metabolic intermediates can be synthesized from acetyl CoA and distributed later to other parts of the cell for biosynthetic and energy-yielding pathways. Even the bypass sequence is absent in microbodies from green leaf and vertebrate animal cells. Once again we see variations on a major metabolic theme by the versatile microbody.

GLYOXYLATE METABOLISM. Glyoxylate is formed in microbodies by at least three different pathways:

1. The glyoxylate cycle or bypass by *isocitrate lyase* activity.
2. The oxidation of glycolate catalyzed by *glycolate oxidase*.
3. A purine degrading pathway in amphibians and some aquatic animals.

Pathways (1) and (3) are absent in a number of species and cell types, but oxidation of glycolate to yield glyoxylate is relatively widespread.

Although still under intensive investigation, we can make some reasonable suggestions concerning the significance of glyoxylate in eukaryote cell metabolism. In those species in which the entire glyoxylate cycle operates, the reactions of the cycle provide a means for channelling two-carbon acetate, via acetyl CoA, into gluconeogenic pathways. Acetyl CoA can be converted in biosyntheses to carbohydrates and then to all the necessary building

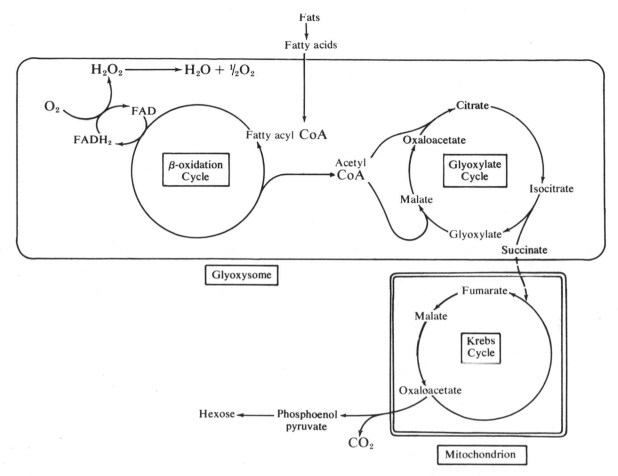

Figure 10.24
Interactions among compartmented metabolic pathways in castor bean endosperm cells. Fats are digested to fatty acids which then enter the glyoxysome and undergo β-oxidation to acetyl CoA units. These units in turn are shunted to the glyoxylate cycle within the same organelles. Succinate produced during glyoxylate cycle reactions is transferred to the mitochondrion where it is converted to oxaloacetate during Krebs cycle oxidations. Oxaloacetate leaves the mitochondrion and is converted in the cytosol to sugars in a reverse glycolysis sequence. This pattern of gluconeogenic conversion of fats to sugars is a major feature of glyoxysome metabolism during castor bean seedling germination and growth. Note that reactions associated with the β-oxidation cycle lead to the production of H_2O_2, which is then disposed of by catalase activity. These latter reactions are typical of microbodies.

blocks for cellular growth and maintenance. Acetyl CoA also serves as an important intermediate leading to energy-yielding Krebs cycle reactions. The two-carbon acetate molecules, therefore, also permit a cell to obtain essential intermediates by which energy-producing reactions occur leading to the formation of ATP, NADH, and $FADH_2$. When only the glyoxylate bypass occurs, the products of the re-

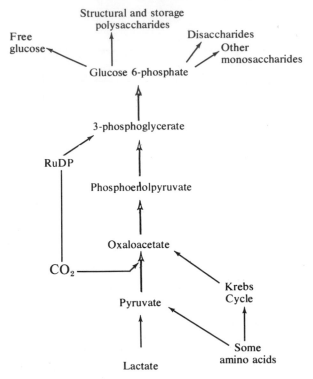

Figure 10.25
A general summary of some gluconeogenic sequences leading to carbohydrate production. Amino acids may serve as precursors by mediation of pyruvate or by Krebs cycle intermediates to form oxaloacetate. Or, pyruvate may be formed from lactic acid and other compounds and then be carboxylated to form oxaloacetate. Sugars may also be formed from RuDP and CO_2 in the dark reactions of photosynthesis. Other pathways exist, such as conversion of fats to sugars through acetyl CoA formation and subsequent reactions in the glyoxylate cycle or the Krebs cycle (see Fig. 10.24).

actions are malate and succinate. These two vital four-carbon metabolites provide an entry into biosynthetic reactions and the Krebs cycle respiration pathway (Fig. 10.26). The glyoxylate cycle or bypass provides, therefore, a more flexible metabolism for cells and also permits the cell to utilize a greater variety of organic compounds to subsidize growth

and reproduction. These advantages are made possible by the compartmentation of the enzymatic activities in microbodies, of both the glyoxysome and peroxisome varieties.

Microbody contribution to leaf cell metabolism is intimately related to activities also taking place in the nearby mitochondria and chloroplasts and in the surrounding cytosol. Although we discussed these systems earlier in Chapter 8, they deserve brief mention again. Under conditions in which photosynthesis is reduced in rate, green cells begin to take up oxygen and release carbon dioxide in the process of **photorespiration.** Photorespiration is an energy-wasting process since energy is directed into oxygen-using, rather than carbon dioxide-fixing photosynthetic activity. Some oxygen consumed during photorespiration is used to form glyoxylate from glycolate in a reaction catalyzed by glycolate oxidase. The hydrogen peroxide which is also produced is degraded by catalase in the microbody. Leaf cells therefore possess peroxisome-type microbodies. These reactions are trivial, however, in relation to the use made of the glyoxylate molecule itself in subsequent pathways.

Under limiting conditions of low CO_2 concentration, glycolate is produced in high concentrations in the chloroplast where it serves little or no useful function. But when glycolate is transferred to microbodies, a series of reactions is initiated leading to a flow of carbon between the Calvin CO_2-fixing cycle and the formation of metabolites that can be diverted into carbohydrate-synthesizing pathways (see Fig. 8.19). In particular, glyoxylate is converted to the amino acids **glycine** and **serine.** Serine in turn is converted to **glycerate** which is energized and used in the synthesis of sugar-phosphates. Sugar-phosphates can be channelled into gluconeogenic sequences, either in the chloroplasts or the cytosol, thus contributing to formation of cellular constituents. These auxiliary possibilities provide a mechanism by which cells whose photosynthetic activities are inhibited can still produce metabolites essential for cell growth. In the leaf cell, microbodies act cooperatively with

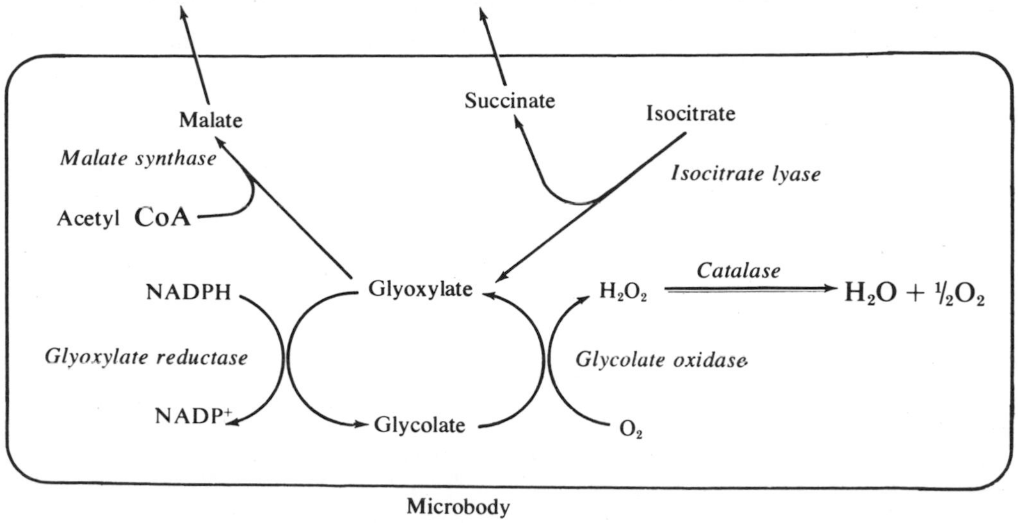

Figure 10.26
Reactions of the glyoxylate bypass in yeast microbodies serve as an anaplerotic pathway to replenish malate and succinate for cell metabolism. Other reactions involving glyoxylate also occur in these organelles, including the expected microbody flavin (glycolate) oxidase – catalase reaction sequence.

chloroplasts, mitochondria, and the cytosol in a stunning display of coordinated activities among different compartments, each of which makes an important contribution to total cell welfare under otherwise inhibiting conditions (Fig. 10.27).

In leaf cells as in vertebrate animals, the glyoxylate cycle or bypass enzymes are absent. While leaf metabolism has been shown to process glyoxylate in other kinds of useful reactions, we know almost nothing about the importance of glyoxylate in vertebrate cells. Because microbodies occur primarily in gluconeogenic organs such as liver and kidney, one would expect microbody activities to be associated with these carbohydrate-synthesizing reactions. Unfortunately, there is no evidence to support this suggestion at present. Except for de Duve's original proposal of a flavin oxidase – catalase system in vertebrate microbodies, little else has been discovered to relate the glycolate or glyoxylate of that system to major metabolic pathways in vertebrates.

OTHER MICROBODY FUNCTIONS. We have mentioned three general functions of microbody metabolism: (1) gluconeogenesis involvement; (2) biosynthesis of the amino acids glycine and serine; and (3) degradation of harmful hydrogen peroxide. There is no question that catalase disposal of hydrogen peroxide is an important microbody activity, but this function is probably a by-product of the fact that hydrogen peroxide happens to be produced in certain microbody reactions, and is less important for the general disposal of cellular peroxide. A number of powerful **peroxidases** are known to occur in the cytosol, and these enzymes efficiently dispose of peroxides produced as byproducts of a variety of metabolic reactions.

Two other functions of microbodies have been substantiated. We mentioned earlier that malate and succinate were products of activity involving the two novel enzymes of the glyoxylate cycle, malate synthase and isocitrate lyase. These two four-carbon

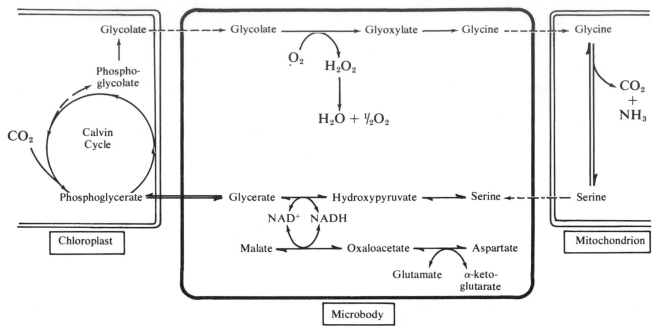

Figure 10.27

Coordination among compartments in the leaf cell. Glycolate, which is manufactured in excess in chloroplasts of photorespiring leaves, is transferred to the microbody where it is processed to the amino acid glycine. Some glycine is converted to the amino acid serine within the mitochondrion, and some serine, in turn, is transferred back into the microbody. Various metabolic intermediates are produced in different enzyme-catalyzed reactions within the microbody, but glycerate may be phosphorylated and enter the chloroplast as 3-phosphoglycerate. In this form, it is one of the critical intermediates of the photosynthetic dark reactions, through which sugars are manufactured. This circuitous metabolic route "salvages" some of the "waste" glycolate formed during photorespiration and allows some sugar production, despite low-efficiency photosynthesis under those conditions. Note that during the conversion of glycolate to glyoxylate in the microbody there is H_2O_2 production by flavin oxidase activity and H_2O_2 disposal by a catalase in typical microbody reactions.

compounds are intermediates in the Krebs cycle, which is essential for energy conservation and aerobic respiration. By replenishing malate and succinate for continued Krebs cycle activities under otherwise limiting growth conditions, glyoxylate enzymes in microbodies provide an **anaplerotic** function. Anaplerotic, or replenishing, activities also provide essential intermediates for gluconeogenesis. A second microbody function briefly alluded to earlier is purine degradation. One of the purine catabolizing enzymes is urate oxidase, but others have been found in microbodies of frogs, chickens, and some protozoa, as well as glyoxysomes of fatty seeds. Some or all of the enzymes of purine catabolism have been found in different species (Fig. 10.28). Purine processing for disposal should therefore be listed among the functions of some microbodies.

Formation of Microbodies

It is widely believed that microbodies bud from the endoplasmic reticulum. Before being pinched off, enzymes synthesized along rough *ER* accumulate in

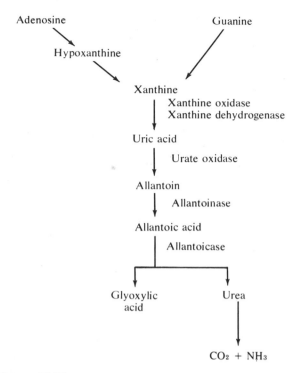

Figure 10.28
Sequence of purine degradation. Different portions of the whole pathway occur in different species and cell types, and many of the enzymes of the pathway have been localized in microbodies.

the *ER* lumen and become sequestered in the microbody as it separates from the *ER*. DNA or ribosomes do not seem to be present in microbodies on the basis of current evidence, so their enzymes must be present when packaging of the organelle takes place. Reports of DNA in microbody preparations have generally been explained on the basis of impurities from other parts of the cell lysates. It is possible that ribosome pockets are incorporated into microbodies as they form from the *ER,* but their role in microbody protein synthesis is uncertain.

Evidence for microbody formation as blebs from *ER* has been obtained from two major kinds of studies: ultrastructural and biochemical. In many electron micrographs microbodies can be seen apposed to and connected with *ER* (Fig. 10.29). This observation is so common that many are willing to accept *ER* formation of microbodies on this information alone. A second line of evidence, however, has strengthened the hypothesis. If radioactively labeled precursors are followed over a period of time after an initial "pulse" exposure, labeled products first appear in rough *ER* and later in microbody-rich fractions isolated from cell homogenates.

In addition to these observations, microbody and *ER* membranes have a very similar lipid composition and at least one enzyme in common. Microbodies have a life span of about 4.5 days in rat liver, so they

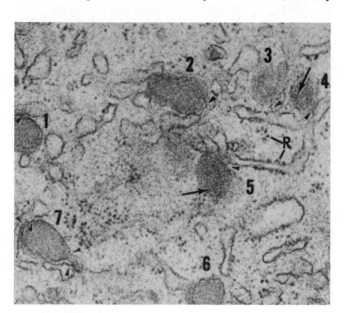

Figure 10.29
Thin section of guinea pig duodenum. All seven microbodies are associated with smooth endoplasmic reticulum and not with rough *ER*. Continuities of microbody membrane and smooth *ER* are shown at arrowheads; arrows indicate wavy, tubulelike structures in No. 4 and 5 microbodies. Ribosomes (R) are present both on the *ER* and apparently free in the cytoplasm. × 45,000. (Courtesy of A. B. Novikoff, from Novikoff, P. M., and A. B. Novikoff, 1972, *J. Cell Biol.* **53**:532–560, Fig. 14.)

must be formed regularly as replacements for organelles that are destroyed.

Several models have been postulated to explain microbody formation and turnover. Experiments designed to follow the decline in radioactivity after an initial "pulse" have been ambiguous and limited in usefulness. It has been found that microbodies may be destroyed at any time during their development; that is, microbodies have an average life expectancy based on statistical studies but not a specified lifetime. One difficulty in interpreting experimental results lies in our lack of information about microbody morphology. If it is assumed that every microbody is a small spheroid or ovoid body and that a certain number exist per cell, one interpretation can be made. If microbodies are less numerous and irregularly larger in size and shape, then the same experimental data can be interpreted in a completely different way. Meaningful studies on the microbody "life cycle" must await information on the size, shape, and number of these organelles in eukaryotic cells.

Evolutionary Origins

Microbodies are a regular component in at least one cell type in almost every eukaryotic species. They can be viewed as simple and relatively inefficient respiratory organelles since hydrogens are accepted by molecular oxygen, as in aerobic respiration. The two kinds of respiratory pathways are profoundly different, however. Microbody reactions do not conserve energy during hydrogen transfer, and hydrogen peroxide is formed when oxygen accepts hydrogens. Toxicity of hydrogen peroxide is prevented by an efficient catalase disposal system. These features, together with the general belief that life was originally anaerobic but became aerobic about 1 billion years ago, have prompted de Duve to postulate a hypothesis of microbody evolution.

According to de Duve, early anaerobic eukaryotes compartmented their primitive respiratory activities in microbodies. Cellular oxidations took place within the microbody, catalyzed by all or most of the enzymes now known to be associated with this organelle. When aerobic species became the predominant organisms in a world with an oxygen-containing atmosphere, the mitochondrion became the main site of cellular oxidations and the microbody declined in capacity and importance in metabolism. Losses of enzymes and metabolic potential have continued to take place in the intervening billion years. These losses vary from species to species and group to group, and we now see a variety of evolutionary differences among eukaryotes. Some activities, such as flavin oxidase—catalase reactions, have been retained in almost all species. Other activities, such as the glyoxylate enzymes in leaves of higher plants and the cells of vertebrate animals, have been lost more recently. Different losses have occurred in different species at different times in evolution.

This hypothesis assumes that all the enzymes and reaction pathways now found in microbodies were present in the ancestral anaerobic species. This assumption is certainly possible, but it is difficult to prove. One profitable approach to the evolutionary questions is to obtain comparative information on enzyme gain or loss in ancestral and descendant lineages. This approach is a standard evolution study method because it is expected that modifications will occur in descendant lineages in direct proportion to the amount of time which has passed since ancestral and descendant lines diverged. From the little information now available, no coherent pattern can be discerned. Differences appear to be associated with different life styles and metabolic requirements, but these are not necessarily coincident with evolutionary relationships. To cite one example of the difficulty in deciding on the correct explanation of microbody evolution, we can look at a plant like castor bean. Its seeds contain microbodies (glyoxysomes) with more kinds of enzymes than are found in any other type of microbody-like organelle. Leaves on these same plants, however, have microbodies without glyoxylate cycle enzymes and with a capacity for synthesizing amino acids, unlike most other microbodies. If the different tissues of a single species can show this range of variability in microbody enzyme repertories, then it would be a major, but not impossible, task to sort out microbodies from different species and decide on a microbody evolution sequence.

Compartmentation is a highly developed subcellular theme in eukaryotic cells. Some organelles, like microbodies, fluctuate in response to growth conditions, fuel sources for carbon and energy, and development. In order to gain some insight into microbody variation (and evolution), M. Müller has examined and compared several groups of protozoa with different life styles. These studies led to the discovery of another microbodylike organelle that had not been known previously.

Müller has studied three different species, each representative of a different and distinctive group of protozoa: (1) *Acanthamoeba castellani,* an acrobic, free-living soil ameba; (2) *Tetrahymena pyriformis,* an aerobic, free-living ciliate; (3) *Tritrichomonas foetus,* a trichomonad flagellate that occurs as a parasite in the genitourinary tract of cattle.

Both *Acanthamoeba* and *Tetrahymena* have conventional mitochondrial structure and functions and microbodies with two or more enzymes from the expected repertory. Catalase is present in microbodies of both species, and *Tetrahymena* also has the two key enzymes of the glyoxylate bypass sequence. So little is known about the other microbody enzymes in these two species that any attempt to define general evolutionary relationships among microbodies would be premature at the present time.

The parasitic trichomonad is a predominantly anaerobic organism but it can tolerate oxygen. It carries on a vigorous respiration in the presence of oxygen but it has no cytochromes and no functional Krebs cycle. *Tritrichomonas* and other trichomonads, like a few other known anaerobic protozoa, have no mitochondria. They do have microbodylike organelles whose single membrane surrounds a granular, unstructured matrix (Fig. 10.30). Since it seemed possible that the microbodylike organelles functioned in cell respiration, in the absence of mitochondria, studies were undertaken. Results reported in 1973–1974 showed that these organelles resembled microbodies in ultrastructure and in their behavior during density gradient centrifugation. The metabolism of these organelles, however, was unlike that of any previously studied microbody.

Catalase and one flavin oxidase were active only in the cytosol and were not localized within the microbodylike structure. If this organelle did serve as a prototype for the ancestral anaerobic microbody as postulated by de Duve, it lacked the common denominator of flavin oxidase—catalase that de Duve suggests for the ancestral structure. This contradiction would appear to rule out the trichomonad organelle as an ancient holdover from primeval times. A unique reaction system was discovered which made it even more unlikely that this organelle was ancestral to aerobic cell microbodies. An electron transfer pathway exists in which a hydrogenase transfers electrons to protons and molecular hydrogen is formed. These organelles are entirely unlike either microbodies or mitochondria since protons act as terminal electron acceptors rather than molecular oxygen. In recognition of the unique chemistry, Müller has proposed the name **hydrogenosome** for this organelle.

Even though the comparative analysis is in an early stage, no particular indication has appeared for an evolutionary progression in microbody types. Protozoa may not be the most suitable group in which to seek patterns of descent. The group is remarkably diversified and seems to have exploited many restrictive ecological niches and life styles. They share some common features of morphology and all are unicellular, but there are profound differences as well. This is the only known group to have different DNA conformations and molecule lengths in mitochondria. Their microbodies may be similarly diverse

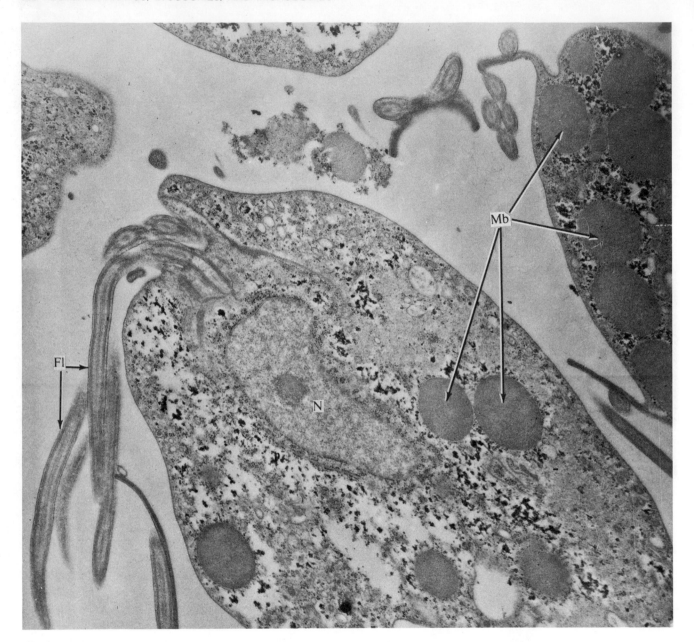

Figure 10.30
Thin section of the trichomonad *Pentatrichomonas*. Both cells contain large, distinctive microbodylike (Mb) organelles, but lack mitochondria. The nucleus (N) contains a prominent nucleolus, and longitudinally-sectioned flagella protrude from the anterior end of the cell. × 25,800. (Courtesy of C. F. T. Mattern and B. M. Honigberg)

in reflection of recent evolutionary modifications that obscure indications of a past common ancestry and lineage.

The hydrogenosome has now been added to the collection of microbodylike structures which includes the glyoxysome and the peroxisome. Until we know more about the relationships among these organelles there is no commonly accepted basis for defining the organelle by any name other than "microbody." Each group of investigators continues to use the term of personal preference. For this reason one should keep all the terms in mind, at least for the present.

PACKAGING AS A GENERAL EUKARYOTIC PHENOMENON

The eukaryotic cell is an amalgam of vesicles, vacuoles, particles, fibers, and complex membrane systems. The structures surrounded by a single membrane, such as microbodies, lysosomes, secretion granules, and Golgi cisternae, are all ultimately related to endoplasmic reticulum and the nucleus. Activation of genes in particular cells leads to the production of RNA transcripts that are translated into proteins along ribosomes at or near the endoplasmic reticulum. Some protein products of gene action accumulate in the *ER* lumen and are included within a membrane-bound organelle. Microbodies are probably formed in this way. Lysosomes may form in this way in certain cells or under certain conditions. Most lysosomes, however, receive their enzyme proteins and surrounding membrane through the mediation of the Golgi apparatus. Proteins synthesized along the rough *ER* are transported through the *ER* lumen to the Golgi apparatus, where they are included in membranous vesicles pinched off from Golgi cisternae; these vesicles become lysosomes if the proteins are acid hydrolases. If the proteins are zymogens or mucus or some other kinds of gene products, we identify the membrane-bound vesicles as zymogen granules, mucus droplets, or other Golgi-formed particles.

The Golgi apparatus itself seems to be formed continuously from *ER,* as its own membranes are used up in packaging exportable proteins. The single membranes of the *ER* and the plasmalemma have the ability to assemble spontaneously from phospholipids, other lipid types, carbohydrates, and proteins. The distinctive features of these membranes reside in large part in their protein molecules, but all the other membrane constituents lend unique properties to membranes in different cells and in different parts of the same cell.

Viewing subcellular compartmentation in general, we find three general classes of organelle:

1. Double-membrane mitochondria and chloroplasts, with their own portion of DNA and ribosomal machinery.
2. Single-membrane organelles and structures that are containers of protein products of nuclear gene action.
3. Organelles lacking their own surrounding membrane, such as ribosomes, centrioles, and chromosomes.

This last group is able to carry out syntheses: protein synthesis at ribosomes, microtubule synthesis at centrioles, and RNA transcript synthesis at chromosomes. Single-membrane-bound structures may modify some molecules, but they are quite limited in their synthetic activities. The double-membrane mitochondria and chloroplasts are very complex and are uniquely able to specify and synthesize some of their own protein constituents.

The *ER,* Golgi apparatus, and plasmalemma all contribute to formation of new membranes, or to reuse of membrane elements for different functions. It is also believed that the nuclear envelope contributes to new membrane systems. Because of these interchangeable and common properties and because all these kinds of membranes may be interconnected at one or more points, they may be considered as morphologically and functionally diversified components of a single **endomembrane system** in the cell. Such a concept has been postulated to explain the dynamic interactions among cellular membranes and their coordinated activities in eukaryotic cells.

SUGGESTED READING

Books, Monographs, and Symposia

Hogg, J. F., ed. 1969. *The Nature and Function of Peroxisomes (Microbodies, Glyoxysomes)*. Annals of the New York Academy of Sciences 168:209–381.

Articles and Reviews

Allison, A. 1967. Lysosomes and disease. *Scientific American* 217(5):62–72.

Avers, C. J. 1973. Peroxisomes of yeast and other fungi. *Sub-Cellular Biochemistry* 1:25–37.

Brady, R. O. 1973. Hereditary fat-metabolism diseases. *Scientific American* 229(2):88–97.

Breidenbach, R. W., and Beevers, H. 1967. Association of the glyoxylate cycle enzymes in a novel subcellular particle from castor bean endosperm. *Biochemical and Biophysical Research Communications* 27:462–469.

Breidenbach, R. W., Kahn, A., and Beevers, H. 1968. Characterization of glyoxysomes from castor bean endosperm. *Plant Physiology* 43:705–713.

Brown, R. M., Jr., Franke, W. W., Kleinig, H., Falk, H., and Sitte, P. 1970. Scale formation in chrysophycean algae. I. Cellulosic and non-cellulosic wall components made by the Golgi apparatus. *Journal of Cell Biology* 45:246–271.

Cunningham, W., Morré, D., and Mollenhauer, H. H. 1966. Structure of isolated plant Golgi apparatus revealed by negative staining. *Journal of Cell Biology* 28:169–179.

de Duve, C. 1975. Exploring cells with a centrifuge. *Science* 189:186–194.

de Duve, C. 1963. The lysosome. *Scientific American* 208(5):64–72.

Frederick, S. E., and Newcomb, E. H. 1969. Cytochemical localization of catalase in leaf microbodies (peroxisomes). *Journal of Cell Biology* 43:343–353.

Kornberg, H. L. 1966. The role and control of the glyoxylate cycle in *Escherichia coli*. *Biochemical Journal* 99:1–11.

Lindmark, D. G., and Müller, M. 1973. Hydrogenosomes, a cytoplasmic organelle of the anaerobic flagellate *Tritrichomonas foetus*, and its role in pyruvate metabolism. *Journal of Biological Chemistry* 248:7724–7728.

Müller, M., Hogg, J. F., and de Duve, C. 1968. Distribution of tricarboxylic acid cycle enzymes and glyoxylate cycle enzymes between mitochondria and peroxisomes in *Tetrahymena pyriformis*. *Journal of Biological Chemistry* 243:5385–5395.

Neutra, M., and Leblond, C. P. 1969. The Golgi apparatus. *Scientific American* 220(2):100–107.

Northcote, D. H. 1970. The Golgi apparatus. *Endeavour* 30:26–33.

Northcote, D. H., and Pickett-Heaps, J. D. 1966. A function of the Golgi apparatus in polysaccharide synthesis and transport in the root-cap cells of wheat. *Biochemical Journal* 98:159–167.

Palade, G. 1975. Intracellular aspects of the process of protein synthesis. *Science* 189:347–358.

Richardson, M. 1974. Microbodies (glyoxysomes and peroxisomes) in plants. *Science Progress* 61:41–61.

Tolbert, N. E. 1971. Microbodies—peroxisomes and glyoxysomes. *Annual Reviews of Plant Physiology* 22:45–69.

Chapter 11

Structures Associated with Cell Movements

Movement is a basic property of living systems. The work of motion requires an input of energy which may be provided in various forms. The most obvious and familiar motions of single cells are swimming and creeping. Many kinds of cells are propelled through liquids by means of **cilia** and **flagella.** These motility organelles act like oars that carry the cell from one place to another. Simple protozoa and algae have cilia and flagella that are basically the same as structures that permit mammalian sperm to swim. The familiar amebae creep along solid surfaces by **ameboid movement;** cells in our own blood system creep along by this type of locomotion; cells grown in culture also move across surfaces by inching forward.

Movement of cell parts occurs commonly and may even be continuous in some cases. Muscles contract, expand, and relax; chromosomes move from one place to another during nuclear divisions; protoplasm streams or churns in most cells. When particles or organelles are carried along with the protoplasmic flow, the streaming movement is called **cyclosis.**

When cells change their positions relative to each other during tissue and organ development, this change constitutes movement because a displacement in space is involved. Growth itself leads to extension from one part of space to another and can be viewed, therefore, as a form of movement. In effect, living systems are in constant motion in at least some portion of the individual during its entire existence.

This universal property of motion is expressed visibly in various ways, some of which were just mentioned. It has become apparent in recent years, however, that the variety of visible movements are actually based on common structural themes within cells. These concepts of cell movement have developed since the 1950s and 1960s from studies using electron microscopy and biochemical methods that were not available earlier. Movements consistently involve protein fibrous structures classified as **microtubules** and **microfilaments.** In some forms of motion only one kind of fiber may be directly or exclusively involved in producing movements. In other cases both microtubules and microfilaments seem to be needed to achieve displacement in space.

Wherever they occur, microtubules are hollow, unbranched cylinders. They vary in length up to tens of thousands of angstroms, but they generally have a diameter of 180–250 Å. On closer inspection, the cylinder wall can be resolved into 13 adjacent filamentous components, each having a diameter of 40–50 Å (Fig. 11.1). Microtubules are widely distributed cell components, but they are apparently absent from certain cells, such as amebae, slime molds, and a few others. Microtubules are most conspicuous in cilia, flagella, the spindle of dividing nuclei, and centrioles, which are usually associated with motility organelles and with the spindle. The tubules are also prominent components of the peripheral cytoplasm in many kinds of cells and within surface specializations of some protozoa (Fig. 11.2).

The second major kind of fibril is the microfilament, which generally occurs in bundles or other groupings rather than singly. Each filament is about 50–60 Å wide, but there are thicker filaments in certain kinds of cells. Microtubules and microfilaments can be

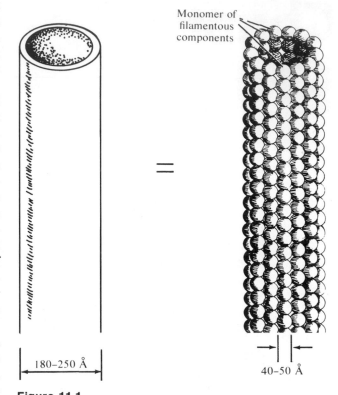

Figure 11.1
The microtubule is a hollow unbranched cylindrical structure (left) formed by 13 adjacent rows of filaments which themselves are linear arrays of the globular protein called tubulin (right).

distinguished on the basis of width, substructure, and their responses to the drugs *colchicine* and *cytochalasin B*. Just as cycloheximide and chloramphenicol selectively affect different ribosomal systems, colchicine and cytochalasin B appear to alter fibrillar systems selectively. Colchicine and its derivatives dismember many microtubular assemblies while cytochalasin B apparently acts on microfilaments. Such discriminating chemical agents are important tools in analyzing relationships between fibrils and specific expressions of cell movement. We will discuss some of these studies later in this chapter.

The most complex and highly evolved system of movement is displayed by muscle. At the same time it is the system that has been studied the longest time and one which provides models that are useful in explaining and understanding simpler expressions of motility. Because of these considerations, we will discuss muscle first. With this foundation it should be easier to understand how other movements have been interpreted and analyzed.

MUSCLE FIBERS

Muscle contraction is work that is subsidized by the chemical energy of ATP. In *striated muscle,* contraction results in voluntary actions by the organism. The structural organization and chemical activities of contraction have been studied for more than 40 years, with increasingly greater understanding during this time. In addition to providing remarkable insight into muscle itself, these studies have allowed us to analyze contractility in nonmuscle systems and to perceive these activities as an outcome of particular evolutionary innovations.

Striated muscle is so named because of its prominent alternating dark and light bands when seen by microscopy. It is also called skeletal muscle because of its close association with parts of the skeleton in vertebrates. The **muscle fiber,** which is a cylindrically-shaped, multinucleate cell, is the whole system (Fig. 11.3). The fiber varies in length but generally has a diameter of 50 to 200 μm. The plasma membrane surrounding the muscle fiber is called the **sarcolemma.** This membrane is bordered by an area of cytoplasm that is called the **sarcoplasm,** which in turn surrounds a bundle of **myofibrils.** Myofibrils are the contractile elements of muscle, running the length of the muscle fiber; each fibril is only about 1–3 μm in diameter. The sarcoplasm contains organized features similar to any eukaryotic cytoplasm. Nuclei, organelles such as mitochondria, and an endoplasmic reticulum (called **sarcoplasmic reticulum** in this system) are all present.

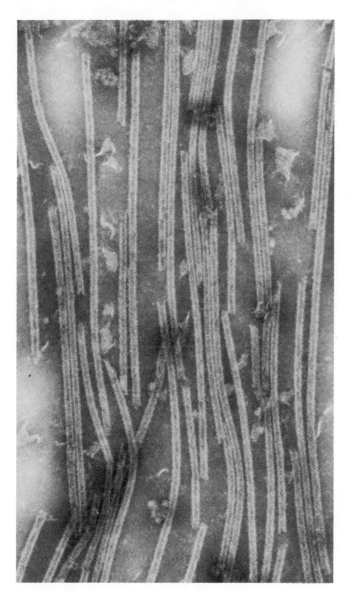

Figure 11.2
A group of negatively stained microtubules from the marginal band just beneath the cell surface of newt (*Triturus viridescens*) erythrocytes. The phosphotungstate stain penetrates and accentuates the lumen of the tubules. × 56,000. (Courtesy of J. G. Gall, from Gall, J. G., 1966, *J. Cell Biol.* **31**:639–643, Fig. 1.)

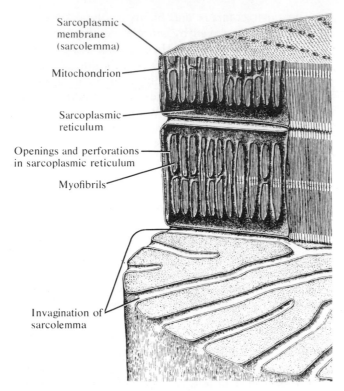

Sarcoplasmic membrane (sarcolemma)

Mitochondrion

Sarcoplasmic reticulum

Openings and perforations in sarcoplasmic reticulum

Myofibrils

Invagination of sarcolemma

Figure 11.3
Cutaway view showing some of the major structured components of part of a muscle fiber. The sarcoplasmic reticulum is perforated and the sarcolemma is infolded at various places, forming a system of invaginations that penetrate through the thickness of the cell (see Fig. 11.10 for further reference).

The striations of muscle myofibrils have been designated by letters of the alphabet in a convention that is readily understood by all investigators (Fig. 11.4). The two major bands are the wider, darker **A-band** next to the narrower, lighter **I-band.** In the center of the A-band there is a lighter region called the **H-zone,** which is bisected by an **M-line.** Each I-band is bisected by a dark, narrow **Z-line.** The region between adjacent Z-lines is called the **sarcomere.** The sarcomere is the contractile unit of the myofibril, and these units are repeated along the length of myofibrils. Each sarcomere occupies about 2.5 μm of space.

Vertebrate smooth muscle, often called involuntary muscle, has no striations. A single nucleus is found in the center of isolated spindle-shaped cells, and the cells are linked together by connective tissue. Considerable progress is being made in understanding smooth muscle, but it is not nearly as well understood as striated muscle systems.

Myofibril Ultrastructure and Chemistry

Myofibrils contain two kinds of **myofilaments** in parallel arrays. These types are thick and thin filaments arranged in specific spatial patterns. Thick filaments, about 150 Å wide, are arranged hexagonally, with about 450 Å of space between pairs of filaments. Thin filaments are about 60 Å in diameter and are regularly arranged between thick filaments so that each thick filament is encircled by six thin filaments. Thick filaments extend from one end to the other of the dense A-band, while thin filaments extend from their connection to the Z-line through the I-band and into the A-band up to the H-zone boundary (Fig. 11.5). When seen in cross section, the pattern varies according to the particular part of the sarcomere in view. Sections through the A-band will therefore have both thick and thin filaments, I-band sections have only thin filaments, and so forth. High-resolution electron microscopy has further revealed the presence of cross-bridges which are disposed along the thick filaments at intervals of 430 Å. These cross-bridges are arranged in pairs in such a way that each pair is rotated about 120 Å from an adjacent pair along the filament. These components have a vital role to play in contraction.

Thick filaments are made of the fibrous protein **myosin.** Each myosin molecule is made up of two identical subunits tightly wound around each other. Myosin molecules aggregate into filaments in a staggered sequence, beginning with the first two molecules lying with their tails side by side and the heads of these asymmetric molecules pointing in opposite directions (Fig. 11.6). This arrangement produces a filament with the globular heads projecting from the central core of the aggregate, except for the central region which lacks these heads. This regularity of projecting heads produces the periodic spacing mentioned above, since the cross-bridges are heads of myosin molecules. These cross-bridges connect the thick myosin filaments to the thin filaments, which are composed of the protein **actin.** The head region of a myosin molecule binds specifically to actin. It is also the part of myosin that has ATPase activity and is, therefore, involved in deriving energy from ATP for contraction.

Actin is different from other proteins of myofibrils since it dissociates into globular monomers (**G-actin**) in the absence of ions, and aggregates into fibrous (**F-actin**) form in the presence of neutral salts. Each G-actin monomer contains one Mg^{2+} ion and one molecule of ATP, both of which are very tightly bound to actin.

Several lines of information clearly show that interactions between actin and myosin are related to contraction in muscle. Actin and myosin combine *in vitro* to form **actomyosin,** a complex that becomes fibrous when placed into distilled water or a dilute solution. Actomyosin fibers contract if ATP is present, but this happens only if the fibers are spread on a surface and are not kept in suspension in liquid. Fibers spread on some surface can lift weights, if ATP is present, while the work of contraction does not take place when fibers are in suspension in liquid. Fibers shorten in all directions in liquid but con-

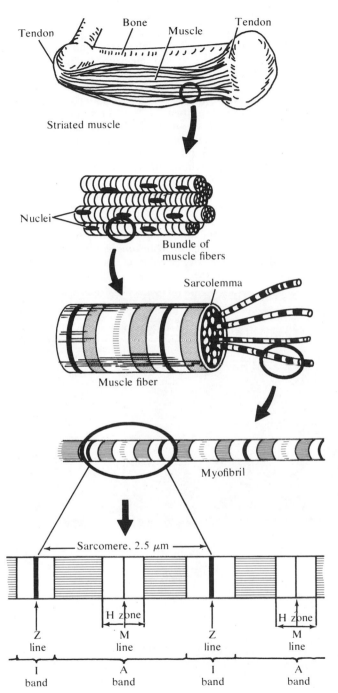

Striated muscle

Nuclei

Bundle of
muscle fibers

Sarcolemma

Muscle fiber

Myofibril

Sarcomere, 2.5 μm

H zone

H zone

Z
line

M
line

Z
line

M
line

I
band

A
band

I
band

A
band

Figure 11.4
The striated muscle system. Striated muscle is made up of cylindrical multinucleated cells called muscle fibers. Each fiber is bounded by a sarcolemma and contains various structured components, including bundles of myofibrils. Each myofibril contains the contractile units, or sarcomeres, of the muscle. Sarcomeres are regions of darker, wider A-bands and narrower, lighter I-bands. Each type of band is bisected by a line. A single sarcomere occupies the area between adjacent Z-lines.

(a)

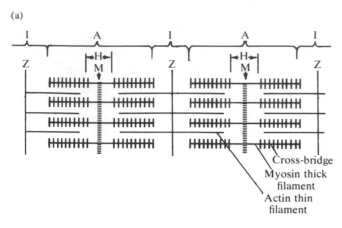

Cross-bridge
Myosin thick
filament
Actin thin
filament

Figure 11.5
Muscle myofilaments: (a) Drawing showing the regular arrangement of myosin thick filaments and actin thin filaments in a sarcomere; (b) electron micrographs of muscle of the fresh water killifish (*Fundulus diaphanus*). (1) longitudinal section of a sarcomere, bounded by two Z-lines and including bands and zones lettered according to the convention described in Fig. 11.4 and the text. (2–6) Cross-sectional views through particular parts of the sarcomere: (2) thick and thin filaments in the A-band region of overlap; (3) thick filaments from the H-zone region of the A-band where no overlap occurs; (4) thick filaments from the A-band immediately adjacent to the M-band; (5) thick filaments from the A-band in the region of the M-band, showing each thick filament connected to each of its six neighboring thick filaments by bridges; and (6) thin filaments from the I-band. (Photographs courtesy of F. Pepe, from *Biological Macromolecules Series: Subunits in Biological Systems.* Marcel Dekker, Inc., New York, 1971, pp. 323–353.)

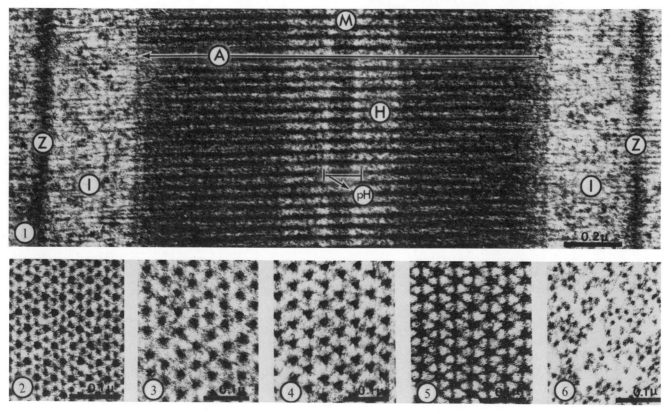

Figure 11.6
A thick filament consists of myosin molecules aggregated such that the molecule heads, which act as cross-bridges during contraction, project from the filament core except in the center, where there are only tail portions of myosin molecule aggregates. The region of "tails" is visible as the H-zone in a sarcomere (see Fig. 11.5).

tract longitudinally in fibers spread on a surface. Even though chemical energy from ATP is used in both cases, this form of energy is not transduced to energy that will directly allow mechanical work such as lifting a weight.

Actin and myosin associate and dissociate during contraction, that is, actomyosin is a complex that is not permanent. Sarcomeres are diminished to about 20–50 percent of their resting length during an episode of contraction. But the individual filaments of myosin and actin do *not* contract. Sarcomeres may increase in length if passively stretched, and again, individual filaments are not stretched. These observations were based on exacting measurements of intact muscle sarcomeres. The A-band width remains constant in contracted, stretched, or resting muscle; therefore, the length of its myosin filaments must also be constant. Actin filaments do not become longer or shorter either. This is clear from the fact that the distance between the Z-line and H-zone does not change during contraction. This is the region in which actin filaments occur. The portion of the sarcomere that alters in length is the I-band, which decreases in contraction and increases in stretching.

Taking all these factors into consideration, A. Huxley and H. Huxley first proposed that changes in sarcomere (hence, muscle) length must be due

to sliding of thick and thin filaments along each other (Fig. 11.7). According to the **sliding filament model,** muscles contract by a mechanism in which myosin and actin filaments slide past each other, driven by longitudinal forces developed by cyclically operating cross-bridges. The Z-lines are drawn together during contraction because the actin filaments attached to a Z-line are drawn in toward the center of the A-band. This event is repeated in each adjacent sarcomere of a myofibril since actin filaments have an opposite polarity on either side of a Z-line. Because of this arrangement, actin filaments in each sarcomere can move similarly toward the center of the A-zone.

Myosin molecules can be selectively fragmented into three regional components by proteolytic enzyme activities (Fig. 11.8). The tail end of the molecule (**light meromyosin**) has no ATPase activity and cannot sponsor binding of myosin and actin. The

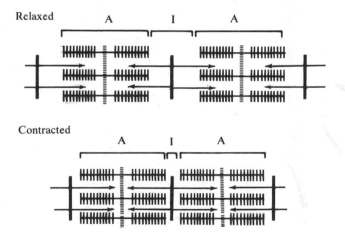

Figure 11.7
Diagram illustrating the sliding filament theory of muscle contraction. The thick and thin filaments slide past each other during contraction but do not undergo change in total length within a sarcomere. The sarcomere itself shortens since adjacent Z-lines are drawn together as actin filaments slide over myosin filaments. Actin filaments on either side of a Z-line have opposite polarity, so that each sarcomere contracts during filament sliding.

opposite end of the asymmetric molecule (**heavy meromyosin**) binds to actin and displays ATPase activity. If heavy meromyosin is digested further, a globular head portion separates from the remainder. The globular head is the specific portion of the molecule that binds to actin and has ATPase activity, and it is the active portion of the myosin molecule.

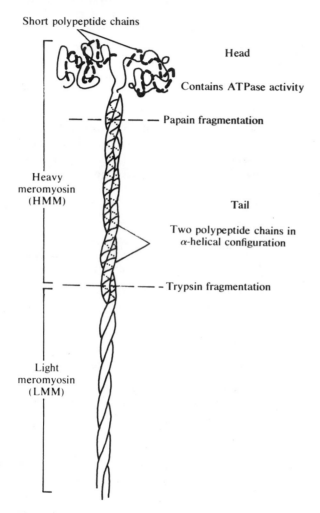

Short polypeptide chains

Head

Contains ATPase activity

Papain fragmentation

Heavy meromyosin (HMM)

Tail

Two polypeptide chains in α-helical configuration

Trypsin fragmentation

Light meromyosin (LMM)

Figure 11.8
The myosin molecule. See text for details.

The globular head portion of myosin is believed to be the cross-bridge that occurs in measurable periodicity along the thick filament, at spacings of 430 Å. These cross-bridges probably break and form continuously during sliding but are never detached from the rest of the myosin molecule. The particular mode of cross-bridging has not yet been resolved, but some models are open to experimental tests.

According to some investigators, the distance involved in an overall cross-bridging movement is 80–120 Å as actin filaments slide over myosin filaments during sarcomere contraction. Because there are two weak regions of myosin molecules which are sites of proteolytic enzyme attack, it is possible that myosin can bend at these two places. If the area between light and heavy meromyosin regions and the area in heavy meromyosin between the globular head and the remainder can provide some flexibility to myosin molecules, then bends could occur. These bends could be involved in the interactions between myosin cross-bridges and actin filaments (Fig. 11.9).

About 80 percent of myofibril protein is myosin and actin. The remaining 20 percent is constituted of four proteins that are all associated with actin filaments: **α-actinin, β-actinin, tropomyosin,** and **troponin.** Tropomyosin accounts for more than 10 percent of myofibrillar protein, and troponin for about 5 percent. Both these proteins have a role in regulating contraction. Troponin consists of three distinct components each of which contributes specifically to actin and myosin association and dissociation. Before considering these regulatory proteins we should have some idea about the stimulus, chemistry, and cellular contributions to episodes of muscle contraction. Afterward these factors can be related to the regulatory processes by which contractions are switched on and off in muscles by control systems.

Coupled Excitation and Contraction

The concentration of ATP in muscle is relatively high during the resting phase. This observation in-

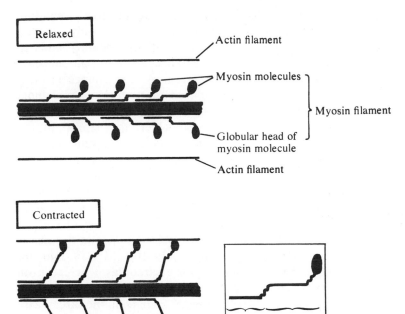

Figure 11.9
Postulated mode of cross-bridging during muscle contraction. The "weak" sites (attacked by trypsin and papain) may provide flexibility to the whole myosin molecule. Bending in these two sites may account for alternating attachment and detachment of myosin heads during contraction and relaxation episodes.

dicates that ATP itself cannot be the factor that initiates and terminates contraction because it would be present in low amounts in resting muscle if its depletion during contraction was the particular signal to shut off contraction. The stimulus that triggers contraction is produced when an electrical impulse from a nerve is received by the muscle and spreads over the sarcolemma. Since there is a higher positive electrical charge outside than inside the sarcolemma, a potential difference exists across this membrane. As the impulse spreads over the sarcolemma, this transmembrane potential difference disappears; this phenomenon is called **depolarization.** Depolarization takes place because the transmembrane potential is discharged as the membrane suddenly becomes permeable to cations such as K^+, Na^+, and Ca^{2+}. The flow of these ions through the membrane leads to depolarization. The extremely rapid spread of the electrical impulse is communicated almost instantly to all the myofilaments of a muscle fiber since they contract simultaneously. The simultaneous contraction occurs even though some myofibrils may be 50 μm deep within the fiber. Simple diffusion of some chemical from the sarcolemma to myofibrils did not adequately explain this phenomenon because the process of diffusion is too slow. The explanation of rapid and instantaneous communication of the impulse finally was discovered in the 1960s. Using electron microscopy, it was found that the sarcolemma is repeatedly invaginated all around the muscle fiber, in such a way that the tubular infoldings of the membrane make contact with most of the muscle myofilaments (Fig. 11.10). This **T system** of transverse tubular infoldings undergoes depolarization, along with the rest of the sarcolemma, upon excitation by an incoming electrical impulse. The T system explains

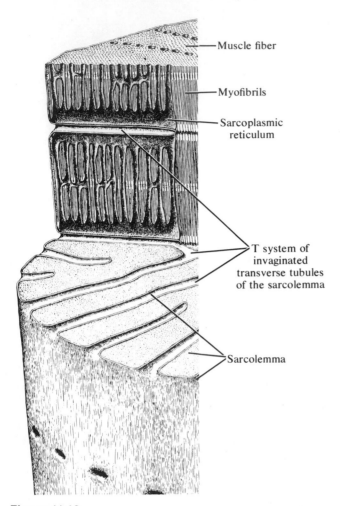

- Muscle fiber
- Myofibrils
- Sarcoplasmic reticulum
- T system of invaginated transverse tubules of the sarcolemma
- Sarcolemma

Figure 11.10
Three-dimensional reconstruction of part of a muscle fiber showing the T system of transverse tubular invaginations of the sarcolemma. (From "The Sarcoplasmic Reticulum" by K. R. Porter and C. Franzini-Armstrong. Copyright © 1965 by Scientific American, Inc. All rights reserved.)

how an impulse can be communicated almost simultaneously to all the sarcomeres of a muscle fiber.

The electrical impulse is translated into chemical or molecular changes in the myofibrils. This translation is mediated by the sarcoplasmic reticulum which envelops the myofibrils and is closely apposed to them. Depolarization of the sarcolemma leads to an increase in permeability of the sarcoplasmic reticulum membrane system, as the change in electrical charge is transmitted through the sarcolemma T system to the cisternae and membranes of the reticulum. Calcium ions, which are stored in the cisternae of the reticulum in resting muscle, then escape rapidly into the sarcoplasm. Once in the sarcoplasm, Ca^{2+} ions trigger the interaction of ATP with the actin and myosin filaments. It has been estimated that 10^{-6} to 10^{-5} M Ca^{2+} is sufficient to initiate a contraction. This concentration of Ca^{2+} is required for activity of myosin ATPase in the sarcoplasm. In resting muscle only 10^{-7} M Ca^{2+} ions are thought to be present in the sarcoplasm, which is an insufficient concentration for enzyme activity. Ca^{2+} ions probably are "pumped" back from the sarcoplasm to the sarcoplasmic reticulum during muscle relaxation, by an enzymelike molecule within the reticulum membrane. This "relaxing factor" is ATP-dependent and can transfer Ca^{2+} ions across the membrane against a Ca^{2+} concentration gradient, using the free energy generated by ATP hydrolysis. On receipt of the next electrical impulse, Ca^{2+} once again is released from its storage vesicles within the sarcoplasmic reticulum, enters the sarcoplasm and another cycle of contraction occurs.

Studies using inhibitors of glycolysis and respiration have shown that muscle contraction can be stimulated repeatedly even if metabolic formation of ATP is prevented. Since there is not enough ATP in muscle to underwrite contraction and since this amount does not decrease during contractions, some other high-energy compounds must contribute to the process. In vertebrate striated muscle, **phosphocreatine** is present in nearly 5 times the concentration of ATP. Dephosphorylation of phosphocreatine by the enzyme **creatine phosphokinase** maintains the cellular concentration of ATP at steady levels. Under conditions known to occur in the sarcoplasm, the reaction leading to ATP formation is favored. In this way, the high-energy phosphoryl group is transferred from phosphocreatine to ADP, forming ATP.

$$phosphocreatine + ADP \xrightleftharpoons[creatine + ATP]{\overset{creatine}{phosphokinase}} \qquad [11.1]$$

Since the terminal phosphoryl group lost from ATP during muscle contraction is rapidly replenished at the expense of phosphocreatine, the sarcoplasmic concentration of ATP remains constant. This level of ATP will decrease in poisoned muscles only after the phosphocreatine supply is exhausted and no new ATP is synthesized in glycolysis, respiration, or by creatine phosphokinase action. An intricate series of feedback controls regulate the rate of respiration and glycolysis and thus the rate of ATP formation in these pathways, as well as controls over the pathways by which ADP is rephosphorylated. Muscle physiology is a subject of widespread interest to biochemists and physiologists, and has been under intensive study for many years.

Troponin and Tropomyosin: Regulatory Proteins

The contraction-relaxation cycles of vertebrate muscle are regulated by the intracellular concentration of free Ca^{2+} ions, under sarcoplasmic reticulum control, but responsiveness to Ca^{2+} involves another component. Interaction of myosin and actin filaments in the presence of Ca^{2+} requires the participation of a tropomyosin–troponin system. Both these proteins are arranged along the actin thin filaments (Fig. 11.11). Troponin consists of three components:

1. A Ca^{2+}-binding element active at concentrations of Ca^{2+} ions from 10^{-7} to 10^{-5} M.
2. An inhibiting component that prevents association of F-actin with myosin.
3. A component that leads to binding of troponin with tropomyosin.

In the absence of adequate free Ca^{2+} ions, troponin components (2) and (3) depress the interaction of

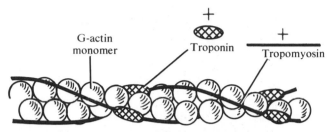

Figure 11.11
Actin filaments in vertebrate striated muscle are composed of two helically wound rows of G-actin monomers. The regulatory proteins troponin and tropomyosin occur in a regular association along the actin filament.

troponin with myosin. Component (3) leads to tropomyosin binding with troponin and component (2) prevents association between thick and thin myofilaments, thereby preventing ATPase activity of myosin. When adequate free Ca^{2+} ions are present, they bind to troponin via its component (1) binding element and a derepression occurs. Myofilaments can now interact, ATPase hydrolyzes ATP, and the muscle fiber contracts.

A great deal has been learned about these regulatory proteins since they were first described in 1964–1965 by S. Ebashi. In addition to exploring their mode of regulating vertebrate striated muscle contractility, these proteins have also been studied in other muscle systems and in various cellular expressions of motility not based on muscle.

As just described, vertebrate muscle contracts when free Ca^{2+} ions interact with actin filaments. When Ca^{2+} is absent or insufficient, these regulatory proteins block active sites on actin thin filaments. The alteration of actin reactivity by troponin is mediated by tropomyosin. A similar system operates in arthropod muscles, such as those in insects, crustaceans, horseshoe crabs, and others. A different regulatory system was reported in 1970, in muscle of clams, scallops, and other molluscs. Molluscan muscle appears to have a regulatory system that involves myosin rather than actin.

In molluscs, Ca^{2+} ions trigger muscle contraction by direct interaction with myosin and not with actin filaments. Tropomyosin is present in actin filaments of all species, but troponin invariably is missing in myosin-linked regulatory systems such as those in molluscs, echinoderms, cephalopods, and some other invertebrate forms. In these systems, tropomyosin seems to have no regulatory function.

Mollusc myosin contains a light-chain component which regulates Ca^{2+} binding by myosin filaments, but the light-chain itself does not bind Ca^{2+}. Purified mollusc myosin binds Ca^{2+} whereas vertebrate myosin does not, and conversely, mollusc actin filaments do not bind Ca^{2+} whereas vertebrate thin filaments do. By appropriate tests, it has been found that actin-linked and myosin-linked systems have some features in common as well as some differences. The two systems have a similar mode of regulation since: (1) absence of free Ca^{2+} ions suppresses ATPase activity and prevents interactions between actin and myosin filaments; and (2) the inhibition can be reversed if 10^{-6} to 10^{-5} M free Ca^{2+} ions are present. The systems differ since: (1) the site of control is on actin filaments in vertebrate systems whereas blocking sites occur on myosin filaments in molluscs and some other invertebrates; and (2) the mechanism by which filament interaction is prevented, the Ca^{2+} effect, is regulated by troponin—tropomyosin mediation in actin-linked contraction but by the light-chain subunit of myosin in the invertebrate systems. Some invertebrate groups seem to have both systems. Insects and crustaceans among the arthropods and earthworms among the annelids seem to have actin-linked and myosin-linked systems regulating muscle contraction. These comparative studies have led to some suggestions concerning the evolutionary pathways leading to muscle regulation. Myosin-linked regulation presumably arose earlier since it occurs in the simpler animal groups. Tropomyosin and actin may not have been involved in muscle regulation in these earlier forms, but they perhaps assumed a regulatory role after the appearance of troponin later in evolution. Further information is needed before these suggestions can be properly evaluated.

With this foundation of information on the components involved in muscle contraction, their organization and regulation, and the mechanisms by which muscle movements take place, we can now proceed to examine other and simpler locomotions. Unlike muscle, in which microfilaments are the underlying fibrillar components involved in movement, some other systems have a microtubule-based system of motility. We will discuss some of these next, and then we will examine systems in which both microfilaments and microtubules contribute to cellular movements.

CENTRIOLES

Centrioles appear as small granules when seen by light microscopy, and structural details are not visible at these magnifications. The granules are associated with spindle formation during nuclear division, and they can be recognized by their behavior, location, and small size. When these granules were examined by electron microscopy in the 1950s, it was realized for the first time that centriole structure was essentially identical in various cell types. Centrioles were about 1600 to 2300 Å in diameter but could vary in length from 1600 to 56,000 Å. Since the limit of resolution of the light microscope is about 2000 Å, it is easy to see why centrioles were not detected in some kinds of cells before electron microscopy was developed. The centrioles would be especially difficult to recognize if they were not in their usual location at the poles of a cell, which is the situation in some cells at some developmental stages.

When seen by electron microscopy in thin sections

of cells, there are 9 sets of tubule triplets in the cross-sectional view of a centriole (Fig. 11.12). Each of the 27 (9 × 3) components is a microtubule about 250 Å wide. Delicate strands appear to connect the triplet groups to each other, and other fine fibrils can often be seen radiating from the central hub of the cylinder to the innermost subfiber of each of the 9 sets. This "cartwheel" configuration is not always present, but if it does occur it is usually localized to the denser proximal end of a centriole (Fig. 11.13). In longitudinal section, the cylinder is seen as a heavy-walled structure with a somewhat denser proximal region in its otherwise translucent and amorphous center (Fig. 11.14).

Centrioles have been found in all eukaryotes except for species which never have a flagellated cell type at any time in development. Species lacking centrioles include certain amebae, unicellular red algae, highly evolved gymnosperms such as pines and their relatives, and all flowering plants. This correlation is interesting in view of the structural identity of the conventional centriole, which acts as a mitotic center, and the **basal body,** from which a cilium or flagellum is produced. All centrioles and

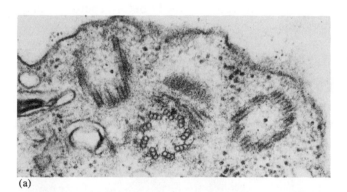

(a)

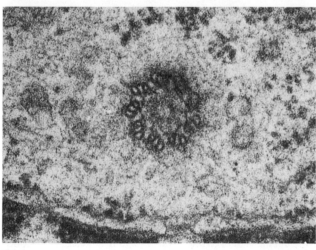

(b)

Figure 11.12
Cross-sectional views of centrioles with the nine sets of triplet tubules. (a) *Chlamydomonas reinhardi,* a unicellular, green flagellate. × 60,000. (Courtesy of U. W. Goodenough) (b) Bone cell of a rat, a mammal. × 82,400. (Courtesy of M. Federman) (c) *Paramecium,* a ciliated protozoan. × 105,-000. (Courtesy of R. V. Dippell). The "cartwheel" is faintly visible in all photographs. Microtubules in the nearby cytoplasm (at arrows in c) are the same in cross-section as the circular microtubules of centrioles.

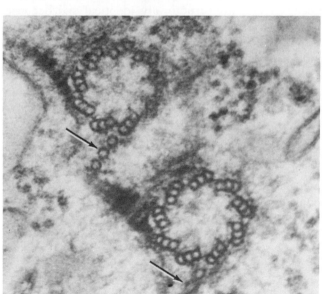

(c)

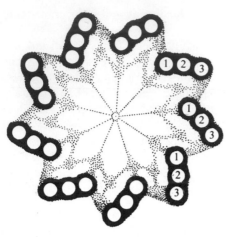

Figure 11.13
Diagram of centriole structure in cross section showing nine subfiber triplets and the faint "cartwheel" pattern of fine fibrils that is sometimes present.

basal bodies have the same fundamental 9 subfiber-triplet organization, but the whole organelles differ in their activities and in their location in the cell. Because they are structurally identical and because a centriole may later become a basal body or a basal body may detach and become a centriole in action and location, the two are considered to be manifestations of the same kind of organelle. A single name identifies the organelles, the preferred term being centriole. We still refer to basal bodies, however, because it is often convenient to use this synonym.

Centriole Formation

There is convincing evidence that shows that centrioles may form in cells which have no existing centriole at the time. Centrioles also may form near to a pre-existing centriole in other kinds of cells. The fact that centrioles can arise *de novo* shows that their formation need not be a replication process. It cannot be replication if there is no pre-existing template or parent structure. *De novo* formation of centrioles further indicates that these structures probably are not genetically endowed. Chromosomes, mito-

chondria, and chloroplasts can only arise from pre-existing equivalent structures. If centrioles specified any of their own properties or carried some genetic component, they should also arise only from pre-existing centrioles. They do not, however, as can be shown by careful studies of various cell types.

If the entire mitotic apparatus, including centrioles, is removed from dividing sea urchin eggs, new centrioles will form and new spindle fibers will appear later. Centrioles are absent from every cell of species like the alga *Nitella*, ferns, and ancient gymnosperms such as cycads, but, centrioles appear *de novo* in particular cells which will give rise to flagellated sperm and they appear shortly before sperm production begins.

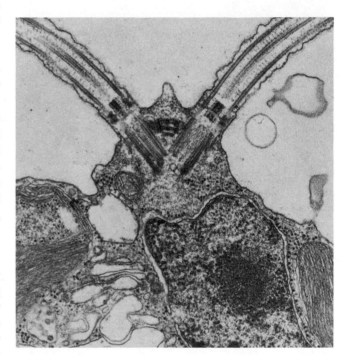

Figure 11.14
Longitudinal section of apical portion of *Chlamydomonas reinhardi*. A heavy-walled centriole subtends each of the two flagella that extrude from the cell surface. × 45,000. (Courtesy of U. W. Goodenough and R. L. Weiss)

Similar *de novo* formation has been described for *Naegleria*, a protozoan that may exist either in an amoeboid or a free-swimming flagellated state. Centrioles are absent in the amoeboid form, but they appear within about 30 minutes during transition to the flagellated form. It is true that tiny centrioles have been overlooked in some kinds of cells and it was mistakenly believed that the first visible signs of a centriole were caused by its *de novo* formation. Apart from such cases, there is reliable information which clearly shows that centrioles can arise spontaneously in some situations and some cell types. In most species, however, centrioles form only in the presence of pre-existing centrioles. Even in these cases we cannot say that the centriole replicates, because the old and new centrioles are always separated by some space, as we will now see.

Stages in structural development of a new centriole were first reported in detailed studies of the unicells *Chlamydomonas* and *Paramecium* in 1968. The sequence is basically the same in both species and probably in other cells as well (Fig. 11.15). The first event in centriole formation is the appearance of a ring of 9 singlet fibers that become the inner *A*-subfibers. Each is circular in cross-sectional outline. The *B*-subfibers assemble onto the *A* singlets, as "horseshoe"-shaped components, rather than as completely circular forms when seen in cross-sectional view. The subfiber-triplets are completed when the *C*-subfibers assemble on the *B* set as incomplete circles in outline. Once the **procentriole** has formed, it grows in length but usually maintains its original diameter during its maturation to become a fullfledged centriole. In *Chlamydomonas,* new centrioles

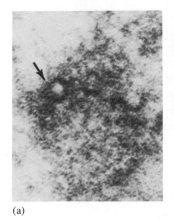
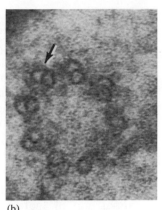

(a)　　　　　　　　　　(b)

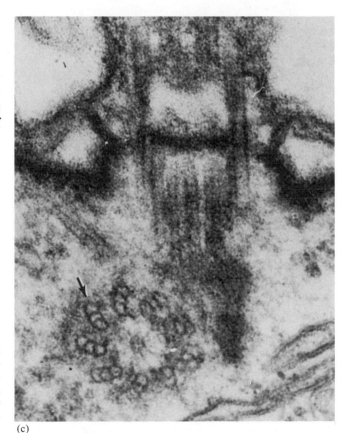

(c)

Figure 11.15
Stages in development of a new centriole at the proximal end of, and at right angles to, the existing centriole in *Paramecium*. (a) Singlet A-subfibers appear first. × 104,000. (b) B-subfibers assemble onto the initial A-set. × 120,000. (c) Triplet sets are completed when the C-subfibers assemble onto the B-subfibers. × 120,000. (Courtesy of R. V. Dippell)

formed in a consistent manner during mitosis, whether or not a pre-existing mature centriole was present at the time. If we generalize on the basis of these observations, it seems that procentriole formation follows the same sequence of events under *de novo* conditions as when a mature centriole is present. Factors that initiate procentriole formation would seem, therefore, to be independent of presence or direction of mature centrioles.

Procentrioles *assemble;* they do not replicate. The procentriole usually assembles in a location which is perpendicular to a mature centriole and at the denser proximal end of that structure (Fig. 11.16). Sometimes, however, procentrioles form parallel to a mature centriole. In these cases, the new centriole may later orient at a tilt to the existing centriole, or perpendicular to it. Centrioles in *Chlamydomonas* develop a tilt after formation and will maintain this tilted aspect even as the centriole pair migrates through the cytoplasm during mitosis and cell division.

J. Gall described the ultrastructure of a cluster of procentrioles in the cycad *Zamia*. Up to 25,000 procentrioles may be present in the cluster which is called a **blepharoplast.** The blepharoplast can attain a diameter of 10 μm in sperm-producing cells, where it develops, so that it had been described much earlier by light microscopists. There is no mature centriole in these cells at the time of blepharoplast formation. Ultrastructure analysis clearly demonstrated the absence of a mature centriole and the procentriolar nature of the blepharoplast. Neither observation was possible by light microscopy. Electron microscopy further verified that each of the 25,000 procentrioles matures to become a basal body from which a flagellum develops.

Centriolar Functions

Centrioles are associated directly or indirectly with two sets of developmental events:

1. Formation of cilia and flagella.
2. Formation of a spindle figure during mitosis and meiosis.

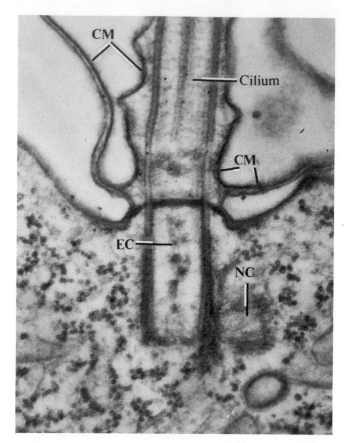

Figure 11.16
Centriole formation in *Paramecium*. The new centriole (NC) forms at right angles to the existing centriole (EC), which is attached to the cilium. A distance of about 700 Å separates the new and existing centrioles at all times. The cell membrane (CM) is continuous around the cell periphery and each cilium of the cell. × 78,000. (Courtesy of **R. V. Dippell**)

Microtubules are a principal component in both systems.

FORMATION OF CILIA AND FLAGELLA FROM CENTRIOLES. Eukaryote **cilia** (shorter appendages) and **flagella** (longer appendages) are produced only from centrioles positioned at the cell periphery. These centrioles may have originated deeper within the cell and later migrated to the cell surface or may have been formed near the surface. A substantial and careful ultrastructural study of centrioles and their associated flagella was reported in 1960 by I. Gibbons and A. Grimstone, in a study using protozoa that live in the gut of termites. These organisms, which are well known flagellates, had been favorite study materials for light microscopy for many years. The 9 subfiber triplets in the centriole were traced to a transition zone just below the shaft of the flagellum. At some distal point in this zone near the flagellar shaft, the *C*-subfibers disappear. The remaining *A*- and *B*-subfibers continue as doublets to a place just below the tip of the flagellum (Fig. 11.17). At the same transition region, a new pair of microtubules appears in the center of the ring of 9 subfiber-doublets. This central pair together with the encircling doublets makes up the so-called "9 + 2" organization found in almost all eukaryote cilia and flagella.

Variations on the "9 + 2" pattern have been found in sperm tails of certain insects. In these species the conventional 9 + 2 arrangement is usually found in other ciliated cells, but 9 + 0, 9 + 1, 9 + 7, or some other variation is found in the sperm flagellum. In the coccid group of insects, the 9 + 2 pattern has been replaced by a totally different microtubular system in the sperm tail.

Several lines of evidence provide support for the direct formation of a cilium or flagellum from an existing centriole. These motility organelles always are an outgrowth of the centriole and never form in its absence. If the appendages are amputated, new microtubules and their surrounding sheath are formed to regenerate the cilium or flagellum from an intact centriole. If the centriole is destroyed or removed from its flagellar attachment, the flagellum will degenerate. A new flagellum will not form unless a new centriole replaces the one that was removed.

In *Chlamydomonas* the centrioles migrate away from the cell periphery during mitosis. The flagellum degenerates after the centriole has detached from it and moved off. At the end of mitosis, new centrioles become established at the cell periphery and they proceed to develop new flagella.

During sperm formation in many species, each flagellum forms from a single centriole produced during the preceding nuclear and cell division. J. Gall showed that a second pair of centrioles formed during spermatocyte division in the snail *Viviparus*. The new pair and the original pair of centrioles then migrated to the apex of the four sperm cells, one centriole per cell, and each centriole gave rise to the one flagellum of the sperm. Aberrant spermatocytes in this species produced multi-flagellated sperm. In these cells the original centriole pair gave rise to a cluster of new centrioles, each of which eventually produced a flagellum of the sperm. The number of centrioles equalled the number of flagella, confirming the 1:1 relationship between these structures. As mentioned earlier, each of the 25,000 flagella of a *Zamia* sperm is produced from a centriole that was part of the blepharoplast in the spermatocyte.

Centrioles display *polarity*. New centrioles form only at the proximal end of a mature centriole, while flagellar outgrowth occurs at its distal end. The organizing events are different in each of these two occurrences. While centriole and flagellum form a continuous structure, there is always a space separating mature centriole from procentriole. In *Viviparus* a distance of 600–800 Å separates the procentrioles from the proximal end of the mature centriole. Sometimes the new centriole moves farther away from the mature centriole, but the centrioles maintain their original orientation to one another.

There is little information on the sites of synthesis of procentriolar components, nor is much known about the pool of precursor materials for centriole assembly. Since centrioles can form spontaneously even in the absence of a mature centriole, little can be said at present about the role of the existing centriole in guiding procentriole assembly.

PARTICIPATION OF CENTRIOLES IN SOME NUCLEAR DIVISIONS. Centrioles are associated with the microtubular displays of the **aster** and **spindle** in dividing

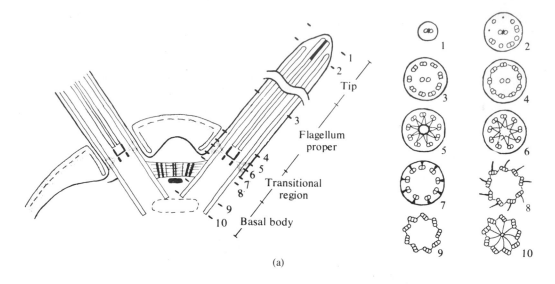

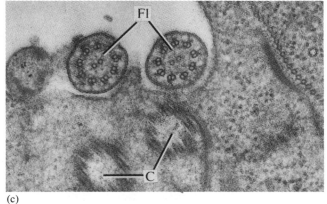

(a)

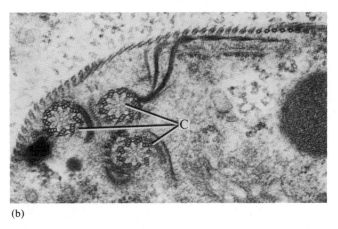

(b)

(c)

Figure 11.17

Centrioles and flagella in unicellular organisms. (a) Schematic drawing of an idealized longitudinal section through the apex of a *Chlamydomonas* cell showing both centrioles and flagella. Four regions of the centriole–flagellum system are shown. A series of cross-sectional views at ten sites along this system shows the subfiber numbers and patterns. (Reproduced with permission from Ringo, D. L., 1967, *J. Cell Biol.* **33**:543, Fig. 30.) (b) Electron micrograph of three centrioles in the protozoan *Hypotrichomonas acosta*, showing a distinct "cartwheel" interior in each. × 49,400. (c) Cross-sectional view of the 9 + 2 microtubule architecture of flagella of the protozoan *Pentatrichomonas*. × 56,000. (Both photographs courtesy of C. F. T. Mattern and B. M. Honigberg) cells (Fig. 11.18). They are not essential for spindle formation or for division. Spindle fibers form regularly in organisms that have no centrioles, such as the flowering plants. There is also experimental evidence showing that spindle formation may proceed in the absence of centrioles, even in species that are normally centriolar. R. Dietz manipulated animal cells to cause the development of an aster at only one pole of the dividing cells. Spindle formation was normal in these cells, even though one pole lacked centrioles and aster. At the end of the division, one daughter cell had centrioles and aster, while the other daughter cell lacked these components, since it formed from the *an*astral half of the mother cell. In the next mitosis, the *a*centriolar, *an*astral cell produced an essentially complete spindle.

Additional lines of evidence also demonstrate that centrioles and asters are not essential for spindle formation. By direct observation, it is clear that spindle microtubules are separated from the centrioles by a space. It is less likely that spindle fibers are assembled by the centrioles if they are never connected. Many lower organisms have centrioles that never participate in nuclear division events but act only in ciliogenesis (formation of cilium or flagellum). In other species, however, centrioles may first act as a mitotic center and later migrate to the cell surface where they produce a motility organelle. According to these observations, centrioles are essential for ciliogenesis but not for spindle formation, even in cells where centrioles regularly participate in nuclear divisions. Some suggestions have been made about the relationship between spindle microtubules and centrioles, but no evidence presently exists to evaluate these ideas.

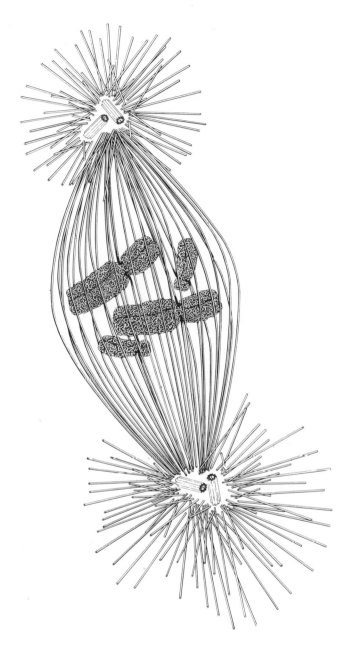

Figure 11.18
An interpretation of mitotic apparatus ultrastructure. A pair of centrioles occupies a clear zone surrounded by short microtubules of the aster at each pole of the spindle. Some microtubular spindle fibers extend from pole to pole, while others extend from one pole to their site of attachment on the chromosome. (Reproduced with permission from DuPraw, E. J., *Cell and Molecular Biology,* 1968, Academic Press. Copyright © 1968.)

Is There DNA in Centrioles?

There have been contradictory reports about the existence of centriolar DNA. Chemical analysis of isolated centrioles has sometimes revealed DNA, but these results have been questioned. It is very likely that the observed DNA was actually contaminating material derived from mitochondria or nuclear sources in these cells. Cytochemical studies have also produced conflicting results. Centrioles show no detectable staining reaction in most cases using either the Feulgen test or acridine orange binding. The 10 μm-wide procentriole cluster in *Zamia* spermatocytes is large enough to be seen by light microscopy and to show positive Feulgen staining if present, but results have always been negative.

Earlier studies of protozoan centrioles showed strong staining with acridine orange, and the appearance of labeled DNA along the ciliary rows of basal bodies according to light microscope autoradiographs (Fig. 11.19). In *Tetrahymena* it was estimated that 2×10^{-16} g of DNA was present at each centriole. This amount is equivalent to about 60 μm of duplex DNA, which is a substantial amount of potential coding length. If true, it further implies that centrioles could have a considerable level of genetic activity. It was, therefore, very important to verify these observations and conclusions.

Two kinds of experimental evidence were later reported, each showing that DNA was absent from centrioles. First, an electron microscope autoradiography study which essentially paralleled the earlier light microscope experiment has been conducted. Since the radioactive label could be placed precisely with identifiable cell structures, it was possible to see whether the label was associated with centrioles. Although centrioles can be identified by location in light microscopy, they can be identified unquestionably in electron micrographs on the basis of unique ultrastructure. In this electron microscopy study it was found that labeled DNA occurred in nearby mitochondria but not in centrioles. Since both organelles are very close in the protozoan cell, it is

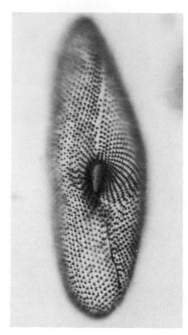

Figure 11.19
The pattern of ciliary rows, each cilium produced from a centriole, is revealed by a silver staining method. This view is of the ventral surface of *Paramecium*. × 700. (Courtesy of R. V. Dippell)

probable that the light microscope autoradiography showed DNA localized in mitochondria but that this result was mistakenly interpreted as showing the existence of centriolar DNA. Other electron microscopic studies also showed that protozoan and metazoan centrioles lacked detectable DNA on the basis of autoradiographic localizations of labeled precursors.

A second type of evidence against the existence of centriolar DNA has been obtained in laser microbeam irradiation studies using an insect species whose centrioles may reach a length of 10 μm. After irradiation with the narrow laser beam, mitochondria were extensively damaged while centrioles remained unchanged. The damaging effect was concentrated in mitochondrial DNA, since the test conditions were designed to sensitize DNA to laser irradiation. If DNA was located in centrioles, it should also have been sensitized and undergone radiation damage. The results of these studies, together with those from electron microscope autoradiography, indicate that centrioles lack DNA. It is therefore unlikely that centrioles genetically control any of their characteristics, and they must, then, be under total nuclear direction. This hypothesis would be confirmed if mutations which affect centrioles could be shown always to be chromosomal and never nonchromosomal.

CILIA AND FLAGELLA

Cilia and flagella are specialized surface structures whose movements propel a cell through a liquid medium. Some ciliated cells are not motile, and in these cells, the cilia sweep materials past the cell surface. We will discuss these and other modified ciliary systems later in this chapter. Flagella are long whiplike appendages that may be more than 150 μm long, while cilia are shorter (5–10 μm, on the average). Cilia occur in relatively large numbers per cell. Flagella are usually fewer on the average, but they are present in the thousands in certain protozoa, some plant sperm, and other cell types. The two kinds of hairlike appendages form from centrioles, and their ultrastructure is basically identical (see Fig. 11.17). As usual in biological systems, some variation in ultrastructural organization does occur in a few species.

The pattern of beating for cilia and flagella or sperm tails is generally distinctive for each type. Cilia beat coordinately in groups or rows of individual extensions, in an action similar to a rowing motion. The beat is transmitted progressively along a row of cilia, rather than occurring synchronously within a group. Motion picture and other studies have shown a similar beating pattern in ciliated cells as evolutionarily divergent as protozoa and vertebrate epithelium. The major differences in beating pattern are in the form and magnitude of the stroke during beating. Flagellar undulation is rather complex and seems to be based on a progressive sharp transition from a straight to a curved position of various regions along the length of the structure. Most information about flagellar motion has come from studies using sperm.

Chemical and Structural Coordinates

Cilia structure and composition are usually studied using electron microscopy and biochemistry, often as joint topics of study. Cilia can be detached from cells by mechanical agitation, as can sperm tails and other flagella, and they can then be collected in sufficient quantities to perform chemical analysis. Ciliary movements can be examined using isolated cilia as well as cells with cilia still attached. The information currently available has been obtained by relating particular chemical constituents to parts of ciliary structure and by analyzing motion on the basis of both kinds of coordinates.

The membrane covering the motility organelle can be removed by detergents, and the remaining fibrous and matrix components of the axial shaft can then be concentrated in separate fractions. Using these materials, further fractionations can then be carried out. It is possible to isolate purified subfiber doublets, singlet microtubules from the center of the axial shaft, and other protein constituents of the system (Fig. 11.20). Once these materials are available in purified form, they can be analyzed chemically, and then each component can be related to ciliary movements.

Three particular protein components have been identified and named, but others have yet to be analyzed chemically. The major protein of the outer doublets and the central singlet fibers is called **tubulin.** The protein has slightly different properties in these two kinds of microtubules. The singlet tubulin is more soluble in some reagents and has a binding site for *colchicine* that is lacking in the doublet protein. The site is normally occupied by

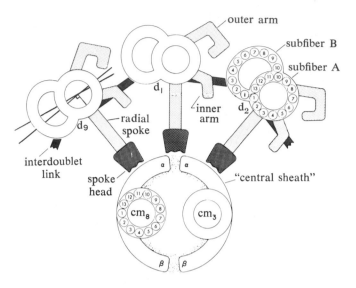

Figure 11.20
Diagram of a cross section through the 9 + 2 axoneme or axial shaft of a cilium or flagellum as viewed from the base toward the tip of the structure. See text for details. (Reproduced with permission from Warner, F. D., and P. Satir, 1974, *J. Cell Biol.* **63**:35–63, Fig. 3.)

guanosine triphosphate (GTP), which helps to explain the differential effect of colchicine on singlet and doublet microtubules. Only the singlets are disrupted in the presence of colchicine; the doublets are unchanged. The disruption of singlets is probably due to interference by colchicine with normal GTP binding on the singlets.

Tubulin resembles muscle actin in amino acid composition and in other important respects. Tubulin and actin differ from each other in at least three ways:

1. Actin responds to Ca^{2+} ions, whereas tubulin does not.
2. Actin binds ATP, while tubulin binds GTP.
3. The two proteins are serologically unrelated, since each responds only to its own antiserum and does not cross-react with the other's.

When purified doublets were extracted, a second kind of protein was identified as a Mg^{2+}-activated ATPase. This protein has been called **dynein.** It had been known earlier that an ATPase existed in the axial shaft of cilia and flagella, and that ciliary motility depended on an energy source. This energy source was presumed to be ATP, and the transfer of free energy to direct ciliary work was further presumed to be catalyzed by an ATPase. Dynein, isolated from a suspension of purified subfiber doublets, therefore fulfilled the ATPase requirement predicted from the earlier studies. The protein was located within cilium substructure by examining doublets after dynein extraction and again after adding back the extracted dynein to the suspension of fibers. When dynein is removed, the arms of subfiber-*A* disappear. These arms are restored after dynein is added back to the extracted fibers. About 50 percent of the ciliary ATPase activity is believed to exist in the dynein arms of subfiber-*A*. The remainder has not been located with certainty, but there is evidence of ATPase activity at the thickened ends of the radial spokes that connect subfiber doublets to the central sheath components surrounding the two singlet tubules.

The third protein constituent is **nexin,** which is located in the links between subfiber doublets. These links are digested by the proteolytic enzyme trypsin, which allows us to identify these interdoublet links as protein. Since the radial spokes attaching subfiber-*A* to the central sheath are also digested by trypsin, the spokes must also be constructed from protein. It is quite possible that the central sheath components are also protein in nature. Additional chemical analyses will provide specific information about these molecules and aid our understanding of their contributions to ciliar motion.

Ciliary Movement: Sliding Microtubule Mechanism
Cilia have the ability to move even when they are detached from the cell. Amputated cilia will continue to swim through a liquid medium until their ATP

supply is exhausted. These observations allow us to draw at least three conclusions:

1. Cells are moved by their cilia, which are the primary organs of motility.
2. ATP is the energy source and its hydrolysis by ciliary ATPase makes free energy available for movement.
3. Ciliary fibers probably serve as the contractile machinery which leads to locomotion.

In addition, we may infer that the source of ATP is within the cell rather than in the cilium itself.

When isolated cilia are placed in the cold for a prolonged period of time in the presence of glycerol (a method called glycerination), the structural integrity of the cilia is maintained but ATP and many other soluble components are removed. Glycerinated cilia can be reactivated if ATP is added back to the extracted nonmotile material. The frequency with which cilia beat is directly proportional to the amount of ATP provided, showing that energy for movement is obtained by ATP hydrolysis. A similar set of events is characteristic of muscle contraction energized by ATP hydrolysis.

If the concentration of Ca^{2+} ions is reduced by some means, the movement of glycerinated cilia supplied with ATP is not inhibited. This result is quite different from the effect of reducing Ca^{2+} concentration on myofibril contraction in muscle preparations. Myofibril contraction is inhibited by removal of Ca^{2+} ions. These observations indicate that the contractility mechanisms are somewhat different in the two systems of movement.

Recent studies by P. Satir and others have provided some remarkable lines of evidence which show that ciliary movement is caused by a **sliding microtubule mechanism** which has some features that resemble muscle sliding filament models (Fig. 11.21). Exacting measurements of many electron micrographs of cuts through different regions of bent cilia made it clear that ciliary subfiber doublets must slide past one another without contracting. As microtubules slide past one another, their movement causes a localized bend of the cilium. Doublet sliding apparently occurs by cycles of attachment and detachment of the radial spokes to the central sheath. In the same way that actin slides along myosin filaments by cyclic attaching and detaching of myosin crossbridges, microtubule doublets slide along the central sheath as the radial spokes make and break contact with the sheath. Individual tubules do not change in length or width which shows that they, themselves, do not contract. Similar observations showed the muscle myofilaments remained constant in length and width, but sarcomeres contracted as sliding took place. The Mg^{2+}-activated ATPase (dynein) of subfiber-A represents only half the ATPase activity of the cilium. There is some evidence that ATPase activity is also concentrated at the swollen ends of the radial spokes where they make contact with the central sheath. If true, this phenomenon would be another similarity with muscle, since muscle ATPase is concentrated in the cross-bridging myosin heads which make contact with actin filaments. Muscle ATPase, however, is Ca^{2+}-dependent.

The sliding microtubule model satisfies conditions for movement in conventional 9 + 2 cilia and flagella and provides a simple explanation for lack of motility in 9 + 0 systems. Without the central sheath and singlet microtubules, there is no place for radial spokes to undergo attachment-detachment cycles. Other variants in structural organization are known, however, especially among sperm of various animal groups, and it remains to be determined whether their motility is based on a similar sliding microtubule mechanism or some modification of this basic model.

The source of ATP used in ciliary movement varies with cell type. Sperm cells generally contain a modified mitochondrion wrapped around the middle-piece region between the sperm head and tail (Fig. 11.22). The usual energy-generating activities take place in this mitochondrion, and ATP is produced. Sperm tails stop moving after their ATP supply is exhausted or if ATP synthesis is inhibited by glycolysis and respira-

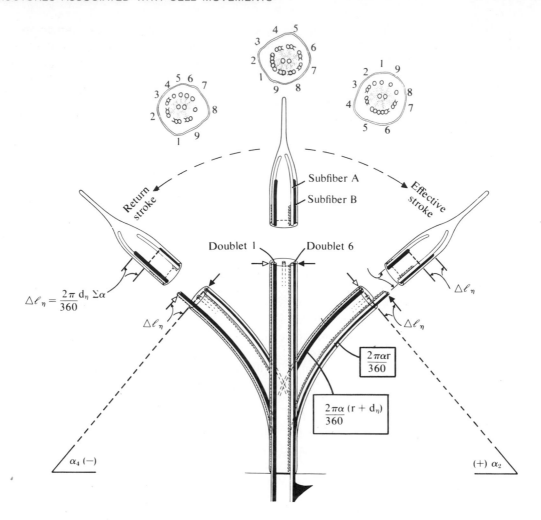

Subfiber A

Subfiber B

Return stroke

Effective stroke

Doublet 1

Doublet 6

$$\Delta \ell_\eta = \frac{2\pi \, d_\eta \, \Sigma \alpha}{360}$$

$\Delta \ell_\eta$

$\Delta \ell_\eta$

$\Delta \ell_\eta$

$\Delta \ell_\eta$

$$\frac{2\pi \alpha r}{360}$$

$$\frac{2\pi \alpha \, (r + d_\eta)}{360}$$

$\alpha_4 \, (-)$

$(+) \, \alpha_2$

Figure 11.21

Diagrammatic illustration of the sliding microtubule hypothesis of ciliary motion. The behavior of subfiber doublets numbers 1 and 6 provides information relating microtubule sliding to cilium bending. Subfiber doublets are shown in cross section near the tip of the cilium. The cilium is in a straight neutral position, bent in the direction of the effective stroke and bent in the return-stroke direction. In the neutral position subfibers-B of all 9 doublets are visible, indicating these terminate at the same level across the tip of the cilium. In the bent ciliary positions, subfiber-B of doublet 1 or of doublet 6 (as well as others) is missing from the outer side of the bend. Since the microtubules are flexible, these cross-sectional views indicate that the microtubule doublet on the inner (concave) side of a bend must slide tipward. When microtubules slide past one another, shear resistance in the cilium changes sliding to bending. Displacement at the tip ($\Delta \ell_\eta$) can be measured using geometrical formulae based on characteristics of the arc produced by bending of the cilium. (Reproduced with permission from Satir, P., 1968, *J. Cell Biol.* **39**:77.)

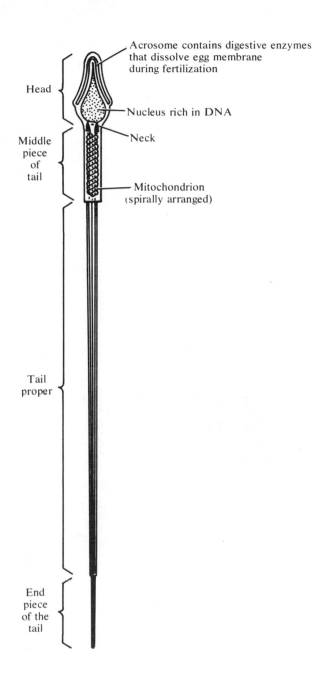

Acrosome contains digestive enzymes that dissolve egg membrane during fertilization

Head

Nucleus rich in DNA

Middle piece of tail

Neck

Mitochondrion (spirally arranged)

Tail proper

End piece of the tail

Figure 11.22
A mature human spermatozoan.

tion poisons. Many cells have mitochondria distributed close to ciliary basal bodies, and ATP could easily be transferred across to the motility appendages.

Bacterial Flagella

The bacterial version of a flagellum is very different from the 9 + 2 eukaryotic organelle. In bacteria each flagellum is made up of 2 to 5 individual filaments which are only 40–50 Å wide but up to several μm long. The width of the flagellum (about 200 Å) depends entirely on the bundle of fibrils. Each filament in a flagellum is made up of a single kind of protein called **flagellin.** Globular monomers of flagellin aggregate end to end to form a filament, like beads on a string. No membrane encloses the flagellum, except in spirochetes, which have an unusual sheathing as well as very different internal components.

Bacterial flagella grow at the tip, whereas eukaryotic flagella grow at their basal end next to the centriole. The bacterial appendage is attached to a granule embedded within the protoplast, but it bears no resemblance to a centriole or other known structure (Fig. 11.23). Flagellar insertion can be readily seen in electron micrographs; flagella remain attached to a protoplast after the bacterial cell wall has been removed. A bacterium is propelled through liquid medium by rotation of its flagellum around its base, unlike the undulating motion of eukaryotic flagella.

Special staining methods are required to see 200 Å-wide flagella in the light microscope where the limit of resolution is 2000 Å. A cementing substance is added first so the flagellum width is increased, and stain is then applied. The organelles require no special treatment to be visible by electron microscopy where 10 Å resolution is easily achieved (Fig. 11.24).

Purified flagellin can be obtained from flagella, which are easily broken off when cells are agitated. As the pH of the medium is varied, globular monomers of flagellin readily aggregate and disaggregate in what is obviously a self-assembly process. Fibrils form spontaneously under proper conditions, but there

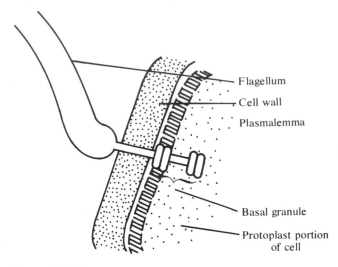

Figure 11.23
Illustration of the probable structure and mode of insertion of the base of a flagellum in a bacterial cell.

must be an organizing site in the structural components in order to produce long polymers. Monomers of flagellin have a molecular weight of 20,000 to 40,000 daltons, which is lower than the weight of eukaryotic tubulin (50,000 to 60,000 daltons).

Flagellin and muscle actin are chemically similar in many ways, especially in having a high proportion of nonpolar amino acids such as aspartic and glutamic acids. No ATPase activity is associated with bacterial flagellar protein, however, while eukaryotic myosin and dynein provide ATPase for motility directly in the structural system which moves.

Modified Cilia and Flagella

It was mentioned earlier that sperm tail ultrastructure varied from the conventional eukaryotic 9 + 2 pattern in some species. Many animal sperm have a 9 + 2 pattern of doublets and singlets but also contain another 9 components that are called **accessory fibers** (Fig. 11.25). These longitudinal components lie outside the ring of doublets and may be dense, massive structures. They are particularly

prominent in mammalian sperm but also occur in some other vertebrates and some invertebrate groups. Little is known of the origin or function of accessory fibers, but they are usually found in species characterized by internal fertilization. Because of this correlation, studies were conducted to see if accessory fibers increased sperm motility in the viscous environment of the female reproductive tract. Evidence that accessory fibers contain an ATPase and a protein similar to one known in striated muscle has provided support for the suggestion that these structures make an essential contribution to sperm motility.

Specialized cells that respond to environmental stimuli such as light, sound, and odor are called **sensory receptors.** A modified cilium, subtended by a

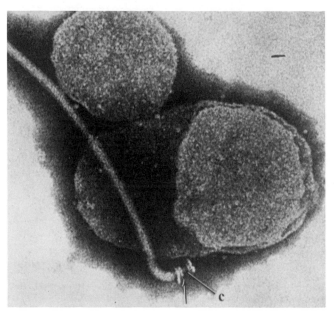

Figure 11.24
Negatively-stained preparation of lysed cells of the bacterium *Rhodospirillum molischianum.* The basal portion of the flagellum consists of two pairs of discs connected by a narrow collar (c) to each other. Globular subunits make up discs, as shown in one of the inner discs (arrow). × 160,800. (Reproduced with permission from Cohen-Bazire, G., and J. London, 1967, *J. Bacteriol.* **94**:458–465, Fig. 3.)

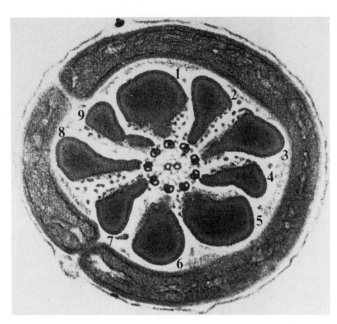

Figure 11.25
Electron micrograph of guinea pig sperm in cross section. The nine dense accessory fibers surround the typical 9 + 2 tubules of the sperm tail. This section was taken from the middle piece of the spermatozoan. × 66,400. (Courtesy of D. W. Fawcett)

centriole, is found in many but not all of these cells types (Fig. 11.26). The central pair of singlet microtubules are usually missing or are extremely short in the ciliary shaft. Visual receptor cells of both rod and cone types have a ciliary structure that connects the outer and inner segments of the cell. A ring of 9 doublets encircles a hollow center where the singlet tubule pair would otherwise be located. In vertebrate visual receptors there may be up to 1000 membrane-limited discs arranged transversely in the outer segment. The discs are apparently formed from extensively folded ciliary membrane, which is derived from the cell plasma membrane just as in any cilium. The folded membrane system in visual receptors provides increased surface area for the conversion of light into electrical impulses. This represents an

evolutionary advance in vertebrate animal eye organization.

Cilia or flagella may serve as both locomotor organelles and sensory receptors in simple eukaryotic species. In the green unicellular flagellate *Euglena,* one of the two flagella is swollen at the base adjacent to the eyespot. The lateral swelling is enclosed by the flagellar membrane but fits neatly into the eyespot concavities. The eyespot itself is a curved

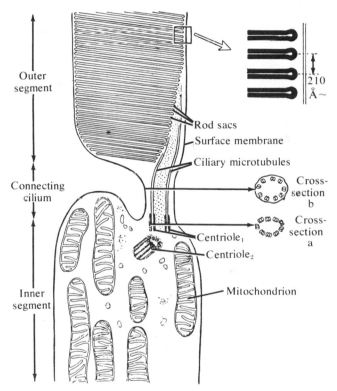

Figure 11.26
Diagram of a mammalian retinal rod cell showing the centriole pair associated with the modified cilium of the outer segment, which contains rod sacs that are transversely arranged membranous discs formed from the folded ciliary membrane. The cross-sectional views through the cilium show (a) the usual subfiber triplets of a centriole and (b) nine subfiber doublets of a ciliary axoneme, but lacking the central pair of microtubules.

plate of lipid material which contains granules of a red pigment. Cells with an intact flagellar swelling are responsive to light, but mutants which lack the swollen region at the base of the flagellum show no movement toward a light source. This observation suggests that the swollen flagellum base has a photoreceptor function, at least in some species. The whole organelle also has an obvious role in cellular locomotion.

THE MITOTIC APPARATUS

Chromosome movement during nuclear divisions will not take place unless there is an organized system of microtubules to form a **spindle.** In addition to the prominent **spindle fibers,** the mitotic apparatus includes other components in animals and many other groups of organisms. An **aster** is found at each pole. The aster surrounds a pair of **centrioles** but is separated from them by a clear zone called the **centrosome.** The spindle and astral fibers are similar in construction, and neither set of microtubules actually touches the centrioles. Species lacking asters have a spindle figure that is different in shape from astral types. Where there are no asters, the spindle fibers are arranged in a cylindrical pattern with a bulging midline, rather than the more spindle- or lens-shaped system which is tapered at the poles (Fig. 11.27). Although it is called a "mitotic" apparatus, similar fibrous configurations are present in meiotic division stages.

Structure

The detailed structural organization and composition of the aster was not described until it was studied by electron microscopy. Many components are below the limit of visibility of the light microscope, but the centrioles can be seen as a barely visible pair of granules separated by a clear zone from radiating astral fibers. In electron micrographs it is possible to see the precise tubular construction of the centrioles, the microtubules that form the astral rays, and various

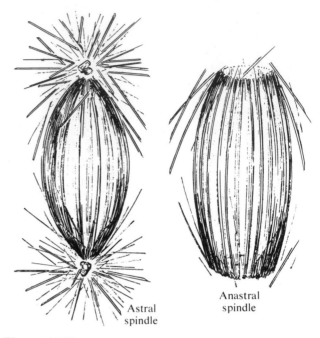

Astral spindle

Anastral spindle

Figure 11.27
Comparative aspects of astral and anastral spindles.

structural components within the "clear" centrosome region.

Ribosomes and small vesicles are dispersed among the astral microtubules and within the centrosome. Numerous amorphous electron-dense bodies called satellites are sometimes found within the centrosome. They are about 1000 Å in diameter, and frequently are connected directly to the aster microtubules. Since satellites are also connected to the centrioles by thin fibrous elements, satellites appear to provide a link between aster microtubules and centrioles. The significance of these connections, however, is not known, and the significance of satellites, themselves, is also unknown since they are only prominent during interphase when aster development is minimal and spindle fibers are totally lacking.

The spindle fibers are conventional microtubules about 180–220 Å wide and many micrometers in length. There may be as many as 3000 microtubules in a spindle figure. The wall of the cylindrical hollow tubules is made from globular **tubulin** monomers arranged into 13 parallel filaments that run the entire length of the microtubule (see Fig. 11.1). According to chemical studies, each tubulin monomer has two binding sites normally occupied by guanosine triphosphate (GTP). In the presence of colchicine, GTP is displaced and its binding site is taken by the drug. These tubules resemble the central pair of singlets in cilia in their construction and their sensitivity to colchicine.

Microtubule monomer units assemble spontaneously *in vitro* to produce recognizable structures essentially identical to spindle fibers that are isolated from various kinds of cells. The entire spindle figure has been assembled *in vitro* and has been shown to be functional since chromosomes moved toward the poles of these fiber assemblies. These observations indicate that formation of functional microtubules in organized patterns takes place by a self-assembly mechanism. Additional evidence also points to a self-assembly process: (1) microtubules will form even when the system is poisoned by cycloheximide, showing that new protein synthesis is not required; and (2) protein monomers must be present in the cytoplasm since materials from unfertilized eggs can be precipitated by antiserum against spindle microtubule protein.

Polymerization of tubulin monomers is inhibited by colchicine because the drug binds to monomer units and blocks the binding sites needed to attach to existing tubulin polymers. The polymers, themselves, are insensitive to colchicine. This particular observation has led to suggestions for a mechanism of chromosome movement during anaphase of nuclear divisions. We will discuss this topic next.

Spindle Function

During metaphase of nuclear division, replicated chromosomes congregate at the midpoint of the spindle. The two parts of each replicated chromosome then separate, and each moves to an opposite pole at anaphase of mitosis. The situation is similar in meiosis, at least for chromosome movement to the poles. The separation of sister chromosomes or half-chromosomes is independent of their subsequent anaphase movement. If the spindle is disrupted, chromosomes will separate but will not move to the poles. The result is formation of a **polyploid** nucleus, having twice the number of chromosomes as the premitotic nucleus.

Before the 1950s there were serious doubts that a spindle figure actually existed. It was considered to be an artifact of cell preparation because well-prepared cells often seemed to have no spindle in dividing tissues. Electron microscopy put these doubts to rest, but evidence in support of the existence of a spindle had also come from studies which used a light microscope equipped with polarizing optics. In this optical system, parallel arrays of fibrous molecules have a brightness due to their property of **birefringence**. Disordered or globular molecules lack this quality. In the early 1950s, S. Inoué showed that birefringence in the spindle region disappeared if colchicine was added to the cell suspension. The birefringence reappeared, however, when colchicine was washed out. Since colchicine had long been known to inhibit anaphase movement of chromosomes and cause the formation of polyploid nuclei, the polarizing microscope studies clearly related chromosome movement to spindle presence and function. The nature of the spindle fibers and their associations with the chromosomes, however, were not clarified until the mid-1960s. At this time important improvements in methods for preparing materials for electron microscopy were developed. As the quality of the images improved, fragile components such as microtubules could be seen consistently to be integral parts of the cell and not artifacts.

Many studies have shown that there are at least two kinds of spindle microtubules (see Fig. 11.18):

1. Those extending from pole to pole.
2. Those attached to a chromosome and terminating near one pole of the spindle.

As early as the 1940s, H. Ris had produced information from careful measurements which showed that spindle fiber lengths changed during anaphase. The chromosome-connected fibers shortened, at the same time that the length of the entire spindle increased. Any model of anaphase movement of chromosomes must account for these simultaneous but opposite changes in the two kinds of spindle fibers. Electron microscopy has verified the occurrence of these changes in lengths of individual spindle fibers.

Energy for chromosome movement is apparently derived by ATPase-catalyzed hydrolysis of ATP. A dyneinlike catalyst has been found in preparations of isolated spindle fibers. Using a specific method which reveals the inorganic phosphate released during ATP conversion to ADP and phosphate, the highest levels of enzyme activity (greatest deposition of phosphate) were found near the spindle poles.

Anaphase Movement of Chromosomes

Early ideas about chromosome movement centered around physical features in the cell, such as electromagnetic field or sol-gel changes. Current notions center on the connections between chromosomes and spindle microtubules and changes these might undergo during anaphase. Three main hypotheses have been proposed to explain the mechanism of anaphase movement: **assembly-disassembly, contraction,** and **sliding microtubules.**

The **contraction mechanism** has the least support. According to this proposal, spindle microtubules would shorten and fold or contract. Instead, the evidence shows that spindle fibers undergo no changes in diameter or in the thickness of the microtubule wall. Another argument against this hypothesis is that contraction fails to explain why pole-to-pole fibers lengthen at the same time that chromosome-to-pole fibers shorten.

The **tubule assembly-disassembly mechanism** was first proposed by Inoué in 1967. It is based on the existence of a pool of tubule monomers and polymerized microtubules of the spindle in an equilibrium mixture. Inoué suggests that chromosomes move toward the poles because spindle fibers disassemble at their poleward end. The disassembly produces shortened fibers and the chromosomes are pulled to the poles. The monomers released from the polar ends of chromosome-connected fibers either return to the cellular pool or add on at the ends of the pole-to-pole fibers, which increases their length. The poles would be pushed farther apart by the lengthening spindle figure made of continuous pole-to-pole microtubules. The "push-pull" system accounts, therefore, for both increasing spindle length and decreasing distance between chromosomes and their respective poles.

There are some difficulties in accepting this hypothesis. A major problem is that the mechanism explains changes in lengths of fibers, but it does not explain how chromosomes move when their fibers are disassembled at the poleward end. It does not necessarily follow that the attached chromosome would be pulled closer to the pole because the fiber shortens at one end. On the other hand, evidence in support of this hypothesis comes from the presence of an active ATPase at the poles, which could liberate the energy to generate the required pulling force. Colchicine effects have also pointed toward an assembly-disassembly mechanism. Weak solutions of colchicine can induce anaphase movement of chromosomes, according to Inoué's experimental results. His interpretation of these results is that colchicine binds to the fibers and leads to partial disassembly which pulls the chromosomes toward the poles. The observed effects of colchicine on chromosome movement may be accurate, but the interpretation of its effect is contrary to known binding properties of the drug. Colchicine binds to monomers and not to polymerized tubule protein. The colchicine effect must be examined, therefore, in relation to its ability to bind to monomers and not to polymers in the system. If we view the system as one in which monomer and polymer protein exist in a dynamic equilibrium, the focus should be on the relative *availability of mono-*

mers and the *accessibility of polymers* for adding and removing monomer units.

When higher concentrations of colchicine are used, intact spindles are disrupted. This event has been shown in spindle studies using electron microscopy or polarizing optics and by light microscopical observations of newly formed polyploid nuclei. Disruption probably occurs because available monomers in the cellular pool bind colchicine and the monomer sites needed for polymerization become blocked. The monomers, therefore, are no longer available to form polymerized tubules. If maintenance of spindle fibers requires a steady input of new monomers to achieve an equilibrium condition directed toward polymer formation, high concentrations of colchicine would cause a shift in this equilibrium. With such a shift, the new equilibrium condition would be in the direction of fiber disassembly on the basis of mass-action physical effects.

When weak concentrations of colchicine are provided, equilibration would be achieved if microtubules underwent partial disassembly. While these explanations could be the basis for normal formation and dispersion of the spindle during nuclear division cycles, they still do not show how fiber changes can lead to chromosome movement.

The **sliding microtubule mechanism** for chromosome movement is derived from analogies with muscle contraction, as is the mechanism for ciliary movement discussed earlier (see Fig. 11.21). According to this hypothesis, chromosomes would move as chromosome-to-pole and pole-to-pole fibers slid past each other. The force required for movement would be generated by alternative breaking and reforming of bridges or chemical bonds during fiber sliding. The spindle fiber system is different in at least one significant way, however, when compared with either the cilium or the muscle systems. No precise spatial relationship between the two kinds of fibers is found in the spindle, while the architecture of the fiber displays in muscle and cilia lend themselves to specific filamentous associations and disassociations.

In some organisms, the chromosome fibers are arranged around the periphery of a "hollow" spindle, while the continuous fibers occur as a separate bundle in the axis of the spindle. In certain protozoa the intact nucleus encircles a bundle of pole-to-pole fibers, while the other spindle fibers extend from each pole to the chromosomes within the membrane-bounded nucleus. Anaphase movement of chromosomes still takes place in these systems even when the two kinds of fibers are in physically separated regions.

While the sliding microtubule mechanism proposed in 1969 by J. McIntosh is an attractive idea, its major advantages are that the mechanism fits the general pattern for ciliary and muscle movements. There is no particular reason to expect all motility systems to have a sliding fiber mechanism. At the present time, therefore, it is difficult to choose between the assembly-disassembly and sliding microtubule hypotheses for anaphase movement of chromosomes.

MICROFILAMENTS AND NONMUSCLE MOVEMENTS

So far we have been discussing systems that are either entirely or predominantly microtubular *or* microfilamentous in underlying structure. Ciliary and chromosomal movements rely primarily on a microtubular apparatus, while muscle contraction seems to work with a system made of microfilaments. Many other forms of cellular motion have been studied, and in most cases these seem to depend on contributions made by both kinds of fibrillar components. In other cases, the movements take place in the total absence of microtubules. This great variety of structural systems has not yet been placed into a single comprehensive framework. We will consider examples of cellular movements that are under study and then look for the common features, if any, that link together seemingly different expressions of motility.

Actin or actinlike filaments have been found in almost all eukaryotic cells. These fibrils are usually 50–70 Å in diameter, they occur in almost all regions that are capable of movement, and they can be recognized by their specific interaction with heavy meromyosin from muscle. Actin filaments can form polarized complexes with heavy meromyosin to produce a "decorated" or "arrowhead" longitudinal display (Fig. 11.28). The more general term micro-

(a)

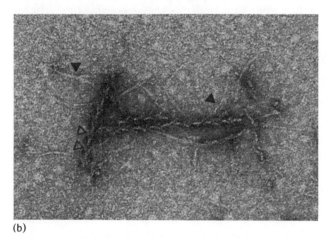

(b)

Figure 11.28
Negatively stained actin and actomyosin filaments from the slime mold *Physarum polycephalum*. (a) Purified actin filaments show an "arrowhead" display after reaction with heavy meromyosin from rabbit muscle. × 218,000. (From Nachmias, V. T. *et al.*, 1970, *J. Mol. Biol.* **50**:83–90, Plate VIc) (b) Purified *Physarum* actomyosin filaments are indistinguishable from those obtained from muscle preparations, once "enriched" for myosin by removal of part of the actin in the preparation. Closed arrows point to long myosin "tails" that project off the "arrowheads"; open arrows point to the visible repeat periods, or cross-overs, on the actin filament. × 92,600. (Photographs courtesy of V. T. Nachmias)

filament has been used because the earlier identifications were based on measured dimensions. Chemical and functional criteria have been used more recently to identify microfilaments of different kinds.

In addition to conventional microtubules and microfilaments, other classes of filaments have been described in certain kinds of cells. Myosin thick filaments are one obvious example of a special kind of filament that differs from the conventional actinlike microfilament in all or most cell types. Many kinds of animal cells contain filaments about 80 to 120 Å in diameter. These filaments can be recognized and distinguished by other criteria in addition to width. They show different responses to physical and chemical agents intended to disrupt microfilaments, and their distribution in the cell is distinctive. In some cases a specific name has been applied to these filament types. For example, filaments found in neurons are called *neurofilaments*. Ameba and slime mold cytoplasm contain *thick filaments* about 150–250 Å wide and about 5000 Å long. These tapered fibers behave, in many ways, like myosin filaments from muscle. Thin actin filaments are also found in the same cells.

All these filamentous components have been related to some movement phenomenon involved in cell locomotion, transport and translocation of organelles and particles within the cell or cell extension, or developmental modifications of cell and organ shape. These relationships have been studied by electron microscopy and other optical methods, effects induced by changes in temperature and pressure, and by responses to colchicine and cytochalasin B. Colchicine is derived from the autumn crocus, a flowering plant, while cytochalasin B is isolated from certain fungi. Both drugs are classified as alkaloids (Fig. 11.29).

Colchicine binds to tubulin monomers and disrupts microtubules under certain conditions. High concentrations of colchicine do not disrupt doublet microtubules in cilia but are effective against the singlet tubules and the spindle fibers in dividing nuclei. The different responses may be related to the mode of

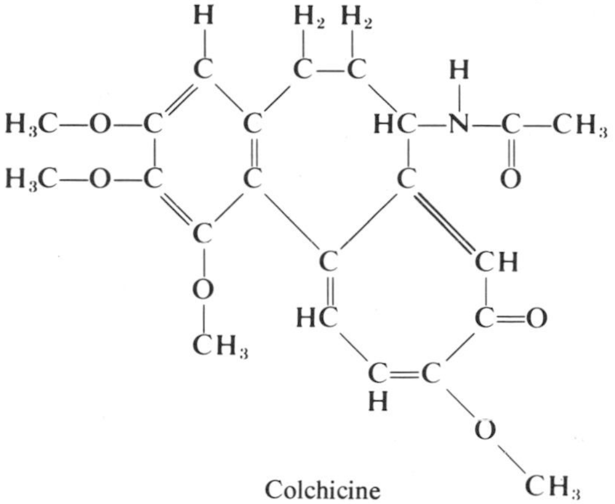

Colchicine
$(C_{22}H_{25}O_6N)$

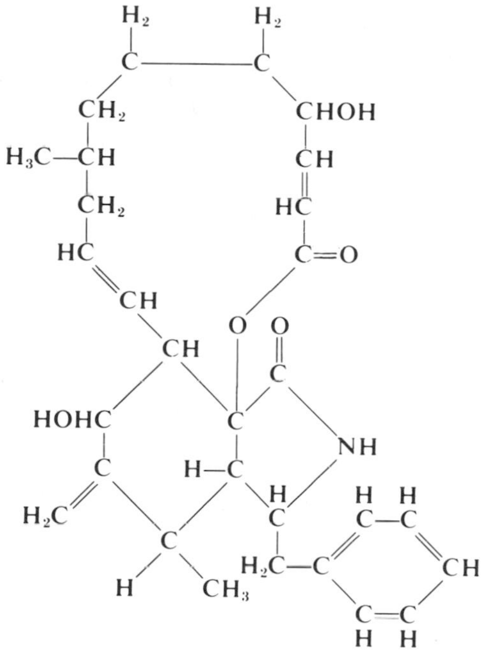

Cytochalasin B
$(C_{29}H_{37}O_5N)$

Figure 11.29
Structural formulae for the alkaloid drugs colchicine and cytochalasin B.

formation of the several kinds of microtubules and the conditions for their maintenance. If continual assembly and disassembly is required to maintain microtubules, colchicine would deplete the pool of available monomers and eventually lead to microtubule disruption. If microtubules reach some stable end point and do not require continued replacement of tubulin monomers, colchicine would have no disruptive effect. At the moment these are only suggestions to explain the differential colchicine effect.

While colchicine has been used for many years to produce polyploid plants with larger flowers or fruits, cytochalasin B only came into widespread use in 1969. T. Schroeder reported at that time that cytochalasin B treatment led to disruption of microfilaments in dividing marine eggs but left the spindle microtubules unaffected. The observations spurred a number of independent studies in various laboratories on the contribution of microfilaments to cell movements, using cytochalasin B as a selective disruptor of these fibrils. Microtubules are insensitive to the drug, but there is some controversy about whether the effects of cytochalasin B are direct or indirect in producing alterations in microfilaments. According to some studies, cytochalasin exerts direct action on actin-like filaments. In other studies the evidence points instead to modifications of membranes, mitochondrial energy yields in respiration, and other indirect routes that eventually may cause microfilament disruption. Since the drug may be used in different concentrations in different experiments, some variation in results can be expected. The current mood is to exercise caution in evaluating results obtained after cytochalasin B treatments.

Movements of Protoplasm

All particles exhibit continuous motion in water because they are randomly bombarded by water molecules. The irregular displacements that result from bombardments depend on the relative amount of kinetic energy of the water molecules which hit particles in suspension. Such random movement is called **Brownian motion,** in honor of Robert Brown who described the phenomenon in 1820. In contrast with undirected Brownian motion, particles, organelles, and the whole cytoplasmic content of cells can move in directed pathways that are far from random. These directed movements are produced by activities of microfilamentous assemblies, or by a combination of microtubules and microfilaments.

SALTATORY MOVEMENTS. Particles suspended in the cytoplasm often undergo directed migrations which result in a displacement of up to 30 μm in a few seconds. The rapid, directed displacement of particles over long distances is called **saltatory movement.** This form of movement has been studied in several species, with different results and explanations. In the giant fresh-water alga *Nitella,* no microtubules are found in the moving regions of the cytoplasm, yet these are the regions in which jumplike saltatory movements have been observed. Since microfilaments are present, these fibrils have been implicated in saltations. Particles adjacent to each other do not necessarily move together. In some episodes some particles may move at a constant rate, while their immediate neighbors are totally unaffected. Various explanations for the way in which microfilaments produce particle saltatory motions have been offered. The major difficulty with all these hypotheses lies in explaining the means by which the motive force is generated. If microfilaments are attached at one end to some cellular structure or region, they may "whip" around and cause particles to be rapidly displaced in some particular direction. An energy supply, which can be transduced to provide the force needed to activate microfilament movement, must be present, however. Little is known about these matters at present.

Various animal cells contain large stores of pigment granules. Under different environmental conditions and stimuli, these granules will disperse or aggregate at extremely rapid rates. They move independently of the protoplasmic flow and can be considered as moving by saltation. In some species of bony fish that were studied few microfilaments were present, but there was an extensive display of microtubules. If the microtubules were disrupted entirely, then pigment granule movement stopped completely. Partial disruption of microtubules led to retarded displacements of these granules. It was suggested that the pigment granules glided over the surfaces of fixed microtubules, in a manner that would be analogous to actin filaments gliding over myosin filaments. The supporting evidence is tentative at present.

Other pigment-containing animal cells have substantial amounts of microfilaments as well as microtubules. In these systems the suggested movement mechanisms involve either one or both of the fibrillar components. The problems in interpreting studies of this kind are numerous. Colchicine and cytochalasin B may have different effects in seemingly similar cell types, and there is relatively little information concerning source and location of ATPase, possible myosinlike filaments, actin-myosin interactions, and other relevant features of these systems. It is quite possible, of course, that different kinds of fibers interact in the different systems that have been studied so far.

PROTOPLASMIC STREAMING. In plants, which have rigid cell walls, the flow of cytoplasm in circular pathways is commonly observed. The large cylindrical cells of the freshwater algae *Nitella* and *Chara* serve as excellent materials for various studies (Fig. 11.30). The outer gel-like layer of **ectoplasm** encloses the fluid sol-like **endoplasm.** In these algae, the chloroplasts are embedded in the relatively stationary ectoplasm, while numerous granules flow with the endoplasmic stream and serve as indicators of **cyclosis.**

Several lines of investigation have shown that the motive force which produces streaming is located at the interface between the stationary ectoplasm and the moving endoplasm. L. Rebhun and others have found twisted bundles of 50–100 microfilaments at this interface parallel to the direction of cyclosis. Micro-

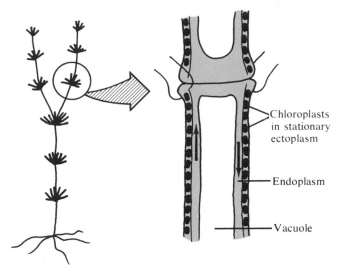

Figure 11.30
Drawing of a *Nitella* plant and a detail of the system showing the distribution of protoplasm and vacuole in the large cylindrical cells of the "stem" region of this alga.

tubules were located in the stationary ectoplasm immediately beneath the plasma membrane. Microtubules seem, therefore, to be unrelated to streaming events.

Thick myosin filaments have not been observed in *Nitella* electron micrographs, but a myosinlike protein must be present since actomyosin complexes can be isolated from these algal cells. Furthermore, using glycerinated materials it was shown that ATPase activity could be demonstrated if Ca^{2+} ions and ATP were provided.

Using cytochalasin B, N. Wessells and colleagues were able to stop cytoplasmic streaming in *Nitella* within 1 hour of treatment. These cells resumed vigorous streaming at original rates once the drug was washed out of the system. Cycloheximide had no effect at any stage in treatment or recovery, indicating that new proteins are not needed for these events. Microtubules also appeared to be unrelated to streaming since colchicine treatment did not stop streaming. Similar results were obtained using oat cells. These

studies further emphasize the participation of microfilaments in plant cell cyclosis.

STUDIES USING THE SLIME MOLD PHYSARUM. Unlike streaming in cells which have rigid walls, protoplasmic flow is rhythmic in *Physarum* and has been called "shuttle streaming." The slime mold is a multinucleate mass of protoplasm lacking any cell partitions. This kind of protoplasmic system is called a **plasmodium.** Endoplasmic streaming is caused by a hydrostatic pressure gradient probably developed by contraction of a fibrillar apparatus in the gel-like ectoplasm behind the flow stream (Fig. 11.31). Prolonged streaming within endoplasmic channels ulti-

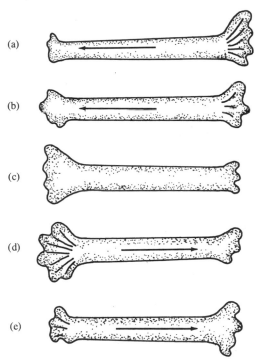

Figure 11.31
"Shuttle streaming" in *Physarum*. Streaming of the endoplasm (a, b) leads to development of pseudopodlike processes (c) that may form at any region ahead of the direction of streaming (d, e).

mately leads to development of a pseudopod-like process which contributes to movement of the plasmodium along a solid surface.

During induced contraction, thick bundles of actin-like filaments develop in the region of ectoplasm which generates the motive force for streaming. *Physarum* yields an actomyosin preparation that behaves very much like actomyosin from mammalian striated muscle. *Physarum* myosin can cross-link with other myosin molecules as well as with actin, further demonstrating its similarities to muscle fibrils. Microtubules apparently are absent in *Physarum*, as they are in muscle.

From these and other studies it would seem that contractility in *Physarum* is based on a fibrillar actomyosin system closely resembling muscle. Contraction is Ca^{2+}-dependent in both systems, and ATPase activity occurs in the presence of adequate Ca^{2+} and ATP. Because the actinlike microfilaments maintain a constant length during episodes of contraction, but myosin filaments are not readily apparent, a modified sliding filament mechanism has been proposed to explain slime mold contraction during streaming. These models are being tested in various laboratories. At present there is little specific experimental information to decide on the precise mechanism leading to contraction in *Physarum* except that it occurs through an actomyosinlike apparatus.

Displacement of Cells

In addition to using ciliary motility, which is based on a microtubular system, single cells can move from one place to another by means of gliding, amoeboid motion, "ruffled membrane" movement, and other ways. Two of these alternatives have been especially well studied. Many amoebae, macrophages, and leucocytes exhibit **amoeboid motion,** while **"ruffled membrane" movement** characterizes locomotion of vertebrate cells in culture. Each of these systems has provided insights into the participation by microfilaments and microtubules in movements.

AMOEBOID MOTION. Amoebae move in a particular direction by forming one or more psuedopods and retracting their posterior region from an attached surface. A continuous forward flow of endoplasm takes place within an enclosure of relatively stationary ectoplasm. Two major hypotheses have been proposed and subjected to more comprehensive tests than have other possible ideas. Each of the two major hypotheses postulates a connection between the forward surge of endoplasm and ultimate cell displacement. The hypotheses differ in the site suggested as providing the motive force for the forward flow of endoplasm.

As endoplasm moves forward, it changes from a gel to a sol state, and when endoplasm is converted back to ectoplasm at the end of the surge, sol-to-gel conversion takes place. Either site could theoretically provide the motive force for endoplasmic flow (Fig. 11.32). According to R. Allen, force is generated at the frontal zone where sol-to-gel change occurs. R. Goldacre, however, has provided evidence showing that the force is generated in the rear of the cell in conjunction with gel-to-sol change in protoplasm consistency. Each model requires contractile fibril assemblies that are coupled to ATP hydrolysis for protoplasmic flow.

Amoebae contain numerous actinlike filaments that complex with heavy meromyosin to produced "arrowhead" displays (see Fig. 11.28). These cells also have occasional thick myosinlike filaments within the cell periphery. Despite the increasing amount of information, the system is extremely difficult to analyze and interpret with confidence. Little evidence which would relate the component steps of the contractile phase with the resulting flow phase of movement has been found. Assembly and disassembly of filaments occurs with astonishing speed and these materials are not easily prepared for the necessary studies using electron microscopy. Cytochalasin does interrupt the forward flow of endoplasm and pseudopod extension, indicating the participation of microfilaments in these aspects of movement.

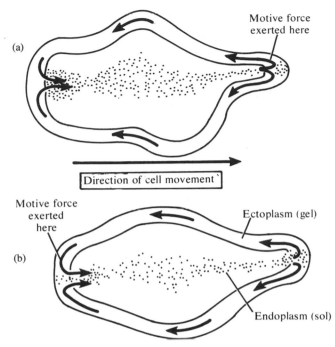

Figure 11.32
Amoeboid movement. The forward surge of endoplasm leads to cell displacement, which may be propelled by a motive force exerted in (a) the frontal zone or (b) the rear of the cell, according to different theories.

RUFFLED MEMBRANE MOVEMENT. The term "ruffled membrane" has been used to describe the appearance of moving mammalian cells when grown in tissue culture in single layers. By time-lapse photography, electron microscopy, and other methods, it has been observed that transient waves of undulating movement by the membrane develop at the leading end of the motile cell. The ruffled membrane is actually a surface protrusion that makes intermittent contact with the glass surface of the tissue culture flask. At these times of contact, the cell moves forward by advancing its unattached membrane regions (Fig. 11.33). A forward flow of cytoplasm is not required for ruffled membrane movement, as it is in amoebae, and protoplasmic streaming seems to occur in the peripheral

regions of these cells rather than in a central channel.

These membranes are also important in intercellular interactions, and in control over cell movements by contact with other cells. The term **contact inhibition** was coined in 1954 to describe the inhibition of ruffled membrane movement in fibroblast cells grown in glass (or plastic) culture vessels. Once the borders of individual cells make physical contact, there is a cessation of movement so that cells do not pile up on one another. The cells are then distributed as a confluent layer, one cell thick. Contact inhibition of locomotion is often confused with inhibition of cell multiplication in many but not all such cell monolayers, but the two inhibition phenomena are separable and may even be based on different mechanisms. Inhibition of cell growth and division in a layer of confluent cells has been called **density-dependent** or **post-confluence inhibition** (see Chapter 13).

It has been posulated that the cell surface membrane is somehow least stable along the leading cell margin, which somehow sponsors ruffling, endocytosis, and other cell surface activities. Perhaps the membrane becomes more stabilized upon contact with

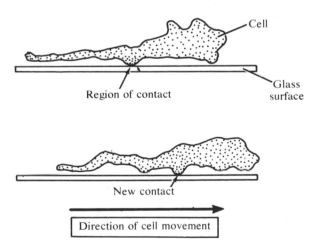

Figure 11.33
Forward movement of cells in culture has been postulated to occur by alternating attachment and detachment of the leading margin ("ruffled membrane") of the cell surface.

another cell surface, causing locomotor inhibition. The experimental evidence provides support for the idea that the locomotor machinery is "turned off" by contact so that traction is no longer exerted at the site of cell contact and the cell stops moving. Since little is known about the machinery of locomotion or the propulsive force for cell movement, the phenomenon of contact inhibition has been difficult to analyze experimentally.

Circulating blood cells do not display contact inhibition *in vivo* or *in vitro*, for reasons that are not known. Nor is it known why some malignant tumor cells lose the capacity for contact inhibition which they possessed before their transformation from normal to neoplastic growth. The basic questions can be answered when more is learned about the properties of the plasma membrane and about the locomotor machinery of the animal cell.

Cultured mammalian cells of all types contain microfilaments and microtubules, and some also have an intermediate class of thicker filaments (80–120 Å diameter). Studies using fibroblast, neuron, dorsal root ganglion, epithelium, cardiac and smooth muscle, and other types of cells have been reported. These studies focused on actinlike properties of microfilaments, isolation of actomyosin complexes, and selective disruption of filaments and tubules by exposure to low temperature, high pressure, colchicine, cytochalasin B, metabolic poisons, and other kinds of agents. All these analyses produced varied results, but the one consistent generalization is that microtubules and microfilaments contribute to movement in different degrees in different cells and situations. Microtubules provide a framework or cytoskeletal component while microfilaments are more directly associated with the contractile machinery itself. Two examples illustrating these points will be discussed briefly.

N. Wessells has reported on a number of experimental studies, including one set of experiments in which embryonic nerve cells were exposed to colchicine and cytochalasin (Fig. 11.34). Cytochalasin stopped cell movement and the leading edge of the

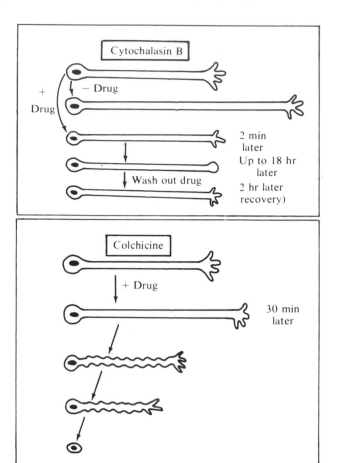

Figure 11.34
Different effects are observed in embryonic nerve cells exposed to cytochalasin B and colchicine. Movement stops in each case but changes in cell morphology vary according to the drug used. See text for details. (Adapted with permission from Wessells, N. K. *et al., Science* **171**:135–143, 15 January 1971, Fig. 5. Copyright 1971 by the American Association for the Advancement of Science.)

undulating membrane was retracted, but the axonal portion of the cell did not change in length. Microfilament assemblies were disrupted, but microtubules and 100 Å filaments were not affected. Cells recovered after the drug was washed out of the medium. In the presence of colchicine, axon growth continued for about 30 minutes. Afterward, the axon gradually shortened until it collapsed into the body of the nerve cell. Neither system was affected by cycloheximide which indicated that protein synthesis was not required in the recovery phase. New fibrils self-assembled from existing monomer pools. These results may be interpreted as showing that locomotion by the ruffled membrane mode of movement at the growing tip was due to microfilament integrity, whereas maintenance of cell shape during growth depended on intact microtubules. The role of the 100 Å neurofilaments was not resolved.

In a similar study, using fibroblasts in cell culture, R. Goldman found that colchicine treatment prevented formation of the long cellular processes of typical fibroblasts (Fig. 11.35) but had no effect on cell attachment or movement. Microfilaments were present in the usual places near the plasma membrane, but microtubules had been disrupted and were not visible. The thicker filaments remained near the nucleus instead of spreading to other parts of the cell. He concluded from these observations that microtubules provided a cytoskeletal framework essential to maintain normal cell shape and to aid in dispersion of the 100 Å filaments within the cell. Since microfilaments were not affected by colchicine and cell movement continued to occur, these fibrils were assumed to be responsible for locomotion. When fibroblasts were treated with low concentrations of cytochalasin B, no effect on microfilaments was observed, but ruffled membrane movement and cell division were inhibited. If higher levels of the drug were provided, microfilaments disappeared. Microtubule and microfilament formation continued to occur during recovery even in the presence of cycloheximide, again indicating that reassembly did not require protein synthesis.

The two sets of experiments produced essentially similar results. But, the differences point out one problem in current studies. There has been little standardization of experimental systems and conditions. One set of results may differ from another because the drug concentration was not the same or the duration of treatment varied. Different types of cells

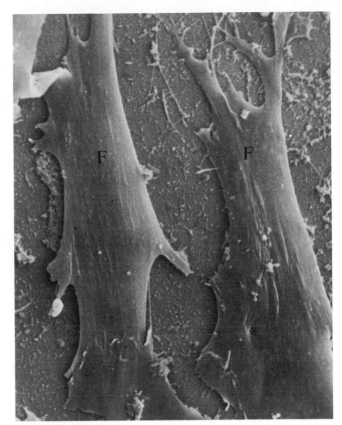

Figure 11.35
Photograph of fibroblasts taken with the scanning electron microscope. Surface features of these non-dividing cells are relatively scarce, but the long processes at the cell periphery can be seen clearly. × 1,740 (Courtesy of K. R. Porter)

may respond in different ways to identical treatments. Until these technical problems are overcome, one set of experiments may not be directly comparable to another set.

MORPHOGENETIC MOVEMENTS. During development the shape of an organ changes, as it proceeds from its embryonic stage to a differentiated form. Changes during morphogenesis usually involve displacement of cells from one relative position in the three-dimensional organ to a different position within that space. Movement of cells which are parts of a tissue or organ, rather than occurring singly, also comes under the heading of cell displacement or motion. The same fibrillar motility components are found at work in these expressions of cell movement, as in the others we have been discussing.

An early clue to the role of microfilaments in changes in organ or cell shape was contained in a study reported by T. Schroeder in 1969. He found that cytochalasin caused cleaving eggs to "round up" as the cleavage furrow disappeared. At the same time, the "contractile ring" of microfilaments in the peripheral cytoplasm just under the furrow also disappeared. When cytochalasin was removed from the medium, the furrow and underlying microfilaments reappeared during the recovery phase. The formation of a furrow has been viewed as an effect induced by contraction of microfilaments, analogous to the drawing of a pursestring (Fig. 11.36). Since microfilaments occupy particular regions in the cell and not others, if one cellular layer is shortened, then the opposite layer, which lacks filaments, will become lengthened.

These observations have formed the basis for the study of cytochalasin effects on embryonic organs grown in culture. Organs which develop lobes have been shown to contain bundles of microfilaments in the areas of depression between lobes, usually lying just under the plasma membrane. The same kind of "pursestring" effect seems to be responsible for cell displacement in the developing organ, as it was in cleaving eggs, but changes in organ shape are a consequence of contractility in this case.

Similar invaginations and evaginations have been reported to occur in several kinds of embryonic organs with microfilaments appropriately positioned just under a kink in an organ. These studies have been interpreted as showing that microfilaments serve as a contractile apparatus in morphogenesis. Further support for contractility has come from studies using calcium ions. When Ca^{2+} ions are injected into young toad embryos using a fine hollow electrode to stimulate the flow of calcium into the system, contraction occurs very rapidly. When these embryos were examined by electron microscopy, it was found that a dense region of microfilaments was present in the place where calcium had entered. If cytochalasin was applied before injection of calcium, contraction did not occur and the dense region of filaments was absent. This response is similar to the response to calcium seen in muscle contraction. On the basis of results in this and other systems, it is thought that the actinlike filaments in the embryo and actin filaments in muscle have similar functions.

SUGGESTED READING

Books, Monographs, and Symposia

Cold Spring Harbor Symposia on Quantitative Biology. 1972. *The Mechanism of Muscle Contraction*, vol. 37. New York: Cold Spring Harbor Laboratory.

Luykx, P. 1970. *Cellular Mechanisms of Chromosome Distribution.* New York: Academic Press.

Sleigh, M. A., ed. 1974. *Cilia and Flagella.* New York: Academic Press.

Articles and Reviews

Abercrombie, M., Heaysman, J. E. M., and Pegrum, S. M. 1970. The locomotion of fibroblasts in culture. II. Ruffling. *Experimental Cell Research* 60:437–444.

Bryan, J. 1974. Microtubules. *BioScience* 24:701–711.

Cande, W. Z., Snyder, J., Smith, D., Summers, K., and McIntosh, J. R. 1974. A functional mitotic spindle prepared from mammalian cells in culture. *Proceedings of the National Academy of Sciences* 71:1559–1563.

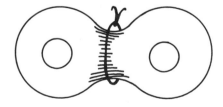

Figure 11.36
Schematic illustration of cell furrow formation as analogous to the tightening of a pursestring around the cell.

Dippell, R. V. 1968. The development of basal bodies in *Paramecium*. Proceedings of the National Academy of Sciences 61:461–468.

Durham, A. C. H. 1974. A unified theory of the control of actin and myosin in nonmuscle movement. *Cell* 2:123–136.

Gibbons, I. R., and Grimstone, A. V. 1960. On flagellar structure in certain flagellates. *Journal of Biophysical and Biochemical Cytology* 7:697–716. (Journal is now entitled *Journal of Cell Biology*.)

Harris, A. 1974. Contact inhibition of cell locomotion. In *Cell Communication,* R. P. Cox, ed., pp. 147–185. New York: John Wiley.

Harris, P. 1962. Some structural and functional aspects of the mitotic apparatus in sea urchin embryos. *Journal of Cell Biology* 14:475–487.

Huxley, H. E. 1969. The mechanism of muscle contraction. *Science* 164:1356–1366.

Inoué, S., and Sato, H. 1967. Cell motility by labile association of molecules. *Journal of General Physiology* 50:259–292.

Ishikawa, H., Bischoff, R., and Holtzer, H. 1969. Formation of arrow complexes with heavy meromyosin in a variety of cell types. *Journal of Cell Biology* 43:312–328.

Johnson, U. G., and Porter, K. R. 1968. Fine structure of cell division in *Chlamydomonas reinhardi*. Basal bodies and microtubules. *Journal of Cell Biology* 38:403–425.

Kubai, D. F., and Ris, H. 1969. Division in the dinoflagellate *Gyrodinium Cohnii* (Schiller). *Journal of Cell Biology* 40:508–528.

Ledbetter, M. C., and Porter, K. R. 1963. A 'microtubule' in plant cell fine structure. *Journal of Cell Biology* 19:239–250.

Murphy, D. B., and Tilney, L. G. 1974. The role of microtubules in the movement of pigment granules in teleost melanophores. *Journal of Cell Biology* 61:757–779.

Murray, J. M., and Weber, A. 1974. The cooperative action of muscle proteins. *Scientific American* 230(2):58–71.

Peachey, L. D. 1965. The sarcoplasmic reticulum and transverse tubules of the frog's sartorius. *Journal of Cell Biology* 25:209–231.

Porter, K. R., and Franzini-Armstrong, C. 1965. The sarcoplasmic reticulum. *Scientific American* 212(3):72–80.

Rebhun, L. I., Rosenbaum, J., Lefevre, P., and Smith, G. 1974. Reversible restoration of the birefringence of cold-treated, isolated mitotic apparatus of surf clam eggs with chick brain tubulin. *Nature* 249:113–115.

Satir, P. 1974. How cilia move. *Scientific American* 231(4):44–52.

Wessells, N. K. 1971. How living cells change shape. *Scientific American* 225(4):76–82.

Part three

The Nuclear Compartment

Chapter 12

Chromosomes and Other Nuclear Components

Eukaryotic cells are distinguished by membrane-bounded compartments. The major membrane boundary within the cell is between the nuclear and cytoplasmic compartments; subcompartments are further delineated within the cytoplasmic volume by a number of membrane systems. No membranes occur within the nucleus, except for transient materials during nuclear divisions. The nucleus is surrounded by two membranes, the **nuclear envelope,** with a space or lumen between the outer and inner membrane systems. The outer membrane is studded with ribosomes and polysomes along its cytoplasm-facing surface (Fig. 12.1). Continuities between this outer nuclear membrane and the endoplasmic reticulum are seen occasionally. The nuclear membranes virtually assure exclusion of cytoplasmic structures from the nuclear compartment, so that mitochondria, lysosomes, microbodies, and other components are not found within the nucleus.

Every cell of a eukaryotic organism possesses one or more nuclei during some stage of its existence, or throughout its lifetime. In certain differentiated cells two or more nuclei are regularly present. Many lower organisms are characteristically binucleate or multi-

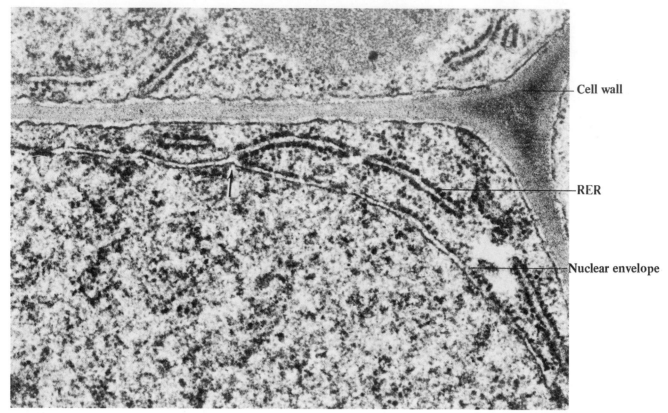

Cell wall

RER

Nuclear envelope

Figure 12.1
Thin-section of root cell of *Lythrum salicaria*. There is a continuity (at arrow) between the ribosome-studded outer membrane of the nuclear envelope and an element of rough endoplasmic reticulum. × 46,000. (Courtesy of M. C. Ledbetter)

nucleate; others are uninucleate throughout their life cycle. In certain mature cell types there may be no nucleus present. The mammalian red blood cell typically loses its nucleus at maturity, and survives for a few months afterward. Some food-conducting phloem cells in flowering plants also lose nuclei at maturity but may continue to function for years afterward. In most species, however, death of the cell causes loss of function shortly after the nucleus disappears.

The nucleus is the site of gene replication and transcription since the **chromosomes,** the major con-centration of cellular DNA, are located within the nucleus (Fig. 12.2). Since the chromosomes do not leave the nucleus, DNA must act as the template for its own replication and for the transcription of RNA within the confines of the nuclear compartment.

Transcript RNAs migrate from their sites of syn-thesis along the chromosomal DNA. Ribosomal RNA participates in ribosome subunit formation within the **nucleolus** of the nuclear compartment, while transfer and messenger RNAs are transported directly to the cytoplasm where they contribute to protein synthesis on ribosomes that also migrate there from

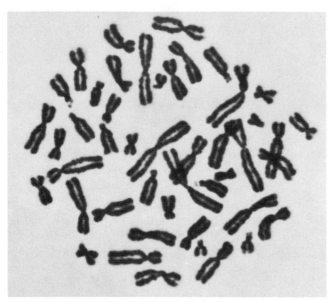

Figure 12.2
Human chromosomes spread out in a metaphase cell that has been treated with colchicine to arrest mitosis at this stage. (Courtesy of L. J. Sciorra)

the nucleus. Eukaryotic cells ordinarily cannot survive for long without at least one nucleolus, since manufacture of new ribosomes is required for continued, long-term protein synthesizing activities. Each nucleolus is formed along the nucleolar-organizing region of a particular **nucleolar-organizing chromosome** (Fig. 12.3). There is at least one such chromosome in a haploid chromosome set, or *genome,* in every eukaryotic species. Many genes other than those for ribosomal RNA are present within the nucleolar-organizing chromosome.

DNA replication and RNA transcription occur at specified times during each **cell cycle,** that is, at a particular interval during the time between one cell generation and the next (Fig. 12.4). During three different intervals in each cell cycle, cellular activities predominate. The fourth time segment is devoted to **mitosis,** at which time daughter chromosomes are segregated into new nuclei.

DNA replication takes place during nonmitotic activity, in the so-called **S (synthesis) period.** By the time mitosis begins, the chromosomes have completed their replication. The usual procession of **prophase, metaphase, anaphase,** and **telophase** stages take place during the very brief mitotic interval (Fig. 12.5). The **interphase** nucleus is therefore the structure that exists during the major, nonmitotic, time period of one cell cycle, the period when most of the significant metabolic activities take place.

Many chromosome studies use metaphase chromosomes because they can be more easily identified, but these are relatively inert metabolically. We will refer to interphase and metaphase nuclei and chromosomes in a large portion of this chapter. Discussion of

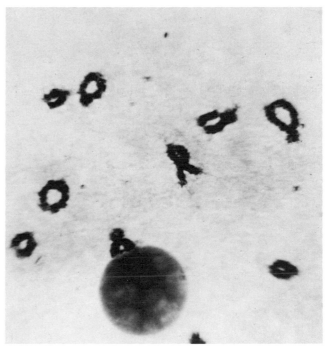

Figure 12.3
Chromosome 6 is the nucleolar-organizing chromosome in maize (*Zea mays*). This is a meiotic cell in the late prophase stage of meiosis, or diakinesis. (Courtesy of M. M. Rhoades)

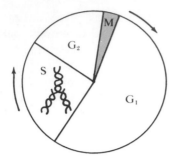

Figure 12.4
The cell cycle. Replication of DNA during the S period and distribution of the chromosomes to daughter nuclei at mitosis (*M*) are separated in time by the G_1 and G_2 phases of interphase nuclear activities. Typical durations of these phases in a mammalian cell are indicated by the proportionate area sizes within the circle.

G

reproduction and its nuclear concomitants will be reserved for Chapter 13.

Chromosomes and nucleoli are bathed in **nucleoplasm,** an amorphous suspension of proteins, particles, fibers, and various organic and inorganic materials. Nucleoplasm is generally distinguished from cytoplasm by its location and structured components, not necessarily by unique biochemical features. "Nucleoplasm" is more a convenient term of reference than a specific functional component.

In contrast with our virtual ignorance about nucleoplasm, a considerable body of information has accumulated over the past 100 years on chromosomes, and more recently on nucleoli. New methods for dissecting these structures by biochemical and microscopical techniques have made possible a new era of **cytology** of the nucleus. The electron microscope has revolutionized cytological study of the cytoplasmic compartments and has aided considerably in exploring nuclear structures. The 1970s will be recorded as the time that molecular approaches to nuclear structure, function, and regulation opened whole new avenues of study. We now stand closer to the threshold of understanding the chromosome than at any time in a century, and this is an exciting time of rapidly developing insights into the fundamental qualities of chromosome organization and activity. Some of these new discoveries will be discussed in this chapter, and others will be explored in Chapters 13 and 14.

THE NUCLEAR COMPARTMENT

The three principal structured components of the nucleus are the **nuclear envelope,** the **chromosomes,** and the **nucleolus.** The major groups of organic compounds are distributed within the nucleus in fairly specific regions. All or most **DNA** is confined to chromosomes, whether these are in an extended or condensed state. **RNA** occurs principally within the nucleolus but also is associated in varying amounts with chromosomes. **Lipids,** found principally as phospholipids, are structural components of the nuclear envelope. **Proteins** of different kinds occur in all parts of the nucleus. Particular attention will be given to some of these types of proteins in subsequent discussions of nuclear structures. Little is known about the carbohydrates of the nucleus.

The Nuclear Envelope

The nuclear materials are separated from the cytoplasm by a double-membrane system generally referred to as the **nuclear envelope.** The outer face of the outer membrane is lined with ribosomes which resemble ribosome groups in the cytoplasm. These ribosomes almost certainly engage in protein synthesis, although we have no information to show that different kinds of proteins are synthesized here than are synthesized by cytoplasmic ribosomes. The inner membrane of the envelope contains no attached ribosomes but is closely associated with condensed chromosomal material (Fig. 12.6).

The intermembrane space is not continuous; it is interrupted by "**pores**" or openings formed by fusion of inner and outer membranes. These openings have a relatively uniform diameter and a highly ordered structure made up of associated nonmembranous

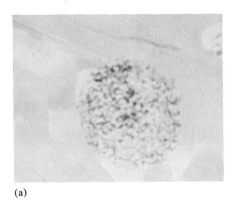

(a)

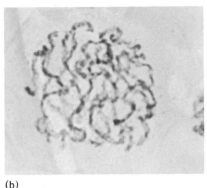

(b)

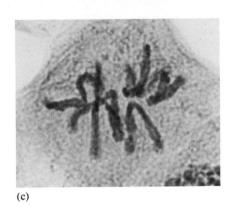

(c)

(d)

(e)

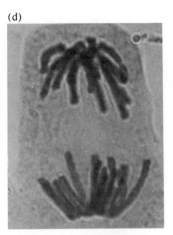

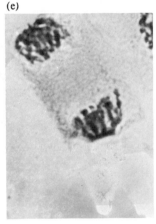

Figure 12.5
Mitotic cells from root epidermis of broad bean (*Vicia faba*):
(a) interphase, representative of G_1, S, and G_2 phases of the
cell cycle; (b) prophase; (c) metaphase; (d) anaphase; (e)
telophase.

components of a distinctive type. A great deal of
attention has been directed toward these pores and
their structure because of their possible involvement
in exchanges of materials between nucleus and cytoplasm. Many studies using isolated nuclear membranes, as well as intact nuclei, have been conducted.

The nuclear envelope is continuous with endoplasmic reticulum (*ER*) at many sites, except in those cells
with minimal reticulum development. Nuclear and reticulum membranes are similar structurally and
chemically. Neither kind of membrane shows the
sharp dark-light-dark staining pattern that typifies the
plasma membrane. The thickness of nuclear and reticulum membranes is essentially the same. The molec-

ular composition of the two membranes is very similar
and freeze-fractured membranes show essentially
similar particles whether the source is nuclear envelope or endoplasmic reticulum. We can view the
nuclear envelope, therefore, as a part of the endomembrane system of the cell, with its closest functional and structural affinities being to the endoplasmic
reticulum. In fact, the nuclear envelope acts like endoplasmic reticulum, since vesicles form from it and
secretory proteins synthesized along its ribosomes
enter the lumen and flow to the Golgi regions for
processing and packaging. In some cells, such as
fungal cells, where there is minimal reticulum, the
nuclear envelope functions somewhat like an *ER*

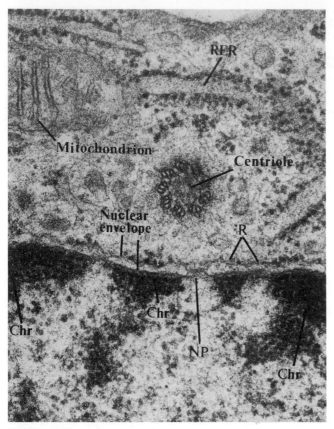

Figure 12.6
Part of a bone cell from the rat. Condensed chromatin (Chr) is closely associated with the inner membrane of the nuclear envelope. Ribosomes (R) are present on the outer surface of the outer membrane of the nuclear envelope, and a nuclear pore (NP) with ill-defined structure is also present. × 51,700. (Courtesy of M. Federman)

system. It would therefore be functionally even more important in minimal-*ER* cell types. In other cell types, however, *ER*-like activities of the nuclear envelope are not considered to provide a major contribution to cellular activities.

THE PORE COMPLEX. Some exceptional cell types have a nuclear envelope which contains no pores, and pore structures similar to those of the nucleus also exist in certain cytoplasmic membranes. By and large, however, pores occur in the nuclear envelope of eukaryotic cells. The opening itself is the site of fusion of two membranes enclosing a cisternal space, so that the lumen on one side of the pore is not continuous with the space between membranes on the opposite side of the pore opening. The pore opening is usually circular in outline and measures about 700–750 Å in diameter for most species (Fig. 12.7).

The perimeter of the pore is accentuated on its upper and lower surfaces by a ringlike system called an **annulus.** These rings are not homogeneous; they have an eightfold radial symmetry caused by the presence of eight symmetrically distributed granular subunits of a nonmembranous nature. The granules of the outer membrane annulus are in register with ribosomes attached to that face of the membrane. The annulus on the inner face of the inner nuclear membrane is similar in structure but has no associations with ribosomes since these are not found along the inner membrane.

The central opening of the pore is often filled with a "plug" of material that sometimes resembles an electron-dense particle. These amorphous materials are distinct from fibrils that can be seen within the pore complexes. Various models of the nuclear pore complex have been proposed. Each model tries to accommodate the different structured components into a coherent framework (Fig. 12.8). Most investigators now agree that both rims of the pore itself are associated with a group of eight granular subunits that are equidistantly spaced within the annulus. Most also accept the other premises of the favored model as including:

1. Eight conical or granular clumps projecting from the pore wall into the interior.
2. Various fibrillar arrangements existing within the lumen of the pore.
3. The fibrils extending into the lumen from the granular subunits of the annulus.

The fibrils were thought to be deoxyribonucleoproteins, but this was disproven by tests which

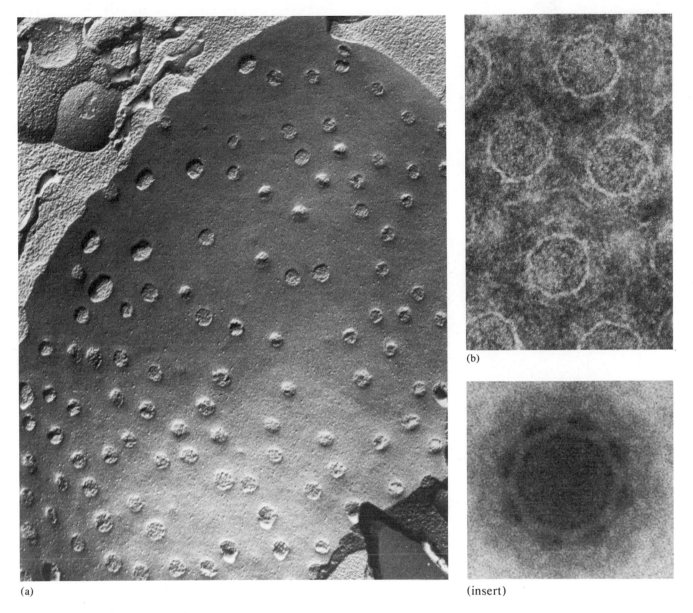

(a)

(b)

(insert)

Figure 12.7

Nuclear pores. (a) Freeze-fracture preparation showing circular pores of the nuclear envelope. × 48,000. (Courtesy of D. Branton) (b) Negatively stained nuclear envelope from oocyte of the newt *Triturus*. × 200,000. *Insert:* 8-fold symmetry of the nuclear pore of the frog *Rana pipiens,* as revealed by a rotation test. × 350,000. (Courtesy of J. G. Gall, from Gall, J. G., 1967, *J. Cell Biol.* **32**:391–400, Figs. 2, 10c.)

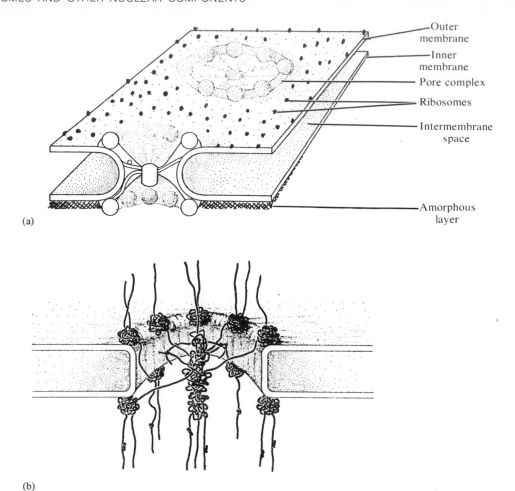

(a)

(b)

Figure 12.8
The nuclear pore complex. Diagrammatic concepts of (a) globular and (b) fibrillar pore complex structures as interpreted from electron micrographs of many cell types. The eight regularly spaced granules of the annulus are found at the pore margin on each surface of the nuclear envelope. Evidence for the occurrence of an amorphous layer of unknown material at the inner surface of the inner membrane has been obtained in experiments reporting the persistence of nuclear shape and nuclear pore complexes after the removal of membrane phospholipids by detergent action. (Reproduced with permission from Franke, W. W., 1970, *Z. Zellforsch.* **105**:405.)

showed the fibrils were not sensitive to DNase digestion. Their ribonucleoprotein nature was shown, however, by their demonstrable sensitivity to RNase and to protease enzyme digestions. Various associations between these ribonucleoprotein fibrils and other structured components within the nucleus have been suggested at one time or another. The precise affiliations remain to be determined. Since nuclear pore complexes are not randomly distributed within the nuclear envelope, there may be particular associations with internal constituents of the nucleus that can be correlated with pore distribution patterns. These relationships are speculative at the present time.

NUCLEAR ENVELOPE CHARACTERISTICS. Nuclear pores have long been thought of as important entryways for nucleocytoplasmic exchanges of molecules and particles in either direction. The pore channel is only about 150 Å wide, however, and cannot easily permit passage of larger materials. The pore channels probably do provide one gateway between nucleus and cytoplasm, but the nuclear membranes themselves almost certainly provide important transport functions in nucleocytoplasmic exchange. A Mg^{2+}-activated ATPase is associated with the membranes as well as with the pore complexes. This enzyme could contribute to active transport of molecules and ions across the membrane (see Chapter 4). There also have been well documented observations of nuclear blebbing. These outpocketings almost certainly are associated with flow of materials between the two major compartments of the cell. In general, the nuclear envelope probably is the more significant component in the control of flow between nucleus and cytoplasm, but pore complexes provide an adjunct system for exchange, especially for ribosomal ribonucleoprotein structures.

Microtubule and microfilament interactions with the nuclear membrane are indicated by experimental evidence. Nuclear migration in some cell types appears to occur in association with such fibrous assemblies, and spindle microtubules are intimately associated with the nuclear envelope in species which maintain an intact membrane system during nuclear division. Some of these features will be discussed in Chapter 13. Little is known, however, about the relationship between nuclear membranes and the events of mitosis and meiosis. Neither the molecular mechanisms nor the biological function of nuclear envelope breakdown during division is known. Breakdown is not required for division since many lower organisms maintain an intact envelope through-

out these reproductive events. Fusions between nuclei are not understood, nor do we know the role played by the nuclear envelope in nuclear fusion. The dynamic activities of the nuclear envelope in mitosis and meiosis will be discussed later, but little more is known than what can be seen in random microscopical observations.

Although nuclear shape has generally been assumed to result from the restrictions imposed by the nuclear envelope and to be modified by microtubule or microfilament associations, this may not be true. R. Aaronson and G. Blobel reported in 1974 that complete removal of the two nuclear membranes led to no change in either shape or ultrastructure of the remaining nucleoplasm and its contents. They verified complete removal of the membranes by showing that virtually no phospholipids remained in their final preparations. These stripped nuclei also retained their arrays of nuclear pore complexes in the same format as they appear in intact nuclei. Even though the membranes were removed by gentle treatment with detergents and other reagents, the pore complexes retained their characteristic locations and characteristic associations with the underlying chromosomal materials. The pore complexes are, therefore, a completely separable component of the nuclear envelope, and they maintain their characteristics by some mechanism that appears to be independent of a nuclear membrane system.

The Nucleolus

The size of the nucleolus usually reflects the physiological state of the cell. The nucleolus is largest in the most active cells and shrinks in inactive cells. There is no membrane around the nucleolus, and two or more may fuse to form a single body in some kinds of cells where nucleolar movement is not impeded. The maximum number of nucleoli seen in a typical cell of a species may indicate the number of **nucleolar-organizing chromosomes.** There is a minimum of one such chromosome per genome, but more than one is usually present in a haploid set of eukaryotic chromosomes. The nucleolus forms at a specific site, called the **nucleolar-organizing region (NOR),** on the particular chromosome(s) in the genome that have this capacity. Even when the nucleolus is not visible, it is possible to recognize the nucleolar-organizing chromosome and region by its different appearance. A constriction in one arm of the chromosome, usually near the free end of the chromosome arm, is the NOR. While some stained preparations appear to show a discontinuity in the NOR, ultrastructural studies have clearly shown that a chromosome fiber runs through this region and is part of the continuous chromosome body.

The nucleolus is a prominent feature of non-dividing nuclei, but it undergoes a cyclic series of changes during mitosis or meiosis in most kinds of cells. As chromosome contraction continues in the early stages of nuclear division, the nucleolus begins to disappear. It is no longer present during the middle stages of nuclear division (metaphase and anaphase), but it reappears at the proper site of the nucleolar-organizing chromosomes during the later stages of nuclear reorganization after chromosome separation in anaphase.

Since the nucleolus is the site of ribosome assembly, it is obviously rich in ribonucleoproteins. Little free RNA is present, and most of the 100 individual kinds of proteins in the nucleoprotein fractions of the nucleolus are undoubtedly ribosomal proteins. Many proteins are nonribosomal but are associated with RNA in precursor ribonucleoprotein particles that are abundant in nucleolus preparations collected after centrifugation of cell-free fractions. RNA polymerase is present, and presumably catalyzes transcription of ribosomal RNA from specific informational DNA in the NOR. We will discuss these genes at greater length shortly.

NUCLEOLAR STRUCTURE AND COMPOSITION. When the nucleolus is viewed with the light microscope, a structureless area and a fibrillar region are seen. The fibrous network has been called **nucleolonema,** or nucleolar threads. The unstructured region has been termed the **pars amorpha.** These terms have little application to nucleolar organization as seen with the electron microscope, since fibrillar material and granular materials are found in both the structureless and fibrillar zones originally distinguished by light microscopy. Although some investigators continue to employ the light-microscope terminology when discussing nucleolar ultrastructure, considerable confusion can be avoided if a different terminology is used.

Four ultrastructural zones can be distinguished in thin-sections through nucleoli (Fig. 12.9). A **granular zone** contains spherical, fuzzy-outlined, electron-dense particles. These particles are about 150–200 Å in diameter and are, therefore, somewhat smaller than mature eukaryotic cytoplasmic ribosomes. The **fibrillar zone** includes somewhat indistinct fibrils, about 50–100 Å in diameter. The granular and fibrillar zones are suspended in a structureless **matrix,** which may be considered a third zone. Faint thread-like connections run through the matrix and appear to join the elements of the granular and fibrillar zones. Finally, 100 Å-wide fibers of low electron density constitute the fourth component, called the **nucleolar chromatin.** These chromatin fibers are a part of the chromosomal complement, but they extend into the nucleolus while the remainder of the chromatin (a term used to refer generally to stainable chromosomal nucleoprotein) is within the nucleoplasm of the nucleus. The relative proportions of these four nucleolar components undergo dynamic changes during the cell lifetime, principally in correspondence with physiological activities. A greater proportion of granular material is often found in nucleoli of more actively metabolizing cells.

The chemical composition of nucleolar components can be analyzed by special cytochemical staining procedures developed for ultrastructure analysis, as well as by conventional methods of analytical biochemistry. In carefully controlled ultrastructural cytochemistry, it is possible to apply digestive enzymes to thin sections and then examine the material to see whether there has been a loss or decrease in electron

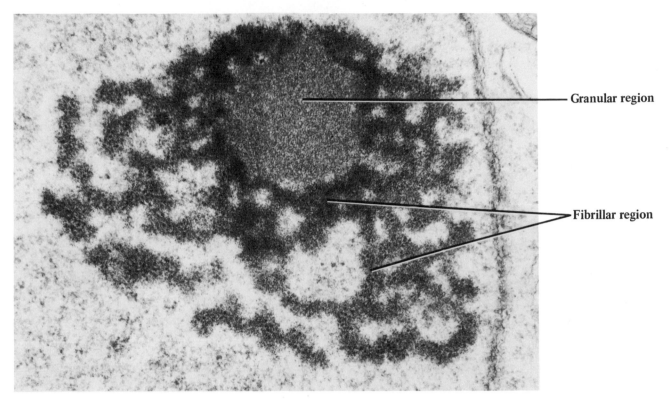

Granular region

Fibrillar region

Figure 12.9
Thin section of nucleolus from spermatogonium of opossum testis. The granular and fibrillar regions were earlier referred to as *pars amorpha* and *nucleolonema,* respectively. × 33,000. (Courtesy of D. W. Fawcett)

density of the nucleolar contents. After digestion with RNase, a striking reduction in granular and fibrillar zone materials is observed. Proteolytic enzymes also reduce the amount of residual material that can be stained. These two tests show that granular and fibrillar zones contain ribonucleoproteins. Nucleolar chromatin, on the other hand, is digested specifically after DNase and protease treatments, as would be expected for chromatin deoxyribonucleoproteins.

Mature ribosomes have not been isolated from nucleoli. A substantial amount of the ribonucleoprotein exists as 45S ribosomal RNA precursor complexed to proteins. Precursor subunits are also found; these contain 28S ribosomal RNA complexed to proteins.

The maturation of precursor particles to functional subunits takes place at some point in time between precursor synthesis in the nucleolus and subunit appearance in the cytoplasm. The final processing may take place at the nuclear periphery, perhaps in association with the nuclear pore complexes of the nuclear envelope, or it may occur on entry into the cytoplasm. It clearly does not take place within the nucleolus itself. Precursor particles of the small ribosomal subunit leave the nucleus so rapidly that little of this material is ever found in isolated fractions of the nucleoplasm and almost none is found in isolated nucleoli (see Chapter 6).

Verification of the chemical composition of nucleolar components can also be obtained from light microscope cytochemistry and from autoradiographs showing the sites of incorporation of radioactive precursors for RNA. Several independent lines of evidence have provided similar kinds of information, and this has led to greater confidence in the accuracy of the molecular localizations (Fig. 12.10). There is less substantial evidence concerning the origin of nucleolar proteins. Many proteins are manufactured in the cytoplasm and are imported by the nucleus and its nucleolus. It is not known whether proteins also are synthesized within the nucleolus. It seems un-

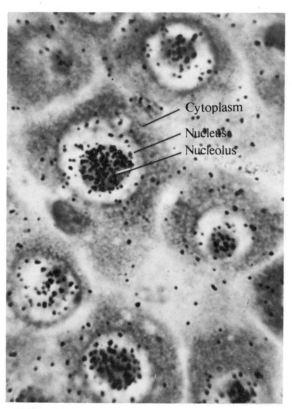

Figure 12.10
Autoradiograph showing localization of labeled RNA precursor in nucleoli of rat liver cells. (Courtesy of S. Koulish, from Koulish, S. and R. G. Kleinfeld, 1964, *J. Cell Biol.* 23:39–51, Fig. 8.)

likely, however, since there are no mature ribosomes on which such proteins might be formed.

GENES AT THE NUCLEOLAR-ORGANIZING REGION. If the nucleolar-organizing region is absent because of mutation or some other traumatic event, nucleoli do not form. This condition is lethal because no new ribosomes can be formed and, therefore, protein synthesis is inadequate to meet the needs of a growing cell or organism. No other part of the chromosome can substitute for the NOR; it is a specific and differentiated region of a particular chromosome. This kind of observation eventually led to a series of studies to determine nucleolar function, using the toad *Xenopus laevis*, other amphibian species, and *Drosophila*, during the 1960s.

F. Ritossa and S. Spiegelman used specially constructed strains of *Drosophila melanogaster* whose chromosomes contained 1, 2, 3, or 4 NORs. Purified DNA from each strain was separated into individual single strands and allowed to hybridize with purified ribosomal RNA to the point of saturation (see Fig. 6.13). If the genes transcribed into ribosomal RNA were localized in the chromosome NOR, then a linear relationship should exist between the amount of RNA hybridized and the number of NORs present in each strain. This relationship is exactly what was found. Twice as much rRNA hybridized with DNA in a strain with two NORs as in a strain with only one NOR, three times as much in a three-NOR strain, and four times as many rRNA–DNA hybrid molecules were found in a four-NOR strain. In other studies it was further shown that increase in NOR numbers did not lead to increasing hybridization when ribosomal 5S RNA or 4S transfer RNA from each strain was used. These two RNA species were thus located elsewhere in the genome and were not present in the nucleolar-organizing regions of the chromosome complement.

GENE AMPLIFICATION IN <u>XENOPUS</u>. Studies using *Xenopus*, which continue in many laboratories today,

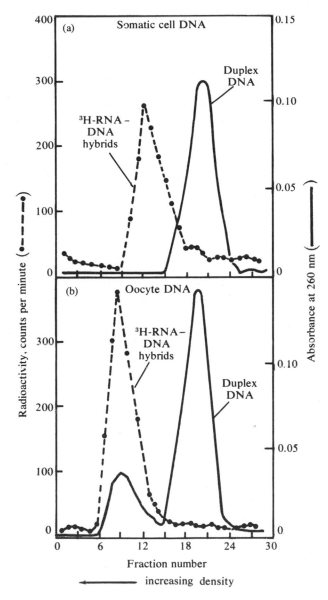

proved even more revealing. To understand the experimental system, we must describe some of the unusual features of nucleolar events in the developing oocyte of amphibians. Amphibian oocytes are very large cells that undergo meiosis to produce the egg cell, which in turn gives rise to the embryo and new individual after fertilization by a sperm. During the early stages of meiosis the oocyte nucleus enlarges considerably, and large numbers of free nucleoli appear. These enlarged nuclei were called "germinal vesicles" in the older literature, a term that is sometimes used today. There may be 600 to 1600 individual nucleoli in an oocyte nucleus, mostly near the periphery. Each nucleolus contains DNA and ribosome precursor RNA.

When DNA of the entire nucleus is purified and analyzed by centrifugation in density gradients of CsCl, a "satellite" DNA fraction is found along with DNA typical of total nuclear or chromosomal DNA (Fig. 12.11). The "satellite" settles in a denser region of the CsCl gradient because of its higher content of guanine and cytosine. When this DNA fraction is hybridized with ribosomal RNA, it is clear that all the ribosomal RNA genes are concentrated in the satellite DNA. For this reason, this satellite DNA has been called the **ribosomal DNA (rDNA) fraction.** If calculations are made from the amounts of rRNA—rDNA hybrid molecules formed in *Xenopus*, it is found that each NOR contains about 450 rDNA genes in a cluster. The rDNA fraction is not present in a *Xenopus* mutant that lacks NOR and nucleoli.

Each of the genes in a cluster is separated by a spacer sequence of DNA. The rDNA genes are not immediately adjacent to each other in the 450-gene

Figure 12.11
DNA-RNA hybridizations of ³H-labeled rRNA and rDNA from *Xenopus* cells: (a) Little or no DNA is detectable in the "satellite" region of somatic cell preparations according to light absorption at 260 nm, but rDNA must be present there since hybrid rRNA-rDNA molecules can be located according to radioactivity measurements; (b) "satellite" rDNA is present in sufficient amount to be observed by light absorption measurements in oocyte preparations. This DNA specifically hybridizes with rRNA according to radioactivity levels. Localizations in oocyte preparations provide support for interpretations of the distribution and meaning of hybrids between rRNA and somatic cell DNA in the "satellite" region of the density gradient.

cluster of repeated rDNA. This determination was verified directly by electron microscopy in studies reported by O. L. Miller and colleagues (Fig. 12.12). The nucleolar nucleoprotein materials were visualized in electron micrographs as feathery, arrow-shaped clusters of fibrils spaced along a thin strand of deoxyribonucleoprotein. Enzyme digestion tests and localization of radioactive precursors within the different regions of the complex showed the molecular nature of the components.

Each rDNA gene transcribes rRNA precursors, which are 40S molecules in *Xenopus*, with many transcripts occurring at each gene sequence. As transcription proceeds from the origin to the terminus of a gene, rRNA transcripts increase in length. The direction of RNA synthesis can thus be seen directly by examining the relative lengths of transcripts along the length of each rDNA gene. The length of the rDNA gene and the length of the longest rRNA transcripts essentially correspond to each other, and to the length expected for a 40S rRNA precursor. The nature and function of the apparently inactive DNA spacers between rDNA genes is under active study at present, but little definitive information is now available except that the spacers contain a very high percentage of guanine and cytosine.

The sequence of events during amphibian oocyte development has been postulated to be as follows:

1. During early prophase of meiosis, rDNA replication takes place while all the rest of the genome remains stable.
2. The extra rDNA clusters, or extra NORs, migrate toward the nuclear envelope where they become enclosed within nucleolar bodies.
3. Transcription of rRNA proceeds within all these nucleoli until the onset of metaphase of the first meiotic division.
4. Further rRNA synthesis stops at this stage and does not begin again until the embryo is about to enter gastrulation approximately two weeks later.

5. Nuclear membrane breakdown takes place by metaphase of the first meiotic division as chromosomes prepare for transport to the poles at anaphase.
6. All nucleoli disappear at metaphase.

This process of **gene amplification** provides a mechanism for synthesis of the enormous numbers of ribosomes that fill the egg cytoplasm and sustain the developing embryo for its first two weeks of existence. New ribosome synthesis begins at gastrulation according to conventional processes, without further amplifications. The extra copies of rDNA are not used after metaphase of the first meiotic division; their activities are restricted to some of the prophase intervals early in meiosis in the oocyte.

Amphibians are unusual in producing extra nucleoli during oocyte development, but other animal types also synthesize rDNA in gene amplification events (Fig. 12.13). In other animals the additional copies of NOR genes are contained within the usual number of nucleoli, but substantial ribosome synthesis can take place as rDNA copies pour forth from the nucleolar-organizing regions. Little is known about the mechanisms and controls that underlie gene amplification in animal oocytes. Amplification of rDNA is a characteristic of animal oocyte differentiation, and it can be viewed as a particular trait that distinguishes one kind of differentiated cell from another. Plant oocyte development apparently lacks rDNA amplification potential, according to recent evidence.

Figure 12.12
Ribosomal DNA transcribing ribosomal RNA molecules, from oocyte nucleoli of the spotted newt *Triturus viridescens*. The tandemly repeated genes are separated by nontranscribing "spacer" segments. Each rRNA gene (= rDNA) is engaged in transcribing many rRNA molecules, which present the appearance of a feathery arrowhead. The newest transcripts are shorter and are closer to the initiation site for each gene. × 27,500. (Courtesy of O. L. Miller, Jr., from Miller, O. L., Jr. and B. Beatty, 1969, *J. Cell Physiol.* **74**, Suppl. 1:225–232, Fig. 3.)

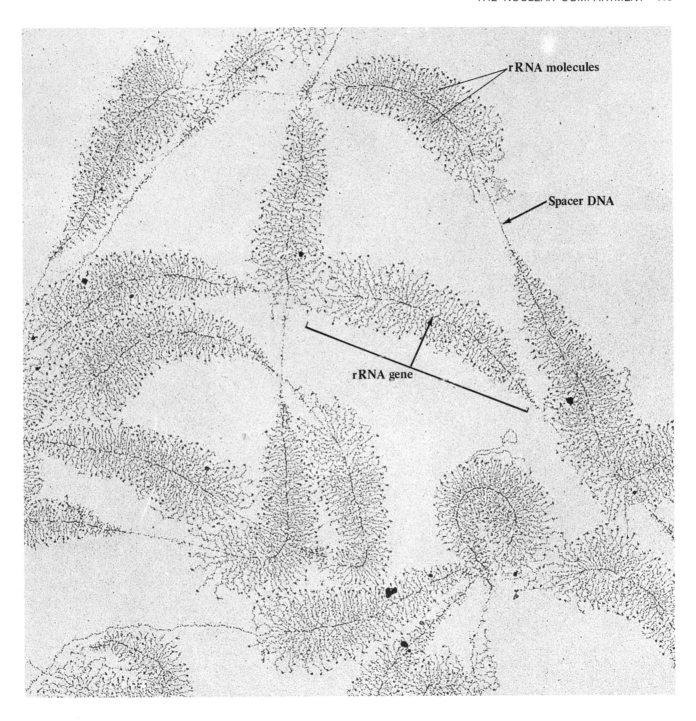

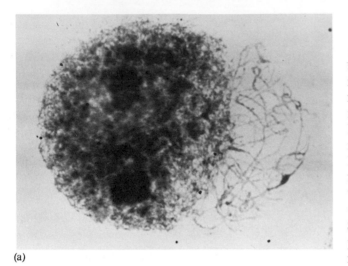

(a)

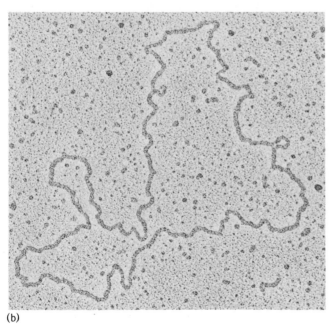

(b)

Figure 12.13

Amplified ribosomal DNA in animals: (a) light micrograph of an oocyte nucleus from the ovary of the Ditiscid beetle, *Dytiscus marginalis*. The large cap of amplified rDNA is to the left and the synapsed meiotic chromosomes are to the right. Approximately 90 percent of the nuclear DNA at this

These studies could be done because of unique and advantageous features of the animal species used. It is now known that rDNA is present in many other species in the NOR of the chromosome; rDNA occurs in clusters of repeated gene sequences specifying ribosomal precursor RNA molecules. One particular method, developed by J. Gall and M. Pardue in 1969, has proven to be especially useful in locating rDNA in chromosomes of a species. In brief, rRNA which has been radioactively labeled is isolated and purified from cells. This labeled rRNA is applied to cells whose condensed chromosomes are well spread out on the slide so that they can be seen most clearly. Some pretreatment is required to cause DNA duplexes to separate into individual strands in certain regions of the chromosome, so that the added RNA can hybridize with single strands of complementary rDNA within the chromosome. The preparation is covered by photographic emulsion and autoradiographs are prepared. Examination of the autoradiographs reveals the location of rDNA by the presence of silver grains produced in the photographic emulsion wherever radioactive rRNA had hybridized.

In this way, rDNA has been shown to occur at the nucleolar-organizing regions of many different species thus far analyzed. Many copies of the rDNA gene must be made at the nucleolar-organizing region to give a sufficient amount of hybridization for autoradiographic detection. One or a few copies of rDNA genes would not be adequate to see the results of hybridization by microscopy. Further discussion

stage of meiosis (pachytene) is rDNA, which is not attached to the chromosomes at this time. × 1,200. (Courtesy of J. G. Gall, from Gall, J. G. and J.-D. Rochaix, 1974, *Proc. Natl. Acad. Sci.* **71**:1819–1823, Fig. 1.) (b) A circular molecule of amplified rDNA from the macronucleus of the ciliated protozoan *Tetrahymena pyriformis*. These rDNA genes, which code for rRNA precursor, exist as free, extrachromosomal molecules with a molecular weight of about 12.6 million daltons. There are about 200 copies of the rDNA gene per genome in the macronucleus of this protozoan. (Courtesy of K. Karrer and J. G. Gall) of rDNA in relation to repetitious DNA and other qualities of the chromosome in eukaryotes will be presented later in this chapter.

THE CHROMOSOMES

Chromosomes were described in the 1880s as vividly-staining, rod-shaped bodies that were always present between the two poles of dividing plant and animal cells. During nondividing intervals, these condensed bodies became greatly extended and appeared as a network of tangled fibers in most kinds of cells. Permanently condensed chromosomes do occur in some simpler life forms, such as the euglenoid and dinoflagellate groups of protists.

T. Boveri established that chromosomes had individuality and persisted from one generation to the next, even though they became dispersed between cycles of nuclear division. Even in the dispersed conformation, specific staining patterns unique to chromosomes continued to be observed in the fibers distributed throughout the nucleus (Fig. 12.14). These stainable materials were called **chromatin,** or chromatin fibers. The term continues to be useful to the present day, although we now know these are strands of DNA complexed with proteins and other organic compounds. The **chromatin fiber** can be considered as the basic structural unit of the chromosome in eukaryotes.

The chromatin fiber consists of a continuous linear DNA duplex strand and associations of basic **histone proteins,** acidic or neutral **nonhistone proteins,** various amounts of **RNA,** and enzymes such as **DNA and RNA polymerases,** among others. While all other components vary in proportionate amounts to DNA, **histones** occur in an equal proportion by weight to DNA in the chromatin fiber. The diameter of the chromatin fiber, as seen by electron microscopy, varies from about 30 Å to 500 Å or more, even within a single strand. Although still somewhat controversial, it is more generally believed now that the typical unit fiber diameter is 100 Å and that greater widths are the result of a fiber folding back upon itself one or more

times. Fibers with diameters of 250 Å or 500 Å can be reduced to fibers with diameters of 100 Å by various treatments that remove Mg^{2+} ions and other ingredients that hold folded regions together (Fig. 12.15). Fibers with diameters less than 100 Å almost certainly represent regions of DNA which contain less protein complexed to the duplex strand. A naked DNA duplex would have a diameter of about 20 Å. The diameter of the duplex increases to 30 Å when a minimum amount of protein is also present on the long molecule in its unfolded state.

The structure and organization of the eukaryotic chromosome and its chromatin fiber currently enjoys a great deal of attention. By concerted efforts in many laboratories, considerable progress has been made in this area in the past few years. We will consider some current points of interest, especially those cases in which methods from microscopy, biochemistry, biophysics, and genetics have converged to clarify chromosome structure and function and the regulation of gene expression.

Chemistry and Structure of the Chromatin Fiber

The absolute amount of DNA varies according to the length of the chromatin fiber. DNA from a large chromosome in *Drosophila* may have a molecular weight of 4×10^{10} daltons, while a typical chromosome from yeast cells has DNA of about 1 to 8×10^8 daltons, which represents a hundredfold difference. Since approximately 2 million daltons of DNA corresponds to a duplex length of 1 μm, the *Drosophila* fiber would measure 20,000 μm (4×10^{10} divided by $2 \times 10^6 = 2 \times 10^4$, or 20,000) from end to end! DNA from a chromatin fiber of yeast would be 400 μm long, if its molecular weight was 8×10^8 daltons. By comparison, the DNA molecule that contains all the genes in *E. coli* is 1300 μm long, or 1.3 mm. Since *Drosophila melanogaster* has 4 chromosomes in its genome and there are 17 or 18 chromosomes in a haploid nucleus of yeast, the total amount of nuclear DNA is vastly greater in eukaryotes than in prokaryotes, in general.

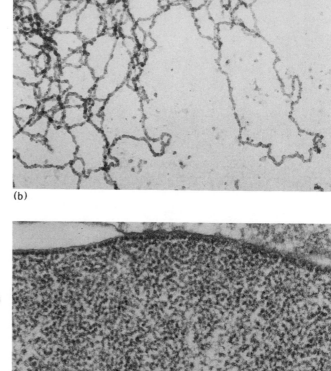

(a)

(b)

(c)

Figure 12.14
The chromatin fiber. (a) Unit fiber, about 100 Å thick, from newt (*Triturus*) erythrocytes, spread on 5 mM sodium citrate (a chelating buffer) and critical point dried. × 160,000. (b) Native fiber, about 200 Å thick, from frog (*Rana pipiens*) erythrocytes, spread on a non-chelating buffer and air-dried. × 36,000. (c) Thin section of a nucleus from frog erythrocyte. The chromatin fibers are about 200 Å thick, which is the same as fibers spread on a non-chelating buffer. × 36,000. Metal ions, such as Mg^{2+}, are present in the 200 Å native fiber but are absent from the 100 Å unit fiber preparations in the chelating buffer. (All photographs courtesy of H. Ris, b and c from Ris, H., 1975, *The Structure and Function of Chromatin*, Elsevier Press, pp. 7–28, Figs. 2 and 3.)

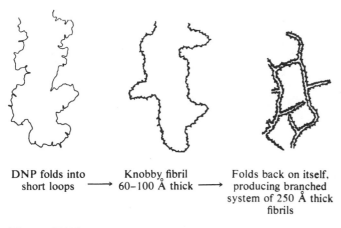

DNP folds into short loops $\longrightarrow$ Knobby fibril 60–100 Å thick $\longrightarrow$ Folds back on itself, producing branched system of 250 Å thick fibrils

Figure 12.15
Interpretative drawings explaining the differences observed in the diameter of chromatin fibers photographed in the electron microscope. The 30 Å-thick deoxyribonucleoprotein thread produces a thicker knobby fibril upon folding into short loops. Additional thickness of the fiber is caused by folding that is maintained in the presence of adequate amounts of Mg^{2+} ions.

It has been estimated that about 13–20 percent of the mammalian metaphase chromosome is DNA and that the remainder consists of RNA and proteins. RNA content appears to vary considerably according to the stage of the cell cycle, with more being present during mitosis than in the interphase between mitotic divisions. Much of the RNA associated with chromosomes isolated at the metaphase stage of mitosis, after breakdown of the nucleolus and nuclear envelope, appears to be ribosomal RNA. This material adheres to chromosomes upon release from confinement in the nucleolus, and when there is no membrane separating chromosomes from the cytoplasm in the middle stages of mitosis. Less ribosomal RNA is associated with chromosomes during interphase.

Some RNA found in the chromatin fiber is used in DNA replication. A short segment of RNA is required as a primer to initiate synthesis of DNA (see Chapter 5). Some RNA may participate in maintaining the chromatin fiber in its folded state. At least it is known that RNA in *E. coli* acts to stabilize the DNA molecule in its compacted form within the bacterial cell. RNA may have a similar function in the eukaryotic folded chromatin fiber. Little else is known specifically about the RNA of the chromatin fiber, except that transcripts for messenger, ribosomal, and transfer RNA must be present in active chromosomes and nuclei isolated during interphase. Since little or no new RNA synthesis occurs during metaphase, isolated chromatin fibers from metaphase chromosomes would be expected to contain the minimum amounts of newly-transcribed RNAs.

A great deal of interest has centered on the chromosomal proteins during the past 15 years. Study concentrated particularly on the histone proteins because they were present in a regular proportion to DNA in the chromatin fiber and because histone synthesis can take place only when DNA is being replicated. Other proteins are manufactured principally when DNA replication is not in progress.

In the early 1960s, James Bonner and colleagues found that chromatin was much more active in transcribing RNA if proteins were removed from the DNA. Native chromatin or reconstituted chromatin whose proteins had been added back to DNA were much less effective in guiding RNA synthesis from DNA templates. Histones in particular had a depressing effect on DNA-dependent RNA synthesis in these systems.

These studies led to the proposal that histone proteins acted as regulatory molecules in gene expression. This notion is not widely accepted today for at least three reasons:

1. Only five major kinds of histone proteins have been found in all eukaryote systems, far too few for specificity in regulating the expression of thousands of different genes.
2. There is little, if any, variation in histone type or content among the many kinds of differentiated tissues of an organism, an unexpected finding for molecules that regulate expression of different genes in different cells.

3. No particular specificity exists for histone binding to different genes, which is unexpected for specificity of regulating gene expression.

Nonhistone proteins, on the other hand, fulfill the requirements not met by histones. They exist in numerous kinds, vary from tissue to tissue in an organism and at different developmental stages, and they are more abundant in active than in inactive chromatin. Gene specificities remain to be demonstrated, but different RNAs are produced by DNA complexed to different sources of nonhistone proteins.

Protein contribution to structure of the chromatin fiber has also been studied vigorously in recent years. In 1953 M. Wilkins presented a model for organization of DNA and protein based on the newly discovered double helix conformation of DNA by Watson and Crick. On the basis of the patterns found in nucleoprotein studied by x-ray diffraction, the protein was postulated to fit into the narrow groove of the DNA double helix.

Similar models, with more specified details, have been presented by Wilkins and others in more recent years. These models are based on complexes which are believed to form between DNA and histone proteins. In 1972, J. Pardon and Wilkins proposed a "supercoil" model for nucleohistone fibers. This model as well as others, relies on a complex between histones and DNA, thought to exist because of the proportionality of the two components, their synchronized synthesis during one interval in interphase nuclei, and the strong affinity between histone and nucleic acid residues. These models can explain the repeat pattern seen by x-ray diffraction, but there have been some differences in interpretation of these patterns. Pardon and Wilkins postulate a double helix of DNA with a "coating" of histones, coiled into a single larger helix. The repeat distance along the axis of this helix is 120 Å, according to x-ray diffraction, and the complex is assumed to measure 100 Å in diameter.

Some objections have been raised by Roger Kornberg, who proposed an alternative model in 1974.

According to Kornberg, an enlarged continuous helix of nucleoprotein would not be sufficiently flexible to allow for the extensive coiling and folding of the chromatin fiber within the condensed chromosome. A fiber that is hundreds or thousands of micrometers in length must be packed into a metaphase chromosome that may be only a few micrometers long. In addition, Kornberg has obtained biochemical evidence that indicates preferential interactions between specific histone types and between these histones and a particular amount of the DNA length.

Kornberg has proposed, in essence, that the repeating unit is made of 200 base-pairs of DNA complexed with two of the major histones (*F2A1* and *F3*) that are themselves complexed into a four-unit package which includes two molecules of each of the two kinds of histones. This repeating unit of 200 base-pairs, plus a histone tetramer, is interspersed along the length of the chromatin fiber with extended segments made up of DNA and the remaining histone types. This model may be compared to a string of beads. The DNA—histone tetramer is the bead, and the space between beads is made up of DNA—histone in a more extended conformation. Electron micrographs show that chromatin fibers are knobby, and Kornberg has suggested that the "knobs" are the repeating units spaced out between the extended regions of the chromatin fiber (Fig. 12.16). A chromatin fiber that consists of many such repeating units would form a flexibly jointed chain, that could easily coil and fold into a very small space.

Figure 12.16
Electron micrograph of chromatin from chicken liver nuclei lysed directly on the metal grid and prepared for microscopy. The chromatin fiber appears to be a flexible chain of spherical particles, about 125 Å in diameter, connected by DNA filaments (at arrows). The spherical particle (nucleosome) contains about 200 base pairs of DNA and an equal weight of four kinds of histone proteins. The same basic structure can be reconstituted *in vitro* from DNA and these four histones (minus the lysine-rich fifth histone, Fl). × 252,000. (Courtesy of P. Oudet and P. Chambon, from Oudet, P. *et al.*, 1975, *Cell* **4**:281–300, Fig. 13.)

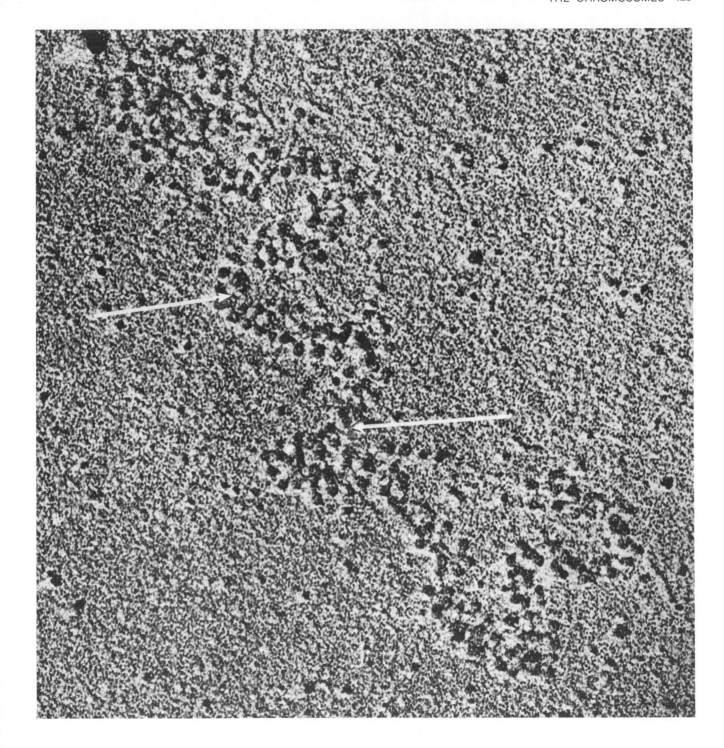

Nonhistone proteins and possibly RNA also may contribute to the structural conformation of the chromatin fiber. None of the nucleohistone models preclude this possibility. There is little information, however, concerning the mechanisms for attachment and detachment of nonhistone proteins and RNA, although this is a constant occurrence during cellular activities.

The Folded Chromosome of *E. coli*

The 1300 μm-long DNA molecule is packed into a nucleoid region only 1 μm in diameter in *E. coli*. How is folding accomplished? The answer to this question has been sought by A. Worcel and others during the 1970s. The interactions that lead to folding must involve only the DNA duplex itself since there is no nuclear membrane to provide additional packaging constraints in prokaryotic cells such as *E. coli*.

When intact nucleoids are isolated free of attached plasma membrane, chemical analysis shows that about 30 percent by weight is RNA and about 1 percent by weight is protein. The major protein is RNA polymerase. Between 60 and 70 percent of the *E. coli* nucleoid is DNA, on a weight basis. Two different observations support the idea that RNA molecules bound to the nucleoids are responsible for stabilizing the condensed state of the DNA:

1. DNA unfolds after brief exposure to RNase, and no other tested enzyme could produce this effect.
2. If cells are grown for a few minutes in the presence of an inhibitor of RNA synthesis before nucleoids are isolated, the DNA unfolds spontaneously.

The second observation further indicates that continual RNA synthesis is required to maintain DNA in its folded conformation in the living cell.

The folded DNA molecule is known to be accessible at all given times to transcription; that is, genes are available for RNA synthesis at any time that cellular conditions and regulation permit. This observation implies that the folding must keep all genes exposed, not masked by tangles of DNA helix. D. Pettijohn and R. Hecht have proposed a model for the folded chromosome that would lead to a structure about 30 μm in diameter after folding and 1–2 μm in diameter if the folded regions are supercoiled (Fig. 12.17). Genes are accessible for transcription in this model, and the calculated size of the molecule corresponds to known nucleoid diameter in the cell. Folding and supercoiling apparently occur by independent means because it is possible to relax supercoils without unfolding and to unfold molecules partially without changing the supercoils.

It is not yet known whether the stabilizing RNA is a unique species or whether it is composed of nascent molecules of an RNA whose mature function is known under other conditions. It is fairly clear that proteins do not serve as stabilizing agents for chromosome condensation in *E. coli* since RNA polymerase is virtually the only kind of protein isolated from nucleoids, and it has no structural function. In this sense, prokaryotic DNA packaging is very different from packaging mechanisms in eukaryotic chromosomes in which histones complex with DNA to stabilize the structure and contribute to its condensation.

Number of Chromatin Fibers per Chromosome

The unreplicated chromosome is difficult to study by microscopy because the interphase structure is extended and tangled. Many observations have been made, therefore, using highly condensed chromosomes isolated from mitotic cells in the metaphase stage. Each metaphase chromosome is obviously a replicated structure whose visible halves are called **chromatids** (Fig. 12.18). When chromatids of a chromosome separate at anaphase, each chromatid is a fully independent chromosome on its own. Knowing these facts about metaphase chromosomes permits

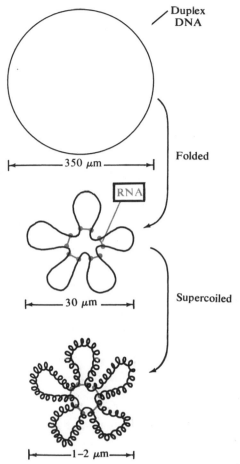

Figure 12.17
Model of the folded chromosome of *E. coli*. The circular DNA is folded into a more compact arrangement that is stabilized by bound RNA molecules. Further compaction is achieved through supercoiling of the major DNA folded regions themselves. (Adapted with permission from Pettijohn, D. E., and R. Hecht, 1973, *Cold Spring Harbor Sympos. Quant. Biol.* **38**:31–51. Copyright 1974.)

study of their chromatin fiber content, and results from these studies can be translated into the situation that exists in the unreplicated chromosome.

Although all will agree that each metaphase chromosome is made up of two distinctive chromatid

halves, there is considerable disagreement over the unity of the chromatid itself as seen in metaphase nuclei. Some observers have claimed to see a "doubleness" to the metaphase chromatid. This observation implies that at least two chromatin fibers are present in the unreplicated anaphase chromosome. By extrapolation, these observers assert that the interphase chromosome is at least two-stranded before DNA synthesis takes place later in interphase. Critics of this point of view suggest that chromatid "doubleness" at metaphase is only an optical illusion brought out by coiling patterns of the single chromatin fiber present in each chromatid.

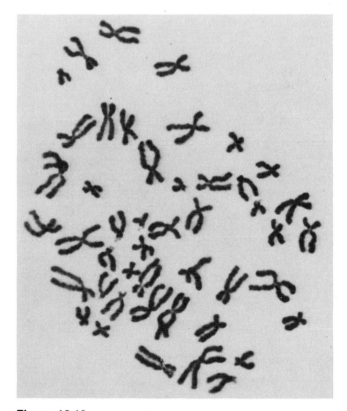

Figure 12.18
The 46 chromosomes of the human diploid nucleus. The two chromatids of each replicated chromosome are joined at the centromere region. (Courtesy of L. J. Sciorra)

One-stranded and multi-stranded models of the chromosome have been in favor alternately. Current information suggests strongly that a one-stranded model is correct. In a one-stranded model there is one chromatin fiber per unreplicated chromosome or one fiber in each chromatid of the metaphase double-structure. The evidence in support of a **unineme** (one thread) model for the chromosome has come from molecular analysis of chromosomal DNA in a variety of eukaryotic species.

If total nuclear DNA is extracted from a known number of cells, it is a matter of simple arithmetic to determine the amount of DNA per haploid nucleus of the species. Additional arithmetic provides the amount of DNA per chromosome in the haploid set. For example, there are approximately 0.84 to 1.2×10^{10} daltons of DNA per haploid nucleus of yeast cells. Since there are 17 or 18 chromosomes in this genome, there would be an average of about 5.4×10^9 daltons of DNA per chromosome. These values can be verified by two independent methods:

1. Carefully isolated yeast nuclear DNA molecules settle in a region of the CsCl density gradient that corresponds to 50S–130S, which in turn correspond to molecular weights of individual DNA molecules between 1×10^8 and 1×10^9 daltons, or an average of about 6×10^8 daltons per molecule. These data support the conclusion that each chromosome contains one continuous DNA molecule (5.4×10^8 daltons of DNA per chromosome and 6×10^8 daltons for a DNA molecule).

2. When molecules of yeast nuclear DNA are carefully prepared for electron microscopy, lengths between 50 and 365 μm can be measured. Since each μm corresponds approximately to 2×10^6 daltons of DNA, the observed molecules would be assigned molecular weights between 1.0 and 7.3×10^8 daltons. This evidence further supports the existence of one DNA duplex per unreplicated chromosome, and the unineme model.

A new method developed by R. Kavenoff and B. Zimm has provided further support for the one-stranded model. The *viscoelastic method* provides information on the rate at which a stretched molecule will recoil during molecule relaxation from its extended state. Only the weight of the largest molecules in the solution can be measured by this method. This feature makes the method even more useful because it can screen out the "noise" of broken molecules also present in the solution of DNA. In 1973 these authors reported the measurement of single molecules of *Drosophila* nuclear DNA which were about 4×10^{10} daltons in molecular weight. Using the equivalence of 2×10^6 daltons to 1 μm of DNA length, these molecules would be 20,000 μm long. These values were verified by two independent methods, including autoradiographs of individual nuclear DNA molecules that were up to 12,000 μm long. This length corresponds to a molecular weight of about 2.4×10^{10} daltons and comes close to the measurements found by the viscoelastic method, which is far more sensitive and reliable than autoradiography of delicate molecules.

Similar information for species other than yeast and *Drosophila* has led investigators to conclude that each chromosome contains a single nucleoprotein fiber which is continuous from one end of the chromosome to the other. The fiber is folded and coiled into a compact, condensed structure that varies in length during the cell cycle but may be only a few μm long during metaphase of mitosis.

The Centromere

Chromosomes are not merely strings of chromatin which condense into visible bodies when seen by microscopy; chromosomes are differentiated along their length into regions that perform unique functions. These functions cannot be taken over by substitute regions located elsewhere along the chromosome, and this is one reason for believing that some parts act differently from others in the genome. This concept can be extended even further since there are whole chromosomes in a genome which are unique and perform functions for which no other chromosome in the complement can substitute.

Two particular differentiated regions of the chromosome are the **centromere (or kinetochore)** and the **secondary constriction,** which is usually associated with the nucleolar-organizing region (NOR). Every chromosome has one centromere that is usually seen as a constricted place. The centromere has also been called the **primary constriction** because of its appearance and its importance in relation to other (secondary) constrictions which have a different function. All other constrictions are secondary, by definition, even if they are not nucleolar-organizing (Fig. 12.19). These constrictions are visible in both light and electron microscopy and are, therefore, not illusions caused by staining.

At one time it was believed that the centromere region lacked chromatin, because it often failed to accumulate detectable amounts of staining agent. Improved preparations revealed visible granules at the centromere, however, and the chromatin fiber runs continuously through this region and the re-

mainder of the chromosome length. Electron micrographs show that the structured components of the centromere region include a part of the continuous chromatin fiber and some special elements that are adjacent to the fiber (Fig. 12.20).

For purposes of greater clarity we will refer to the area of chromosome as the **centromere region** and the visible structures in this region as the **centromere.** These concepts are often ignored and both the region and the structures are often referred to simply as the centromere, a condition which may be nearer the true situation in many cases.

The shape of the centromere is consistent for a species group but may vary among eukaryotes. It is often a disclike structure in animals, lying closer to the outer part of the chromosome thickness. During metaphase, spindle fibers (microtubules) can be seen to be inserted into a three-layered centromere. The outer layer is moderately dense and is called the **centromere plate;** the inner dense layer is formed from compacted chromatin fiber; and the middle layer is the most transparent of the three.

The outer dense material of the centromere is probably the more significant component since it is present even when more organized centromere layered structuring is undefined. For example, in some of the insects of the order Hemiptera (bees, wasps, and others), there is a diffuse centromere region rather than one localized site. Spindle fibers attach at many places along the length of these chromosomes, in areas covered by a dense centromere-platelike substance. Certain groups, such as flowering plants, have a ball-shaped outline in the centromere region which lacks obvious layered structure. In these plants, present evidence indicates that the spindle fibers emanate directly from the tangle of chromatin fiber (Fig. 12.21). Plant chromosomes have a primary constriction that is similar to the centromere region in animal chromosomes, even though the centromere structure varies. Diffuse centromere chromosomes, however, lack a primary constriction region. The structural differences have only been defined by electron microscopy. Modified methods may reveal that

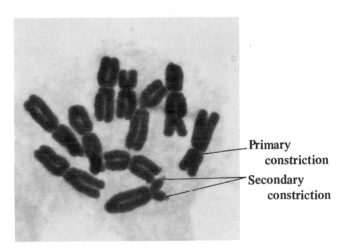

Primary constriction

Secondary constriction

Figure 12.19
Ten of the 16 metaphase chromosomes from onion root tip cell. There is an obvious primary constriction in each replicated chromosome (centromere region), and one chromosome has a secondary constriction at its nucleolar-organizing end. × 4,000.

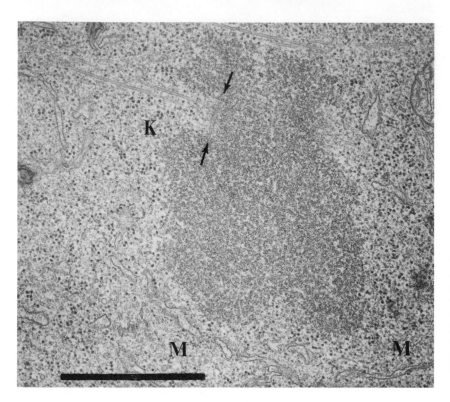

Figure 12.20
Thin section of Chinese hamster chromosome showing the centromere (kinetochore) region (K). A disclike three-layered centromere passes through the primary constriction (arrows) with a layer of fine fibers around it. Membrane profiles (M) are also present. × 39,000. (Courtesy of E. Stubblefield, from Stubblefield, E., 1973, *Internat. Rev. Cytol.* **35**:1–60, Fig. 22.)

plant chromosomes have a more organized centromere, since the structure went unobserved in animal materials until the mid-1960s.

Whether or not a centromere plate or its dense-layer equivalent is present, chromosome movement to the poles at anaphase will usually not take place unless a functional centromere region is present. Except for the unusual diffuse centromere cases, spindle fiber attachment takes place only at the centromere of a chromosome. Up to 150 fibers may be inserted at the ball-shaped centromere in plants, but about 5–20 fibers are usually found attached to the animal centromere. Chromosomes that lack a centromere because of some traumatic event do not move to the poles under their own power. They may be dragged along with surrounding chromosomes, but more often they lag behind and may not be included in the new nucleus

that organizes during telophase. In some cases a chromosome may come to have two centromere regions. If each centromere moves toward an opposite pole at anaphase, a bridge-chromosome is observed (Fig. 12.22). This observation provides additional evidence linking anaphase movement of chromosomes to centromere activity. The relationship between microtubules (spindle fibers) and anaphase movement was discussed earlier in Chapter 11.

The **arms** of a chromosome vary in relative length depending on the location of the primary constriction. When the centromere is median in location, the equal-armed chromosome is termed **metacentric. Submetacentric** chromosomes have arms of somewhat unequal length while **acrocentrics** have very unequal arm lengths (Fig. 12.23). Chromosomes with a terminal centromere have only one arm and are called

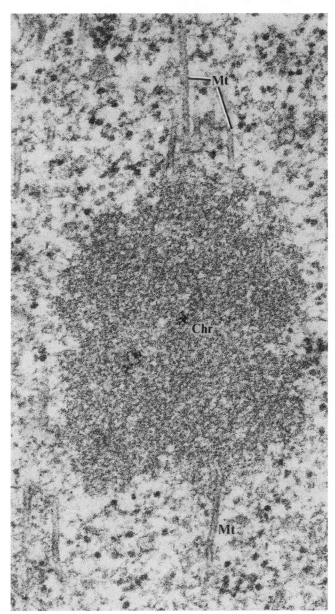

Figure 12.21
Mitotic cell of the African violet (*Saintpaulia ionantha*). Microtubules (Mt) are inserted into the ill-defined centromere region of dividing chromosomes in flowering plants. × 81,000. (Courtesy of M. C. Ledbetter)

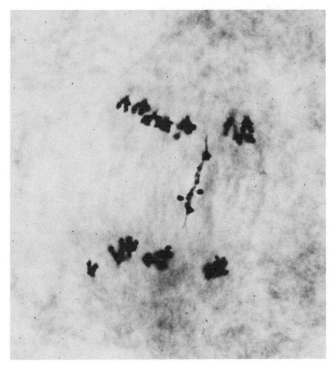

Figure 12.22
Anaphase bridge and two acentric chromosome fragments in the first meiotic division of maize (*Zea mays*). (Courtesy of M. M. Rhoades)

telocentrics. Until they were shown to exist by electron microscopy, it was not believed that telocentric chromosomes existed normally. Changes in chromosome morphology and number have been studied in relation to evolutionary events during species formation. Some of these topics will be discussed in Chapter 14.

The gross morphology of the chromosomes in a set provides the primary means of identification for the individual members of the genome. In most cases, however, chromosomes have very similar form and length so that precise identity is often impossible, and this situation is especially true for species with small and numerous chromosomes. New methods for staining chromosomes, such as chromosome banding, have

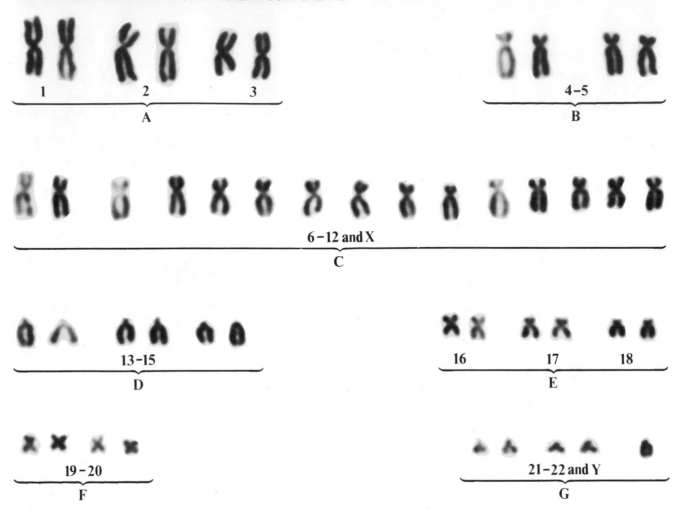

X Chromosomes occupy the second position in the C group

Figure 12.23
The human male diploid chromosome complement arranged according to size and centromere location of each chromosome pair. Chromosomes 1–3 are large *metacentrics*, chromosomes 4 and 5 are large *submetacentrics*, and chromosomes 13–15 are large *acrocentrics*. There are no telocentric chromosomes in the human nucleus. It is clear from this display that chromosomes of the same shape may vary considerably in size (compare metacentric chromosome 1 and chromosome 19).

overcome these problems and permit investigators to identify the different chromosomes of the complement. Chromosome banding will be discussed in a later section of this chapter.

The chemical composition of the centromere materials is unknown, and not much is known about the fundamental characteristics of centromere activity and its function in chromosome movement. The forces that hold the metaphase chromatids together at the centromere region have not been identified (Fig. 12.24) and the relationship between chromosome structure and spindle fiber formation and attachment is not well understood. Centromere materials must be present for spindle fibers to attach to the chromosomes, but this vital region must have more than a mere holdfast function. Far less is known about the non-DNA portions of the chromosome, mostly because specific models to test and sensitive methods for analysis are lacking.

HETEROCHROMATIN AND EUCHROMATIN

In 1928 E. Heitz proposed the term **heterochromatin** to refer to chromatin that remained in a condensed state during interphase. That portion of chromatin which became extended during interphase was called **euchromatin** (Fig. 12.25). Increasingly detailed studies since 1960 have shown several other fundamental differences between these two states of chromatin organization.

Genetic studies showed that heterochromatic regions were relatively inactive parts of chromosomes. Few mutations were found in heterochromatic segments when they were compared with equivalent lengths of euchromatin in the same chromosome or in different chromosomes. Since gene action is expressed

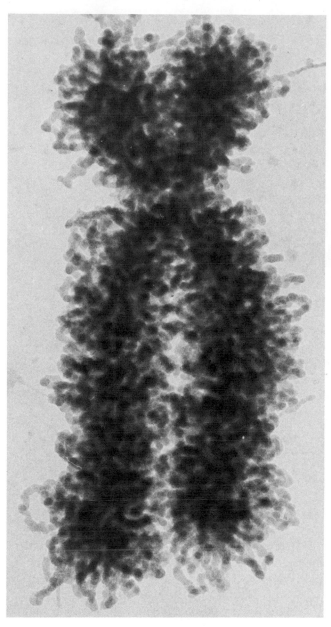

Figure 12.24
Electron micrograph of a whole mount of human chromosome 12. The two chromatids are joined together at the centromere region. × 60,400. (Courtesy of E. J. DuPraw, from DuPraw, E. J., 1970, *DNA and Chromosomes,* Holt, Rinehart & Winston, Fig. 9.10, p. 144.)

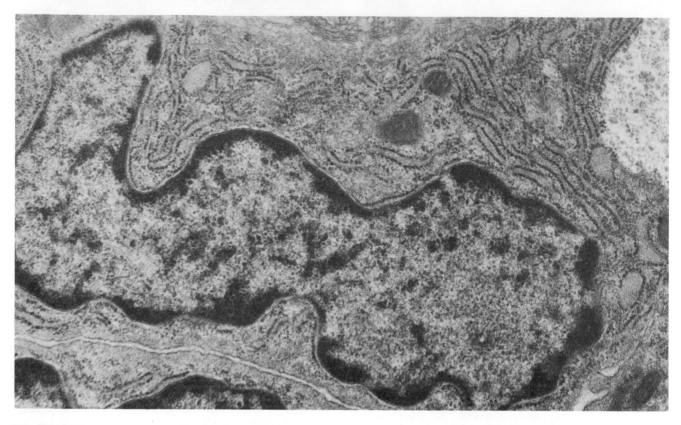

Figure 12.25
Thin section of rat osteoblast. The condensed heterochromatin is principally dispersed at the nuclear envelope, and euchromatin fills the remainder of the interphase nucleus. × 24,000. (Courtesy of M. Federman)

by transcription of messenger RNA during interphase, it seems that little or no messenger RNA is synthesized along these regions of heterochromatin. Even if mutations were present, they could easily go undetected if their altered genetic information was not transcribed and thus remained unexpressed in cell development.

When cells of various eukaryotes are incubated with radioactive precursor during DNA synthesis and examined by autoradiography, different chromosomes and regions of individual chromosomes are found to replicate at different times during the hours required for the whole complement to be duplicated in the S-period of the cell cycle. As first shown by J. H. Taylor in 1960, the latest chromosome regions to begin replication were invariably heterochromatic (Fig. 12.26).

Heterochromatin has thus come to be viewed as chromatin which is condensed during interphase, relatively inactive in gene expression, and late-replicating. During the 1960s, it also became apparent that there were two basic types of heterochromatin. Genetic and cytological studies of mouse chromosomes showed that female cells contained one X chromosome in the euchromatic state and the second X chromosome in the condensed, genetically-inactive

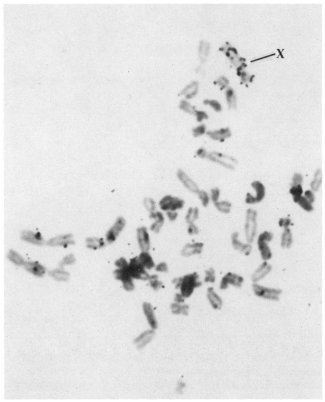

Figure 12.26
Autoradiograph showing silver grains over chromosome regions that replicate late in the S period of the human cell cycle. These regions are largely heterochromatic in nature. (Courtesy of L. J. Sciorra)

heterochromatic state. During the early stages of embryo development from the fertilized egg, both X chromosomes are euchromatic, but one X becomes inactivated soon afterward. Mary Lyon showed that either X chromosome could become condensed and genetically inactive, on a random basis. Similar X-inactivation has since been shown to characterize most other mammalian species, including the human species. Regardless of the number of X chromosomes in the cell, in both normal and aberrant individuals only one X remains euchromatic and all the others

condense. Once heterochromatization has been decided, the inactivated chromosomes remain permanently condensed for the life of the individual. A similar situation has been demonstrated for some insects in which an entire chromosome complement is inactivated in the male diploid cells.

The concept of **facultative** and **constitutive** heterochromatin has been derived from these and other studies. Facultative heterochromatin, such as the heterochromatin in the mammalian X chromosome inactivation situation, contains active genes but may become condensed and genetically inactive in response to physiological and developmental processes. These chromosomes may later be euchromatized, a process which obviously occurs during egg formation when even the inactive X chromosome becomes functional during gametogenesis. Constitutive heterochromatin, on the other hand, is believed to remain condensed during interphase in all cells. As shown by studies since 1970, constitutive heterochromatin seems to be present at the nucleolar-organizing region and around the centromere in many, but not all, species, and it may also occur in a whole chromosome or chromosome arm. These discoveries depended on the development of new and very sensitive methods of detection and identification.

Chromosome Banding

Beginning in 1969, new methods were developed to stain chromosome regions differentially. These methods produced distinctive bands of stained chromatin interspersed with unstained interband areas. The detailed images of relatively small and numerous chromosomes, which were obtained with these stains, have been of immense practical value in human chromosome analysis and of substantial theoretical importance. Chromosome banding studies have opened a whole new avenue to explore the structure and function of chromosomes and their nucleoprotein fiber components.

Two main categories of chromosome banding patterns have been described:

1. Bands of presumptive heterochromatin distributed along the arms of the chromosomes have been revealed by staining with fluorescent dyes such as quinacrine (**Q-bands**) or Giemsa and other nonfluorescent stains (**G-bands**). A reverse band (**R-bands**) pattern occurs when the method is varied slightly, producing R-bands that are reciprocals of Q-bands and G-bands.

2. **C-bands** that distinguish constitutive heterochromatin regions, particularly around the centromere, are visualized by Giemsa staining after special pretreatments that include HCl and NaOH denaturing steps. Giemsa merely contrasts the C-band regions that have responded differently than nonbands to pretreatment; it has no other particular specificity.

Giemsa-stained preparations can be examined by ordinary light microscopy using a white light source, but fluorescent Q-bands can only be seen when ultraviolet light is employed in conjunction with special optics of a fluorescence microscope (a modified form of light microscope).

According to current interpretations of experimental evidence, some generalizations about mechanisms of banding can be made, although these interpretations may change. Q-banding appears to result from interactions between quinacrine and regions of DNA that are rich in adenine and thymine. Guanine and cytosine quench fluorescence, so regions rich in these bases would be present in the unstained interbands (Fig. 12.27). Since G-band and Q-band patterns are very similar, an underlying feature common to these two banding mechanisms must exist. The pattern similarity can hardly be caused only by the DNA since Giemsa is not specific for DNA, much less for adenine—thymine groups in nucleic acids. It is possible that the pretreatments cause Giemsa to interact with nonhistone proteins complexed to DNA, but more evidence is needed. Since treatment with the protein-digesting enzyme trypsin is required for G-banding, some relationship between G-bands and the protein

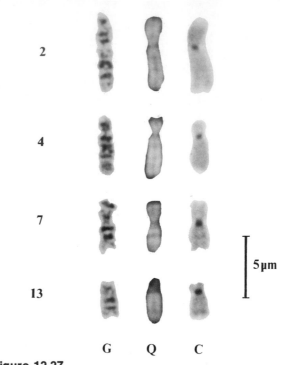

Figure 12.27
Comparison of banding patterns in selected human chromosomes (identified at the left). G- and Q-banded patterns are relatively similar although the poor resolution of fluorescent staining makes this difficult to see clearly. C-bands occur primarily at the centromere region. (Courtesy of C. Hux)

components of the chromosomes must exist. A composite picture of each chromosome requires superposition of Q-, G-, and R-bands since there often are slight variations among these patterns. Q-bands and G-bands are not necessarily identical. C-bands, however, are quite distinctive in location.

C-banding apparently occurs because a greater concentration of DNA remains in C-banded regions than in other regions of the chromosome after specific pretreatments. DNA and protein are extracted preferentially from non-C-band regions, so these regions stain less with Giemsa, and an interband effect is produced. The relatively higher amounts of nucleoprotein in C-band regions than in interband regions causes the differential staining effect with Giemsa. Very little nucleoprotein extraction occurs during pretreatment for Q-, G-, or R-banding, so these staining patterns are unrelated to concentration of remaining DNA. The mechanisms of banding must be better understood, however, before banding can be related to chromosome organization.

At least two kinds of observations support the conclusion that both major kinds of banding pattern (C- *versus* Q-, G-, and R-bands) are somehow related to the presence of heterochromatin in the bands:

1. The condensed state of interphase chromatin corresponds with bands, and not with interbands, on chromosomes whose structure and composition are better known.
2. Regions corresponding to bands are the same ones that have been shown to replicate late in autoradiographic analysis.

Banding patterns provide much greater resolution of heterochromatic regions along each chromosome than either the study of condensed interphase chromatin or autoradiographic study of replication time. If their reaction specificities can be determined, banding methods promise to be very important in revealing the underlying differences in composition of the differentiated parts of the chromatin fiber.

Heterochromatin Ultrastructure

Most information on heterochromatin ultrastructure has come from electron micrographs of thin sections through nuclei. Areas of condensed chromatin are often found near the nuclear envelope, but the heterochromatic and euchromatic chromatin fibers seem to be continuous and therefore parts of the same underlying structure. These observations suggest that the chromatin fiber may occur in either a condensed or an extended state during interphase but not as separate kinds of structures or independent fibers, in most cases. In suitable cells, such as blood lymphocytes, which can be induced to undergo mitosis from their usual interphase state, heterochromatin clearly alters to a euchromatic state before the mitotic divi-

sion actually begins. The conversion of heterochromatin to the euchromatin conformation is further evidence that parts of the chromatin fiber may undergo different degrees of condensation at different times in the cell cycle or lifetime of the cell.

More detailed photographs of interphase nuclei show that condensed chromatin fibers are highly folded and collapsed against other folded regions. Euchromatic parts of the chromatin fiber are more extended and can be followed along greater distances in a continuous strand. Conformationally, therefore, heterochromatin regions of the chromatin fiber are highly folded into compacted areas that appear condensed and deeply-stained in light microscope preparations. Since the euchromatic parts of the fiber are less folded, there is less stainable material per unit area, and euchromatin appears lightly stained by conventional methods of light microscopy. The causes for differential folding remain to be determined, most probably by biochemical methods in conjunction with microscopy.

Repetitious DNA at the Centromere

C-bands occur regularly at the centromere regions of all chromosomes in a complement of many species and in a few other regions that vary from one species to another. These areas of constitutive heterochromatin are permanently condensed at all times in the life of the organism. In 1970 M. Pardue and J. Gall showed that the highly repetitious "satellite" DNAs from mouse nuclei specifically occurred at the centromere regions of all 20 chromosomes of the haploid complement. It was their observation, in fact, that led to the development of the Giemsa banding methods by T. C. Hsu and others in 1971.

When nuclear DNA is centrifuged in cesium chloride density gradients, one or more "satellite" peaks may be present, along with the bulk nuclear DNA fraction (Fig. 12.28). Even greater resolution can be obtained in some cases when DNA is centrifuged in cesium sulfate gradients that contain silver or mercury ions. If a DNA component can bind pref-

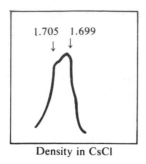

1.705   1.699

Density in CsCl

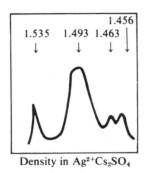

1.535   1.493   1.463   1.456

Density in Ag^{2+}Cs$_2$SO$_4$

Figure 12.28
Tracings of nuclear DNA separated in a density gradient of CsCl compared with greater separation and resolution of the same DNA preparation centrifuged in cesium sulfate gradients containing silver ions. The ρ values are different because they are determined relative to the buoyant density of the gradient material in which the DNA equilibrates.

erentially to Ag$^+$ or Hg$^+$ ions, this affinity creates a denser fraction that will settle in a different region of the gradient than will DNA not bound to these heavy metal ions. By using a combination of methods, as many as five satellite DNA fractions have been recovered for some species.

Satellite DNAs are composed of highly reiterated sequences, as shown by their very rapid rates of renaturation from the single-stranded melted DNA state (see Chapter 5). The more rapid the rate of reannealing of single strands, the more repetitious the DNA, or, the longer the stretch of DNA molecule containing the same base sequence repeated along the length of the molecule. In some cases 10 million

or more repeats of the same sequence may occur; 1 million repeats is about average for these rapidly-renaturing repetitive DNAs.

Using a method they had first developed to locate nucleolar-organizing repetitious DNA in *Xenopus* oocytes and other cells, Pardue and Gall looked for the chromosomal sites containing satellite DNAs in mouse cells. After appropriate pretreatment to melt DNA so that single strands would be available for hybridization directly on the microscope slide preparation (*in situ* hybridization), radioactively labeled DNA or RNA was applied and located later by autoradiography (Fig. 12.29).

In the DNA–DNA hybridizations, radioactive single strands of melted satellite DNA were applied to the chromosomes on the microscope slides. Wherever complementary base-pairing occurred, radioactive strands could be seen by development

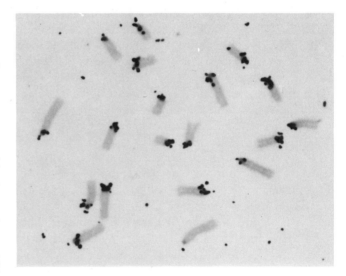

Figure 12.29
Mouse tissue culture chromosomes hybridized with radioactively-labeled RNA copied from mouse satellite DNA. Only centromeric heterochromatin is labeled in these acrocentric chromosomes, as seen by the distribution of dark silver grains against the lightly stained chromosomes. × 1,420. (Courtesy of M. L. Pardue and J. G. Gall)

of silver grains in the photographic emulsion that covered the preparation. Silver grains were found at the centromere regions.

For RNA−DNA *in situ* hybridizations, isolated satellite DNA was placed in an artificial system in which RNA could be transcribed from the mouse DNA. This RNA, with incorporated radioactive label, was then applied to pretreated chromosomes on slides. Wherever complementary satellite DNA was found, radioactive transcript RNA was bound, and it was revealed by silver grains in the emulsion developed by autoradiography. Silver grains were located over the centromere regions in this case too. Repetitious satellite DNAs had clearly been derived specifically from centromere regions of all mouse chromosomes. Pardue and Gall also found that Giemsa staining of pretreated chromosomes produced highly stained centromere regions whether or not hybridization was later carried out. This particular observation led to the new techniques of Giemsa-banding beginning in mid-1971, as mentioned earlier.

In situ hybridizations have shown similar centromere localizations for highly reiterative satellite DNAs in a number of species. A few exceptions have been found, but most species show the expected pattern. These are the same regions that produce C-bands and contain constitutive heterochromatin. In some species whole chromosome arms contain constitutive heterochromatin; for example, whole chromosome arms in the Syrian hamster, but not in the Chinese hamster, contain heterochromatin.

In recent studies of *Drosophila virilis,* Gall was able to determine the base composition and sequence of the repeat unit in each of three major satellite DNAs. This 7-nucleotide repeat may occur up to 10 million times. It is clear that these repetitious simple-sequence DNAs are not genetically informational. Very few codons occur, and these contain mostly adenine and thymine, with very little guanine and cytosine represented. Furthermore, centromere satellite DNA does not seem to be transcribed in the living cell. When whole-cell RNA is isolated and used for *in situ* hybridization, none of the labeled RNA binds to the centromere regions. Since transcription does not take place, satellite DNA must have some function other than carrying structural-gene information. Whether it regulates gene action, contributes to chromosome structural organization, or has some other possible function remains to be determined. In view of the astonishing rate of progress in this research area, answers may well have been obtained by the time this book appears. One aspect of satellite DNA which has already become clear is that these repetitious simple-sequence DNAs are conserved in evolution. There seems to be very little difference in composition and base sequence for centromeric DNAs from many unrelated species.

Other Repetitious DNAs

Simple-sequence centromere DNA may be repeated 1–10 million times. Repetitious DNA of more complex sequence, such as nucleolar-organizing DNA, may occur in hundreds or thousands of sequence repeats. We have already described nucleolar-organizing DNA, which is present in segments of about 450 repeats in *Xenopus laevis.* In the nucleolar-organizing chromosomes of maize (corn, or *Zea mays*) as many as 5000–8000 repeats have been discovered. In each case, the repeated sequences are transcribed into the large ribosomal precursor RNA that subsequently is packaged in ribosomal subunits (18S and 28S in animals, 18S and 25S in plants). Repetitious DNA in the nucleolar-organizing region differs from repetitious DNA at the centromere region in at least three ways:

1. Actual length of the repeat nucleotide sequence.
2. Number of repeats in a localized region of chromosomal DNA.
3. Activity in transcribing RNA.

In addition, nucleolar-organizing DNA has much more variable base composition than centromeric DNA. Both types, however, are heterochromatic.

At least three other kinds of repetitious DNA have

been identified in addition to sequences at the centromere and nucleolar-organizing regions: (1) genes for ribosomal 5S RNA, (2) genes for transfer RNAs, and (3) histone genes. These types of repetitious DNA have been studied, particularly in animal species, but they are probably not fundamentally different in plant chromosomes.

In *Xenopus laevis,* M. Birnstiel and colleagues have shown that genes for 5S RNA of the ribosomes occur in tandem repeated sequences. Using the *in situ* hybridization technique, radioactively labeled 5S RNA was found at the ends of the longer arm in all, or most, of the 18 chromosomes in the complement. In *Drosophila melanogaster,* however, 5S RNA genes are restricted to one site on chromosome-2. The 200 repeated 5S RNA sequences represent a stretch of about 24,000 nucleotides (120 nucleotides per 5S RNA × 200 repeats = 24,000).

In situ hybridizations involving 4S transfer RNA have been more difficult to pinpoint among the chromosomes, but they seem to be scattered in separated repeat groups within the chromosome complement in *Xenopus* and *Drosophila.* An estimated 750 genes are transcribed into transfer RNAs. Whether these genes are present in equal, or unequal, numbers per RNA type is unknown. Newer and more sensitive radioisotope detection methods are now available. These methods use radioactive iodine atoms tagged on to RNA, and better localizations are expected in the near future.

Localization of genes encoded for histone proteins has been carried out using labeled messenger RNA in *in situ* hybridizations. Histone messengers are essentially alike, regardless of the species from which they are obtained, so studies can be conducted using histone messengers from one species that will hybridize to histone DNA from the same or a different species. In *Drosophila melanogaster,* about 200–300 repeats of the histone sequence have been localized to one limited region of chromosome-2. Since five major types of histone proteins exist, it is still uncertain how the five kinds of genes are distributed among the repeated histone-specifying segments.

The five specific kinds of repetitious DNA were thus well characterized and located on chromosomes by combined methods of genetics, biochemistry, and cytology. These repetitious DNAs represent some of the heterochromatin of the chromosome complement, but banding studies show that there must be other heterochromatic DNAs. Except for C-bands of constitutive heterochromatin, all, or most of, the remaining heterochromatin is assumed to be facultative and therefore able to contribute to gene expression under some physiological and developmental conditions.

In a much broader context, a current major problem in cell biology concerns the overall organization of repeated and nonrepeated DNA sequences in the eukaryotic genome. The relationship of DNA sequence organization to gene expression can only be determined by fine dissection of chromosome structure and organization. Repetitive and nonrepetitive DNA arrangements and interactions are believed to hold some of the clues to understanding the eukaryotic genome.

POLYTENE CHROMOSOMES

In most cells, chromosome replication is synchronized with cell division so that the two sets of daughter chromosomes become segregated into separate cells. Other kinds of cells, however, may undergo chromosome replication(s) without accompanying cell divisions. This process is called **endoduplication** when the replicated chromosomes accumulate in the same nucleus, rather than in separate nuclei. If these replicated structures become separated chromosomes, the nucleus is considered to have undergone **endopolyploidy,** that is, to have multiples of the usual number of chromosome sets characteristic of the species.

If the original and replicated strands remain tightly

associated in a single chromosome structure, then **polytenization** is said to have occurred. **Polytene chromosomes** occur regularly in some protozoa, dinoflagellates, and flowering plants, but they are not as distinctive as the giant chromosomes typically found in some tissues and organs of most two-winged insects (members of the order Diptera). The best known polytene chromosomes are those from cells of the salivary glands in dipteran larvae. For this reason polytene chromosomes are often referred to as "salivary gland chromosomes." Similar polytene chromosomes are found in other larval structures, but they are not as easily obtained or handled as salivary gland structures.

Polytene chromosomes from dipteran larvae may have undergone as few as four replications of the diploid chromosome complement or as many as fifteen endoduplications. The amount of DNA, which is measured by chemical or staining methods, is the usual basis for determining the number of replicates that occur in polytene chromosomes. *Drosophila melanogaster* salivary chromosomes contain 1000 times more DNA than do ordinary diploid somatic cells; this increase in DNA content indicates that ten replication cycles have taken place ($2^{10} = 1024$). Some dipteran species have 32,768 times the ordinary diploid DNA content, having undergone fifteen replications without subsequent separation of the individual strands ($2^{15} = 32,768$). These precise amounts, of course, are usually not obtained in the measurements, but the values are so close to the theoretically expected ones that we can safely accept the theoretical amounts and the corresponding number of DNA strands these amounts indicate in the polytenized nuclei.

Polytene chromosomes are much thicker than ordinary chromosomes from somatic cells because of lateral increase in the number of strands. Polytene chromosomes are also very long because they occur in the extended conformation of the interphase nucleus. The long, thick interphase chromosomes are further enlarged because homologous chromosomes are closely paired from end to end along the length of the chromosome. When nuclei from the diploid larval cells are examined, only the haploid chromosome number seems to be present. In special strains it is possible, however, to see that each chromosome actually is a pair of closely aligned individual structures. The identification of paired regions is made relatively simple because of conspicuous *banding* which is a special quality of dipteran polytene chromosomes. Each chromosome has a unique pattern of bands and interband regions, and the chromosomes of different species can be distinguished on this basis as well as the individual chromosomes within the nucleus (Fig. 12.30). Banded polytene chromosomes have been used to locate genes, and chromosome maps are generally produced from combined genetic and cytological studies of these species.

Organization of Polytene Chromosomes

The bands on polytene chromosomes are DNA-rich sections, and the interbands contain far less DNA per unit area or length. The nucleoprotein fibers are continuous throughout the length of a polytene chromosome, so there must be greater packing of chromatin in the bands to account for their higher concentration of DNA. The number of bands per genome varies between 2000 and 6000, according to the insect species. The midge *Chironomus tentans* has about 2000 bands, while *Drosophila melanogaster* has about 5000 bands distributed among the four chromosomes of the haploid complement.

There is no correlation between band number and either chromosome number or total DNA content in a genome. There is, however, a constancy of band number and pattern for every tissue in an individual or species. This correspondence is one line of evidence that is used to support the suggestion that each band is the equivalent of one gene. The average DNA content in one band is about 30,000 base-pairs in *D. melanogaster* and near 100,000 base-pairs in

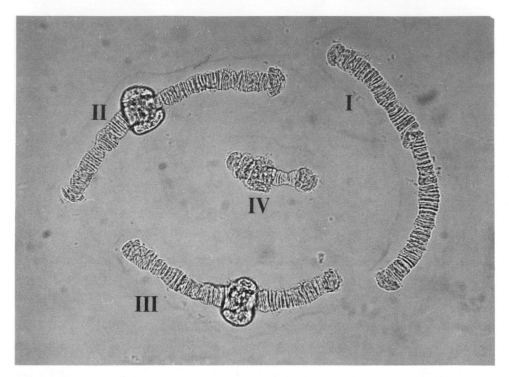

Figure 12.30
Phase contrast light micrograph of the chromosome complement from salivary gland nuclei of the midge *Chironomus tentans.* The chromosomes are identified by the roman numerals. Each chromosome is recognizable by its morphology and banding display. The 8 polytene chromosomes are very closely paired in the diploid cell, which gives the impression of only four chromosomes. × 375. (Courtesy of B. Daneholt)

Ch. tentans. This amount is 50 to 100 times more DNA than is required for an average gene coding for a protein made of 300 amino acids (300 amino acids × 3 base-pairs per codon = 900 base-pairs). About 1000 base-pairs are calculated to be in the stretched DNA segment of an interband, which has an average length of 0.3 μm. This length is a good fit for gene length, but most of the available evidence shows that active genes are within bands and not interband regions. The problem of DNA content and gene number has been the subject of considerable study.

The number of bands and the number of genes shows a 1:1 correspondence, according to genetic analysis. The size, number, and positions of bands are all under genetic control since these features are consistent within a species but differ from one species to another. The more closely related two species are, the more similar their banding number, size, and pattern. This evidence points to a similarity between bands of polytene chromosomes and the knobs, called **chromomeres,** of meiotic chromosomes that are likewise genetically determined and species-specific. The number of chromomeres seen in extended chromosomes during early meiosis is approximately the same as the number of bands counted in interphase polytene nuclei. In species lacking polytene chromosomes, the number of chromomeres still approximates the number of genes believed to occur in a particular genome. Although J. Belling had postulated the equivalence of genes to chromomeres in the late 1920s and early 1930s, support for this idea has only recently been obtained. According to current hypothesis, the bands of polytene chromosomes are equivalents of chromomeres seen in meiotic chromosomes.

Considering the vast amount of DNA per band (or chromomere), considerable amounts of repetitious DNA would be expected to be present in a genome that has only about 5000 genes but has 50–100 times more DNA than is required to code for these genes. This is not the case, however. Renaturation studies have indicated that there is as little as 5–20 percent repetitious DNA in *D. melanogaster* nuclei, and a large part of this repetitious DNA can be accounted for by rRNA, 5S rRNA, tRNA, and histone repeated sequences. It seems that most (80–95 percent) of the *Drosophila* genome, and chromosomes of other dipterans, is made up of unique-copy DNA; that is, each gene is present in one sequence that is not repeated in the chromosome complement. Unique-copy DNA reanneals very slowly during renaturation of melted strands, and these results make the question of band DNA content versus number of genes per band even more puzzling.

Some insights have been provided by recent studies of *Ch. tentans* by J.-E. Edström, a Swedish biologist. One particular band on chromosome-4 of this species is known to code for only a few salivary secretion proteins, yet it has enough DNA to code for 20–30 proteins. The RNA is transcribed from all the DNA in this band, but only a small fraction of the transcribed RNA is actually translated into protein. The rest of the transcript remains in the nucleus and is eventually degraded there. This observation suggests that most of the band DNA is not involved in coding for protein but may be utilized either in regulating or in transporting the messenger portion of the transcript.

A large proportion of transcribed RNA, presumably from many genes, is similarly retained in the nucleus and eventually is degraded there. This **heterogeneous RNA** is complementary to most of the DNA in the genome, as is shown by *in situ* DNA-RNA hybridization tests. It is not ribosomal or transfer RNA since it does not specifically hybridize with known DNA regions encoding these particular types of RNA. It thus seems that a great deal of the unique-copy DNA in the chromosome is transcribed, but a relatively small region of the transcribed RNA is translated into protein. It is very clear from these considerations that the amount of DNA per genome cannot be equated with gene number, even when it is unique and not repetitious. The amount of DNA per band or chromomere may be 20–30 times more than is needed to code for a protein, so that it is very likely that most DNA functions in some other way than simply providing information for an amino acid sequence in a translated protein.

One band may sometimes contain many copies of a particular gene sequence, however. Clustered repeat genes for histone protein have been localized to a single region near the centromere on chromosome-2 in *D. melanogaster,* according to *in situ* hybridization using labeled histone messenger RNA. In an earlier study, 5S ribosomal RNA repeated genes had been localized in another band of this same chromosome. On the basis of these studies and earlier evidence for clustered rDNA at the nucleolar-organizing region in *Drosophila,* it is apparent that generalizations cannot be made about the number of active gene sequences per band in a polytene chromosome or chromomere in any chromosome. Each case must be tested and after enough information has been obtained for various species, it may be possible to make some significant generalizations. We are only at the beginning of these studies now.

Puffing and Gene Expression

The chromosome band may exist in either of two alternative states: (1) compact, as wide as the adjacent interbands, and showing no signs of RNA transcription; or (2) looser-stranded, swollen, and forming a **puff** in which RNA and nonhistone proteins accumulate (Fig. 12.31). A puff is basically a localized decondensation of a chromosome band, but if the puffed region is unusually large and well defined, it is referred to as a **Balbiani ring,** in honor of the Italian biologist who first described these phenomena in 1881 without knowing their nature or significance (Fig. 12.32).

Although these giant dipteran bodies were not recognized to be chromosomes until 1933–1934, a considerable amount of genetic and cytological study soon established their nature and activities. W. Beer-

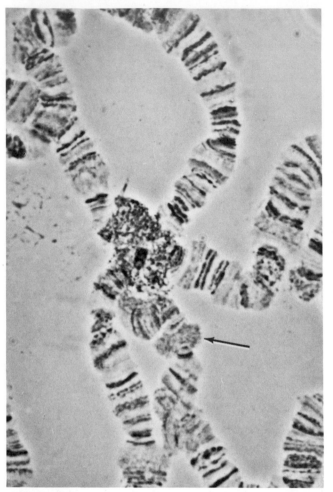

Figure 12.31
Light micrograph of a region of the salivary gland chromosomes of *Drosophila virilis*. A puff is visible (arrow) in the right-hand chromosome below the center. The granular region in which all the chromosomes converge is called the *chromocenter,* and it includes most of the heterochromatin of the complement. × 1,600. (Courtesy of J. G. Gall, from Gall, J. G. *et al.,* 1971, *Chromosoma* **33**:319–344, Fig. 5.)

mann in Germany has been prominent in polytene chromosome analysis and has contributed a number of important and substantial lines of information. In the 1940s and early 1950s, Beermann and others had

established that puffs were specific for particular bands of each chromosome and that they appeared and disappeared in conjunction with cellular differentiation events. Using specific stains for DNA, RNA, and proteins, as well as data from incorporation of radioactive precursors of these macromolecules, Beermann and others showed that DNA replication usually did not occur during puffing, whereas RNA and proteins were synthesized and accumulated during puff development. When inhibitors were present, puffing was prevented, or regressed if in progress, and no new RNA or protein appeared at the puff sites.

This evidence clearly indicated that puffing was an expression of gene action, with RNA transcripts formed from the extended DNA in this region. Proteins that are synthesized in the cytoplasm are transported to the puffed sites in the nucleus, although some proteins may also be synthesized in the nucleus. Beermann's model, proposed in 1952, is still widely accepted. He suggested that a puff arose by unfolding of DNA at the chromomere (band), and that this folding made these DNA looped regions accessible to transcription (Fig. 12.33).

Only a few dipteran species have been studied

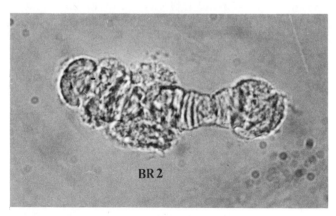

Figure 12.32
Chromosome IV of *Chironomus tentans* salivary gland nuclei, showing the well developed Balbiani ring known as BR 2. × 900. (Courtesy of B. Daneholt)

extensively, and *Chironomus tentans* has provided exceptionally fine material for various investigations. The Balbiani rings are particularly well developed in *Chironomus* and are not typical of all polytene chromosomes during their active intervals. Even in *Ch. tentans,* only about 15 percent of the 2000 bands have been seen to incorporate radioactive RNA precursors. The other bands apparently remain inactive during larval stages of development that have been analyzed.

The evidence suggested a relationship of DNA to gene expression, but a link with particular protein products of gene action was not obtained until 1961. At that time Beermann showed conclusively that a particular Balbiani ring on chromosome-4 was responsible for synthesis of one or several proteins of salivary secretion material in *Chironomus*. In the absence of this puff, there was no synthesis of the particular proteins. In salivary glands that did produce this Balbiani ring, only four cells of the entire gland were seen to form a Balbiani ring and only these cells synthesized the proteins (Fig. 12.34). In a hybrid formed between two *Chironomus* species, only one of which normally synthesized these proteins and showed development of the specific Balbiani ring, only one of the two homologous paired polytene chromosomes produced a Balbiani ring and only half as much protein was synthesized. The correspondence between puffing, gene action, and formation of a specific product of this gene action clearly showed that puffing accompanied transcription of encoded information into RNA transcripts. In a study reported in the 1970s, J.-E. Edström showed that all DNA at this band was transcribed into heterogeneous RNA and that only a small part of this RNA was apparently messenger RNA which was used in translation of the salivary secretion proteins.

Because most chromosomes are too difficult to

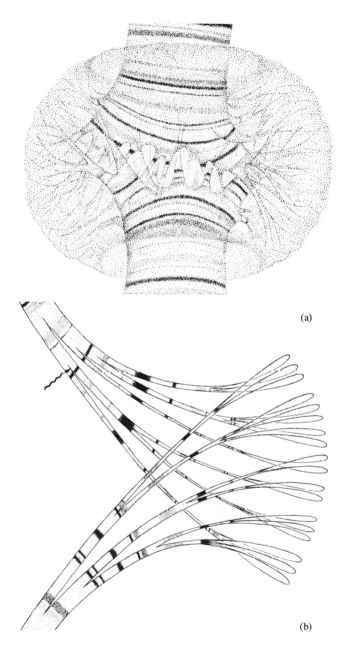

(a)

(b)

Figure 12.33
Diagram showing (a) the formation of a puff, as caused by (b) local unfolding of DNA at a band of the salivary gland chromosome. (From "Chromosome Puffs" by W. Beermann and U. Clever. Copyright © 1964 by Scientific American, Inc. All rights reserved.)

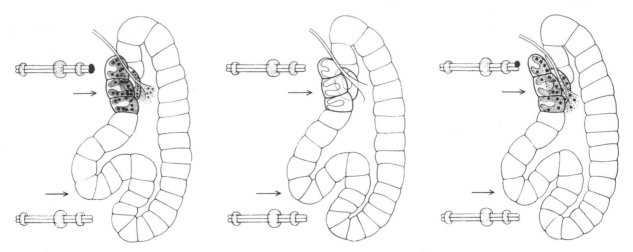

Figure 12.34

Diagrammatic summary of Beermann's experiments with *Chironomus* hybrids. There are four cells in the salivary gland of *Ch. pallidivittatus* (left) that produce a granular secretion. A puff is formed at one end of chromosome-4 only in these particular cells. The corresponding cells in *Ch. tentans* (center) produce a clear secretion that lacks these protein granules. No puff forms at the end of chromosome-4. In hybrids (right) between these species, half the amount of protein granules is found in the secretion as in the *Ch. pallidivittatus* parent. Only the chromosome derived from that parent forms a puff in the critical four cells of the gland. Each chromosome-4 contributed by a parent to the hybrid was identifiable by its banding pattern. (From "Chromosome Puffs" by W. Beermann and U. Clever. Copyright © 1964 by Scientific American, Inc. All rights reserved.)

identify and observe in the active interphase stage, it is not known whether puffing is a general phenomenon or only a particular aspect of cellular differentiation during larval development in dipteran insects and perhaps a few other species groups.

Puffing provides one mechanism for amplification of gene action products even when only one gene copy may be present in the genome. For example, in the silkworm *Bombyx mori* one gene copy is encoded for the silk fibroin protein, but during four days of larval development as many as 1×10^{15} molecules of the protein may be synthesized in the polytene cells of the silk glands. The production of many messenger RNA molecules during puffing provides a basis for the synthesis of an enormous number of proteins during translation of the messengers. Even though messenger RNAs may have a short lifetime, their abundance assures the synthesis of tremendous amounts of a particular protein in a very short interval during development.

More stable forms of messenger RNA also are known for eukaryotic cells, such as the mammalian red blood cell. Differentiation in that case involves synthesis of hemoglobin over the lifetime of the cell from messengers that can be used over and over again for weeks or months. Amplification of less stable messengers represents one mechanism employed during differentiation, as in polytene situations in larval cells, while stable messengers provide an alternative means for gene expression during cell differentiation in other cell types.

Regulation of Puffing

In the early 1960s, U. Clever showed that the molting hormone *ecdysone* was directly involved in puff formation in *Chironomus*. When the hormone was withheld from developing cells, the usual puffing pattern did not appear and the affected cells did not differentiate. When ecdysone was injected into such affected cells, the same puffing pattern took place as that occurring during the normal molting process in the larvae preparing to pupate and undergo metamorphosis.

In subsequent studies, Clever further showed that the formation of the earliest puffs depended on hormonal control, but that later puffs depended on the formation of the first ones in the developmental sequence and also on new RNA and protein synthesis. The first two puffs formed even in the absence of RNA and protein synthesis, but ecdysone had to be present. The chromosome appears to be one of the primary sites of hormone action, but many kinds of regulatory events and components interact during differentiation of complex tissues and organ systems.

The mechanism of puffing has been explained by two alternative hypotheses:

1. The primary event in the induction of gene action involves detachment of associated proteins from DNA in the chromatin fiber, making the exposed DNA accessible for the initiation of transcription. Puffing is seen in this hypothesis as the first event in gene action leading to transcription.
2. Puffing itself may be a result of transcriptional activity of chromosomal DNA in a particular band or chromomere site.

It remains to be determined whether either of these alternatives or another, as yet unspecified, hypothesis is correct.

SUGGESTED READING

Books, Monographs, and Symposia

Cold Spring Harbor Symposia on Quantitative Biology. 1973. *Chromosome Structure and Function*, vol. 38. New York: Cold Spring Harbor Laboratory.

DuPraw, E. J. 1970. *DNA and Chromosomes*. New York: Holt, Rinehart, and Winston.

Yunis, J. J., ed. 1974. *Human Chromosome Methodology*. New York: Academic Press.

Articles and Reviews

Aaronson, R. P., and Blobel, G. 1975. Isolation of nuclear pore complexes in association with a lamina. *Proceedings of the National Academy of Sciences* 72:1007–1011.

Beermann, W., and Clever, U. 1964. Chromosome puffs. *Scientific American* 210(4):50–58.

Bishop, J. O. 1974. The gene numbers game. *Cell* 2:81–86.

Britten, R. J., and Kohne, D. E. 1970. Repeated segments of DNA. *Scientific American* 222(4):24–31.

Brown, D. D., and Dawid, I. B. 1968. Specific gene amplification in oocytes. *Science* 160:272–280.

Brown, S. W. 1966. Heterochromatin. *Science* 151:417–425.

Burkholder, G. D. 1975. The ultrastructure of G- and C-banded chromosomes. *Experimental Cell Research* 90:269–278.

Daneholt, B. 1975. Transcription in polytene chromosomes. *Cell* 4:1–9.

Delius, H., and Worcel, A. 1974. Electron microscopic visualization of the folded chromosome of *Escherichia coli*. *Journal of Molecular Biology* 82:107–109.

Gall, J. G. 1969. The genes for ribosomal RNA during oogenesis. *Genetics* (Suppl.) 61:121–132.

Hamkalo, B. A., Miller, Jr., O. L., and Bakkens, A. H. 1973. Ultrastructure of active eukaryotic genomes. *Cold Spring Harbor Symposia for Quantitative Biology* 38:915–919.

Kavenoff, R., Klotz, L. C., and Zimm, B. H. 1973. On the nature of chromosome-sized DNA molecules. *Cold Spring Harbor Symposia for Quantitative Biology* 38:1–8.

Kornberg, R. D. 1974. Chromatin structure: A repeating unit of histones and DNA. *Science* 184:868–871.

Lewin, B. 1974. Sequence organization of eukaryotic DNA: Defining the unit of gene expression. *Cell* 1:107–111.

Oudet, P., Gross-Bellard, M., and Chambon, P. 1975. Electron microscopic and biochemical evidence that chromatin structure is a repeating unit. *Cell* 4:281–300.

Pardue, M. L., and Gall, J. G. 1970. Chromosomal localization of mouse satellite DNA. *Science* 168:1356–1358.

Perry, P., and Wolff, S. 1974. New Giemsa method for the

differential staining of sister chromatids. *Nature* 251: 156–158.

Petes, T. D., Newlon, C. S., Byers, B., and Fangman, W. L. 1973. Yeast chromosomal DNA: Size, structure, and replication. *Cold Spring Harbor Symposia for Quantitative Biology* 38:9–16.

Stein, G. S., Stein, J. S., and Kleinsmith, L. J. 1975. Chromosomal proteins and gene regulation. *Scientific American* 232(2):46–57.

Stubblefield, E. 1973. The structure of mammalian chromosomes. *International Review of Cytology* 35:1–60.

Yunis, J. J., and Sanchez, O. 1973. G-banding and chromosome structure. *Chromosoma* 44:15–23.

Chapter 13

Mitosis and Meiosis

During the nineteenth century it was firmly established that life comes from pre-existing life, and cells come from pre-existing cells. Each new generation of cells or individuals is the result of **reproduction.** Since progeny resemble their parents, there must be mechanisms that ensure faithful increase and transfer of genetic information. Increase is essential because more copies of the genetic instructions must be made if the progeny are to get all the information they need to grow up and produce their own offspring in turn. Once the genes have multiplied, they must be transferred from parent to progeny if an independent generation is to grow and develop. Both processes, *increase* and *transfer* of genes, must be accomplished with considerable fidelity, or progeny would not resemble their parents, as they do.

The increase in genetic material occurs when DNA replicates. Since we know that new DNA is faithfully copied from template DNA in parent cells, we have the basis for understanding how progeny and parents continue to be more like each other than like unrelated individuals. Transfer of the genetic information in eukaryotes is accomplished by **mitosis** or **meiosis.** These nuclear division processes include accurate systems for distributing chromosomes to progeny nuclei. If reproduction is *asexual,* mitosis is the only mechanism which ensures that daughter cells will receive equal and identical copies of chromosomes from the parent cell.

In *sexually-reproducing* systems, mitosis also takes place, but it is not primarily responsible for the unique sets of chromosomes that will be present in progeny cells. Two processes absent from asexually-reproducing forms but present in sexual life are **meiosis** and **fertilization.** The division of parental nuclei by meiosis leads to halving of the chromosome number in the **gametes.** When two gamete nuclei fuse to form a single **zygote** nucleus, fertilization has been accomplished and the chromosome complement has been returned to its former number.

Sexual cycles are punctuated by meiosis and fertilization episodes (Fig. 13.1). Asexual cycles are maintained by mitosis as the only nuclear division event, and nuclear fusions are not necessary to initiate the next generation. Sexual forms cannot initiate a new generation until nuclear fusion has produced a zygote which has information from two different sources.

Sexual reproduction is therefore often called *biparental,* while asexual reproduction is *uniparental.* These terms are accurate for species where sexes are

Reproduction

Asexual

Sexual

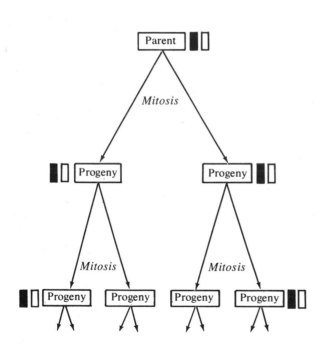

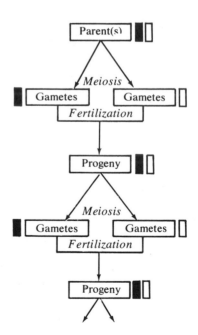

Figure 13.1
Schematic diagram illustrating the basic differences between asexual and sexual reproduction.

separated into different individuals, usually males and females. Many sexual forms, however, have both kinds of sex organs so that one individual may produce both **sperm** and **eggs,** the common forms of gametes. Because of many variations on the basic themes, we should view reproduction by asexual means as a system in which one nucleus can lead to a new generation, whereas sexual generations can only be initiated by two nuclei that fuse to form the initial product.

Asexual forms produce offspring exactly like themselves. Sexual species produce progenies that are somewhat different from their parents and from each other. The differences between sexual generations are a reflection of genetic events that lead to *new combinations of genes.* The **segregation** of genes in meiosis and the **recombination** or **reassortment** of genes at fertilization produce and maintain higher levels of variation in sexually-reproducing species. Since evolutionary changes are drawn from gene pools in populations, increased genetic variability is expected to provide increased opportunities for beneficial evolutionary developments. The significance of sexual reproduction in evolution cannot be overemphasized. The focus of this chapter is on the nuclear division processes that lead to new cell generations and on other cell activities that accompany the reproductive events.

THE CELL CYCLE

We have known about the visible events of mitosis for 100 years. The gradual condensation of chromosomes, their separation to opposite poles of the cell, and their reorganization into new daughter nuclei can be observed by ordinary light microscopy. These dramatic events were a focus for many cell reproduction studies until the early 1950s. At this time the concept of DNA as the genetic material was being established, and new methods were developed and applied to analyze growth and reproduction at the cellular level.

In particular, methods to measure the amount of DNA in nuclei that were stained with the Feulgen reagents were developed. It had long been known that this staining method revealed DNA by its magenta coloring, but it was now possible to determine how much DNA was present by measuring the amount of stain that was bound to the preparation. Another important innovation in the 1950s was the development of autoradiography. The time and place of DNA replication could be monitored by examining silver grain distributions, which indicated newly-synthesized DNA containing incorporated radioactive precursors. Both experimental methods clearly showed that DNA replicated hours before there was any visible sign of mitosis.

With this new information, attention was directed toward the seemingly quiet **interphase** stage between mitotic divisions. DNA replicated during interphase. Other studies soon showed that interphase was also the time in which synthesis of proteins and RNA took place. In fact, the most active and vital time in mitotic cells was between divisions and not during mitosis itself. As more information was obtained about the **cell cycle,** a gap in time was discovered between the end of mitosis and the beginning of DNA replication. Another gap of time separating the end of replication from the onset of mitosis was also noted (Fig. 13.2).

Following a convention first suggested in 1953, the phases of the cell cycle have been labeled in the following way: The two time gaps are called G_1 and G_2; the period of DNA replication is called $S;$ and the time devoted to mitosis is called the M phase. These four phases constitute the nuclear portion of the cell cycle. In most eukaryotic systems, there is an accompanying division of the cell when the new nuclei are enclosed in their own separate cell boundaries. Cell division, or **cytokinesis,** usually begins during the telophase stage of mitosis. While it is not an integral part of mitosis itself, cytokinesis usually

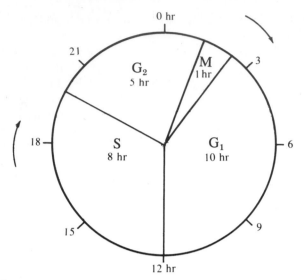

Figure 13.2
Stages of the cell cycle and their proportional duration in an average 24-hour mammalian cell cycle.

is synchronized with mitotic events. We will discuss cell division in another section of this chapter.

Variations in Cell Cycle Phases

Autoradiographic studies showed that DNA replicated in the S phase and provided information on the length of the S phase. Since mitosis can also be visualized by microscopy, the time spent in each of the four phases could be measured. S and M phases were measured directly, and G_1 and G_2 phase durations were inferred by calculating the amount of time which elapsed between M and S (G_1 phase) or S and M (G_2 phase).

Mammalian cells in culture are favorite materials for cell cycle studies. These cells have relatively similar cell cycles in general, but cycles vary somewhat from one cell type to another. The whole cycle takes 18–24 hours. In a typical adult mammalian cell cultured from human tissue, G_1 lasts 8 hours, DNA is synthesized for 6 hours in the S phase, G_2 continues for about 4.5 hours, and mitosis is completed within

1 hour. The largest variation in cell cycles of similar cells is usually found in the duration of G_1. The length of the S period also varies, but the combined time for S and G_2 shows the least change in response to external conditions.

Cell cycle measurements have also been made for cells from higher plants, where a similar 10–30 hour cycle characterizes mitotic cells. Mitosis generally takes more than 1 hour in plant cells, and S is also slightly different in duration in plant cells in comparison with animal cells.

Embryonic cells and many lower organisms show variations of the cell cycle that provide important insights into the general nature of the sequence of events. Many animal embryos undergo rapid division, with smaller and smaller cells produced after each of the divisions. The rate of DNA synthesis is about 100 times faster in these nongrowing embryonic cells than it is in adult cells from the same species, and there is usually no G_1 phase. DNA synthesis begins therefore during, or immediately after, mitosis is completed.

Adult *Xenopus* cells have a long S period lasting about 20 hours, while embryonic *Xenopus* cells carry out DNA synthesis for most of the 25 minutes of their cell cycle. In embryonic *Xenopus* cells, there is no G_1 phase, G_2 is very brief, and the S phase begins before mitosis has been completed. In sea urchin embryos, a 70-minute cell cycle is divided into M lasting about 40 minutes, less than 15 minutes of S (which begins during telophase of mitosis), and 20 minutes of G_2. These examples are fairly typical of vertebrate and invertebrate embryo cell cycles.

Many protozoa, fungi, and other lower organisms have no G_1 phase in their cell cycle. Like embryonic cells, the cycle lasts a relatively brief time. Other simple organisms, of course, may have a typical G_1 phase. The occurrence of G_1 seems to be correlated with the amount of growth and biosynthesis going on in the cell. If growth is minimal, as in animal embryos, or very rapid, as in some lower organisms grown under optimum conditions, then G_1 is brief or absent altogether. Since DNA synthesis may even begin before mitosis has been completed, the requirement for G_1 seems the most dispensable in a cell cycle. There usually is no continued cycling without S and M phases, however, and these rarely occur without an intervening G_2.

These observations point out that mitosis takes place after DNA synthesis has occurred. If DNA synthesis stops, the cell will not undergo mitosis; it becomes noncycling, rather than being "stuck" at some point in the cycle. Mitosis cannot begin immediately after DNA replication since there is always a G_2 phase, generally of relatively short duration. Other preparations for mitosis must be made, in addition to synthesis of DNA. If protein synthesis is inhibited during G_2, the cell will not divide. Very little is known about the particular proteins needed by the cell to enter mitosis, but they probably include structural proteins as well as enzymes. Histones are known to be synthesized at the same time as DNA, during the S phase, but nonhistone proteins are made at various times during interphase.

Turning to another part of the cycle, some cells are not ready to begin DNA synthesis when a previous mitosis has ended. These cells have a G_1 phase. Other cells have everything they need and dispense entirely with G_1. The preparations required for DNA synthesis are not known, but various proteins must be needed, including structural, catalytic, and regulatory proteins for DNA replication and chromosome formation.

The critical transitions in the cell cycle are from the G_1 phase to the S phase when replication begins and from G_2 to M when mitosis begins and chromosomes will be distributed to daughter cells. There is a point of "readiness" in cells that enter S phase. When S-phase cells are fused with G_1-phase cells in specially-treated cultures, the two nuclei of a fusion cell remain separate and each can be identified by its chromosomes. The G_1 nucleus begins to synthesize DNA much earlier than it normally would. This experiment shows that something is present in S-phase

cells which triggers or allows DNA replication to take place.

Other experiments involving fusions between cells in different phases of the cell cycle have been done to test the nature of the second crucial transition from G_2 to M. In these cases the M-phase nucleus causes condensation of chromosomes in G_1, S, or G_2 nuclei in the same fusion cell. G_1 chromosomes condense even though they have not yet replicated, and they appear as single chromosomes. G_2 chromosomes have already replicated, so they have the conventional organization as they enter mitosis. When S-phase and M cells are fused, however, small fragments of condensed chromosomes appear from the S nuclei. Some feature of M cells, therefore, leads to chromosome condensation, even in unprepared nuclei. Unprepared nuclei do not proceed normally through a mitotic division, however, if they are prematurely condensed in a fusion cell. We will review some aspects of DNA replication next. DNA replication was also discussed in Chapter 5.

Replication of DNA

$E.$ $coli$ cells growing at a maximum rate will double their numbers every 30 minutes. DNA is continually synthesized in such cultures. Autoradiographic studies by J. Cairns showed that $E.$ $coli$ cells synthesized their new duplex DNA at a rate of about 40 μm per minute. At this rate, only 30 minutes would be needed to replicate the entire 1300 μm-long DNA molecule. This prokaryotic DNA has only one point of origin at which each cycle of replication begins (Fig. 13.3). New semiconserved duplexes are formed as replication proceeds in both directions away from the point of origin. The resulting image is similar to the Greek letter $theta$ (θ), so molecules like this are referred to as replicating θ-forms.

A new replication cycle can begin before the previous one is finished; this can be seen in autoradiographs which show more than one **replication fork.** When the amount of radioactivity at each replication fork is determined, it is clear that the forks did not

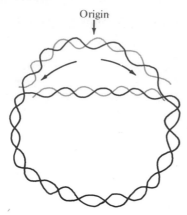

Origin

Figure 13.3
Bidirectional replication of the circular DNA molecule of
E. coli. Newly-synthesized strands are gray.

begin to form at the same time. Since almost all the
cell activities are taken up with DNA synthesis,
while other syntheses are going on simultaneously,
many investigators do not think that prokaryotes
have sequentially-ordered cell cycle activities.
Prokaryotes are usually excluded from cell cycle
studies, therefore.

Eukaryotic chromosomal DNA also replicates
bidirectionally. But, electron micrographs of replicating DNA isolated from chromosomes show multiple
points of origin (Fig. 13.4). Each of these "bubbles"
along the duplex DNA represents a bidirectionally-
replicating loop, according to autoradiographic analysis.

DNA is synthesized at a slower rate in eukaryotes than in bacteria. The estimated rate for eukaryotic chromosomal DNA synthesis is about 1–2
μm per minute in adult cells in culture. Since the
average S phase lasts about 6–8 hours in these vertebrate cell types, only 720 to 960 μm of DNA
would be replicated at a rate of 2 μm per minute (2
μm $\times$ 480 minutes = 960 μm in 8 hours). There may
be 100–200 times this amount of DNA in a chromosome complement. To explain replication of such a
large amount of DNA during the S phase, it has been

suggested that 100–200 different points of origin for
replication must exist among the chromosomes. These
replicating segments, called **replicons,** can be seen
directly in electron micrographs as regions with
"bubbles" or loops. Other methods have also been
used to estimate the number of replicons in a chromosome complement, as well as the average length of
duplex DNA per replicating segment.

Some regions of DNA begin to replicate at the
beginning of the S phase, while others begin at various
times afterward. Euchromatin regions generally
begin to replicate earliest, and heterochromatin
replicates latest in the S phase of a cell cycle. It is
believed that heterochromatic DNA is duplicated
at a faster rate so that it completes replication along
with all euchromatin DNA by the end of the S phase.

Many species have been analyzed since J. Taylor's
first important studies in 1960. All have shown the
same features of late-replicating heterochromatin
and earlier-replicating euchromatin. Confirming evidence for these generalizations has been obtained
by banding studies and from the *in situ* hybridizations
discussed earlier, in Chapter 12. Time of replication can be used as one identifying feature for heterochromatin or euchromatin in unknown cell systems
and as a means of identifying specific chromosomes
in a set.

The triggering events that initiate DNA synthesis
in the cell are not known. Certain proteins are obviously needed if replication is to begin and continue,
but external factors are also important in controlling
internal events. For example, some kinds of mammalian cells grown in culture form only one layer
of cells attached to the glass dish. These cells become
noncycling once the confluent monolayer of cells
has covered the glass surface of the dish. If some cells
in the monolayer are scraped away, cycling begins
again in the cells adjoining the opened space in the
dish until the space is refilled. After this, the cell
cycle shuts down again.

The shutdown of cell multiplication in confluent
cell monolayers has been called **density-dependent** or

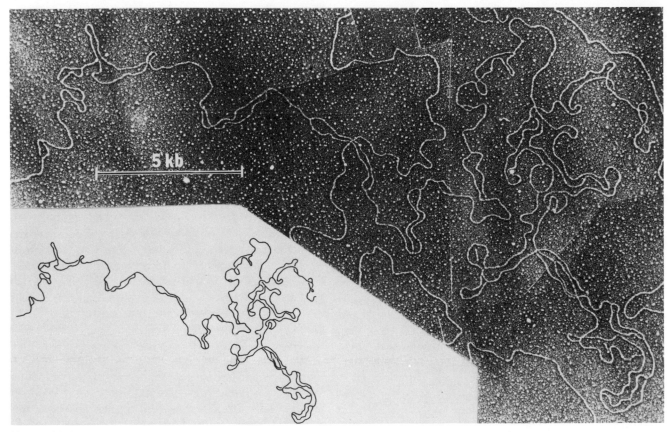

Figure 13.4
Electron micrograph and interpretive drawing of replicating chromosomal DNA from nuclei of cleaving fertilized eggs of *Drosophila melanogaster*. The portion of the molecule shown here is 119,000 base-pairs long and contains 23 "eye" forms. A kilobase (kb) is a unit of length equal to 1000 bases or base pairs in single- or double-stranded nucleic acids, respectively. (Courtesy of H. J. Kriegstein and D. S. Hogness, from Kriegstein, H. J., and D. S. Hogness, 1974, *Proc. Natl. Acad. Sci.* **71**:135–139, Fig. 1.)

postconfluence inhibition. This phenomenon is separable and probably distinct from contact inhibition of cell locomotion in many such monolayer cultures, but the two kinds of inhibition have often been confused when the single term "contact inhibition" is applied to both phenomena (see Chapter 11). A number of specific signals must be involved in cell cycling because many kinds of cultured mammalian cells continue to grow in masses and do not shut down after a single-layer sheet is formed. Hormone-like agents, ions, and small molecules in the system also exert controlling effects on different kinds of cells under different conditions.

In some cases there appear to be communication sites in adjoining membranes of those cells that stop cycling after forming a confluent single-layer sheet. Cells that go on to form multi-layered masses do not have these **junctions** between cells. The nature of the junction ultrastructure and other features of plasma membranes make it very likely that signals can be communicated among cells of a population in some cases (see Chapter 4).

Cancer cells grow and divide without restraints. They may be cells which are noncycling in normal development but which have reinstituted their cell cycle after transformation to neoplastic growth. There are many variations involved in tumor cell phenomena, since different cells act and respond in different ways to internal and external conditions. At present it seems that in cancer cells the restraint over entering the *S* phase is lost. The restraint cannot be restored at this time. The best therapy that can be applied in cancer situations, at present, is to destroy or remove the aberrant cells as selectively as possible. If viruses are found to be the primary causative agent of cancer induction, then perhaps more specific therapies, in the form of antisera and vaccines, could be applied. The main basic research problem, however, will continue to be focused on the nature of the cell cycle controls and the ways in which these controls are overridden or swamped by external and internal influences.

MITOSIS

The continuous sequence of events in mitosis is divided, by convention, into arbitrary stages (Fig. 13.5): **prophase** (*pro:* before), **prometaphase, metaphase** (*meta:* between), **anaphase** (*ana:* back), and **telophase** (*telo:* end). The morphological features of mitotic division are quite well known, but the underlying molecular and biochemical mechanisms which produce the visible events of mitosis are being studied only now. One point, which is known from cell cycle studies mentioned earlier, is that nuclei normally will not enter mitosis unless there has been a prior replication of chromosomal DNA. It is also known that the cycle of chromosome condensation during mitosis can be induced in nuclei from any other part

of the cell cycle, since G_1, S, or G_2 nuclei undergo chromosome contraction if they are present in a fusion cell along with an *M*-phase nucleus.

The capacity of cells to undergo DNA replication and mitosis varies across a broad spectrum. Muscle and nerve cells do not divide; bone cells stop dividing in the adult; cell division in skin and blood-forming tissues takes place throughout the life of the organism but is modulated to compensate for losses of old cells; and other tissues divide only in response to some external stimulus such as wounding. Higher plants that live for many years have *meristematic* tissues that grow and divide as long as the plant lives. In these divisions, some cell products remain meristematic and others go on to differentiate into the mature tissues of the roots and stems. Onion or broad bean root-tip mitosis is studied in most biology classes. The dividing cells are part of the root-tip meristem that produces epidermis, vascular tissues, and other parts of the root at the same time that a reserve of meristem cells is retained.

DNA replication produces the new genetic material for the next cell generation. Mitosis is a mechanism for distributing this material to the daughter cells. It is a remarkably accurate process that works equally well for a few chromosomes as for a few hundred. Occasionally a mistake may be made during distribution; for example, sister chromosomes may fail to disjoin (Fig. 13.6). **Nondisjunction** leads to added or missing chromosomes in the daughter nuclei. In most cases this causes cell malfunction or even death. The consequences of nondisjunction during mitotic or meiotic divisions will be discussed in Chapter 14.

Stages of Mitosis

The first obvious signs of mitosis, signaling the beginning of **prophase,** are condensations of the chromosomes. As prophase continues, the chromosomes become shorter and thicker. Their morphological characteristics are clearer and individual chromosomes can be recognized. They can clearly be

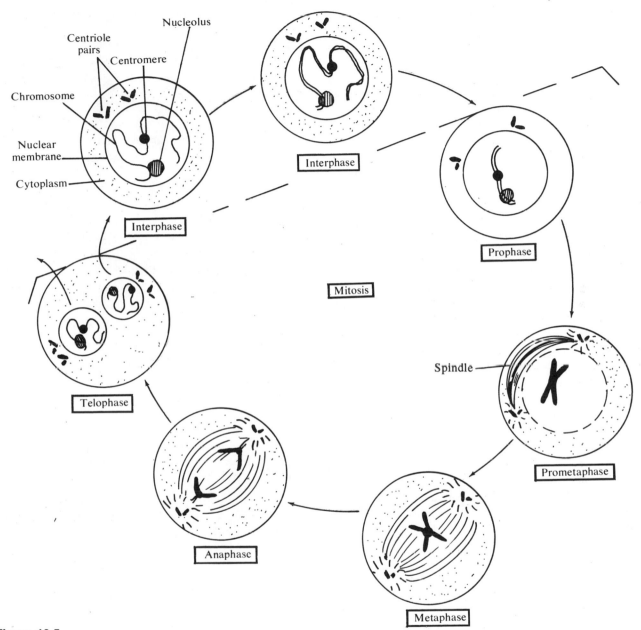

Figure 13.5
Stages of mitosis. Chromosome replication takes place during interphase, and chromosome distribution to daughter nuclei is accomplished during mitosis. The chromosome number in daughter nuclei is the same as it was in the original parent nucleus.

Figure 13.6
Nondisjunction of sister chromosomes or chromatids leads to an extra chromosome at one pole and a missing chromosome at the other.

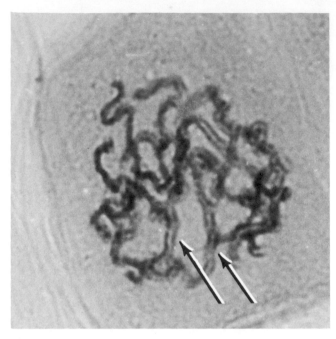

Figure 13.7
The dyad structure of mitotic prophase chromosomes from broad bean (*Vicia faba*) indicates that these are replicated chromosomes.

seen as double structures by about mid-prophase (Fig. 13.7). Toward the end of prophase the nucleolus and nuclear envelope disappear, their components becoming dispersed and generally indistinguishable from other parts of the cytosol.

Centrioles, which doubled in number in the preceding interphase, now separate with one pair migrating to the opposite pole of the cell. The spindle microtubules become evident next to the nuclear envelope. After the nuclear envelope disappears, the spindle occupies a more central location, between the two pairs of centrioles. Cells that normally have no centrioles will still have a spindle that fills the position which had been occupied by the intact nucleus. If the nucleus was centered, so is the spindle. Eccentrically placed nuclei are replaced by off-center spindles, that lead to an asymmetric division in which one daughter is larger than the other (Fig. 13.8).

During **prometaphase,** the contracted chromosomes move toward the equatorial plane of the spindle. Movement is somewhat erratic, as seen in time-lapse photography; the chromosomes do not move unwaveringly toward their ultimate positions. Some individual chromosomes streak across the plane of the spindle while other chromosomes stay put or even jiggle about aimlessly. Prometaphase *congression* of chromosomes is poorly understood

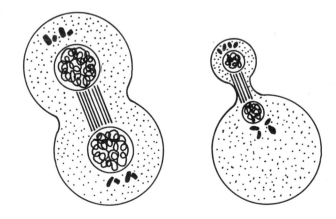

Figure 13.8
The position of the spindle leads to equal- or unequal-sized daughter cells at the completion of cytokinesis following symmetrical or asymmetrical mitosis, respectively.

since spindle fibers are not yet attached. Finally, as though a signal has been given, all the chromosomes line up by their **centromeres** at the equatorial plane of the spindle figure.

Each **metaphase** chromosome is a replicated structure, made up of two *chromatids*. These chromosomes are aligned so that the centromere of each chromatid faces the opposite pole of the cell. By this time the chromosome-to-pole fibers are inserted at the centromere of each chromatid, so it is even easier to see that sister centromeres face opposite poles at metaphase (Fig. 13.9). Continuous pole-to-pole spindle fibers pass right by the metaphase chromosomes and are not attached to them in any way. The forces which keep the two chromatids of a chromosome together from the time of replication in interphase until the end of metaphase are not known. It is particularly puzzling since each chromatid can be separated from its sister in cells treated with colchicine. This drug prevents formation of the spindle or disrupts one that has already formed. In the presence of colchicine, mitosis is interrupted at the metaphase stage and goes no further. The sister chromatids separate passively and a new nucleus organizes with double the number of chromosomes of the original parent cell (Fig. 13.10). This process must be a normal occurrence in some differentiated tissues, such as liver, where tetraploid or octaploid cells are commonly found. Cells in higher plants and animals are usually **diploid,** a condition characterized by the presence of two sets of **homologous** (genetically similar) chromosomes in the cell nucleus.

The relatively brief metaphase is followed by **anaphase** which begins when chromatids of each chromosome separate (Fig. 13.11). Sister chromatids move to opposite poles of the cell, since their centromeres were aligned in this way at metaphase and their fibrous attachments point toward only one pole. When chromatids separate from each other, each becomes a full-fledged and independent chromosome, no longer acting together with its sister.

Once the chromosomes have arrived at their respective poles, **telophase** begins. During this final stage of mitosis, the condensed chromosomes begin to unfold and gradually assume the appearance they had during interphase (Fig. 13.12). Nuclear reorganization takes place, including organization of new nucleoli and nuclear envelope. Nucleoli form specifically at nucleolar-organizing regions of particular chromosomes in the complement and nowhere else. The new nuclear envelope first appears around individual chromosomes. These membranous elements eventually coalesce to form the continuous nuclear envelope surrounding all the contents of the nucleus. Interphase has then begun.

Consequences of Mitosis

At the end of mitosis each daughter nucleus contains an identical set of chromosomes, which is also identical to the original parent nucleus. By the processes of mitotic distribution, each cell is assured of receiving a complete set of genetic instructions, since DNA was duplicated in the *S* phase prior to

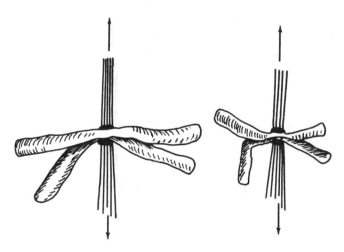

Figure 13.9
When metaphase chromosomes align at the equatorial plane of the mitotic cell, centromeres of the sister chromatids of a chromosome face opposite poles. This orientation determines that their separation will be to these opposite poles at anaphase.

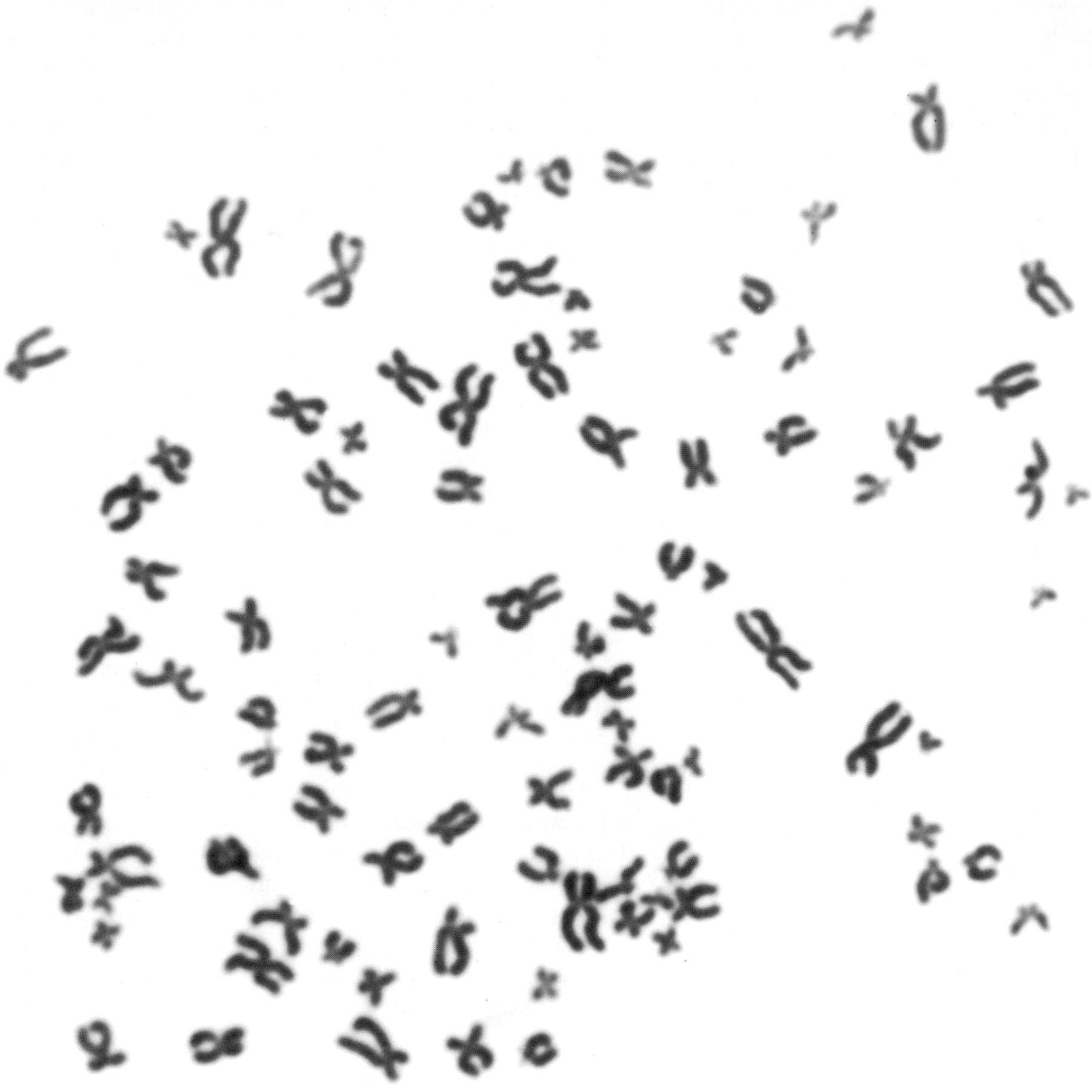

Figure 13.10
Polyploid metaphase mitotic nucleus from human cells isolated from amniotic fluid. There are 92 chromosomes in this tetraploid nucleus, instead of the diploid complement of 46 chromosomes. (Courtesy of C. Hux)

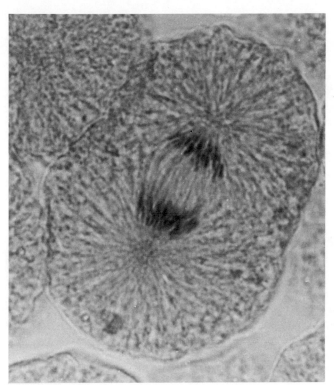

Figure 13.11
Mitotic anaphase in a blastula cell of whitefish.

mitosis. The same distribution mechanism operates successfully for cells that are haploid, diploid, or polyploid; or normal or aberrant in chromosome number. This mechanism ensures identical chromosome complements in continuing generations. The fidelity of distribution is responsible for genetic constancy in mitotic generations. Variations can be introduced by mutations or other random modifications of chromosomes, but these variations in turn will be transmitted faithfully to all descendants of the modified cell. Asexually reproducing species are therefore genetically rather uniform. Variability exists between populations rather than within populations in asexual species, depending on chance mutations that occurred and were incorporated into population gene pools.

Each body (somatic) cell in a multicellular organism arises by mitosis, whereas gametes or spores sooner or later will form after meiotic divisions in sexual species. The billions of somatic cells in a human being have the same genetic content, all derived originally from the fertilized egg and all its cell lineages produced by mitosis during development and differentiation. The differences in cell appearance, function, and activity are due to systems of regulation that control gene action, and to other internal as well as external influencing factors. Mitosis delivers the chromosomes, but other processes then direct the expression of genetic potential into the variety of cells, tissues, and organs of the individual.

The appearance of mitosis in evolution was a significant step forward. It permitted increase in chromosome number, which has advantages in relation to gene reassortment during sexual reproduction.

Modifications of Mitosis

As we should expect for biological systems, mitosis in some cells and organisms may be accomplished in a somewhat different way from the average format we just described. Only a few general examples will be described, some because of inherent interest and others because they provide some clues to evolutionary patterns.

One of the commonest variations is *intranuclear* mitosis, which is mainly encountered in some of the lower eukaryotes. The nuclear envelope remains intact throughout mitosis, so that the chromosomes are always compartmented. Apart from this feature, there are other nuclear differences, depending on the species. For example, centrioles may or may not play an active role in spindle formation or in any mitotic events. In yeast, centriolar plaques occur at each pole, with spindle microtubules in between (Fig. 13.13). *Chlamydomonas* centrioles are active during cell division but not during mitotic division of the nucleus.

The location of spindle microtubules varies. Sometimes it is formed within the intact nucleus, while at

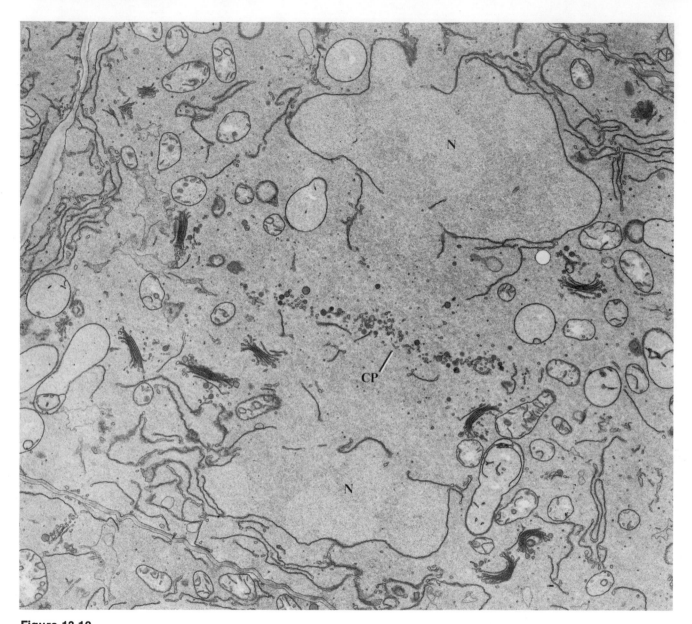

Figure 13.12

Part of a telophase mitotic cell from root epidermis of maize. The membranes of the two reorganizing nuclei (N) are closely associated with the chromosomes (pale outlines in permanganate-fixed preparations). The two daughter nuclei are separated by the beginnings of a new cell plate (CP). Microtubules, chromosome fibers, and other fibrous structures are not preserved in cells fixed with permanganate salts. × 10,000. (Courtesy of H. H. Mollenhauer)

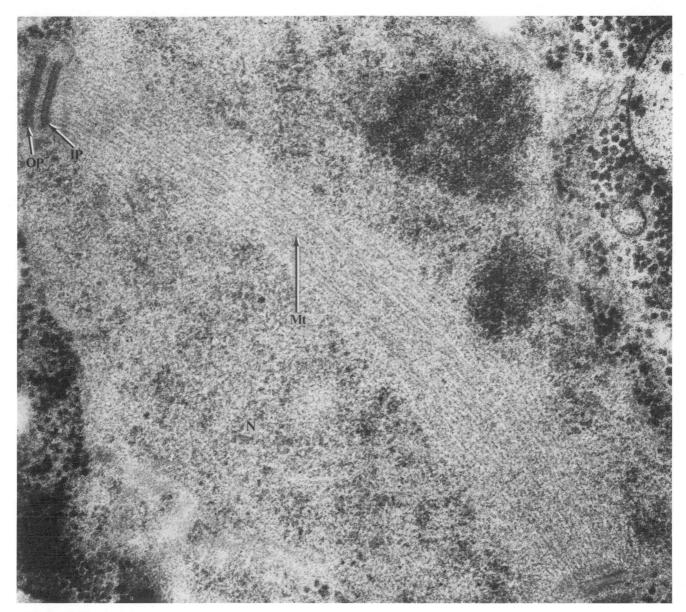

Figure 13.13
Spindle plaques in anaphase I meiotic nucleus (N) of bakers' yeast (*Saccharomyces cerevisiae*) during ascospore formation. The inner plaque (IP) at each pole is a differentiated part of the nuclear envelope, and microtubules (Mt) are inserted directly into the plaque. The outer plaque (OP) is in the cytoplasm immediately adjacent to the nucleus and inner plaque. × 72,000. (Courtesy of P. B. Moens)

other times, it is not. In the latter case, there often is a connection between the spindle fiber in the cytoplasm and the intranuclear chromosome. The connection is generally made at the centromere which is embedded in the nuclear envelope (Fig. 13.14). Even though the nuclear envelope persists, the nucleoli may or may not be retained during mitosis. Chromosomes may be permanently condensed, as in *Euglena* and other species, or they may undergo the usual cycle of unfolding and folding that is typical of most mitotic systems.

In one species of dinoflagellate, an interesting

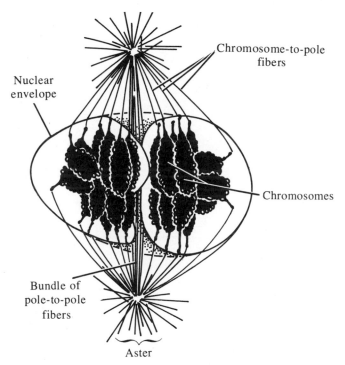

Figure 13.14
Drawing of an anaphase nucleus in the protozoan *Barbulanympha* showing the pole-to-pole fibers located outside the nuclear envelope and the chromosome-to-pole fibers attached to the chromosomes within the nuclear envelope where the centromeres are attached. (Redrawn from Cleveland, L. R., 1953, *J. Morphol.* **93**:371–403, Plate 59, Fig. 6. With permission of The Wistar Press, Philadelphia.)

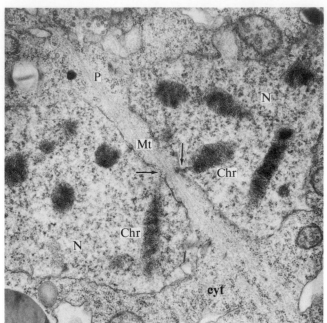

Figure 13.15
Electron micrograph of longitudinal section through one lobe of a late division nucleus of the dinoflagellate, *Gyrodinium Cohnii*. Microtubules (Mt) occur within the cytoplasmic channel between the lobes of the nucleus (N) and extend for a short distance into the cytoplasm at the polar (P) end. Chromosomes make contact (arrows) with the nuclear envelope bounding the cylindrical channel of cytoplasm (Cyt). × 18,700. (Courtesy of H. Ris, from Kubai, D. F., and H. Ris, 1969, *J. Cell Biol.* **40**:508–528, Fig. 24.)

mitotic variation has been described in the microtubular system associated with chromosome movement in anaphase (Fig. 13.15). During mitosis channels form in the cytoplasm and penetrate through the body of the intact nucleus without disturbing the nuclear membrane continuity. Bundles of microtubules appear within the cytoplasmic channels and lend rigidity to these regions. Chromosomes within the nucleus are attached by their centromeres to the nuclear membrane at the borders of the rigid channels, but they are never connected through the membrane with the microtubules. As the nuclear membrane lengthens, the chromosomes are carried to opposite poles, undoubtedly given leverage by the presence of the rigid channels on the outside.

This pattern seems to be a compromise between conventional anaphase movement of chromosomes which are directly attached to microtubules, and the separation of DNA which occurs during fission in bacteria. In both the dinoflagellate and bacteria, the DNA is carried passively to opposite ends of the cell or nucleus by the lengthening membrane. In bacteria, the plasma membrane is held taut by the rigid cell wall; in the dinoflagellate, the nuclear membrane is made more taut by the microtubule-filled cytoplasmic channels. It may be coincidental, but dinoflagellate chromosomes have no proteins complexed to the extensively folded DNA.

Ciliated protozoa such as *Paramecium* and *Tetrahymena* have two kinds of nuclei in the cell. The smaller micronucleus does not undergo mitosis whereas the larger macronucleus does (Fig. 13.16). The macronucleus is a compound structure that contains many copies of the chromosome set. It is the metabolically active component in ciliates. The micronucleus may be lost without causing immediate damage. It is the component which undergoes meiosis. An amicronucleate cell is therefore asexual. New cells are produced by fissions, but a gradual depletion of the macronuclear material occurs and such amicronucleate strains cannot be maintained indefinitely.

The polytene chromosomes in larval cells of dipteran insects and in some other species furnish striking examples of permanent interphase systems. DNA replication takes place in repeated rounds, but there is no subsequent separation of the strands in a mitotic division because the *M* phase is missing in this cell cycle.

In other kinds of cells there are also rounds of **endoreplication,** or, **endoduplication,** but the new chromosome strands separate from each other. Since the separated chromosomes remain within the confines of the same nuclear envelope, the process is

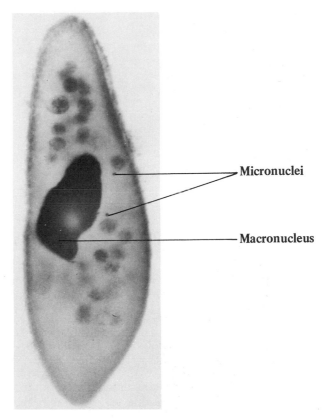

Figure 13.16
There are two tiny micronuclei near the large macronucleus in this species of *Paramecium.* × 700. (Courtesy of R. V. Dippell)

called **endomitosis** (*endo:* within). During endomitosis, there is an increase in chromosome sets, so the products of endomitosis are **endopolyploid** nuclei (Fig. 13.17). Cells of some tissues, such as mammalian liver, commonly have endopolyploid nuclei with 4 or 8 sets of chromosomes, rather than the usual 2 sets present in diploid cells in other tissues of the organism.

It may not be possible to see the chromosome number directly in typical interphase cells, but the amount of DNA per nucleus can be measured accurately by the amount of Feulgen stain which becomes bound to the DNA. Endopolyploidy can be

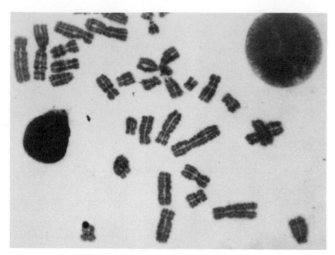

Figure 13.17
There have been two rounds of replication without separation of the chromatids (endoreplication). Such aggregates are called *diplochromosomes*. Chromosomes have been stained using a G-banding method that produces a unique band pattern for each of the human chromosomes in the haploid complement. (Courtesy of C. Hux)

or cytokinesis, may not take place at all in some organisms. Slime molds have multinucleated protoplasm without a single partition present. Other lower eukaryotes may have one, two, several, or many nuclei per cell, sometimes in regular patterns and sometimes erratically. In the higher eukaryotes cytokinesis and mitosis are usually synchronized, and uninucleate cells are produced. Some tissues do vary, but the general situation is uninucleate.

Animal cells typically divide by **furrowing**, also called **cleavage** (Fig. 13.18). This pattern of division is also found in a few kinds of plant cells and among some groups of protists. The first sign of cell division is a constriction or furrowing, in a line with the midpoint of the spindle figure. The furrow becomes progressively deeper, until the cell assumes an exaggerated dumbbell shape. Finally, the cell pinches into two individual and separate daughter cells. Depending on the position of the spindle in the cell, the daughter cells may or may not be equal-sized.

Most higher plant cells divide by **cell plate formation** rather than furrowing. The cell plate begins to

induced at will by applying colchicine to the mitotic system. Since spindle formation is prevented, the replicated chromosomes separate but are not carried to opposite poles of the cell. When telophase reorganization occurs, the new nucleus has double the usual number of chromosomes. Higher levels of polyploidy can be induced by regulating the amount of colchicine applied and the time it remains in the system. Treatments of this kind do not lead to polytene chromosomes, however. It is not known why certain chromosomes regularly become polytene and others do not.

CYTOKINESIS

Depending on inherent properties of the species and cell type, division of the cell contents may or may not be synchronized with mitosis. Cell division,

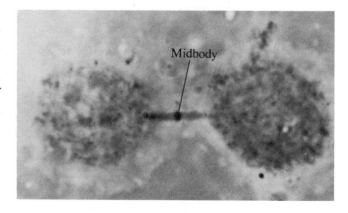

Midbody

Figure 13.18
Phase contrast light micrograph of section of HeLa cell in culture. The constricted region has elongated considerably, and the midbody is evident in the bridge connecting the sister cells (see also Fig. 13.20). × 2,500. (Courtesy of B. Byers, from Byers, B., and D. H. Abramson, 1968, *Protoplasma* **66**:413–435, Fig. 8.)

form in the center of the spindle midpoint, and material is added at both ends until it stretches completely across the width of the cell. Furrowing begins at the cell periphery and moves inward to the cell center; cell plate formation moves outward toward the edges of the cell (Fig. 13.19). Cell plate materials are provided in part by secretion vesicles made at the Golgi apparatus and partly by membranous elements of the endoplasmic reticulum. After these have been incorporated into the developing cell plate, other substances are added to make up the new cell wall. Cell wall carbohydrates are deposited only on the outside surface of the new plasma membranes which separate the two daughter cells.

Algae and fungi generally divide by a process of cytokinesis that resembles bacterial cell division. In these eukaryotes, the plasma membrane usually invaginates around the cell midline, and new cell wall forms alongside. As the plasma membrane pinches in closer toward the center of the cell, the new wall is extended until the two growing ends of the wall join at the center and completely separate the daughter cells. Since the separation proceeds

from the outside inward, this process is similar to animal cell furrowing. The two mechanisms, however, are entirely different in their basic features.

Furrowing

The general impression gained by observing dividing animal cells is of a constricting "pursestring effect" (see Fig. 11.36). Because of this activity, it has been suggested that a "contractile ring" of filaments causes more and more constriction until the furrow cuts through and the two daughter cells have been separated. The suggestion has gained some strength from evidence showing that cytochalasin B can reverse the pursestring effect, presumably due to its disruptive influence on bundles of microfilaments in the furrow region. There is some difference of opinion about the contractile ring proposal. Although there is no dispute about the experimental observations, the interpretation of these observations is surrounded by some controversy. One argument against the contractile ring proposal is that furrowing proceeds most dramatically in uncrowded cell groups. Much less vivid constriction occurs in dividing cells that are densely packed into a solid tissue or mass, but cytokinesis still occurs by a gradual furrowing process. There is little indication of a pursestringlike contraction in these densely packed cells.

Electron microscopic studies have contributed more details of the structural aspects of furrowing. The first ultrastructural change takes place at the periphery of the spindle, in the midline opposite the first inward point of furrowing. Dense material collects at the peripheral spindle fibers along the midline and gradually increases in amount until it fills the entire equatorial plane of the dividing cell. This dense region is called the **midbody**. As constriction proceeds toward pinching the cell in two, the midbody becomes smaller (Fig. 13.20). It usually disappears before cleavage is completed but may persist somewhat longer before disappearing in certain cell types.

The midbody is a characteristic structural feature

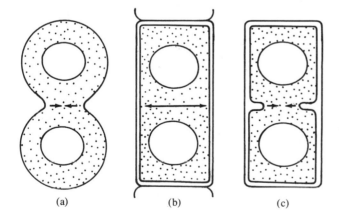

Figure 13.19
Drawings showing the differences between (a) furrowing and (b) cell plate formation in higher eukaryotes and (c) new wall formation as it occurs in many algae, fungi, and bacteria.

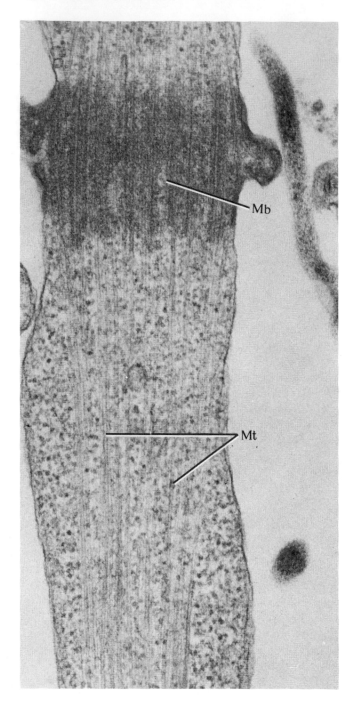

of animal cell division. It is made from spindle fiber components, vesicles, and dense material of unknown composition. Its architecture and function are not known. Midbody development from late anaphase to the end of telophase can be seen by light microscopy as well as electron microscopy. The more detailed image from electron microscopy has not yet aided our understanding of the functional role of the midbody in cytokinesis of vertebrate and invertebrate cells.

Cell Plate Formation

Cell division in plants shows a common pattern of cell plate formation. Pollen-forming meiotic cells in some flowering plants divide by furrowing, so the process is not found exclusively in animals.

Cell plate formation usually begins during late anaphase or early telophase, a characteristic timing in most cell types with synchronized mitosis and cytokinesis. In plants, the spindle fibers begin to disappear at the poles during telophase, but they remain at the midpoint of the spindle figure. These spindle microtubules usually *increase* in number at the midpoint during formation of the new cell plate and wall. The first signs of the new plate appear as

Figure 13.20
Electron micrograph of thin section through the elongated bridge in furrowing HeLa cell (see Fig. 13.18). The midbody (Mb) is a distinctive region of the furrow at this late stage in cell division. × 67,500. (Courtesy of B. Byers, from Byers, B., and D. H. Abramson, 1968, *Protoplasma* **66**:413–435, Fig. 9.)

Figure 13.21
Cell plate development in plants: (a) Thin section of maize epidermal cell fixed in permanganate. Plate development proceeds from the center toward the periphery of the cell (left to right) and incorporates membranes and assorted vesicles. × 23,500. (Courtesy of H. H. Mollenhauer) (b) *Arabidopsis thaliana* root cell. Note greater development of the phragmoplast (Phr) at the left, compared with more recently formed cell plate nearer the cell periphery (to the right). × 37,000. (Courtesy of M. C. Ledbetter)

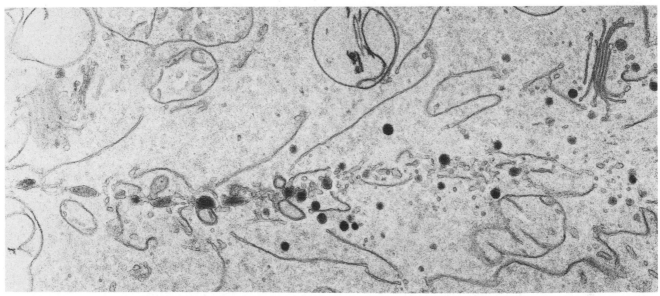

(a)

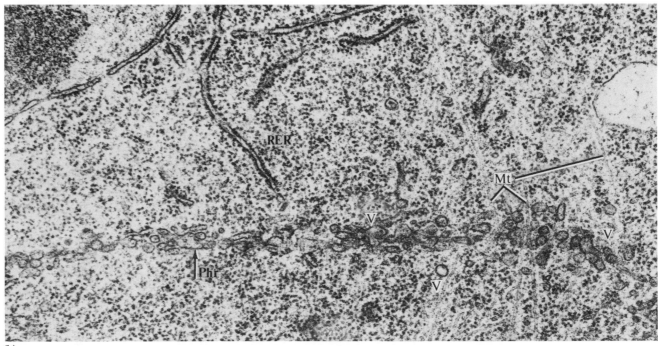

(b)

small vesicles filled with electron-dense material collect in the equatorial plane of the spindle figure. These vesicles are formed from the cisternae of the Golgi apparatus and are filled with carbohydrate and lipid materials that have been processed and packaged within the Golgi region (Fig. 13.21).

Membranous elements of the endoplasmic reticulum also congregate in the vicinity of the newly forming cell plate, along with vesicles and microtubule components. These all coalesce to form the new plasma membrane and outer primary wall coating, which together are called the **phragmoplast**. The phragmoplast is doughnut-shaped at first, when seen from a top view, but it grows outward toward the cell wall as more materials are added and fuse with existing phragmoplast structure. Materials are also added to the phragmoplast in its center "hole" region, until the structure is a solid layer touching at all points of the surrounding cell wall. At this time the division is finished, and daughter nuclei have been separated into individual cell cytoplasms.

Continuities between daughter cells are maintained by fine fibrous strands called **plasmodesmata** (Fig. 13.22). These fibrous connections extend through the new cell plate and cell wall at various places and presumably act as a communication channel. Since plasmodesmata are present throughout a row of cells, there is a passageway that can extend along a considerable length of cells in a vertical file.

Once the daughter cells are separated by the completed cell plate, more wall materials are deposited on the plate surface outside the cells. Wall materials are never deposited on the inner face, since this is plasma membrane. The layers of primary and secondary cell wall are made of secretions produced in the cell. The ways in which wall fibers are arranged contributes to the relative level of flexibility in the cell wall. Primary cell wall fibers are deposited in a

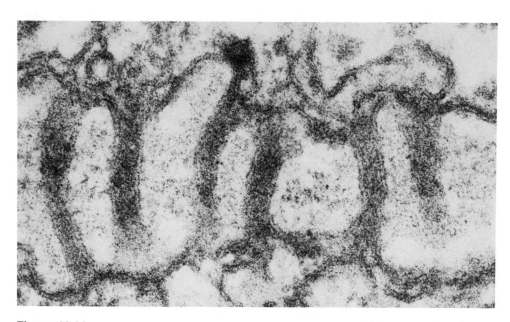

Figure 13.22
Plasmodesmata in *Potamogeton natans* root cells. These continuities between cells provide a passageway for transfer of materials from one cell to its neighbor. × 231,000. (Courtesy of M. C. Ledbetter)

more random pattern, providing greater flexibility for the cell to grow and extend the cell wall at a rate that keeps up with the increasing size of the protoplast. Secondary walls are much more rigid and are usually deposited on cells that no longer grow in length or width. Secondary wall fibers are more highly organized in pattern, and seem to be deposited in a more defined architecture. There is a region of oriented microtubules immediately adjacent to the developing cell wall (Fig. 13.23). The relation be-

tween cytoplasmic microtubules and the deposition of wall fibers is uncertain.

MEIOSIS

Almost every kind of eukaryotic cell can carry out mitosis at least sometime in its existence, but meiosis is a nuclear division process that is highly restricted both to cell type and to time of occurrence in a cell lineage. Only cells of sexually reproducing species

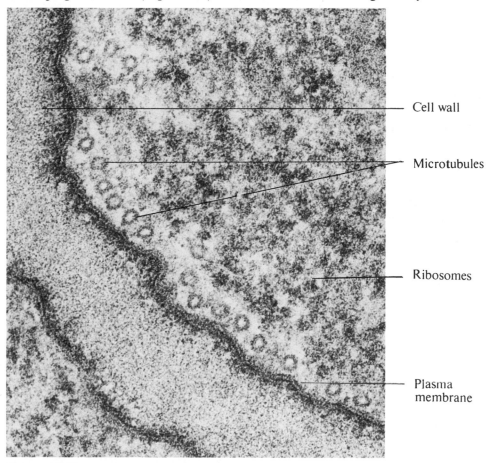

Cell wall

Microtubules

Ribosomes

Plasma membrane

Figure 13.23
Circular cross-sectional views of microtubules arranged next to the cell wall in *Phleum pratense* root tissue. × 180,000. (Courtesy of M. C. Ledbetter)

have the capacity to undergo meiosis, and only special cells in the multicellular individual switch from mitosis to meiosis at specified times in a life cycle (Fig. 13.24). Since meiosis is a *reduction division,* whose end products have half the number of chromosomes as the original mother nucleus, we should suspect that meiosis is related to **gamete** production. In

animals, some lower plants, and various protist and fungus groups meiosis leads directly to formation of gamete nuclei. These nuclei can fuse with other gamete nuclei to produce the **zygote,** which is the initial cell representing the new generation. The fusion nucleus has double the number of chromosomes since one set from each of the two gamete nuclei has merged into a single zygote nucleus.

Higher plants regularly produce **spores** as direct products of meiosis. These spores undergo different degrees of development, depending on the group of plants, and they eventually develop into some gamete-producing system or structure. From such systems, and only from such systems, will gametes then be produced after ordinary mitosis. Plants, therefore, have an intermediate stage in the life cycle, but gametes are eventually formed only from cells that can originally be traced back to a meiotic division. Many variations on the life cycle theme have been described in algae, fungi, and protists.

Every sexual species is recognized by the two decisive events of its life cycle: **meiosis** and **nuclear fusion** to produce a zygote. The kind of fusion called **fertilization** usually refers to union between a larger egg and a smaller sperm. The term has been used in a more general sense, however, and we will use it synonymously with other descriptions of sexual nuclear fusion.

Organs or other structures in which gametes are produced are called **gonads.** The female gonad is an **ovary** or **oogonium** in which eggs are produced. The male equivalent is a **testis** or **spermatogonium** where sperm are formed. The **oocyte** undergoes meiosis to produce one functional egg and three abortive cells, while the **spermatocyte** produces four functional sperm after meiosis is completed (Fig. 13.25). In plants, meiosis takes place in **sporocytes** in a **sporogonium** and spores are the meiotic products (Fig. 13.26). Spores will ultimately develop into egg-producing, or sperm-producing, systems. Because oocytes, spermatocytes, and sporocytes have the unique property of meiosis, all can be considered as a

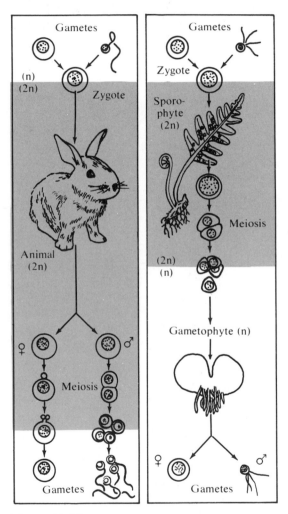

Figure 13.24
Comparison of animal and plant life cycles showing the differences in the time of meiosis in sexual reproduction.

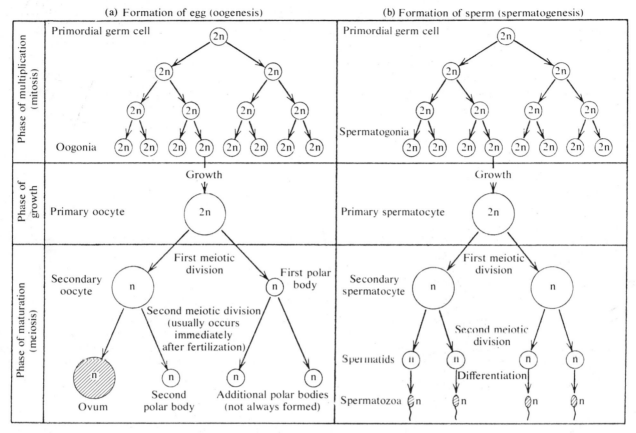

Figure 13.25
In higher animals (a) oogenesis leads to one functional egg per oocyte, whereas (b) spermatogenesis typically results in four functional sperm per spermatocyte at the conclusion of meiosis.

variety of **meiocyte.** It is very convenient in general discussion to avoid the specific term and use only the more general description of a cell as a meiocyte. The sequence of events during meiosis is basically the same in all sexual species and in most meiocytes. There are always some variations, however, as might be expected.

The significant features of meiosis are: (1) reduction of the chromosome number by one-half in cells produced by meiosis, and (2) formation of genetically different gametes from a meiocyte or population of meiocytes. The mechanics of meiosis clearly show how chromosome reduction is accomplished, and we will discuss these features first. Afterward we will consider in some detail the processes of chromosome *synapsis, assortment,* and *crossing over* which lead to a *segregation* of alleles into different meiotic products. All these events contribute to potentially high genetic variability in sexually reproducing species, and to long-term evolutionary advantages.

The Two Divisions of Meiosis

The usual sign of a switch from mitotic to meiotic division activity in a meiocyte is enlargement of the

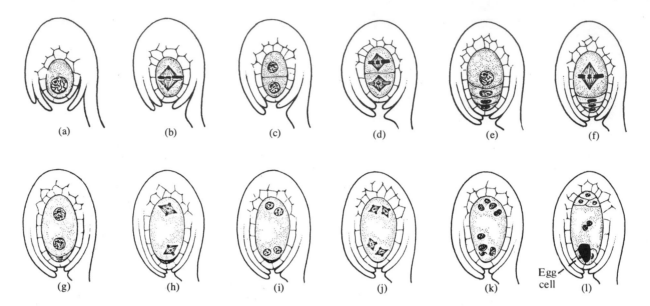

Figure 13.26
Development of the egg cell in flowering plants: (a-e) Meiosis occurs in the megasporocyte within the female structures of the flower; (f-k) three successive mitotic divisions take place in the one functional meiotic cell product, leading to eight nuclei; and (l) one of these eight nuclei is enclosed in the egg cell, while the others may or may not proceed to develop when fertilization has been accomplished. Once the chromosome number has been reduced by one half during meiosis, there is no further change in the haploid number until the egg is fertilized and the diploid zygote has been produced.

nucleus. Beadlike **chromomeres** become especially noticeable along the finely dispersed chromatin, whereas they were much less visible in any preceding mitotic cycle (Fig. 13.27). The S phase of the meiotic cell cycle is about 3 times longer than the S phase of prior mitotic cell cycles in the population. The significance of this observation is uncertain, but there is good evidence that the longer period of DNA synthesis is due to the presence of fewer active replication forks in the chromosome set and not to a slower rate of replication per μm of chromosome length.

DNA replication takes place during premeiotic interphase, in the S phase of the cell cycle, according to autoradiographic and other evidence. Cells which are haploid normally do not undergo meiosis, whereas cells with two or more sets of chromosomes have meiocyte potential. In a typical diploid species, a diploid cell has meiocyte potential. This diploid or 2X amount of DNA doubles to 4X during S phase. There is no further substantial increase in the amount of DNA during the two divisions that make up meiosis. The two divisions may proceed in immediate sequence or after a period of delay in the first of the two divisions. For example, some meiocytes may remain suspended in prophase of the first division and not continue through to the end of the entire process for months or years. In these cases a stimulus is required for the process to resume. For oocytes to resume meiosis, penetration by the sperm stimulates continuation of the oocyte division. After the egg nucleus has formed at the end of meiosis, it fuses with the sperm nucleus that had entered earlier.

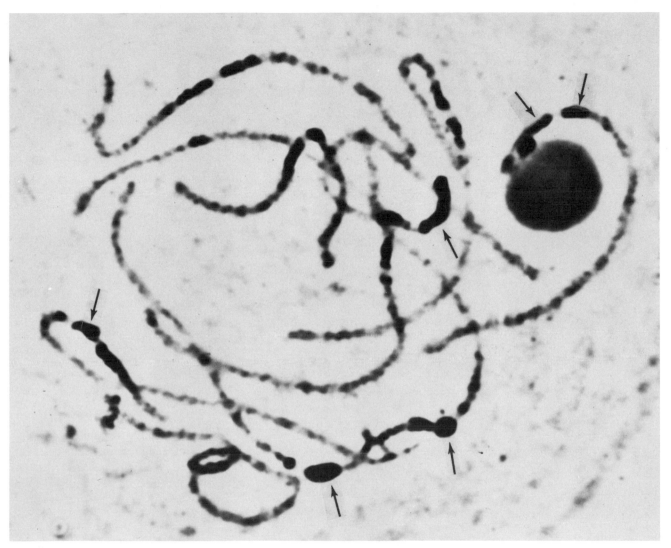

Figure 13.27
Pachytene chromosomes of castor bean (*Ricinus communis*).
Beadlike chromomeres occur along these paired chromo-
somes, along with large heterochromatic knobs (at arrows).
Chromosome 2 is the nucleolar-organizing chromosome
in this complement of ten chromosomes. (Courtesy of
G. Jelenkovic)

Since there is a 4X amount of DNA in the replicated meiocyte nucleus, the meiotic products would have 2X DNA if division stopped after one round. These would still be diploid cells, and they would be equivalent to any body cell in DNA content. By going through a second division, the meiotic products are reduced to the 1X DNA level and are truly haploid cell products. When such 1X cells fuse later, they restore the conventional 2X state of diploid cell DNA. The DNA content of the species remains constant generation after generation because of the cycle of meiosis and fertilization. If meiosis did not precede fertilization, the chromosome number would double endlessly, and this does not happen. Even before meiosis was described in the 1880s, it had already been predicted that a compensating reduction process would be found to explain the constancy of chromosome number in sexual species.

THE FIRST MEIOTIC DIVISION (MEIOSIS I). Meiocyte nuclei proceed through a sequence of prophase, metaphase, anaphase, and telophase intervals during the continuing series of events (Fig. 13.28). We arbitrarily designate stages because of convenience in analysis and discussion. Because the prophase of Meiosis I is the most complex, protracted, and significant interval, it has been further subdivided. The substages of leptotene, zygotene, pachytene, diplotene, and diakinesis are generally recognized as parts of prophase I.[1]

Leptotene. This earliest substage of prophase I is difficult to isolate. The nuclei are highly hydrated and usually suffer some alteration during preparation

---

[1] The suffix -tene designates an adjective, whereas -nema is used in noun forms of a word. There is some preference for leptonema, zygonema, pachynema, diplonema, and diakinesis among biologists because these are more correct grammatically in references to the names of the stages. They retain leptotene and other adjectival terms to describe structures or activities of each stage. For example, "leptonema is the first substage of prophase I and leptotene chromosomes are finely dispersed threads of chromatin." I think it will be less confusing to use the -tene form throughout, even if it is less correct from a grammatical standpoint.

of the material for microscopy. Where it has been possible to fix and stain leptotene nuclei, the chromatin threads are seen as a jumble of tangles. It is almost impossible to follow any single thread for even a micrometer of its length without losing it in a maze of other threads. The replicated nature of the chromosomes is not apparent by light microscopy (Fig. 13.29).

Nucleoli and nuclear envelope resemble the interphase structures, but recent electron micrographs of pollen-forming meiocytes in wheat have shown that striking changes take place in the nature of the nuclear envelope. These changes may be related to aligning of the chromatin at the nuclear envelope and may aid in pairing of chromosomes, which occurs in the following substage. We will mention this aspect later in discussion of chromosome pairing (synapsis). In the light microscope, therefore, leptotene cells are recognized by enlarged nuclei in which there is finely dispersed chromatin. The chromatin characteristically has beadlike chromomeres spaced along the threads, and the threads themselves look unreplicated.

Zygotene. The continuation of prophase I into zygotene is signalled by close pairing of homologous chromosomes. This specific associating process is called **synapsis.** Both sets of chromosomes in a diploid meiocyte have the same gene complements, except for the X and Y or other **sex chromosomes** present. The nonsex chromosomes (**autosomes**) and any kind of sex chromosome present in pairs will begin to synapse at various places along the chromosome length. Synapsis does not seem to begin at a particular part of a chromosome, nor do certain chromosomes pair before the others.

As synapsis takes place a complex structure becomes organized in the space between paired chro-

Figure 13.28
The stages of meiosis outlined in diagram form for a meiocyte containing three pairs of chromosomes. (Adapted from Lewis, K. R., and B. John, 1963, *Chromosome Marker.* With permission of J. & A. Churchill, Ltd., London.)

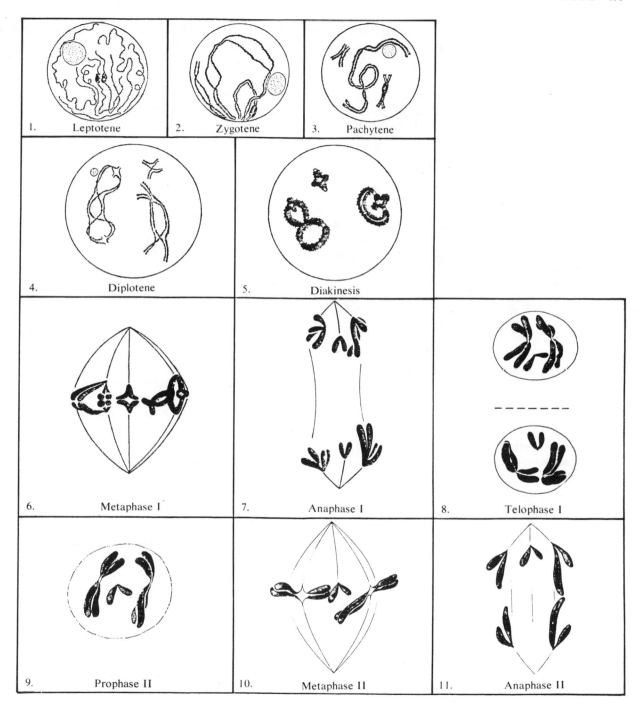

1. Leptotene
2. Zygotene
3. Pachytene
4. Diplotene
5. Diakinesis
6. Metaphase I
7. Anaphase I
8. Telophase I
9. Prophase II
10. Metaphase II
11. Anaphase II

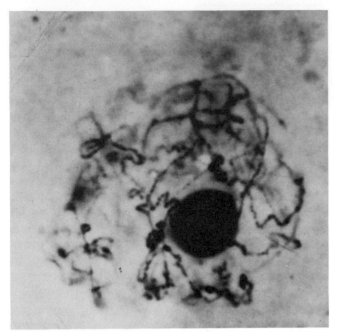

Figure 13.29
Leptotene stage of meiosis in maize. (Courtesy of M. M. Rhoades)

mosomes. This structure is called the **synaptinemal complex,** and it is found uniquely in meiotic cells during early prophase I (Fig. 13.30). When pairing is complete and the synaptinemal complex is fully developed, zygotene is also ended. The paired chromosomes still appear as two single structures since microscopy does not resolve the doubleness of the strands.

Pachytene. This substage is recognized by completely paired homologous chromosomes. The chromosomes have condensed to a greater degree, but do not yet show their replicated condition at this stage. A chromosome pair is called a **bivalent,** and each chromosome of this pair is made up of two replicated halves called **chromatids.** These two terms are very useful in describing the significant events of diplotene and later stages of meiosis.

As pachytene continues the bivalents condense further, but they remain as closely synapsed pairs (Fig. 13.31). In favorable materials the pachytene chromosomes can even be individually identified according to relative length and other features. Far more importantly, there is evidence that exchanges between homolgous chromosome segments, a process called **crossing over,** takes place during pachytene. We will consider crossing over and related events in a later part of this chapter.

Diplotene. This part of prophase I begins with *opening-out* of synapsed chromosomes at various places along their length. The homologues remain associated only at certain places called **chiasmata** (sing.: **chiasma**), which are considered to be sites where previous crossovers had occurred (Fig. 13.32). In favorable materials such as insect spermatocytes, it is clear that *each* chiasma involves only two of the four chromatids at that site of the bivalent. Other chiasmata may involve the same or any other two of the four chromatids, so the total effect is of exchanges involving all parts of the bivalents. Only the individual chiasma is traced to a two-chromatid exchange event.

The diplotene stage may last for weeks, months, or even years in some meiocyte types. Large oocytes from amphibians have an especially striking diplotene chromosome complement (Fig. 13.33). These **lampbrush chromosomes** were so named because of their superficial resemblance to the lampbrush that was a common utensil in the 19th century. The chromatin fibers loop out from a central axis. They are matched pairs of actively transcribing genetic regions. The pattern of loops is constant for each chromosome of a species and serves as one basis for chromosome mapping.

Diakinesis. Chromosomes continue to condense during diplotene and are at their most contracted length by the time of diakinesis. Bivalents are short and thick, chiasmata are evident, and the individual bivalents are relatively well spaced within the nucleus. They are very easily counted at this stage (Fig. 13.34). As diakinesis reaches its conclusion, the nucleoli and nuclear envelope disappear and spindle microtubules are visible within the chromosome area.

Prometaphase. Chromosomes move somewhat erratically, but congression at the midpoint of the spindle eventually takes place. By the conclusion of

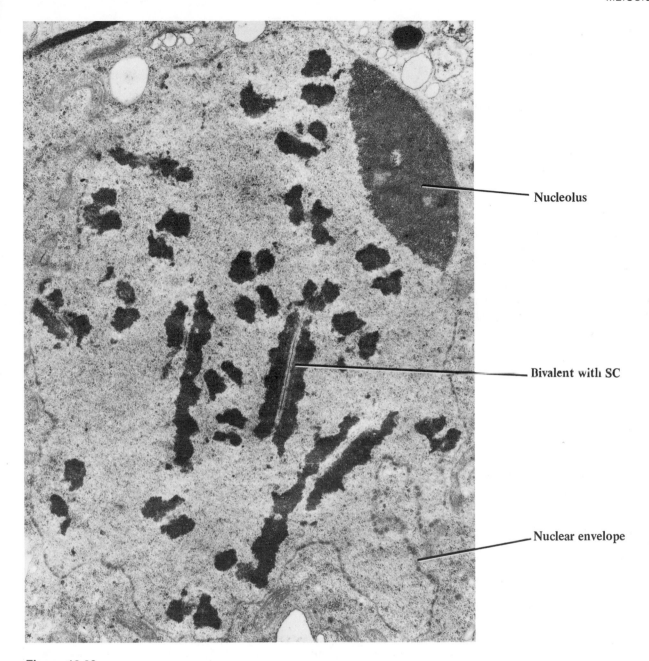

Nucleolus

Bivalent with SC

Nuclear envelope

Figure 13.30
Electron micrograph of thin-section of a meiotic cell of the ascomycetous fungus *Neottiella*. One of the pachytene chromosome pairs has been sectioned favorably to display a synaptinemal complex (SC) in the space between the paired chromosomes of the bivalent. × 16,000. (Courtesy of D. von Wettstein, from Westergaard, M., and D. von Wettstein, 1970, *Compt. Rend. Lab. Carlsberg* **37**:239–268, Fig. 1.)

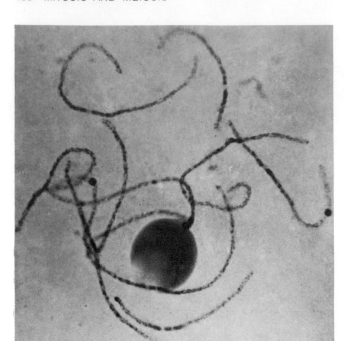

Figure 13.31
Pachytene stage of meiosis in maize. The ten pairs of chromosomes can be identified in such favorable material. (Courtesy of M. M. Rhoades)

this preliminary phase, the bivalents are aligned along the equatorial plane of the spindle, and metaphase I has begun.

Metaphase. In meiotic metaphase the ends of the bivalent chromosome arms are positioned on the spindle equator, and the centromeres of the homologous chromosomes are as far apart as physically possible (Fig. 13.35). This situation is the opposite of chromosome alignment during a mitotic metaphase, when centromeres are aligned at the spindle midpoint, and the chromosome arms wave about in all directions on either side of this zone.

Each chromosome of the bivalent is made of two chromatids, and the centromeres of each sister chromatid pair remain closely associated on the same side of the spindle midpoint. They face the same pole, while the sister chromatids of the homologous chromosome of the same bivalent both face the opposite pole. When anaphase begins, the whole duplicated chromosome separates from its homologue. The duplicated chromosome is called a **dyad,** which describes its double nature and indicates it is composed of two chromatids.

Anaphase. Anaphase begins when homologous chromosomes separate and move toward opposite poles. Each dyad has the 2X amount of DNA, so reduction to haploidy is not yet achieved if we view it from the standpoint of DNA content. On the basis of chromosome number, however, half the number of chromosomes are present at each pole after anaphase as were in the original meiocyte nucleus (Fig. 13.36). Completely haploid nuclei, in terms of DNA content as well as chromosome number, will not be formed until the second meiotic division when individual chromatids of each dyad are separated into individual nuclei. In some species the anaphase I nuclei may enter almost immediately into the second meiotic division. In other species there is a telophase stage beforehand.

Telophase. Chromosome reorganization takes place if this stage occurs in a meiotic sequence. Chromosomes unfold to greater lengths, nucleoli and nuclear envelopes reappear, and daughter nuclei are then evident. Telophase is abbreviated in many species, and nuclei may go directly to prophase or even metaphase of the second meiotic division. In other cases there may be an extended interphase before the reorganized nuclei divide in Meiosis II. An accompanying cytokinesis may or may not occur at the end of Meiosis I. In most cases, cytokinesis is delayed until both divisions are finished, at which time four separate cells are formed to enclose the four nuclear products of meiosis.

THE SECOND MEIOTIC DIVISION (MEIOSIS II). There is nothing particularly striking or unusual about Meiosis II. Nuclei proceed through conventional

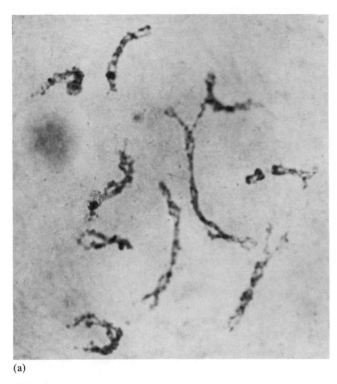

(a)

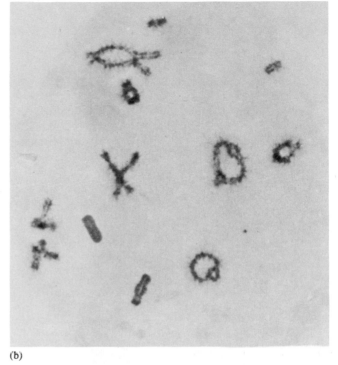

(b)

Figure 13.32
Diplotene stage of meiosis is signaled by opening-out of the paired homologous chromosomes, except at chiasmata.

(a) Maize. (Courtesy of M. M. Rhoades) (b) Grasshopper. (Courtesy of N. V. Rothwell)

prophase, prometaphase, metaphase, anaphase, and telophase, or at least the last few stages of this division (Fig. 13.37). There is considerable variation among species in their M I-M II transition. Meiosis II is usually described as a "mitotic" division because its stages are similar to conventional chromosome distribution events of a mitosis. It is not mitosis, however, but it is the second set of events in a total meiotic division cycle. The main result of Meiosis II is separation of the dyads into individual chromatids, at which time these become full-fledged chromosomes in their own nuclei. At the end of Meiosis II, the DNA content of each nucleus is reduced to the 1X level which matches the haploid number of chromosomes

also present. At this time, each meiotic product has half the DNA content and half the chromosome number of the original meiocyte nucleus. Reduction division is complete only when this point is reached.

Once telophase and cytokinesis are completed, each meiotic product can make its unique contribution to the life cycle. A gamete must first fuse with another gamete of a compatible type before a new individual can develop. Spores, on the other hand, can develop directly into new individuals without prior fusion. In fact, the major distinction between a spore and a gamete is the ability of the spore to develop by itself into a new individual. It doesn't matter whether the spores and gametes are produced by

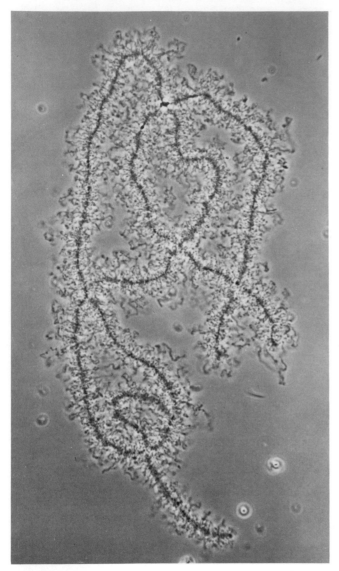

Figure 13.33
Phase contrast photograph of unfixed lampbrush chromosome (diplotene bivalent) from the newt *Triturus viridescens*. Three chiasmata hold the paired chromosomes together. × 440. (Courtesy of J. G. Gall)

mitosis or meiosis, their potential for development is based on characteristics other than nuclear mode of origin.

Meiosis as a Source of Variability

Variability is the hallmark of living systems. New genetic information arises by **mutation** of existing genes, and each gene can therefore occur in alternative forms called **alleles.** The particular combination of alleles in an individual is called its **genotype,** and almost every individual in some populations may have a different genotype. These enormous numbers of genotypes do not arise by mutations in individuals in each generation; they are produced by reshufflings of

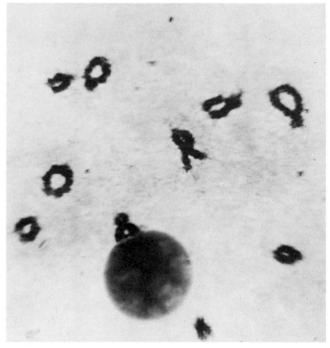

Figure 13.34
Diakinesis in maize meiocyte. Note the nucleolar-organizing chromosome (No. 6 in the complement of ten) associated with the nucleolus. (Courtesy of M. M. Rhoades)

Figure 13.35
Metaphase I in maize. The centromere is oriented closest to the pole for each homologous chromosome of a pair (note spindle fiber attachments). (Courtesy of M. M. Rhoades)

existing allele combinations during sexual reproduction.

All the progeny of an asexual generation are identical since the same genotype has been transmitted by mitosis from parent to offspring. The progenies of sexually reproducing species have different genotypes because there are mechanisms that regularly lead to new combinations of alleles. The principles of **segregation** and **reassortment** of alleles were first postulated by Mendel in 1866 to explain the results of breeding studies. If progenies have genotypes that are different from those of their parents, the inheritance factors must first be segregated and then recombined in different sets (Fig. 13.38). We know that meiosis is the time that homologous chromosomes are segregated into different nuclei. If there were different alleles on one or more pairs of homologous chromosomes, alleles would be segregated into different haploid nuclei at meiosis.

When gametes carrying different alleles fuse to form the new generation, **reassortment** of alleles takes place and the progeny may have different combinations of alleles (genotypes) than were present in their parents. The Mendelian rule of **independent assort-**

Figure 13.36
The dyad (replicated) structure of each chromosome is evident during homologue separation at anaphase I of meiosis in maize. (Courtesy of M. M. Rhoades)

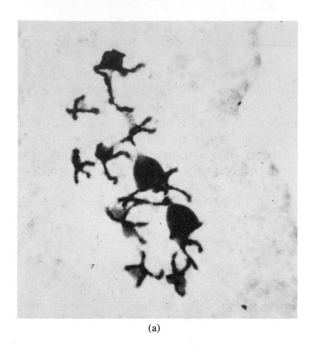

(a)

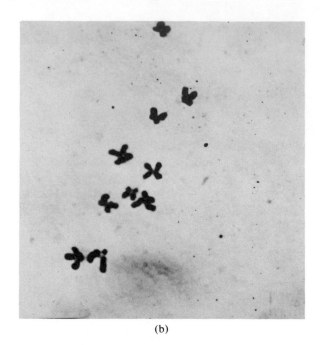

(b)

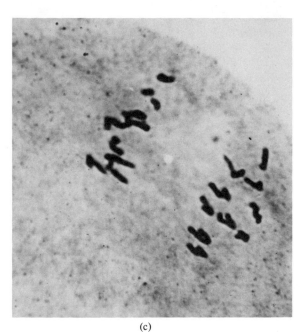

(c)

Figure 13.37

Second division stages of meiosis in maize. Only one of the two nuclei is shown in each photograph, but the other nucleus produced after the first meiotic division would be engaged in identical activities. (a) Ten dyads at prophase II. (b) Ten dyads at metaphase II. (c) Ten monads to each pole at anaphase II. (All photographs courtesy of M. M. Rhoades)

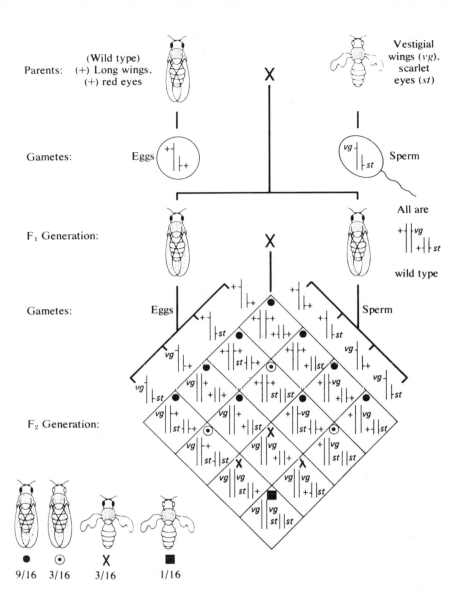

Figure 13.38
Independent assortment. Mendelian inheritance of two different genes each located on a different chromosome leads to the familiar 9:3:3:1 ratio of F_2 progeny phenotypes when mutant alleles act as recessives to their wild-type alternatives.

ment of alleles states that the proportion of each genotype in the progeny depends on random recombinations. This rule has been verified in every case for alleles of genes carried on different chromosomes. As the chromosomes assort at random, so do the alleles they carry. Depending on the numbers of different pairs of alleles and chromosomes, considerable diversity may occur *regularly* in sexually reproducing species. Alleles are segregated regularly at each meiosis and reassorted regularly at gamete fusion.

There may be 5,000 to 10,000 different genes in a species, but there are usually fewer than 50 different chromosomes in the haploid set. How do alleles on the same chromosome become separated and recombined into new genotypes? We know that they do appear in recombinant genotypes that are different from the parental combinations. Not only do **linked** genes (those on the same chromosome) recombine, but they do so with predictable frequency and regularity. Some mechanism must exist that first separates and then recombines alleles originally present within the same chromosome. This mechanism must be a regular feature of the meiotic nucleus since meiosis is the time that alleles are segregated into haploid nuclei.

We know that the mechanism responsible for recombining linked genes is **crossing over,** which involves exchange of homologous chromosome segments (Fig. 13.39). Recombination of genes on different chromosomes takes place by independent assortment, but recombinations between linked genes depend on crossing over. In the next section we will deal specifically with questions concerning the nature, timing, mechanism, and consequences of crossing over between linked genes in sexually reproducing species. Appropriate information about crossing over in viruses and prokaryotes must also be included since we have obtained some very important insights from these systems.

CROSSING OVER

Crossing over between linked genes leads to recombinant genotypes in progenies. We can ask some general questions about crossing over:

1. What is the chromosomal basis for crossing over?
2. When does crossing over take place?
3. What is the mechanism responsible for crossing over?
4. What kinds of evidence indicate that crossovers have occurred?

Once we have discussed these questions we can turn to two other questions about the process:

1. What is the ultrastructural framework for crossing over?
2. Why is crossing over a regular feature of meiosis but not mitosis?

As we answer these major questions, more detailed questions will be raised. Recombinations between linked genes were noted in the first years of this century, but the cytological and biochemical aspects of these events have come under more intensive investigations only during the past 20 years.

Exchange Between Homologous Chromosomes

Since linked genes recombine, some physical basis for the new arrangements of alleles must exist. In studies using chromosomes marked by many known alleles, it seemed that whole blocks of chromosome length must be involved in crossing over. In 1931 two independent reports showed that genotype recombinations were the result of recombined or exchanged chromosome segments. Special stocks of *Drosophila* were used. These stocks had modified X chromosomes that were immediately recognizable by microscopy and gene markers that could be followed in progeny by genetic analysis. C. Stern provided **cytogenetic** (combined cytological and genetic) evidence for the chromosomal basis of crossing over by doing experiments with these special *Drosophila* strains. A very similar study using special strains of corn with cytological and genetic chromosome markers was reported by H. Creighton and B. McClintock. Since results were similar for an animal and a plant species, the conclusions were accepted as being generally applicable to other plants and animals.

Each *Drosophila* female parent in Stern's experiments had two modified X chromosomes; one was very short, and the other was longer because a piece of another chromosome was attached to it. Each X chromosome contained different alleles for the two genes that were to be followed in the genetic part of the analysis:

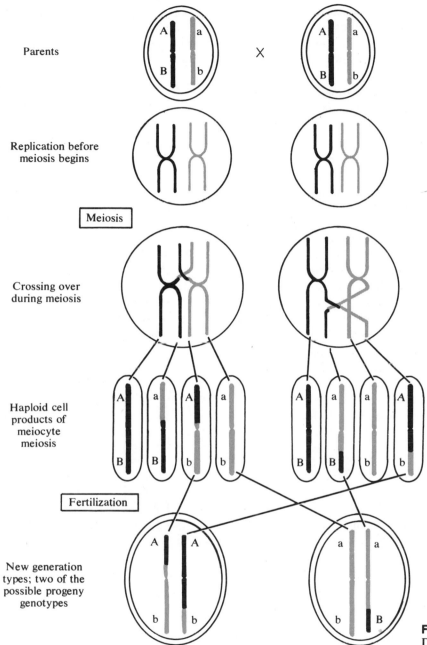

Parents

Replication before
meiosis begins

Meiosis

Crossing over
during meiosis

Haploid cell
products of
meiocyte
meiosis

Fertilization

New generation
types; two of the
possible progeny
genotypes

Figure 13.39
Diagram illustrating recombination of
linked genes as the result of crossing
over during meiosis.

1. The short X contained the recessive allele for carnation eye color (*car*) and the dominant mutant allele for bar eye shape (*B*).
2. The long X contained the wild-type alleles for both genes (++).

The male parents had normal length chromosomes, and the X was marked by *car* and the wild-type allele (+) for eye shape. The four kinds of genotypes in their progenies were examined for chromosome constitution in relation to alleles carried on these chromosomes. Since each female receives one X from each parent, they all must have received the one normal *car* + from the male parent. If an unchanged X had been received from the mother, then the daughters with the *carnation, bar eye* phenotype would have one short X, and the daughters with the *wild-type* phenotype would have one very long X chromosome (Fig. 13.40). Daughters that had *bar-eyed* or *carnation* phenotypes had the recombinant maternal X chromosome.

When these genetic recombinant females were examined cytologically, each had an X chromosome that could only have arisen in their mother if homologous chromosome segments had undergone an exchange. Crossing over thus involves an exchange of homologous chromosome segments and leads to genetically recombinant progeny.

Timing of Crossing Over

The studies by Stern and Creighton and McClintock did not indicate the time of crossing over, however, and this was an important question then as now. Recombinant chromosomes could have been formed before, during, or after the chromosomes had replicated. It was not known in those days that DNA replication took place in premeiotic interphase, or even that DNA was the genetic material. The question was technically difficult to approach because progeny were random samples of meiotic products which had undergone many stages of change and development before they were examined either genetically or cytologically.

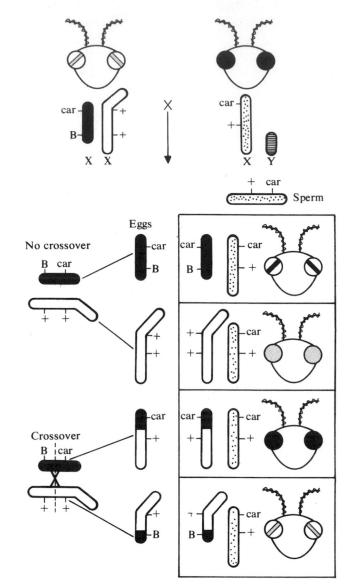

Figure 13.40
Summary diagram of Stern's classic cytogenetics experiments relating chromosome exchange to genetic recombination in *Drosophila*. See text for details.

An experimental approach to determining the time of crossing over became possible in the 1940s. *Neurospora* had come into wide use as an experimental organism for genetic studies, and a number of alleles were known for the species. The enormous advantage of *Neurospora* as an experimental organism was that all the products of a single meiocyte could be identified, and many sets of individual meiocyte products could be analyzed in a single genetic cross. This fungus belongs to the Ascomycetes, a group which produces **ascospores** by meiosis in the sexual phase of the life cycle (Fig. 13.41). Since the meiotic products are contained in a sac, called an *ascus* (plural: *asci*), it is a relatively simple matter to see what the genotypes of each ascospore set may be.

The experiment was designed to see whether crossing over takes place before or after DNA replication is completed. If DNA replicated first, then there would be four chromatids per bivalent at meiosis. It was known that only two chromatids were involved in a single crossover event, so asci were examined to see if they included *both* parental and recombinant spores (Fig. 13.42). If crossing over took place before DNA replication, then asci would contain *either* recombinants or parentals, but not both. Since the results of this experiment showed that both

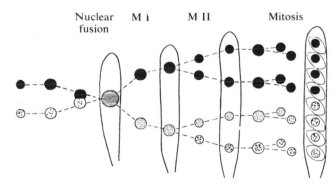

Figure 13.41
Ascospore formation in *Neurospora crassa*. Each of the four haploid nuclei produced by meiosis undergoes a mitotic division which results in eight nuclei, each enclosed in an ascospore within the ascus.

parental and recombinant spores could be present in a single ascus, it appeared that chromosomes had replicated and therefore crossing over had occurred in the *"four-strand" stage*. This term is widely used to describe a bivalent with replicated chromosomes, that is, with four chromatids.

It was discovered later that DNA replication was completed in the haploid nuclei of the fungus long before these nuclei fused to form the diploid meiocyte nucleus. In other words, crossing over during meiosis *must* take place after DNA replication in these ascomycetes, because the chromosomes are already double structures at the time of the sexual fusion, which precedes meiosis. Such information, however, was not available in the 1940s.

The Breakage and Reunion Mechanism

Two major hypotheses to explain crossing over had been proposed in the 1930s. In 1931 J. Belling proposed a **"copy-choice" mechanism** which stated that the beadlike chromomeres "split" and were joined together in different combinations depending on the way in which the new connecting strands switched back and forth during chromosome replication (Fig. 13.43). In 1937, however, C. Darlington proposed that chromatids were broken during diplotene. Then, the rejoining of the broken ends would produce recombined chromatids, if these were closest together. Both hypotheses required that chromosome replication take place during mid-prophase of meiosis, which was the time they could first see that chromosomes were double structures. We now know these ideas were incorrect with regard to timing, but aside from the detailed inaccuracies of each hypothesis, each could be revised enough to be used again in a modified version in the 1950s and later.

The copy-choice mechanism has enjoyed relatively little general acceptance, but it has been used again and again to interpret unusual recombination situations. Any copy-choice mechanism requires the occurrence of DNA synthesis *during* crossing over. It cannot be the major mechanism, if it occurs at all,

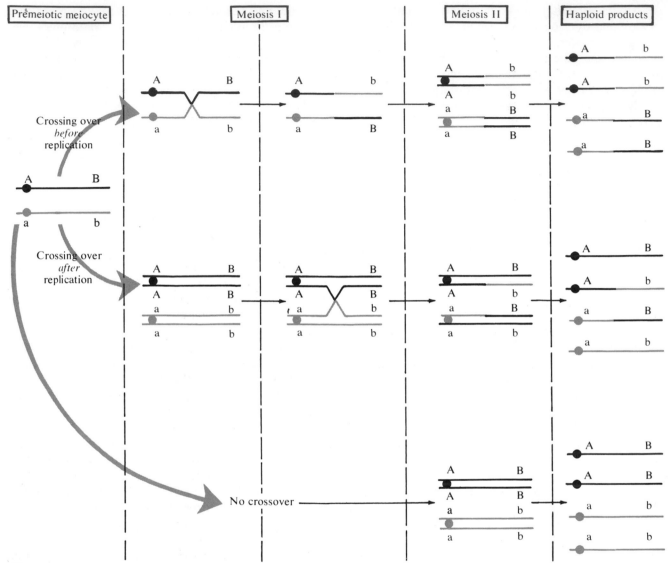

Figure 13.42

Diagrammatic summary of *Neurospora* experiments designed to determine the time of crossing over relative to the time of DNA replication during meiosis. Asci containing only parental (noncrossover) combinations would be formed according to either hypothesis, but asci containing both parental and recombinant types were also found, as predicted by the hypothesis that crossing over occurred after DNA replication.

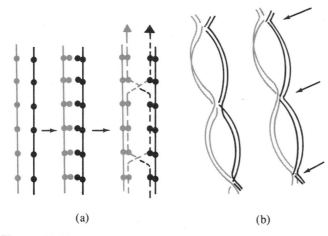

(a)   (b)

Figure 13.43
Diagrammatic illustration of two postulated mechanisms of crossing over: (a) copy-choice, according to Belling; and (b) breakage-and-reunion (at arrows), according to Darlington.

because there are some species whose haploid nuclei are already replicated before sexual fusion takes place. Crossing over must take place after replication in these species and in other meiocytes as well, because it is known that DNA replicates in premeiotic S phase of diploids.

THE PHAGE LAMBDA EXPERIMENTS. Major experimental support for crossing over by the mechanism of **breakage and reunion** was provided in 1961 by M. Meselson and J. Weigle. They used strains of phage lambda (λ) which were genetically marked with alleles of two genes and chromosomally marked with isotopes of carbon and nitrogen. The progenies could, therefore, be examined cytogenetically since parallel observations of genetic recombinants and recombinants with marked chromosomes could be made. In this case, unlike studies of *Drosophila* or corn in which chromosomes are examined by microscopy, "chromosomes" were identified by the centrifugation behavior of phage DNA in density gradients of cesium chloride. These two experimental situations are entirely comparable, however.

One phage lambda strain was the double mutant *cmi*, and the other had the two wild-type alleles, ++. Experiments were done reciprocally, so that the mutant strain was grown in $^{13}C^{15}N$ media, while the wild-type was grown in normal $^{12}C^{14}N$ media, or vice versa. We will follow the $++^{13}C^{15}N \times cmi^{12}C^{14}N$ crossing experiment, although results were similar in the reciprocal cross. One other feature of the experimental situation was that the two genes were situated near each other far to one end of the chromosome. This feature provided an important reference point for subsequent interpretation of the crossing over experiments.

Once the phage strains were prepared, they were introduced by multiple infection into their *E. coli* host cells which were growing in normal $^{12}C^{14}N$ media. These media will be called the "light" media, or isotope combination, and $^{13}C^{15}N$ will be called "heavy" in further references. Each of the two possible mechanisms for crossing over predicts different and mutually exclusive outcomes for the nature of the recombinant phage progeny. If crossing over is by copy choice, then crossing over occurs during DNA replication. *All* recombinants should be "light" in this case, since all would be produced from new DNA made in "light" medium. The recombinant strands would be products of switching back and forth between the two parental DNA strands.

If breakage and reunion produces recombinants, then at least some recombinants should have parts of both parental strands and their corresponding isotope labels. In this experiment genetic recombinants with recombined isotopically-marked DNA were found (Fig. 13.44).

The results point very strongly to breakage and reunion, but some interpretations of the results can be considered ambiguous. Phage replication takes place in repeated rounds during infection, so the progeny included some phage with essentially unreplicated DNA, some with one strand of the duplex from the original parent and the other strand of newly synthesized DNA (semiconservatively replicated

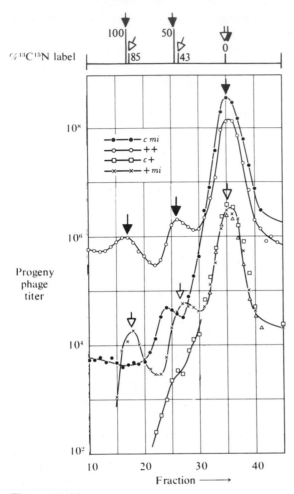

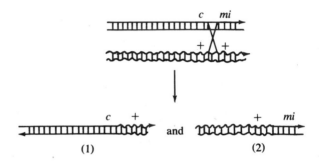

Figure 13.44
Results of the cross between $++^{13}C^{15}N$ and $cmi^{12}C^{14}N$ using phage lambda. Arrows at selected DNA peak fractions in the gradient display indicate the percentages of $^{13}C^{15}N$ label incorporated in the duplex DNA molecules. These arrows are also shown in the numerical scale at the top of the diagram. All four combinations of alleles occur preponderantly in the unlabeled fractions, having arisen after repeated rounds of replication in unlabeled medium. The 100-percent and 50-percent labeled peak fractions of the $++$ class represent unreplicated and semiconservatively replicated first generation molecules, respectively, as do the 85-percent and 43-percent labeled peak fractions of the $+mi$ recombinant class. See text for details. (Reproduced from Meselson, M., and J. J. Weigle, 1961, *Proc. Natl. Acad. Sci.* **47**:857–868, Fig. 4.)

duplex DNA), and most with both strands from newly polymerized DNA. Most progeny were, therefore, totally unlabeled, whether they had parental or recombinant DNA. The crucial progeny classes are the two recombinants $+mi$ and $c+$, whose DNAs have retained some "heavy" parental strands. If a break occurred between the c and mi gene loci and rejoining produced a recombinant duplex molecule, then the result would be

Because the genes are near one end of the chromosome, the break and rejoin would produce $c+$ recombinants, with about 15 percent "heavy" label, and $+mi$ recombinants, with about 85 percent label. Some $+mi$ recombinant DNA was found. The $c+$ recombinant could not be clearly measured for density, but it appeared in a low-density part of the gradient (see Fig. 13.44). In addition, $+mi$ recombinants with somewhat less than 50 percent "heavy" label were found in a well-defined density peak.

These two classes are significant. The 85 percent "heavy" $+mi$ recombinants must have been formed *without* significant DNA replication, since any new DNA would be "light." The remainder of the "light" component could have come from DNA of the parental cmi, which rejoined after the initial breakage (recombinant chromosome #2 shown above). Recombinants with somewhat less than 50 percent "heavy" DNA must have undergone one semiconservative replication in "light" medium. The duplex DNA with one original recombined $+mi$ strand and one newly-synthesized complement would have half the density of an unreplicated recombinant duplex DNA.

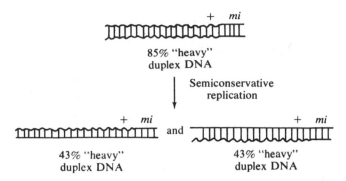

85% "heavy" duplex DNA

Semiconservative replication

43% "heavy" duplex DNA and 43% "heavy" duplex DNA

Even though phage systems are different from those in eukaryotic chromosomes, genetic and linkage activities are similar. There is no reason to believe that recombination of linked genes would occur by any basically different crossing over mechanism in phages, prokaryotes, or eukaryotes. The results from the lambda experiments were accepted as showing that the probable mechanism of crossing over in any genetic system was breakage and reunion.

DNA REPAIR-SYNTHESIS. The understanding of DNA replication and its requirements of enzymes, unwinding protein, RNA primer, and other factors leads to the belief that a similar set of requirements must exist for breaking and rejoining DNA in crossover events. Breaks in DNA are caused by **nuclease** action, addition of new stretches of nucleotides requires **polymerases,** and sealing of broken ends calls for **ligase** activity (Fig. 13.45). If single-stranded regions are to be held straight enough for synthesis to occur at a replicating fork or in another polymerizing situation, an unwinding protein would be needed to keep the strand from kinking up. Evidence that DNA synthesis occurs during crossing over could be obtained by assaying for these proteins or their activities, as well as by looking for incorporation of radioactive precursors into polymerized DNA segments. All the requirements for DNA synthesis have been found in meiocytes of lily plants, the most intensively investigated system to date. Similar components of the DNA synthesis system have also been found in meiocytes from mammalian and other animal species. A great deal of the understanding of this system has come from studies by H. Stern and his associates.

Lily buds provide ideal material for biochemical studies of meiosis because they can be grown in quantity under controlled conditions, and their meiotic activities are essentially synchronized in all male meiocytes of a bud. From previous studies, it was known that lily flower buds of a particular length are engaged in a particular stage of meiosis. Enough bud material was collected to conduct biochemical studies of the individual stages of meiosis and of different times during one of these stages.

About 0.3 percent of total nuclear DNA in lily bud cells is synthesized during zygotene, after 99.7 percent of the DNA replication has occurred in the premeiotic S phase. Sensitive tests indicate that this **zygotene-DNA** is synthesized in long pieces, and it hybridizes with DNA in all the chromosomes. These observations indicate there is no one unique place for DNA formation during zygotene. If zygotene-DNA synthesis is inhibited, pairing of homologous chromosomes will not take place, and crossing over is therefore prevented. In this situation, no chiasmata are seen in later prophase I, an observation which shows crossing over has not occurred.

The function of zygotene-DNA is undetermined, as yet, but it is isolated in a complex together with nuclear membrane lipoproteins. This complex is very transient and disappears by pachytene. Stern has suggested that zygotene-DNA may play an active role in initial stages of chromosome synapsis, but there is little evidence to support this idea at present.

Synthesis of DNA also occurs during pachytene, but it is different from zygotene-DNA patterns. **Pachytene-DNA** is made in very short pieces, and its synthesis is balanced by degradation of existing DNA. No *net* synthesis of DNA occurs, since the amount formed is balanced by the amount lost during pachytene. If pachytene-DNA synthesis is inhibited, the

(a)

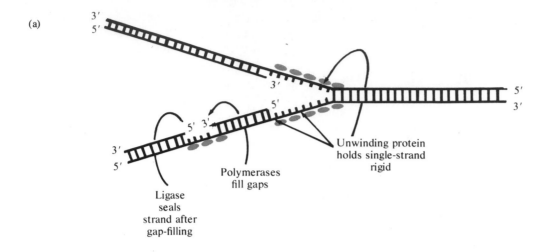

3′
5′

3′

5′

3′
5′

5′
3′

5′ 3′

Unwinding protein
holds single-strand
rigid

Polymerases
fill gaps

Ligase
seals
strand after
gap-filling

(b)

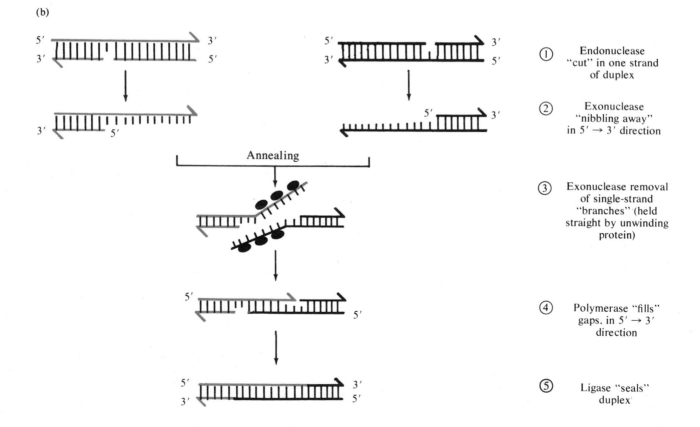

5′　　　　　　　　　3′
3′　　　　　　　　　5′

5′　　　　　　　　　3′
3′　　　　　　　　　5′

① Endonuclease
"cut" in one strand
of duplex

3′　　　　　　5′

5′　　3′

② Exonuclease
"nibbling away"
in 5′ → 3′ direction

Annealing

③ Exonuclease removal
of single-strand
"branches" (held
straight by unwinding
protein)

5′　　　　　　　　　5′

④ Polymerase "fills"
gaps, in 5′ → 3′
direction

5′　　　　　　　　　3′
3′　　　　　　　　　5′

⑤ Ligase "seals"
duplex

chromosomes fragment. This DNA is therefore essential to maintain or restore the integrity of the chromosome. Synthesis of pachytene-DNA resembles synthesis of **repair-DNA** during gene recombination in other systems or during repair of radiation-damaged DNA in *E. coli* and eukaryotic cells.

One particular deoxyribonuclease has been found exclusively in meiocytes. This enzyme "nicks" one strand of duplex DNA molecules, causing breaks in the sugar-phosphate backbone of DNA. The enzyme is first detected in late zygotene. It reaches a peak of activity in pachytene and is not detectable by the end of pachytene.

Nucleotide incorporation into DNA during pachytene has been shown by radioactivity studies. The filling in of gaps in broken strands of DNA requires polymerases, and the broken ends must be joined with newly-synthesized stretches of DNA. This sealing action is carried out by a ligase which has been found in lily meiocytes during pachytene.

The short fragments of DNA found during pachytene have disappeared by the end of this prophase substage. These pieces presumably have been made part of repaired DNA and been sealed into a continuous strand of duplex DNA by ligase action, which could account for their disappearance.

One protein isolated from lily buds is very similar to the unwinding protein first studied in phage T4 systems (see Chapter 5). The T4 protein is essential for both replication and recombination in T4-infected bacterial cells, and the eukaryote protein seems to perform similar functions in lily and mammalian meiocytes. The eukaryote protein appears first during leptotene, increases to a steady level in zygotene, and

Figure 13.45
Diagram illustrating one of the postulated mechanisms to explain breakage and reunion events during crossing over in virus and prokaryote DNA molecules: (a) Composite diagram showing some of the proteins required for DNA synthesis; (b) flow scheme showing breakage and reunion events followed by DNA-repair synthesis during crossing over leading to genetic recombination.

disappears by the end of pachytene. Since this protein is found in association with nuclear membrane fractions isolated from meiocytes, Stern considers its function to be closely related to pairing of DNA during synapsis in zygotene. We will, therefore, mention the unwinding protein again when we consider the role of synapsis in crossing over.

The evidence from lily studies provides very strong support for the activity of a repair-DNA synthesis system in the breakage and reunion mechanism that finally produces recombinant DNA. Whether the repair is a "cut-and-patch" or a "patch-and-cut" operation remains to be determined (Fig. 13.46). General opinion favors a "cut" by endonuclease action, followed by the filling in of gaps in DNA, which are finally sealed by ligase. The "cut-and-patch" scheme is therefore the more likely of the two alternatives.

Chromosome Synapsis: Prerequisite for Crossing Over

Exchanges of homologous chromosome segments take place regularly in meiocytes but not in mitotic cells. In seeking some way of explaining the basic distinctions between these nuclear division systems, one immediately obvious difference is that synapsis occurs between meiotic chromosomes but does not happen in diploid nuclei of somatic cells. It would only be logical to expect that some mechanism must bring together the homologous chromosomes, stabilize their paired association, and activate the breakage and reunion mechanism of crossing over. The regularity of synapsis matches the regularity of crossing over occurrences. Synapsis begins in zygotene and ends at the beginning of diplotene. It is precisely during zygotene and pachytene that crossing over is thought to take place. The chiasmata seen at diplotene are considered to be evidence of prior crossovers between synapsed chromosomes of pachytene bivalents.

We have mentioned several times that chiasmata are sites of previous crossovers, but what evidence is there to support this interpretation? One convincing kind of evidence comes from documented

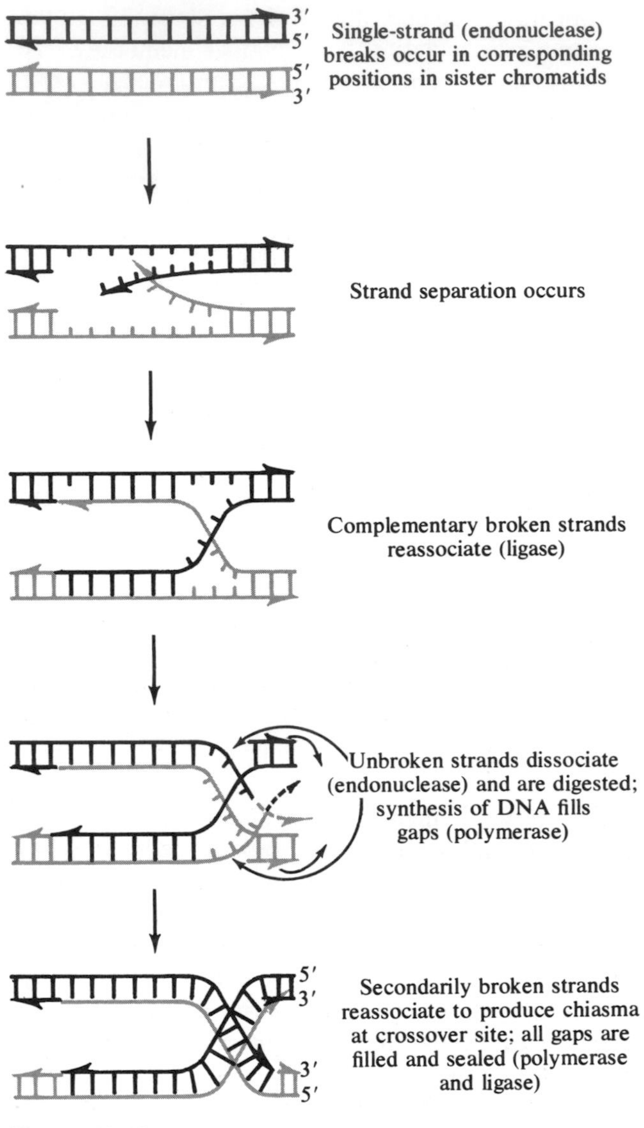

Single-strand (endonuclease) breaks occur in corresponding positions in sister chromatids

Strand separation occurs

Complementary broken strands reassociate (ligase)

Unbroken strands dissociate (endonuclease) and are digested; synthesis of DNA fills gaps (polymerase)

Secondarily broken strands reassociate to produce chiasma at crossover site; all gaps are filled and sealed (polymerase and ligase)

Figure 13.46
Molecular model for crossing over and chiasma formation in eukaryotic bivalent chromosomes, according to R. Holliday. Enzymes involved in breakage and subsequent rejoining by DNA-repair synthesis are indicated in parentheses.

observations of many normal and abnormal meiocytes. In almost every case where synapsis is a regular prophase I event, chiasmata are seen at diplotene to metaphase I (Fig. 13.47). A few important exceptions to this phenomenon are seen among some insect species.

Chiasmata are absent in mutants which show no chromosome pairing (*asynapsis*). Unpaired chromosomes (**univalents**) can be easily identified. Even if loose pairing is observed, no chiasmata are formed in the paired regions of a bivalent. Mutants that undergo synapsis at zygotene but then become *desynaptic* also have no chiasmata. The pairing apparently does not last long enough for crossing over, or it does not help in subsequent chromosome exchanges.

Wherever it is possible to analyze genetic recombination along with cytological observations of chiasmata and synapsis, the analysis shows that gene recombinations are produced only from meiotic systems in which synapsis and chiasma formation are normal. The classic studies of *Drosophila* and corn in 1931 (see Fig. 13.40) clearly showed that chiasmata in paired X chromosomes were correlated with the production of gene recombinants. A more striking example is furnished by comparisons between *Drosophila* males and females. Crossing over takes place only in females and not in males, and chiasmata are visible only in oocytes and not in spermatocytes of this insect. Although chromosomes in spermatocytes are paired, their association is very loose and is not the same as that in bivalents in oocyte nuclei.

A direct correlation between chiasma formation and crossing over by breakage and reunion was demonstrated in 1966 by P. Moens. He followed the pattern of radioactive label in individual chromatids of diplotene bivalents in grasshopper spermatocytes. Chromosomes were fully labeled with tritiated precursor *two* cycles before meiotic DNA replication in the *S* phase. The chromosomes were then allowed to replicate for the next two cycles in unlabeled medium. Assuming semiconservative replication occurred, meiotic bivalents should have only one

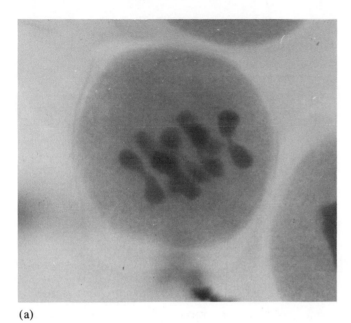

(a)

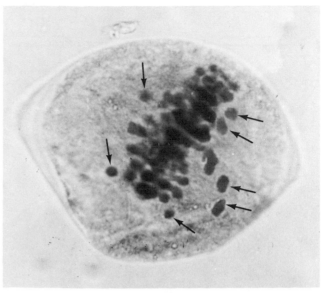

(b)

Figure 13.47
Metaphase I in meiocytes of spring beauty (*Claytonia virginica*) flower buds: (a) cell from a diploid plant, with eight bivalents about to disjoin. (b) Cell from a polyploid plant, with about ninety-eight chromosomes. Most of the chromosomes are bivalents or multivalents, but some un-paired univalents (arrows) have failed to align on the equatorial plane of the spindle. × 1,200. (Courtesy of N. V. Rothwell)

radioactively labeled chromatid in each chromosome pair (Fig. 13.48). By direct observations of diplotene bivalents, it was possible to decide whether a chiasma was due to simple overlapping of one chromatid on another or to prior breakage and reunion. The particular distribution of radioactivity clearly showed that chiasmata were not the result of chromatid overlaps but were caused by physical exchange of chromatid segments. A chiasma is, therefore, the site of a previous exchange event, and it can serve as evidence of crossing over in meiosis.

The Synaptinemal Complex

In 1956 M. Moses first described the **synaptinemal complex** as a three-layered structure that was present in bivalents of crayfish meiocytes (Fig. 13.49). This structure has since been found in pachytene nuclei of all the major eukaryote groups. It is only visible with the electron microscope, and its formation and fate have been followed by ultrastructure analysis. One **lateral element** is produced by each homologous chromosome. The entire synaptinemal complex is formed when these lateral elements are joined and held in register by a less dense **central region** in which there may be a **central element** and a number of fine strands that cross the 1000 Å width of this region. Variations in widths and patterns of the lateral elements and central region have been described for various species, but the basic construction is the same.

Many observations point to the prerequisite of proper formation of synaptinemal complex for chiasma formation and, by implication, for crossing over, which leads to chiasmata. Only some of the evidence will be mentioned here:

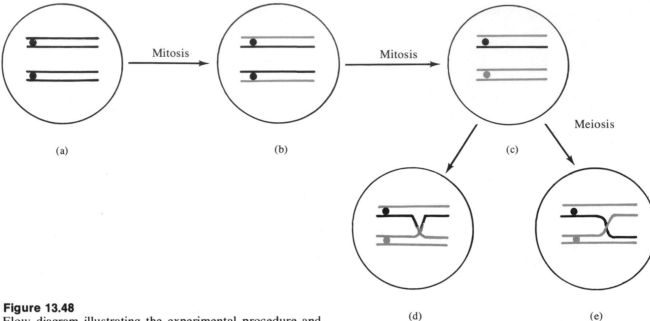

(a) (b) (c)

Mitosis Mitosis Meiosis

(d) (e)

Figure 13.48
Flow diagram illustrating the experimental procedure and results reported by P. Moens that demonstrated that a chiasma was the site of a previous crossover event by a breakage-and-reunion mechanism. (a) Cells were grown in labeled medium until DNA was fully-labeled. (b) After one generation and semiconservative replication of DNA, all the chromosomes are labeled in one of the two chromatids (half-labeled). (c) After a second generation and another round of DNA replication the radioactive label is present only in one chromatid out of the four in a bivalent (quarter-labeled). (d) Predicted appearance for a diplotene bivalent if a chiasma is the site of a breakage-and-reunion event. (e) Predicted appearance of a diplotene bivalent if a chiasma is due to chance overlapping of chromatids. (f) Autoradiograph of grasshopper diplotene bivalents. The bivalent with the chiasma shown at the arrow shows distribution of radioactive label as predicted according to breakage and reunion (see d). (Photograph courtesy of P. B. Moens, from Moens, P. B., 1966, *Chromosoma* **19**:277–285, Fig. 5.)

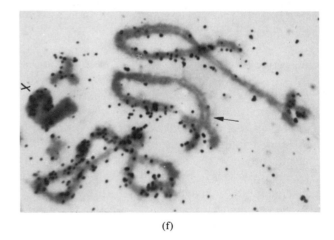

(f)

1. The synaptinemal complex (SC) forms during zygotene, is continuous from one end of the bivalent to the other by pachytene, and is shed from bivalents at diplotene, except for modified segments that remain at the chiasmata.
2. Asynaptic mutants which show no recombination or chiasmata are likewise lacking in SC formation.
3. Inhibition of protein synthesis during zygotene stops SC formation, synapsis does not take place, and chiasmata are not formed.
4. Colchicine treatment during leptotene or very

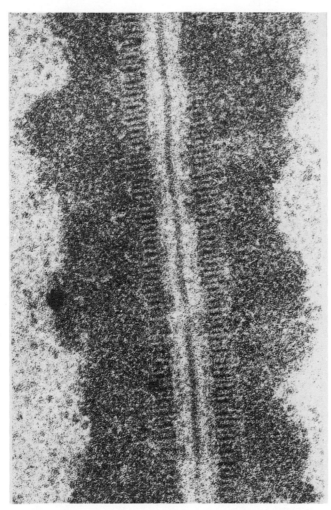

Figure 13.49
Longitudinal section through bivalent with synaptinemal complex at pachytene of meiosis in *Neottiella*. Sandwiched between the chromatin of the two homologous chromosomes is a banded lateral element associated with each chromosome and a lighter central region in which a dense central element is situated. × 90,000. (Courtesy of D. von Wettstein, from von Wettstein, D., 1971, *Proc. Natl. Acad. Sci.* **68**:851–855, Fig. 1.)

early zygotene inhibits SC formation, even though lateral elements have already appeared on the unpaired chromosomes, and prevents synapsis and chiasma formation.

5. If inhibitors are present until late zygotene or early pachytene and then removed, SC formation is not reinitiated, pairing does not take place, and chiasmata are not formed.

Lateral elements form during leptotene and become parts of the SC that acts to stabilize pairing between homologous chromosomes. If SC integrity is destroyed, paired chromosomes separate from each other. Once begun, pairing must be maintained if crossing over and chiasmata are to be found. The SC itself *does not initiate* the pairing process, but it does *stabilize* the bivalent after homologous regions have become associated. The presence of the SC does not guarantee that crossing over will take place. In *de*synaptic mutants, SC formation takes place, but the bivalents separate precociously and chiasma frequency is sharply reduced or completely absent. A species of mantid (related to grasshoppers) has been found to form perfectly normal SCs for all bivalents, but chiasmata do not form. Male *Drosophila*, on the other hand, form no chiasmata in their spermatocytes, and one report stated that loosely paired chromosomes had no SC structures.

In general it would seem that SC formation is necessary to stabilize closely paired chromosomes during pachytene. SC formation usually provides the physical closeness for crossing over to take place, but there must be other factors that trigger crossing over in pachytene bivalents since presence of SC does not guarantee that crossing over will take place. The desynaptic mutants and the achiasmatic mantid species provide evidence for separable processes that lead to SC formation and to crossing over afterward. In most cases the two processes are synchronized, or perhaps normal SC formation triggers normal crossing over events. We are just beginning to find answers to some of these puzzling questions.

The regularity of crossing over in meiocytes and erratic occurrences in somatic cells is directly correlated with presence and absence of SC, respectively. Even when chromosomes pair in normal and intimate association in somatic cells, as in the polytene nuclei of *Drosophila* and other dipteran insects, there is no evidence of SC structures. Chromosome pairing, by itself, is no guarantor of crossing over, since some kinds of somatic cells show no chromosome pairing and still have low but constant frequencies of crossing over.

For example, crossing over in somatic cells of the fungus *Aspergillus* has been detected in about 2 percent of the mitotic cells analyzed. These crossovers are common enough for construction of chromosome maps, which match those obtained by conventional genetic analysis of meiotic cells. The results of crossing over would therefore appear to be the same in mitotic and meiotic cells, but frequencies differ and the ultrastructural correlates of crossing over are also different. These observations strengthen the idea that the *regular* occurrence of crossing over in meiocytes is largely due to synaptinemal complex development. The SC stabilizes the paired chromosomes at a constant distance of about 1000 Å and may even activate crossing over under appropriate conditions.

One asynaptic mutant of *Drosophila* shows no chromosome pairing, synaptinemal complex formation, chiasmata, or genetic recombination in meiotic cells. Crossing over in somatic cells of this mutant is unaffected, since somatic cell recombinants are produced in the usual low frequencies. This mutant provides additional evidence that the processes of synapsis and crossing over in meiocytes are under different controls from somatic cell processes that lead to gene recombinants.

ALIGNMENT OF HOMOLOGOUS CHROMOSOMES. The synaptinemal complex is the final product of chromosome synapsis, but it is formed *after* chromosomes have begun to pair. The mechanisms responsible for bringing chromosomes into alignment and for matched pairing must be separate from processes that contribute to the final synaptinemal complex. How do homologous chromosomes "find" each other? What are the initial steps in chromosome pairing? Some tentative answers have been suggested in the past few years.

All major groups of eukaryote meiocytes have been found to have pachytene bivalents firmly attached to the inner nuclear membrane (Fig. 13.50). In most cases, the two ends of each bivalent are attached all around the membrane, but some animals have a polarized region of attachment with all the ends of the bivalents at a part of the nuclear membrane that is opposite the centrioles (Fig. 13.51). This display has been responsible for calling the pachytene stage of these animals a "bouquet" stage. Sometimes a chromosome end is embedded in the nucleolus, especially if it is the nucleolar-organizing end of the chromosome. The other end is attached to the nuclear membrane, however.

Unpaired leptotene chromosomes are also attached by their two ends to the nuclear envelope, with the lateral element attached at the same time and place. How does attachment occur? One 1974 study of pollen-forming meiocytes from wheat plants showed that bundles of microtubules formed in the nuclei during premeiotic interphase, before lateral elements or synaptinemal complexes could be seen. The microtubules disappeared during pachytene, after synaptinemal complexes were formed and stabilized.

These observations prompted the suggestion that movements of chromosome ends to the nuclear membrane might be aided by the microtubular bundles. Various changes took place in the nuclear membrane during this time, including rapid formation of many nuclear pores from an essentially nonporous nuclear envelope. Although many details remain to be worked out, it is reasonable to assume that an organizational framework for aligning chromosomes at the nuclear envelope may involve transient bundles

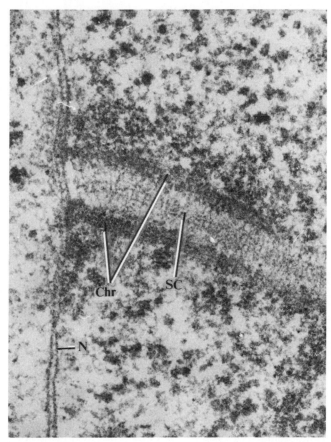

Figure 13.50
Longitudinal section through pachytene bivalent of *Locusta migratoria*. Attachment to the nuclear membrane (N) is clearly evident. There is a well-developed synaptinemal complex (SC) between the homologous chromosomes (Chr). × 100,000. (Courtesy of P. B. Moens)

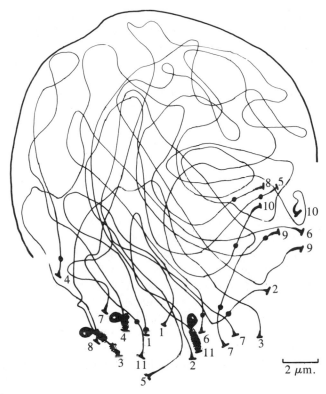

Figure 13.51
Reconstruction of all synaptinemal complexes in an early pachytene spermatocyte of *Locusta*. Each line represents a synaptinemal complex of a bivalent; the dot on the line represents the centromere in these acrocentric chromosomes; the larger dark areas beside the complexes for chromosomes 3, 4, and 11 represent the nucleoli and the dotted areas nearby represent the associated heterochromatin of these nucleolar-organizing regions. The line perpendicular to the end of the complex indicates the attachment point at the nuclear envelope. Both ends of each complex are attached at the polar side of the nuclear envelope. In addition to the eleven autosome pairs, there is one unpaired X chromosome in these spermatocytes, but it is not shown because no synaptinemal complex is formed at this univalent chromosome. Chromosomes 7 were still partially unpaired in the cell at the time of fixation and sectioning so that there are three ends present. (Reproduced with permission from Moens, P. B., 1973, *Cold Spring Harbor Sympos. Quant. Biol.* **38**:99–107, Fig. 3. Copyright 1974.)

of microtubules that act as aids or guides in wheat. Whether or not this situation also prevails in other species is not known.

Reconstruction studies of prophase nuclei indicate that movement of chromosome—nuclear membrane sites takes place, bringing ends of homologous chromosomes to within about 3000 Å of each other. Since synaptinemal complex formation sometimes is initiated internally rather than at the ends of zygotene chromosomes, pairing cannot be initiated exclusively at the nuclear envelope.

DNA SEQUENCE MATCHING. Complementary base pairing is the one mechanism we know that can explain the precision of meiotic synapsis, down to the smallest region of the chromosome. But chromosomes are highly condensed structures whose DNA is tightly folded from lengths of thousands of μm to only a fraction of this length. For *Drosophila* it has been calculated that 61,000 μm of DNA is condensed into 110 μm of total pachytene chromosome length. Only 0.2 percent of the genome is therefore accessible for pairing or for recombination.

One promising line of information on the initial DNA pairing problem has come from biochemical studies of lily meiocytes reported by Stern and others. The zygotene-DNA synthesized during the pairing stage is generally distributed over all the chromosomes, according to DNA–DNA hybridization studies. Zygotene-DNA is synthesized in reasonably long pieces, and an unwinding protein is synthesized during this same zygotene stage. Zygotene-DNA is also specifically part of a lipoprotein complex which is part of the nuclear membrane. This DNA is part of the membrane only during zygotene, because it is found at later stages in nuclear DNA which is isolated and purified of membrane residues. All these observations point to some unique activity of zygotene-DNA during zygotene. This DNA may be the special pairing strands that bring homologous chromosomes together initially and specifically. If the unwinding protein is active in this process, it would allow zygotene-DNA strands to be extended and make precise base-pairing more efficient at this stage. If DNA synthesis is inhibited during leptotene (which prevents synthesis of zygotene-DNA), then pairing will not take place since the zygotene stage is not achieved. Zygotene-DNA is essential if pairing is to take place initially.

Other suggestions have been made to explain the first pairing reactions, but little evidence is now available to make evaluations. At present it seems most likely that at least three distinct processes are involved in chromosome synapsis:

1. Premeiotic alignment of chromosomes by some nuclear membrane-mediated mechanism.
2. Base pairing of DNA regions to bring homologous chromosomes into initial precise register during zygotene.
3. Formation of the synaptinemal complex which stabilizes the paired chromosomes and keeps them in register during pachytene.

Once the synapsed chromosomes have been stabilized, crossing over is presumed to take place. The enormous problem of how matched strands from different homologous chromosomes can have access across the synaptinemal complex, undergo crossing over, and produce recombinant chromatid DNA remains to be solved. Various suggestions have been made, many of which are rather complex. It is quite possible that the fine strands seen traversing the synaptinemal complex are chromatin fibers that are near enough to recombine if they are homologous to one another in a specific region. The geometry of the problem is vexing. It must be solved, however, if the relationships between molecular and structural components in producing gene recombinations during meiosis are to be understood.

SUGGESTED READING

Books, Monographs, and Symposia
Mitchison, J. M. 1971. *The Biology of the Cell Cycle.* Cambridge: Cambridge Univ. Press.
Swanson, C. P., Merz, T., and Young, W. J. 1967. *Cytogenetics.* Englewood Cliffs, N.J.: Prentice-Hall.

Articles and Reviews
Bajer, A., and Molé-Bajer, J. 1969. Formation of spindle fibers, kinetochore orientation, and behavior of the nuclear envelope during mitosis in endosperm. *Chromosoma* 27:448–484.

Bennett, M. D., Stern, H., and Woodward, M. 1974. Chromatin attachment to nuclear membrane of wheat pollen mother cells. *Nature* 252:395–396.

Gillies, C. B., Rasmussen, S. W., and von Wettstein, D. 1973. The synaptinemal complex in homologous and nonhomologous pairing of chromosomes. *Cold Spring Harbor Symposia for Quantitative Biology* 38:117–122.

Hanawalt, P. C., and Haynes, R. H. 1967. The repair of DNA. *Scientific American* 216(2):36–43.

Hartwell, L. H. 1974. Genetic control of the cell division cycle in yeast. *Science* 183:46–51.

Martz, E., and Steinberg, M. 1972. The role of cell-cell contact in "contact" inhibition of cell division: A review and new evidence. *Journal of Cell Physiology* 79:189–210.

Mazia, D. 1974. The cell cycle. *Scientific American* 230(1):54–64.

Meselson, M. S., and Radding, C. M. 1975. A general model for genetic recombination. *Proceedings of the National Academy of Sciences* 72:358–361.

Meselson, M. S., and Weigle, J. J. 1961. Chromosome breakage accompanying genetic recombination in bactriophage. *Proceedings of the National Academy of Sciences* 47:857–868.

Moens, P. B. 1968. The structure and function of synaptinemal complexes in *Lilium longiflorum* sporocytes. *Chromosoma* 23:418–451.

Moens, P. B. 1973. Quantitative electron microscopy of chromosome organization at meiotic prophase. *Cold Spring Harbor Symposia for Quantitative Biology* 38:99–108.

Nicklas, R. B. 1974. Chromosome segregation mechanisms. *Genetics* 78:205–213.

Stern, H., and Hotta, Y. 1973. Biochemical controls of meiosis. *Annual Reviews of Genetics* 7:37–66.

Stern, H., and Hotta, Y. 1974. DNA metabolism during pachytene in relation to crossing over. *Genetics* 78:227–235.

Taylor, J. H. 1965. Distribution of tritium-labeled DNA among chromosomes during meiosis. I. Spermatogenesis in the grasshopper. *Journal of Cell Biology* 25:57–67.

Westergaard, M., and von Wettstein, D. 1972. The synaptinemal complex. *Annual Reviews of Genetics* 6:71–110.

von Wettstein, D. 1971. The synaptinemal complex and four-strand crossing over. *Proceedings of the National Academy of Sciences* 68:851–855.

Chapter 14

Cytogenetics: Classical and Human

Cytogenetics is a hybrid science which is based on parallel observations of gene behavior and chromosome or DNA behavior. Each set of observations reinforces the other, and the combined information usually provides more powerful support for hypotheses than either could do alone. The term cytogenetics is derived from cytology (study of cells by microscopy) and genetics (study of genes by breeding and pedigree analysis).

Cytology traces its beginnings to the seventeenth century when the first important observations with primitive light microscopy were reported. Genetics was officially born in 1900 with the "rediscovery" of Mendel's 1865 genetic studies. Cytogenetics is therefore a twentieth century science, which goes back to the very early 1900s.

Combining the newly discovered methods of genetics with the improved cytological methods developed in the late 1800s, W. Sutton formulated the Chromosome Theory of Heredity in 1902. Experimental support for this theory was collected primarily by T. H. Morgan and his colleagues working with *Drosophila melanogaster*. By 1915 they could provide evidence showing that genes were within chromosomes. Maps of chromosomes showing gene locations continued to verify the basic postulates (Fig. 14.1).

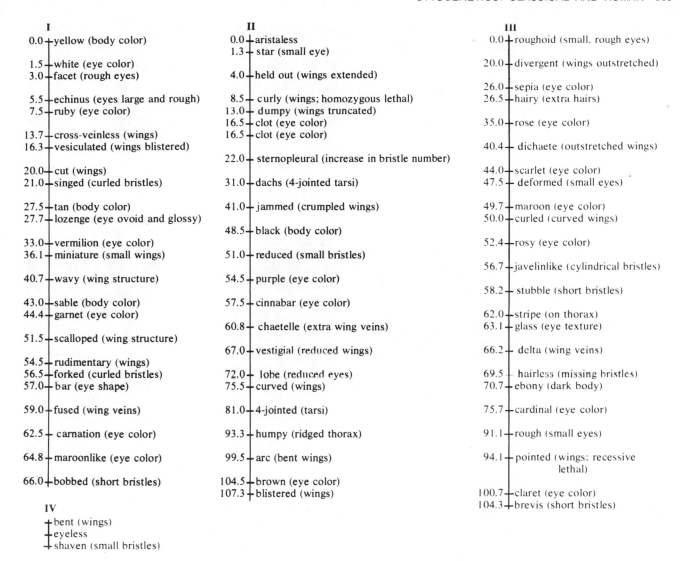

I

0.0 — yellow (body color)

1.5 — white (eye color)
3.0 — facet (rough eyes)

5.5 — echinus (eyes large and rough)
7.5 — ruby (eye color)

13.7 — cross-veinless (wings)
16.3 — vesiculated (wings blistered)

20.0 — cut (wings)
21.0 — singed (curled bristles)

27.5 — tan (body color)
27.7 — lozenge (eye ovoid and glossy)

33.0 — vermilion (eye color)
36.1 — miniature (small wings)

40.7 — wavy (wing structure)

43.0 — sable (body color)
44.4 — garnet (eye color)

51.5 — scalloped (wing structure)

54.5 — rudimentary (wings)
56.5 — forked (curled bristles)
57.0 — bar (eye shape)

59.0 — fused (wing veins)

62.5 — carnation (eye color)

64.8 — maroonlike (eye color)

66.0 — bobbed (short bristles)

IV

bent (wings)
eyeless
shaven (small bristles)

II

0.0 — aristaless
1.3 — star (small eye)

4.0 — held out (wings extended)

8.5 — curly (wings; homozygous lethal)
13.0 — dumpy (wings truncated)
16.5 — clot (eye color)
16.5 — clot (eye color)

22.0 — sternopleural (increase in bristle number)

31.0 — dachs (4-jointed tarsi)

41.0 — jammed (crumpled wings)

48.5 — black (body color)

51.0 — reduced (small bristles)

54.5 — purple (eye color)

57.5 — cinnabar (eye color)

60.8 — chaetelle (extra wing veins)

67.0 — vestigial (reduced wings)

72.0 — lobe (reduced eyes)
75.5 — curved (wings)

81.0 — 4-jointed (tarsi)

93.3 — humpy (ridged thorax)

99.5 — arc (bent wings)

104.5 — brown (eye color)
107.3 — blistered (wings)

III

0.0 — roughoid (small, rough eyes)

20.0 — divergent (wings outstretched)

26.0 — sepia (eye color)
26.5 — hairy (extra hairs)

35.0 — rose (eye color)

40.4 — dichaete (outstretched wings)

44.0 — scarlet (eye color)
47.5 — deformed (small eyes)

49.7 — maroon (eye color)
50.0 — curled (curved wings)

52.4 — rosy (eye color)

56.7 — javelinlike (cylindrical bristles)

58.2 — stubble (short bristles)

62.0 — stripe (on thorax)
63.1 — glass (eye texture)

66.2 — delta (wing veins)

69.5 — hairless (missing bristles)
70.7 — ebony (dark body)

75.7 — cardinal (eye color)

91.1 — rough (small eyes)

94.1 — pointed (wings; recessive lethal)

100.7 — claret (eye color)
104.3 — brevis (short bristles)

Figure 14.1
Chromosome map of *Drosophila melanogaster*. Relative distances between genes are shown to the left of each of the four chromosomes of the haploid complement. Chromosome I is the X chromosome, and II, III, and IV are the autosomes. No genes have been mapped on the Y chromosome, so it is not shown here.

The golden era of "classical" cytogenetics extended from the late 1920s to the early 1940s. Many species were studied, and many new insights were obtained by investigations of unusual or anomalous chromosome and gene behavior. Each test verified parallels between genes and chromosomes and gave new and broader significance to cytogenetic trends in evolution of eukaryotes. With the beginning of molecular genetics in the 1950s, cytogenetic analysis could be extended to the level of chromosomal DNA activities. Molecular cytogenetics has enjoyed a steady increase in attention and forms an indispensible part of the methods for modern study of the genetic apparatus in viruses, prokaryotes, and eukaryotes.

Our special interest in human cytogenetics extends from our understandable desire to know more about ourselves and from the clinical usefulness of cytogenetic analysis. Human cells can be cultured outside the body under controlled conditions (Fig. 14.2). These cells are a major source of material for mapping genes on chromosomes and for analyzing the biochemical and regulatory features of human gene action in producing cellular characteristics. Human chromosome studies actually lagged far behind those for other species. The first reported human chromosome counts were made in 1912. It was reported that men had 47 chromosomes and women had 48, with the sex-determining mechanism based on one X chromosome in males and two X chromosomes in females.

This totally inaccurate report was contradicted in the early 1920s when the very small Y chromosome was detected in males. XY males and XX females were the rule, and the sex-chromosome distinction was correctly evaluated on this basis. Human chromosome number was counted as 48 in human somatic cells, however. This count was published in 1923 and was accepted for the next 33 years. In 1956 the correct number of 46 chromosomes in somatic cells was reported. The development of improved methods for preparing and staining chromosomes made it possible to conduct accurate studies of human chromosome materials.

Continued innovation and improvement in cytological methods have led to the present status of understanding and to identification of the human chromosome complement. Each chromosome can now be recognized. Its parts can also be discerned by using new banding techniques for staining (see Chapter 12). Once it was possible to identify each

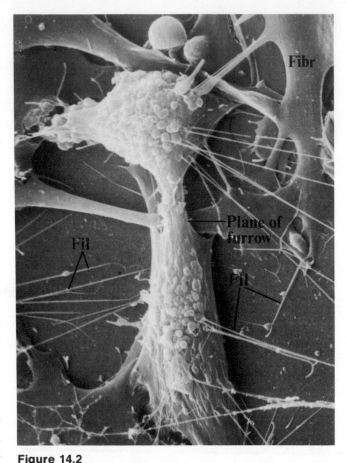

Figure 14.2
Photograph of normal rat fibroblasts growing in explant culture, taken with the scanning electron microscope. Cell in center is just completing cytokinesis. Blebbing of surface is characteristic of cells entering G_1 of the cell cycle. Other cells in the neighborhood are in a later stage of interphase and have relatively smooth surfaces. Slender filopodia connect the dividing cell to the substrate below. × 2,700. (Courtesy of K. R. Porter)

chromosome arm with certainty, it was a simpler job to assign gene locations to chromosomes from studies of cultured cells. We will discuss some of these studies, as part of the discussion of human cytogenetics, in this chapter. The first part of the chapter, however, will deal with classical cytogenetic analysis, which is the basis for newer investigations. Throughout this book we have referred to molecular cytogenetic studies, which also have a classical foundation on which interpretation and experimental approach are based.

CLASSICAL CYTOGENETICS

Changes Involving Chromosome Structure

Since genes are within chromosomes, modifications of chromosome structure can be expected to have detectable genetic effects if the altered region includes known gene loci. Cytogenetic analysis would show, therefore, that an alteration in gene behavior is paralleled by a predictable, structural change in the chromosome carrying the gene. Four general types of chromosome modification lead to a change in the linear order of genes on the aberrant chromosome: (1) deletions (also called deficiencies), (2) duplications, (3) inversions, (4) translocations. Each of these structural changes is caused by *breaks* in chromosome continuity and *rejoining* in an altered order from the original. All these kinds of aberrations may be *intra*chromosomal, but in some cases the changes may involve two or more chromosomes in the complement.

DELETIONS. A **deletion** involves the loss of a portion of the chromosome. The lost piece may be from an end of the chromosome or from some internal location (Fig. 14.3). A **terminal deletion** happens when a single break occurs, with later "rehealing" of the broken end. An **interstitial deletion** is produced by two breaks, loss of the piece between breaks, and rejoining of the broken ends of the chromosome remainder.

Cytological evidence of deletions can be obtained in different ways, depending on the species. When giant polytene or lampbrush chromosomes are available, such as those in dipteran insects and amphibians, respectively, direct examination of the chromosome bands or loops can be made. By comparison with the

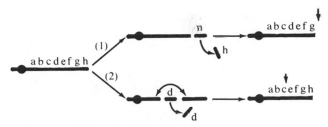

Figure 14.3
Origin of a deletion that is (1) terminal or (2) interstitial in its location on the chromosome.

standard polytene or lampbrush chromosome map, the missing piece can be identified by missing bands or loops in the chromosome preparations. Since polytene somatic chromosomes are paired, as are the homologous diplotene lampbrush chromosomes, the missing piece can be located first by looking at hybrid cells in which one chromosome of the pair is structurally normal and the other is aberrant. The normal chromosome often shows a looped-out region with no partner region for synapsis on the aberrant chromosome.

The same kind of looping-out is observed in conventional hybrid meiocytes during pachytene, if the deletion is large enough to be visible. Deletions in virus DNA can be detected in exactly the same way in electron micrographs of duplex DNA containing one normal strand and one deleted strand (Fig. 14.4). Observations of these types confirm the precision of synapsis between chromosomes and of exact base-pair matching between the two strands of duplex DNA molecules.

Genetic analysis of deletion mutants has been revealing. Most species cannot tolerate extensive loss of gene blocks. *Drosophila* mutants that lack as few as 50 bands (about 1 percent of the total in the haploid chromosome set) are usually lethals. Even if the homologous chromosome is intact, a deletion of this relatively small size in the aberrant homologue leads to death. Deletions of parts of chromosomes usually have deleterious effects, even if they are not lethal (Fig. 14.5). In humans the deletion of part of the

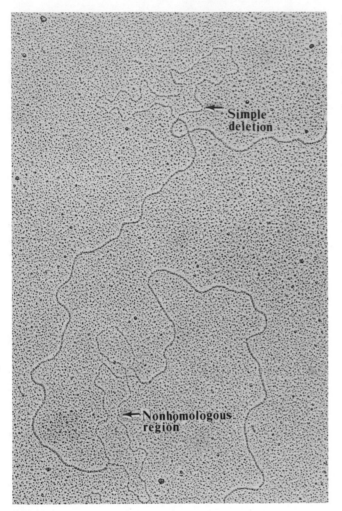

Figure 14.4
Electron micrograph of a platinum-shadowed heteroduplex DNA molecule made by annealing melted single strands from two different strains of phage lambda. The DNA of one strain has a deleted region plus another deletion into which a short piece of nonhomologous DNA has been substituted for the native longer region. The simple deletion can be detected by a single-stranded loop. The unpaired, nonhomologous region is evident by an opened area with one of the single strands longer than the other. (Courtesy of B. Westmoreland and H. Ris, from Westmoreland, B. *et al.*, 1969, *Science* **163**:1343–1348.)

short arm of chromosome 5 leads to patients with **cri-du-chat (cat-cry)** syndrome. These individuals have a number of body abnormalities and show severe mental retardation. While the other abnormalities lessen in adulthood, mental retardation remains severe and is the most conspicuous problem in the life of the patient. These individuals have one normal chromosome 5 along with the partially deleted homologue.

Deletion mutants usually behave as *recessives* in genetic crosses; that is, they are similar to mutants which produce defective or no gene product. If linked genes on either side of a deletion can be mapped by genetic tests, they are shown to be closer together in the mutant, than in a wild-type strain. These genetic

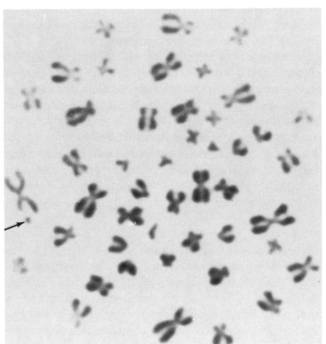

Figure 14.5
A spontaneous break in the long arm of chromosome 1 has produced a deletion. The deleted piece (arrow) is visible nearby. The deleted piece would probably be lost from the cell during subsequent mitotic divisions since it has no centromere to ensure accurate anaphase movement to the poles. (Courtesy of T. R. Tegenkamp)

tests as well as observations of the characteristics of the deletion mutants go hand in hand with the chromosomal analysis and confirm the diagnosis.

Since known deletion mutants are either lethal or deleterious types, it would seem that gene losses have little ongoing evolutionary benefit. There are many species groups, however, in which parts of chromosomes have been lost in the more recent as compared with the ancestral species. In many cases where it can be properly studied, the lost chromosome parts are almost entirely heterochromatin in composition. Since constitutive heterochromatin usually contains little or no genetic information (see Chapter 12), it may be a more dispensable component of the chromosome complement. Too little is known about heterochromatin function, however, to evaluate its importance or lack of importance in evolutionary episodes.

Deletions have a low probability for transmission during reproduction since they often lead to death of the gametes or the early embryo. Spontaneous abortion in both plant and animal species is a common consequence of deleted chromosomes carried in the developing embryo. In plants, a few seeds may be produced, if any, or various malformations may occur in the young seedlings. In animals, death most often occurs before hatching or birth if substantial deletions are present in even one of the homologous chromosomes of the pair.

DUPLICATIONS. As the name implies, a **duplication** involves the presence of an extra piece of a chromosome somewhere in the chromosome complement of a cell. It may be attached somehow to a chromosome that already has one copy of the genetic region involved, or to another chromosome in the complement. In some cases, the duplicated fragment may exist as a separate piece not connected to other chromosomes in the nucleus.

Duplications may arise in various ways. In some cases, at least two breaks must have occurred within a chromosome, so that the duplicated piece could be inserted and then sealed into the structure.

In other cases, some copying error may have happened during DNA replication, and extra regions may have been produced somewhere along the chromosome. In one of the better studied examples of duplication, a process of *unequal crossing over* has been suggested as the cause of the *bar eye* duplication in *Drosophila* (Fig. 14.6). Cytological examination of the *16A* region of the X chromosome shows that a particular region containing 4–5 bands is repeated in tandem in *bar* mutant males. Since progeny of *bar* mutants frequently segregate into normal as well as into exaggerated *bar-double* types with three tandem repeats, unequal crossing over has been invoked as the most likely explanation for the origin of the duplication. The eye shape (normal, *bar*, and *bar-double*) corresponds to the banding pattern of region *16A* in the X chromosome in every case.

Detailed studies of banding patterns in polytene chromosomes of species of *Drosophila* and other dipterans have shown a relatively high incidence of duplications in natural populations. While similar studies cannot be carried out with most plant and animal species, there are reports of synapsed regions occurring in haploid meiocytes of unusual haploid derivatives of conventional diploid plants. These paired regions have been interpreted as evidence of the presence of duplications since synapsis is almost always highly specific for homologous gene sequences.

Duplications are more likely to survive in populations since they do not necessarily have lethal or harmful effects on the individual. In fact, according to current ideas, duplications may be responsible for

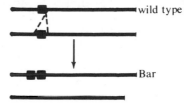

Figure 14.6
Assumed origin of the Bar duplication in *Drosophila melanogaster* by unequal crossing over.

gene increase during species evolution. If originally duplicated genes were present and were retained in the population, they would be expected to undergo different mutations with time, since mutation is a random event. Different random changes could lead eventually to modified products of the duplicated genes. Ultimately such completely different products would occur that they would have to be called different genes. The original duplicates would appear as totally distinct genes at the present time, whether or not their past history was known. Since we have actually observed very few genes in the process of evolving or at some intermediate, recognizable stage, gene duplication leading to gene increase remains a hypothetical but attractive possibility.

The whole problem is even more complicated today because we know there is repetitious DNA and unique-sequence DNA (see Chapter 12). The origins of these forms of DNA are not known, and theories about their evolution are just now being formulated. There are actually two questions: "how did genes increase in number during evolution?" and "how did repetitious DNA originate in eukaryotes?" According to some current studies there may not be many more unique-sequence genes in the highest eukaryote life forms than there are in *E. coli*. *E. coli* has been estimated to have about 4000 genes and no repetitious DNA, while eukaryotes may have 5000–10,000 genes and variable amounts of repetitious DNA.

Duplications can be detected genetically by the absence of a recessive phenotype in diploids which are heterozygous for the duplicated region. Appropriately constructed genetic stocks are required to do the breeding analysis, but the results are consistent and different from those in crosses between conventional dominant and recessive gene mutants. The results are also different from genetic tests using deletion mutants, in which dominant phenotypes are not produced in heterozygotes containing one normal chromosome with a recessive allele present. Further details can be found in texts on genetics or cytogenetics.

INVERSIONS. **Inversions** are the commonest structural aberration encountered in natural populations of higher plants and animals. One reason for this is their easier detection and identification; another is that they rarely have lethal effects on individuals carrying such rearranged chromosomes in either the heterozygous or homozygous state. An inversion arises when a piece of the chromosome has been cut out by two breaks and is then reinserted in reverse order from its original sequence (Fig. 14.7). The usual inversion involves only one arm of a chromosome (**paracentric** inversion). When the inverted region includes the centromere, it is described as **pericentric**.

The simplest genetic test for inversions is to compare the linkage maps for noninverted and inverted homozygous strains. The gene order will be altered in the chromosome with the inversion, so that *abcdefghijk* may show up as *abcgfedhijk*. The gene block *defg* has been inverted and shows different linkage distances with *abc* on one side and *hijk* on the other. The genetic test is straightforward but time-consuming.

If cytological examination of the meiotic bivalents can be made, an inversion "loop" in pachytene bivalents of individuals heterozygous for the inverted chromosome is looked for (Fig. 14.8). Since somatic chromosomes in salivary gland cells of dipterans are also paired, inversion loops are clearly evident in polytene chromosome preparations from hetero-

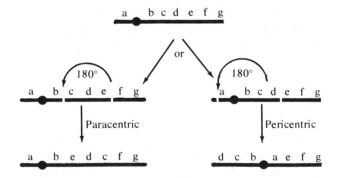

Figure 14.7
Origin of inversions of the paracentric and pericentric types.

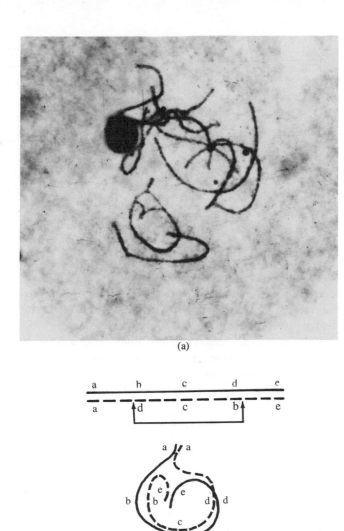

(a)

(b)

Figure 14.8
Pachytene nucleus from maize (*Zea mays*) meiocyte heterozygous for an inversion in chromosome 2. (a) Inversion loop. (Courtesy of M. M. Rhoades) (b) Interpretive diagram illustrating the inverted and noninverted chromosomes and their alignment after pairing at pachytene.

zygotes. Homozygous individuals have the same gene order on both homologous chromosomes, so inversion loops would not be found for either the normal or the inverted gene order. Heterozygotes must be examined cytologically for diagnostic loops. Banding patterns of salivary gland chromosomes will show inverted order even in homozygotes, of course, but all the chromosomes must be searched through and tedious comparisons with the normal chromosome banding maps must be made.

Inversion heterozygotes sometimes show sterility or reduced fertility. These effects can be traced to crossing-over events during meiosis and the kinds of chromosomes that become incorporated into the spores or gametes which are produced (Fig. 14.9). Depending on the location and size of the inversion and whether or not it includes the centromere, the number of crossovers per bivalent and the total number of chromatids involved, as well as other factors, meiotic products with chromosomes containing deletions and duplications may be formed. The relative effect on fertility depends on how severe the chromosome aberrations turn out to be in each case.

Long-term studies of natural populations of *Drosophila* species have shown not only that inversions may be quite common, but that certain inversions confer a selective advantage on the population. For example, T. Dobzhansky has shown that there are three main gene arrangements in chromosome 3 of *Drosophila pesudoobscura* populations in the western United States. The *Standard* (*ST*) arrangement is commonest in low-altitude populations during warm weather; the *Arrowhead* (*AR*) arrangement is found more often in populations from higher altitudes and cooler temperatures; and the *Chiricahua* (*CH*) inverted chromosome is commoner in cooler weather but is equally frequent at any altitude.

The whole species is maintained in favorable numbers by shifting balance among the three inversion types, in response to altitude and temperature. Its success is due, in large part, to its ability to modify its populations according to the environmental conditions at different times of the year and in

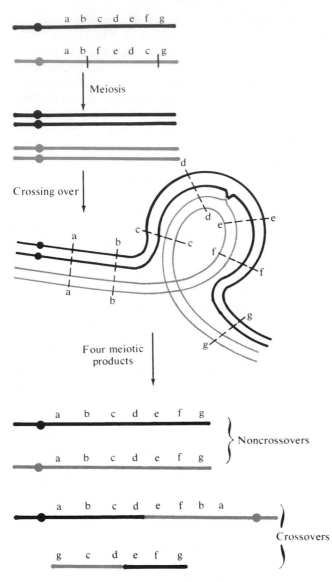

Figure 14.9
Diagram illustrating the origin of chromosomes containing deletions and duplications as the result of a crossover event within an inverted region in a meiocyte of an inversion heterozygote.

different locations in the species range. Inversions would not only be tolerated in such populations, they would be maintained by positive selection pressures.

TRANSLOCATIONS. A **translocation** occurs when a block of chromatin is broken off and relocated to another place on the same chromosome or to a different chromosome (Fig. 14.10). If one broken piece is involved, it is a **simple** translocation; if parts of two nonhomologous chromosomes are broken off and exchanged, it is called a **reciprocal** translocation. If the process has not involved removal or addition of a centromere, the translocated chromosomes will act just like normal chromosomes at meiosis. If a centromere has been removed, the **acentric** chromosome remainder cannot move to the poles at either mitosis or meiosis and may be lost from the cell. This behavior could lead to cell death, if the chromosome fragment contains genes that are vital to growth and development. A translocation that produces a **dicentric** (two-centromere) chromosome can also cause problems at nuclear division if the two cen-

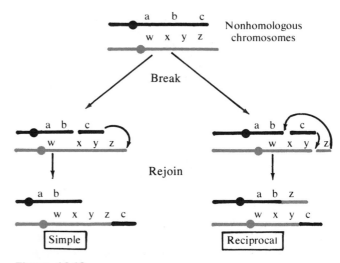

Figure 14.10
Origin of simple and reciprocal translocations by exchange between nonhomologous chromosomes.

tromeres move toward opposite poles and the chromosome bridge between them is broken (Fig. 14.11).

Translocations may be identified in genetic tests by a change in linkage relations for the genes of the exchanged segments. Genes that once assorted independently (on separate chromosomes) may now show linked inheritance (on the same chromosome), and vice versa. Cytological tests for translocations can take advantage of meiotic or mitotic chromosome preparations. In heterozygotes, where translocated and nontranslocated chromosomes are both present, pachytene chromosome figures reveal characteristic shapes showing that more than two chromosomes are synapsed in the total configuration. Only two chromosome segments synapse at any one specific region, however. These translocation configurations are very

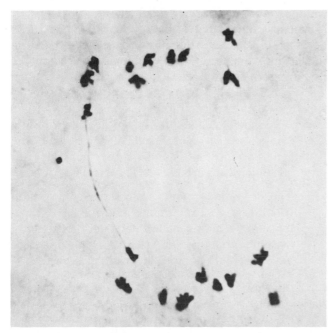

Figure 14.11
Chromosome bridge between disjoining halves of a dicentric chromosome during anaphase I in maize meiocyte nucleus. A fragment lacking a centromere lies alongside the bridge. (Courtesy of M. M. Rhoades)

specific and provide undeniable evidence of homologies and rearrangements (Fig. 14.12).

Mitotic chromosomes will show altered morphologies if either simple or reciprocal translocations have occurred. The simple translocation is easiest to identify because one chromosome of the set will be shorter than usual (lost a piece) and another will be longer than usual (gained a piece). Chromosome morphology studies have been especially useful in identifying chromosomal aberrations in humans. The aberrations have also been useful in mapping genes on human chromosomes, as we will discuss in a later section of this chapter.

Translocations may lead to reduced fertility in some populations, if irregularities in pairing at meiosis cause deletions and duplications in the resulting spores or gametes. The degree of effect depends on many factors, including the positions of the translocated segments, the number of locations of crossovers at meiosis, and the orientation of the chromosomes at metaphase I of meiosis (Fig. 14.13).

A striking example of naturally-occurring translocations has been described in species of *Oenothera*, the evening primrose. In most cases all 7 chromosomes of the haploid set have undergone reciprocal translocations. Because of an incompatibility system that operates in these self-pollinating species, the only surviving members of each sexual generation are translocation heterozygotes. Meiosis in these heterozygotes is remarkably regular, even though 14 chromosomes (7 from each parent) are synapsed together in a complex figure at meiosis. By metaphase I, the 14 chromosomes have become aligned at the equatorial plane of the spindle such that alternating chromosomes face the same pole and adjacent chromosomes face opposite poles. In this way each daughter nucleus receives a complete set of 7 chromosomes with all the genes intact (Fig. 14.14). Since these dyads undergo normal disjunction at anaphase II, the meiotic products are viable.

Translocations have been an important factor in *Oenothera* species evolution, providing permanent

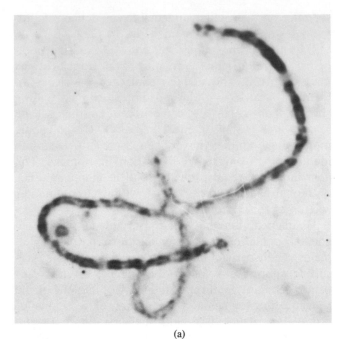

(a)

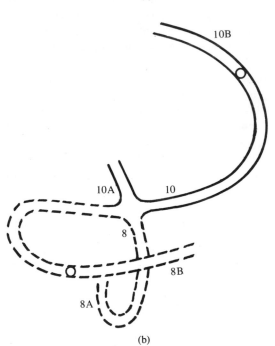

(b)

"hybrid vigor" to the populations. A few *Oenothera* species have the conventional 7 pairs of chromosomes without translocations. These invariably are very restricted in geographic distribution and are actually relict, or ancient remnants of, populations of former widespread species. The translocation-heterozygote species are widespread, remarkably prolific, and successful weeds for the most part, and are the most recent populations to have appeared during *Oenothera* evolution. They have largely replaced their 7-paired ancestors in many localities and have spread far beyond the ancestral species ranges in a brief period of geological time. The combination of unusual features, including reciprocal translocations of all the chromosomes, has provided a unique pathway for evolutionary success in this flowering plant group.

Changes Involving Chromosome Number

Mitosis and meiosis are remarkably accurate processes, but occasionally a mistake or failure occurs in a nuclear division. Cells may come to have an extra chromosome, to lack a chromosome, or even to have more than two sets of chromosomes. Some of these mistakes can be traced to meiosis, and others can be assigned to failure in mitotic distribution of chromatids to the poles. Whatever the origin of the change in chromosome number, two general categories of numerical variants are recognized:

1. Those variants with a somatic complement which is some exact multiple of the basic chromosome set are called **triploids, tetraploids,** or some other specific kind of **polyploid,** if they have a **euploid** number higher than diploid.

Figure 14.12
Pairing during pachytene in maize meiocyte nucleus heterozygous for a reciprocal translocation: (a) Four chromosomes are engaged in synapsis, but only two chromosome segments are synapsed at any one point along the complex; (b) interpretive drawing of the four-chromosome complex involving chromosomes 8 and 10 and reciprocally translocated chromosomes 8A/10B and 8B/10A. (Photograph courtesy of M. M. Rhoades)

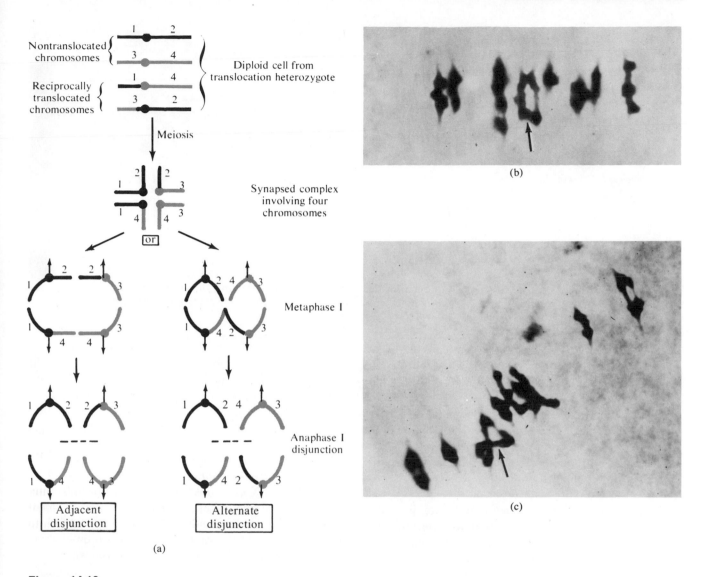

(b)

(c)

Figure 14.13
Chromosome alignment and disjunction in meiosis: (a) Diagrammatic illustration of the effects of chromosome alignment at metaphase I and mode of disjunction at anaphase I on gene content in the resulting meiotic products produced by a translocation heterozygote. Chromosome arms are numbered to indicate genetic regions. Duplications and deficiencies result after adjacent disjunction, but a complete set of genes is distributed to each pole after alternate disjunction. (b) Alignment of a ring of four chromosomes at metaphase I in maize which will result in adjacent disjunction. (c) Alignment of a ring of four chromosomes at metaphase I in maize which will result in alternate disjunction. (Photographs courtesy of M. M. Rhoades)

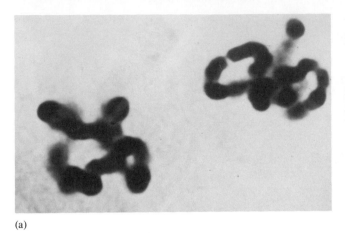

(a)

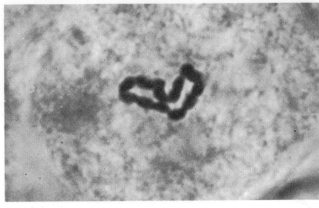

(b)

Figure 14.14
Metaphase I in meiocytes heterozygous for reciprocally translocated chromosomes involving the entire complement: (a) babe-in-the-cradle (*Rhoeo discolor*) with all chromo- somes linked together in a ring. (Courtesy of N. V. Rothwell) (b) Circle of fourteen chromosomes in *Oenothera*, the evening primrose.

2. Cells or individuals with a somatic chromosome complement which is an irregular multiple of the basic set are called **aneuploids.**

Just as there may be specific designations of polyploids, there may also be terms that describe an aneuploid more accurately. For example, an aneuploid may have three copies of one of its chromosomes present, rather than having the usual two copies found in the diploid condition. This type of aneuploid is called a **trisomic;** the aneuploid state is called **trisomy.** If there is only one copy of one chromosome and two copies of all the others in a complement, the aneuploid is described as **monosomic,** and the aneuploid state is called **monosomy.** The base of reference is the somatic diploid state in which there are two copies of each chromosome. This state is called **disomic.** These variants have been studied in the higher plants and animals where the adult organism is conventionally diploid. A few studies along similar lines have been conducted with predominantly haploid species of the lower eukaryote groups.

ANEUPLOIDY. Anaphase separation of homologous chromosomes in meiosis may be abnormal, so that spores or gametes may be formed with one chromosome missing or may have an extra chromosome (Fig. 14.15). The failure to separate properly is called **nondisjunction,** and it is a frequent cause of aneuploid meiotic products.

In *Drosophila* it was shown that unusual genetic results in sex-linked inheritance could be traced to eggs with two X chromosomes or no X chromosome, in addition to the regular haploid number of autosomes. These cases usually are caused by the failure of synapsis and crossing over of X chromosomes in oocyte meiosis. The unpaired X chromosomes do not become properly oriented in metaphase, and both may arrive at the same pole by random motion. Other anomalous events have also been seen and suggested as explanations for gametes with the $n + 1$ or $n - 1$ chromosome number in a species, where n is the haploid number of chromosomes. If these gametes are viable, they may fuse with a normal haploid gamete and produce an aneuploid individual which

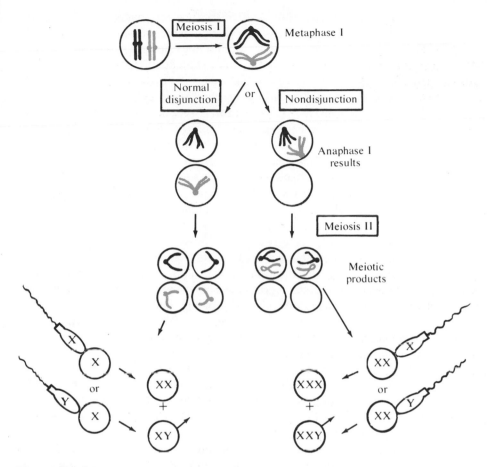

Figure 14.15
Flow diagram illustrating the origin of aneuploid cells or individuals as the result of nondisjunction at meiosis (right), as compared with normal disjunction (left).

is trisomic $(2n + 1)$ or monosomic $(2n - 1)$ for the X or some other chromosome.

Aneuploidy may arise in somatic cells when a chromosome lags behind at anaphase and is not included in the new daughter nucleus. This $2n - 1$ nucleus may continue to divide by mitosis without further malfunction and give rise to a lineage of cells made up entirely of aneuploid descendants of the original aberrant. Mosaic patches of diploid and aneuploid tissues have been studied in plants and animals and can usually be traced back to the time of anaphase lag in the original abnormal mitosis. The chromosome that lagged behind may be eliminated altogether from further mitoses. This event is more common than the alternative in which the lagging chromosome is included in the second daughter nucleus, which would then become $2n + 1$, or trisomic.

Many detailed studies of trisomic flowering plants

and *Drosophila* have been made. The extra chromosome usually produces some visible specific change in the organism, so that the particular chromosome can be predicted, just by looking at the mature or juvenile individual. Transmission of the extra chromosome is usually through the egg rather than the sperm, probably because of the more protected nature of the egg within the female, and its large nutritional reserves. The egg is more likely therefore to tolerate gene unbalance than the sperm.

Chromosome loss is more harmful in diploid organisms than the addition of a chromosome. This effect may be the result of unmasked harmful or lethal alleles on the monosomic chromosome, which are not expressed in diploids or in trisomics. Monosomics are rare in every species that has been studied. A number of trisomies are known to occur in humans, but the only monosomic condition that survives is the one where the second X chromosome is missing. There are no autosomal monosomic conditions in humans, although it is suspected that some spontaneous abortions are caused by lethal monosomic chromosome complements. We will discuss aneuploidy involving autosomes and sex chromosomes in humans later in this chapter.

The first clinical condition discovered to have a chromosomal basis in humans was **trisomy-21,** also known as **Down's syndrome.** In 1959, three years after the human chromosome number had been established as 46 in somatic cells, patients with 47 chromosomes were described. The extra chromosome was No. 21, one of the smallest autosomes in the set (Fig. 14.16). Other trisomies have since been described.

CENTRIC FUSION. Chromosome numbers that differ from the usual diploid complement can also arise without a change in the amount of chromatin. The usual way this happens is by **centric fusion,** in which two smaller chromosomes fuse to form one larger one. In the usual situation a large **metacentric** chromosome is derived from two smaller **acrocentrics** (Fig. 14.17).

It originally had been thought that one of the centromeres was lost during centric fusion, together with some of its surrounding heterochromatin. Ultrastructure studies of metacentric chromosomes, however, have shown that the single centromere region contains two centromere plates. Most acrocentric material is apparently retained in the metacentric chromosome, including the two centromere plates now present in a single region.

This feature of centric fusion was suspected to occur because studies using various invertebrates had shown that chromosome *numbers* varied between individuals in some populations while the number of chromosome *arms* was the same. Centric fusions produced fewer chromosomes, and dissociation of metacentrics to form two acrocentrics also occurred in these populations. This meant that both centromeres were present in the metacentric chromosome, if it could later dissociate into two functional acrocentric chromosomes. Neither the mechanism for centric fusion nor the mechanism for dissociation are known.

Centric fusion has played an important role in chromosome evolution in many animal groups, and in some of the flowering plants. In a particularly well documented study of *Drosophila* species, evolution has followed the pattern of a reduction in the haploid chromosome number from 6 to 3 (Fig. 14.18). The trend to fewer and larger chromosomes is especially clear in species groups of *Drosophila* and other organisms that are known to be in a direct evolutionary lineage. Centric fusions can be verified in *Drosophila* by comparing banding patterns of the salivary gland chromosomes in the acrocentric and derived metacentric chromosomes of the complements. On the basis of many studies, it has been concluded that acrocentric chromosomes are an indication of a more "primitive" evolutionary status in animals, while metacentrics indicate a more "advanced" status. This generalization may be true for many groups, but it cannot be accepted entirely for every species comparison.

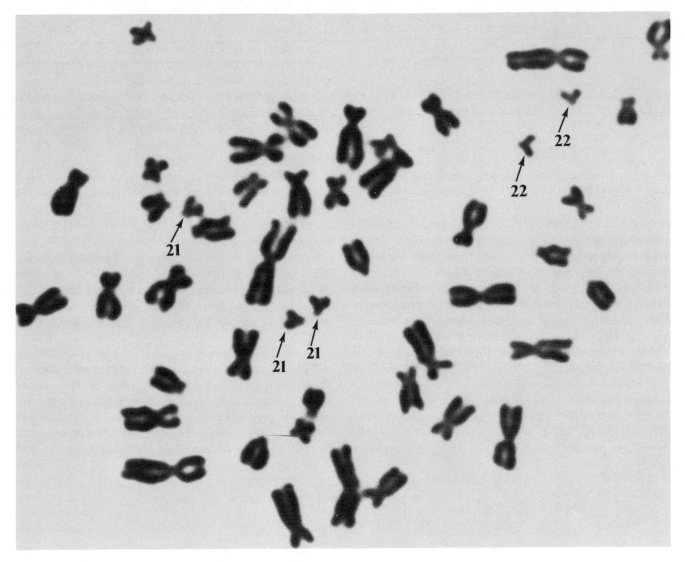

Figure 14.16
Chromosome complement of a female patient with Down's syndrome, or trisomy-21. There are forty-seven chromosomes, five of which are from the small acrocentric 21–22 (G) group. The extra chromosome is easily detected and identified in such a well-spread chromosome preparation. (Courtesy of C. Hux and T. R. Tegenkamp)

Figure 14.17
One large metacentric chromosome from two equal-sized acrocentrics (top) or one large submetacentric from un-equal-sized acrocentrics (bottom) may originate after centric fusion events.

POLYPLOIDY. Haploidy is the predominant state in many sexually-reproducing lower eukaryotes, with a very transient diploid state that may only be represented by the zygote. Diploid adults are the predominant mode in higher eukaryotes, with a brief haploid state that may only include the gamete cells. The two chromosome sets (genomes) in diploid cells are homologous, since they have the same gene loci but may have different allelic forms of these genes on the chromosomes.

Polyploids may have multiples of the same genome (**autopolyploids**) or there may be different genomes in the same cell (**allopolyploids**), depending on the origin of the particular polyploid individual or species. In natural populations, polyploids whose genomes are a mixture of auto- and alloploid chromosomes are usually found.

Autopolyploids are rare. Autopolyploid organisms would have problems during meiosis because synapsis would involve more than two homologous chromosomes (Fig. 14.19). These **multivalents** segregate erratically at anaphase I, unlike the regular separation of bivalents, and gametes would be formed with different degrees of chromosomal unbalance. These gametes usually are not viable.

Allopolyploids are known, and in some cases their origin can be traced back to two different species which hybridized and then experienced a doubling in chromosome number. If the hybrid was a diploid with two different genomes, synapsis would be irregular or largely absent and the hybrid would be sterile. When such a diploid becomes tetraploid after some event, either spontaneous or artificially induced, each dissimilar chromosome has a partner, and normal meiosis will occur in the new allotetraploid (Fig. 14.20).

Natural populations of successful polyploids have an even number of chromosome sets. They are tetraploid, hexaploid, and octaploid. Each of these types forms bivalents at meiosis, so segregation of chromosomes is regular, and gametes have a reduced but complete chromosome constitution. Polyploids with an odd number of genomes, such as triploids and

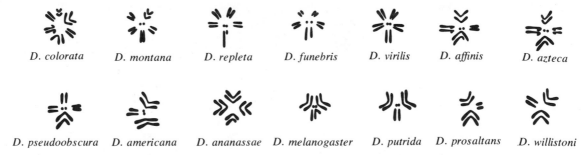

D. colorata　　*D. montana*　　*D. repleta*　　*D. funebris*　　*D. virilis*　　*D. affinis*　　*D. azteca*

D. pseudoobscura　*D. americana*　*D. ananassae*　*D. melanogaster*　*D. putrida*　*D. prosaltans*　*D. willistoni*

Figure 14.18
Chromosome complements of fourteen different species of *Drosophila* showing the aneuploid series varying between three and six pairs of chromosomes in the diploid cells. The X and Y chromosomes of these male genomes are shown at the bottom of each drawing.

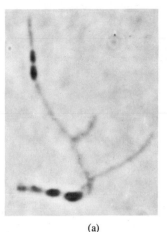
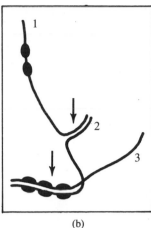

(a)   (b)

Figure 14.19
Synapsis involving three homologous chromosomes in castor bean (*Ricinus communis*): (a) trivalent chromosome No. 9; (b) interpretive drawing showing two-by-two pairing among the chromosomes, with one of the chromosomes (2) paired with parts of the other two homologues (1, 3). Crossing over and chiasma formation can occur in synapsed regions (arrows). The distinctive heterochromatic knobs identify this chromosome in the complement and further show that the three synapsed chromosomes are identical in this trivalent. (Courtesy of G. Jelenkovic)

pentaploids, are sterile. Their chromosomes form multivalents, bivalents, and univalents at meiosis, and unbalanced chromosome numbers are found in spores or gametes. If these polyploids are found naturally or are grown as a cultivated species, they must be propagated by some asexual mechanism. They may increase through **parthenogenesis** (development of an individual from an unfertilized egg) or some other substitute for sexual reproduction.

Many cultivated plants are polyploids. Wheat occurs as diploid, tetraploid, and hexaploid types. Common wheat contains 3 pairs of genomes, each of which can be traced back to particular wild diploid relative species. Its hexaploid nature was determined by cross breeding with other species, since the regular formation of 21 bivalents, by itself, would not show there were three different sets of 7-chromosome genomes. Cultivated tobacco, cotton, potatoes, and others are polyploids whose ancestor species are known to be diploids.

Breeding programs to introduce desirable genes from different genomes into a hybrid may be arranged, and polyploidy may then be induced in the hybrid strain by colchicine treatment. Regular meiosis takes place, and the polyploid becomes a higher-yielding, fertile strain. Larger fruit, seed, and flower size in polyploids make these variants more valuable economically. In fact, orchids and other plants raised for their flowers, rather than for their fruit, are deliberately polyploidized to produce sterile, odd-numbered genome types. Triploids and pentaploids produce large flowers that continue to bloom for longer times because they do not form seeds and fruits. Sterile odd-numbered polyploids are therefore valuable under certain conditions of the marketplace. Of course, there are many diploid cultivated plants, such as tomatoes, corn, and other flowering plants.

Polyploidy has played a significant role in plant evolution, if its wide distribution and high incidence are any indication (Fig. 14.21). About 50 percent of known flowering plants are polyploid, particularly among some families such as the grasses. Ferns and other ancient land plants are usually high polyploids, but many gymnosperms (pines, cycads, and relatives) are still diploid for the most part. The success of polyploids may be due to their greater hardiness and vigor, and their ability to exploit new habitats. Studies of related species distributed over glaciated and non-glaciated regions of Europe and North America have shown that the older diploid relatives remained in the unglaciated areas while their polyploid descendants took over the glaciated areas after the warming in Pleistocene and Recent geological times. Polyploidy provides selective advantages, particularly when environmental changes are in progress.

Polyploidy is rare in animals because the balanced autosome—sex chromosome mechanism for sex determination and differentiation would be upset by variable numbers following meiosis or genome in-

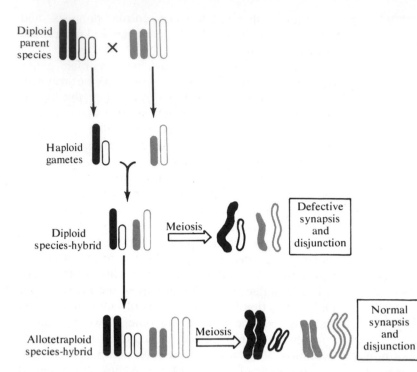

Figure 14.20
Diagram illustrating the origin of a sterile diploid species-hybrid and the beneficial effects of doubling the chromosome number on meiotic pairing in the allotetraploid.

creases. Some invertebrate groups are polyploid, but these are almost all species that reproduce by parthenogenesis. This asexual mode of reproduction provides a bypass around the problem of hormonal unbalance created by disturbances in chromosome number and balance.

HUMAN CYTOGENETICS

Interest in human genetics and the chromosome complement has increased steadily with the introduction and development of new methods for study. The improved methods for obtaining well-spread out mitotic chromosomes, using cells in culture or from direct samples, has led to greater emphasis on chromosome analysis. After J. Tijo and A. Levan showed, in 1956, that there were 46 chromosomes in human somatic cells, the complement was described and each

chromosome was identified as belonging to one of seven groups that differed in total length and in centromere position (Fig. 14.22). Newer staining methods that produce distinct patterns of bands and autoradiography techniques that show early- and late-replicating chromosomes permit the recognition of each of the 24 chromosomes specifically. Every somatic cell has 22 pairs of autosomes, and one pair of sex chromosomes. Females have two X chromosomes and males are XY.

Gene locations on the 24 kinds of chromosome (autosomes 1–22, X, and Y) are proceeding rapidly now that human somatic cells in culture can be fused with mouse, rat, and other cell sources. The new field of **somatic cell genetics** has expanded tremendously since the first reports of somatic cell fusions in 1960. We will discuss some of the methods, discoveries, and implications of these studies later in this section.

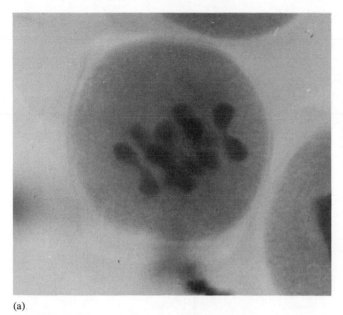

(a)

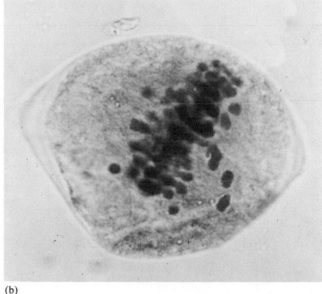

(b)

Figure 14.21
Cells from different populations of spring beauty (*Claytonia virginica*). (a) The diploid plants have eight pairs of chromosomes while the polyploid plants in this unusual species may have sixteen or more pairs of chromosomes. There is considerable aneuploidy in the group, as indicated by the variable numbers and chromosome pairing. (b) A number of univalents are evident among the ninety-eight chromosomes in this high polyploid cell. × 1,200. (Courtesy of N. V. Rothwell)

Medical genetics is also a rapidly expanding field of human study. In addition to the correlations between genes, chromosomes, and clinical defects in patients, considerable progress is under way in the areas of *prenatal diagnosis* of inherited afflictions. Samples of amniotic fluid taken from the mother during pregnancy contain cells and products of fetus growth and metabolism. By analyzing these materials, it is sometimes possible to determine whether or not the developing fetus has a genetic or chromosomal defect.

The Human Chromosome Complement

Two human cell types commonly studied in culture are **lymphocytes** (a type of blood cell) and **fibroblasts** (a type of connective tissue cell). These cells have a relatively similar cell cycle with the usual sequence of phases, of which G_1 is the most variable phase. Chromosomes from mitotic metaphase are the ones usually studied since these are the most contracted, morphologically distinct, and easiest to collect. If colchicine is added to cells that have been stimulated to divide, metaphase nuclei accumulate because there is no spindle and no anaphase separation of the chromatids of each chromosome.

Metaphase nuclei of these cells are photographed, and the individual chromosomes are cut out and positioned according to an established convention. The largest chromosomes are first in the numbered sequence, and the arrangement proceeds through intermediate-size chromosomes to the smallest in the complement. This completed picture represents the

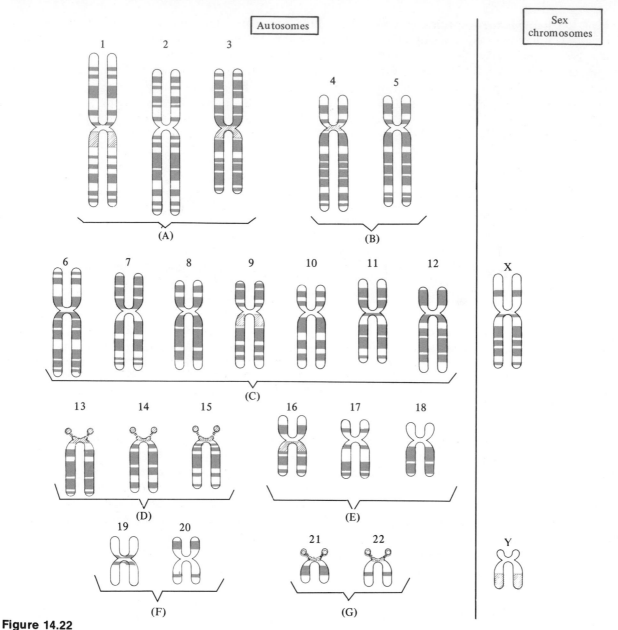

Figure 14.22
The twenty-two autosomes and two sex chromosomes of the human chromosome complement, arranged according to size and centromere location into seven groups (A to G) and numbered individually. The satellited chromosomes 13, 14, 15, 21, and 22 are all nucleolar-organizing.

karyotype of the individual or species (Fig. 14.23). When the photographed chromosomes of a karyotype are drawn diagrammatically to emphasize their features, the diagram is called an **idiogram.** One feature that has been noted is the presence of **satellites** on all five acrocentric chromosomes belonging to the D (Nos. 13, 14, and 15) and G (Nos. 21 and 22) chromosome groups. These satellited regions are often associated together in a preparation; they are the nucleolar-organizing regions attached to a common nucleolus that was present earlier in interphase and prophase. Nucleoli are no longer visible in metaphase nuclei, but the 5 pairs of acrocentric chromosomes often retain their associative grouping.

Certain regions of the chromosomes contain **heterochromatin,** a form of chromatin that remains

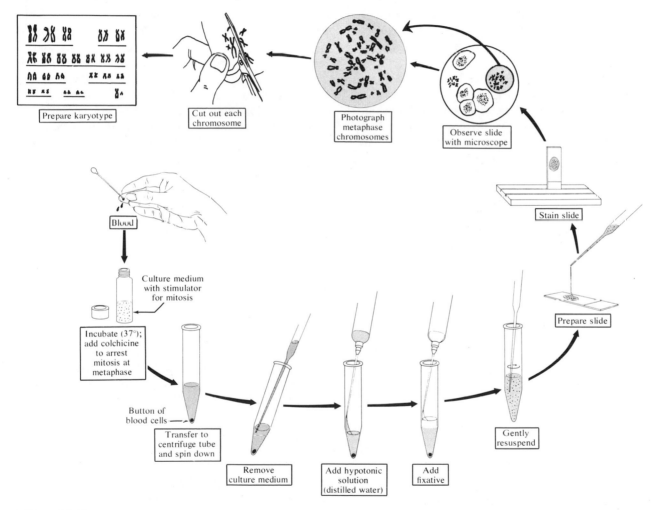

Figure 14.23
Flow diagram illustrating the procedures for karyotype preparation.

condensed during interphase, while **euchromatin** is greatly dispersed (see Chapter 12). Heterochromatin is not readily detected in metaphase chromosomes when all the chromatin is condensed to a maximum, but it can be recognized in earlier stages of mitosis by its staining properties. It is also recognizable in autoradiography studies which show heterochromatin regions to be late-replicating, while euchromatin regions replicate earlier in the S phase of the cell cycle. Most heterochromatin is localized around the centromeres and at the ends of chromosomes (**telomeres**). These regions contain highly repetitious simple-sequence DNA.

One heterochromatic structure has been identified as **sex chromatin** or the **Barr body** (Fig. 14.24). This structure is the condensed, inactive second X chromosome in XX female cells. Males have no Barr body since their only X chromosome remains euchromatic at all stages of development. The two X chromosomes of female cells are euchromatic during meiosis and very early embryo development. The second X becomes condensed and genetically inert afterward. According to the genetic analysis either X may be inactivated on a random basis.

The Sex Chromosomes

Most information on sex determination systems came from studies of insect chromosomes, which have genetic material that is easy to obtain and easy to study cytologically. There are several different sex-determining mechanisms among insects, but species with XX females and XY or XO (pronounced X-"Oh") males were believed to be models for understanding human and other mammalian sex chromosome activities. The *Drosophila* chromosome complement has been especially well studied, and different combinations of autosomes and sex chromosomes were introduced into strains to see what pattern might emerge (Table 14.1).

The variations all pointed to a pattern which showed that the ratio between number of X chromosomes and number of autosome sets was the critical factor in sex determination. If the ratio was 1.0 (XX : AA), a female developed; if the ratio was 0.5 (X : AA), a male would develop from the fertilized egg. Since *Drosophila* males can be either XY or XO in diploid laboratory strains, the Y chromosome seemed irrelevant to sex determination. There are genes on the Y chromosome, however, so XO males were genetically different than those with the XY complement. Intersexes were produced when the X : A ratio deviated from the norms of 0.5 and 1.0 for males and females, respectively. In other insects, such as grasshoppers, the males are normally XO, because there is no Y chromosome in those species. This evidence further supported the **Sex Balance Theory** of sex determination.

When human and other mammalian chromosomes became readily available for cytological study, it soon was found that a different sex-determining role was played by the mammalian Y chromosome than had been found in insects. In mice, cats, humans, and other mammals, the X chromosome is female-determining and the Y-chromosome is male-determin-

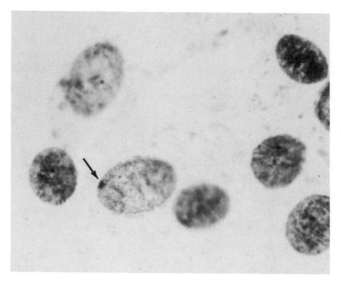

Figure 14.24
Barr body, or sex chromatin, in the human female. One of the two X chromosomes remains condensed at interphase, and is located at the periphery of the nucleus (arrow). (Courtesy of T. R. Tegenkamp)

Table 14.1 Chromosomal basis for sex determination in *Drosophila*

SEX CHROMOSOMES PRESENT	NUMBER OF AUTOSOME SETS	RATIO OF X : A	SEX OF INDIVIDUAL
XX	2	1.00	Female
XY	2	0.50	Male
XXX	2	1.50	Metafemale (sterile)
XXXX	3	1.33	Metafemale (sterile)
XXY	2	1.00	Female
XXX	4	0.75	Intersex
XX	3	0.67	Intersex
XY	3	0.33	Metamale (sterile)
X	2	0.50	Male (sterile)

ing (Fig. 14.25). More specifically, if a Y chromosome is absent then female development takes place in the presence of 1, 2, or more X chromosomes (Table 14.2). If there is a Y chromosome, it is male-determining even if as many as 4 X chromosomes are also present. In mammals, XO individuals are female, and a Y chromosome is present in every male. The sex balance mechanism cannot apply to mammals as it does to insects, since XO insects are males.

There is no known mammalian organism with both X chromosomes missing, and no case of a male with a Y chromosome but no X. The X chromosome must be vital to cell and organism development since it is always present. The Y chromosome is dispensable, since all females lack this chromosome. The proper development and differentiation of sex organs and secondary sex characteristics will only take place in human beings if there are at least two sex chromosomes, either XX or XY. Females with only one X chromosome have various clinical problems, and are sterile since sex organs remain undeveloped in the adult. Mouse XO females, however, are perfectly normal and fertile.

The sex-determining genes on the X and Y chromosome are believed to act only during the early months of human embryo and fetus development.

Table 14.2 Anomalies involving sex chromosomes in humans

INDIVIDUAL DESIGNATION	CHROMOSOME CONSTITUTION*	SEX	FERTILITY
Normal male	46, XY	Male	+
Normal female	46, XX	Female	+
Turner syndrome	45, X	Female	−
Triplo-X	47, XXX	Female	+
Tetra-X	48, XXXX	Female	−
Penta-X	49, XXXXX	Female	?
Klinefelter syndrome	47, XXY	Male	−
Klinefelter syndrome	48, XXXY	Male	−
Klinefelter syndrome	49, XXXXY	Male	−
Klinefelter syndrome	48, XXYY	Male	?
Klinefelter syndrome	49, XXXYY	Male	?
XYY-male	47, XYY	Male	+

* The two-digit number indicates the total number of chromosomes in diploid cells, followed by the exact number and kinds of sex chromosomes in this complement.

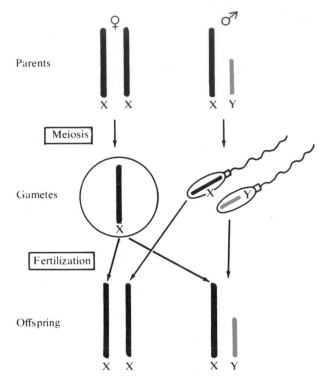

Figure 14.25
The chromosomal basis for sex determination in human beings and other mammalian species.

By about 6 weeks of age, the embryo begins to produce male hormones if a Y chromosome is present. These hormones influence differentiation of unspecified gonads so that they begin to develop into testes. External genitalia differentiate in the fetus at 3 months of age. If development continues normally in the fetus and after birth, male internal reproductive tract and other secondary sex characteristics will appear. Apart from these sex-determining genes, there is no unambiguous genetic evidence that other genes are located on the Y chromosome.

If the embryo has no Y chromosome, its gonads do not begin to differentiate until about the twelfth week of age. At this time the rudimentary gonads differentiate into ovaries. The internal reproductive tract then develops, as do the external genitalia, and female primary (gonad) and secondary sex characteristics continue to differentiate under hormonal influences until late adulthood. *Determination* of sex is chromosomal. It is expressed by development of the **primary sex characteristics**—testes or ovaries—from the pair of embryonic gonads. *Differentiation* of internal and external **secondary sex characteristics** is under the dual influence of gene and hormonal factors before and after birth.

Many genes on the X chromosome are completely unrelated to sex or sexual development. In this way the X is quite different from the Y. Occasional evidence from pedigree studies has indicated a slight degree of homology between human X and Y chromosomes. Inheritance patterns predicted or observed for exclusively X-linked or exclusively Y-linked genes are different from each other and different from patterns of genes that would be on both the X and Y chromosomes. While the genetic evidence is scanty and controversial, cytological studies have shown that some homology does exist. Meiotic cells from the testis have been studied to determine the nature, if any, of synapsis between X and Y chromosomes in spermatocyte prophase I. A small region of the Y and X does synapse and a synaptinemal complex can be seen. Whether or not there is any crossing over between these two chromosomes must await evidence of genetic recombination.

Medical Cytogenetics

We have already mentioned the presence of an extra chromosome in most cases of trisomy 21 (**Down's syndrome,** formerly called "mongolism"). This was the first disorder to be associated with a chromosome anomaly, but others have since been found. Other aneuploid complements involving autosomes or sex chromosomes are being found increasingly as clinical studies intensify. In addition to trisomy and monosomy, a number of clinical conditions are associated with deletions and translocations of chromosome arms or parts of arms. Some chromosomal variations cause no detectable symptoms in the individual. These have been discovered from general samplings of newborns or institutionalized populations, or from family members of some patient with a chromosomal anomaly.

ANEUPLOIDY. About 96 percent of patients with **autosomal trisomies** are trisomy 21 cases. Other trisomies that have been reported involve chromosomes 8, 13, and 18. Statistical information on frequency of trisomies in live births shows about 1 in 650 is trisomy 21, and 1 in 4000–8000 is trisomic for chromosome 13 or 18. Mental retardation is characteristic of all these individuals, along with a variety of physical malformations. Lower life expectancy is the general rule in individuals with these abnormalities. There is no known case of **autosomal monosomy** (45 chromosomes), although some reports of monosomy involving a G-group chromosome (Nos. 21 or 22) have appeared. These reports have not been verified and are believed to represent substantial deletions or translocation of G-chromosome to some other chromosome in the set.

A greater variety of aneuploidy involving the sex chromosomes has been found (see Table 14.2). The two main types of sex chromosome anomalies are *45, X* females with clinical symptoms of **Turner's syndrome** and *47, XXY* males with **Klinefelter's syndrome.** Almost all *45, X* conceptions result in miscarriages, but the incidence of Turner's syndrome live births is 1 in 2500. In most cases the missing sex chromosome is caused by loss of the paternal X or Y, probably at fertilization and not by meiotic nondisjunction. According to studies of a blood group gene marker on the X chromosome, almost 80 percent of the cases showed the X chromosome in *45, X* patients was of maternal origin.

Slightly more than half the cases diagnosed as Turner's syndrome have the standard *45, X* monosomic karyotype (Fig. 14.26). The remainder have been found to have X chromosome deletions or mosaic tissues with various karyotypes in different patches. Mental retardation does not necessarily occur; some *45, X* females have scored very high on IQ tests.

Women who are *45, X* have no Barr body in somatic cells. Their single X chromosome remains euchromatic during interphase. Males with *47, XXY* karyotypes, on the other hand, show one Barr body in somatic cells. These simple diagnostic tests provide preliminary information that relates sex chromosome anomaly to clinical symptoms. Analysis of the chromosomes generally confirms the predictions. Women with two Barr bodies instead of the normal one have also been reported. These *47, XXX* karyotypes have been discovered in samplings of newborns or other groups, and they represent about 1 in 1500 live births. There are no clinical symptoms; fertility and intelligence levels are the same in individuals with these karyotypes as in the general population. However, *48, XXXX* females are mentally retarded.

Males with Klinefelter's syndrome account for 1 in 1000 newborns, or 1 in every 500 male infants every year. Karyotypes vary between *47, XXY* and *49, XXXXY* extremes, with about 20 percent of the total caused by mosaic tissues or more than two X chromosomes. The *47, XXY* karyotype is found in 80 percent of Klinefelter males (Fig. 14.27). These men are usually sterile, but they are not necessarily mentally retarded. The degree of retardation increases with increasing number of X chromosomes.

Until 1965 there were only 10–20 known *47, XYY* males. At present the statistics from various samplings show that about 1 in 650 live male births is *47, XYY*. This karyotype has no general effect on fertility or intelligence, and there is probably no correlation between the karyotype and abnormal behavior, as has been suggested from biased samplings of men in penal or mental institutions. Studies of "antisocial" behavior in *47, XYY* males are not fully evaluated as yet.

DELETIONS AND TRANSLOCATIONS. Three general kinds of deletions have been found in human chromosomes:

1. Deletions of part of the short arm of a chromosome, designated *p* −.

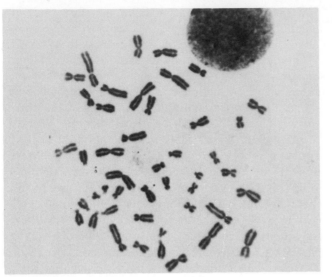

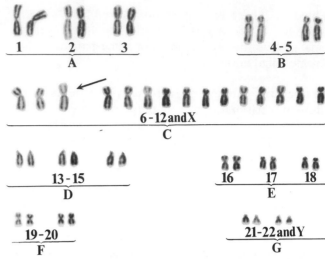

1   2   3   4-5

A   B

6-12 and X

C

13-15   16   17   18

D   E

19-20   21-22 and Y

F   G

X Chromosomes occupy the second position in the C group

Figure 14.26
Chromosomes of a *45,X* female with Turner's syndrome: (a) metaphase chromosome spread; (b) karyotype prepared from the metaphase chromosome photograph shown in a. (Courtesy of T. R. Tegenkamp)

2. Deletions of part of the long chromosome arm, designated $q-$.
3. Deletions of both telomeres of a chromosome with subsequent fusion of the broken ends to form a ring chromosome, designated as r.

A deletion of part of the short arm of chromosome 18 would be designated $18p-$, $18q-$ designates a deletion of part of the long arm of chromosome 18, and $18r$ labels chromosome 18 as a ring chromosome because of telomere deletions and later fusion of the two broken ends.

Since these karyotypes have only been reported after 1963, and chromosome analysis is more difficult to perform for structural than for numerical variations, there is little statistical information on their frequencies. Deleted chromosomes from the B, C, D, E, and G size groups have been reported (Fig. 14.28). Almost all result in severe physical malformations and mental deficiency, and many lead to death in infancy or childhood. Most deletion types have no more than one-third of an arm missing. Because a variable amount of genic material is lost, the symptoms vary somewhat among patients with similar chromosomal anomalies.

More translocation anomalies are being discovered as cytogenetic studies continue and expand. The best known anomaly involves translocation of part of chromosome 21 with either a D or a G group chromosome (Fig. 14.29). If a translocation chromosome involving No. 21 is present, it may be part of a 46-chromosome karyotype of a patient with Down's syndrome or part of a 45-chromosome karyotype in a normal person.

The 46-chromosome patient would have the usual two No. 21 chromosomes plus part of another chromosome 21 attached to either a D or a G chromosome in the group. The chromosome count would be 46, but the karyotype would have two unpairable chromosomes: the normal chromosome and the translocated chromosome which is partly homologous with the normal one. In 45-chromosome karyotypes, only one

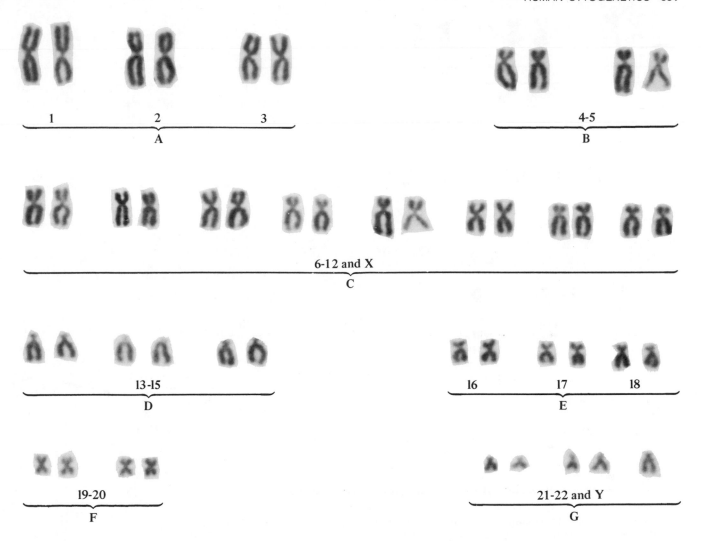

1   2   3
A

4-5
B

6-12 and X
C

13-15
D

16   17   18
E

19-20
F

21-22 and Y
G

X Chromosomes occupy the second position in the C group

Figure 14.27
Karyotype of a human male with Klinefelter's syndrome.
There is an extra X chromosome, making the total count
47 instead of the normal 46. (Courtesy of C. Hux)

chromosome 21 would be detectable, together with
a longer translocated chromosome carrying the rest
of the second No. 21. The 46-chromsome individual
has three copies of chromosome 21 genes and develops Down's syndrome, while the 45-chromosome
individual has only two copies of chromosome 21
and is normal.

Where cases of trisomy 21 occur repeatedly in a

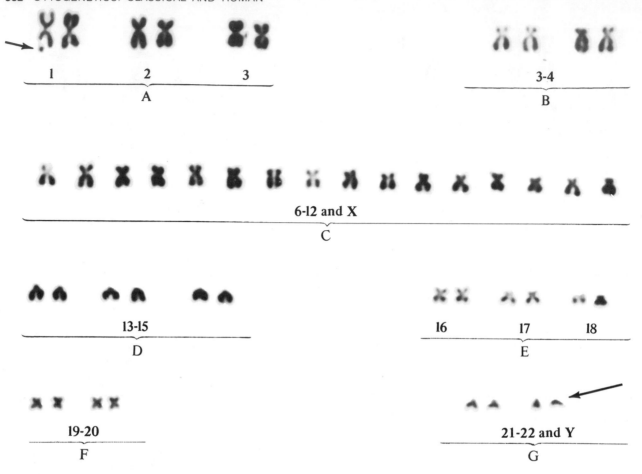

X Chromosomes occupy the second position in the C group

Figure 14.28
Karyotype of a female patient with a deletion of part of the short arm of chromosome 22 (called the "Philadelphia" chromosome). The deletions are incurred spontaneously during the lifetime of the individual and are not transmitted to offspring. There also has been a break in one chromosome 1, with the deleted segment near the broken long arm of the chromosome (see Fig. 14.5). (Courtesy of T. R. Tegenkamp)

family, it is usually because the translocated chromosome is transmitted through the gamete to the new generation. If the gamete carries one No. 21 and one translocated chromosome with No. 21 genes, it will cause the development of a trisomy 21 individual because the other gamete will have its usual complete set of chromosomes, including its own No. 21. This type of gamete fusion produces three copies of chromosome 21 genes in the new individual that develops from the fertilized egg (Fig. 14.30). About 3 percent of trisomy 21 patients carry a translocated chromosome in a 46-chromosome karyotype.

Figure 14.29
Translocation involving chromosome 14 of the D group and chromosome 21 of the G group in a female patient with Down's syndrome. There are 46 chromosomes in the nucleus, but in addition to two chromosomes 21 there is a third set of chromosome 21 genes attached to the 14/21 translocated chromosome. This patient is trisomic for chromosome 21, and she displays the typical clinical symptoms of Down's syndrome. (a) Metaphase chromosome spread. (b) Karyotype prepared from the metaphase chromosomes shown in a. (Courtesy of T. R. Tegenkamp)

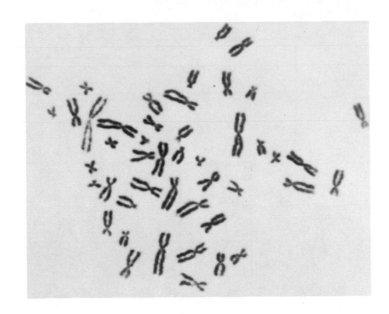

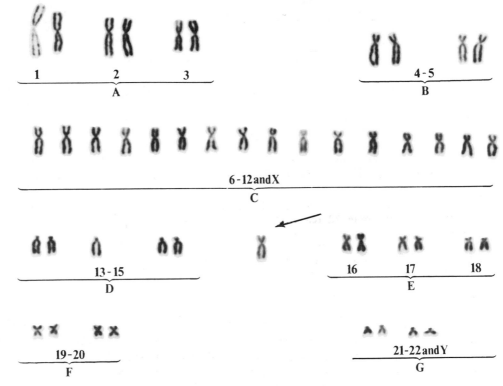

X Chromosomes occupy the second position in the C group

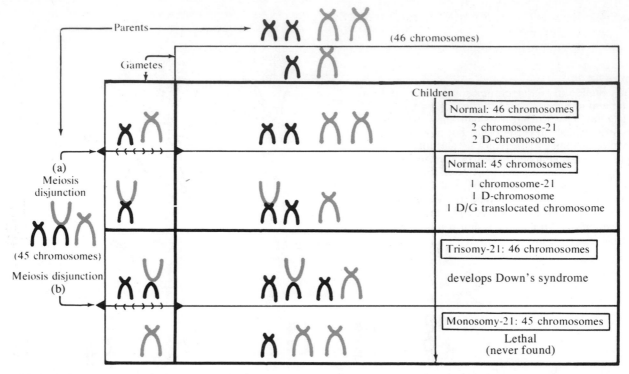

Figure 14.30

Summary chart showing the basis for cases of hereditary transmission of Down's syndrome. The parent with 46 chromosomes contributes one chromosome 21 and one D chromosome to each child through the gametes. The parent with 45 chromosomes, including a D/G translocated chromosome, may contribute any one of four kinds of chromosome combinations to the child, depending on the disjunction of chromosomes at meiosis. (a) Chromosome 21 plus a D chromosome may go to one pole and the remaining D/G translocated chromosome may go to the opposite pole at meiosis I. In either case the child would be normal. (b) Chromosome 21 and the D/G chromosome may go to one pole and the D chromosome may go to the opposite pole at meiosis. The combination produces a trisomy-21 child with Down's syndrome, but the alternative monosomic combination must be lethal since it has never been observed even in aborted embryos or fetuses.

PRENATAL DIAGNOSIS. In some cases, it is helpful to the family and physician to know the chromosome constitution of the fetus. If one of the prospective parents has a family history of trisomy 21, there is a higher risk that the fetus will carry extra copies of the chromosome. Even if there is no history of the condition in the family, older women have a greatly increased probability of giving birth to trisomy 21 children. The risk is about 1 : 2000 for a 20-year-old woman and 1 : 50 for a woman over 45 years of age. The family may wish to know if the fetus is trisomic, regardless of the decision they may make if the tests prove positive.

Another situation where chromosome analysis is desirable is in cases involving risk of transmitting sex-linked genetic afflictions. Males usually develop sex-linked diseases or malformations since they have only one X chromosome (Fig. 14.31). If there is a family history of the condition, or if previous boys have been born with the disease, the parents may wish to know if the unborn child will be a boy. This in itself will not say that the male fetus carries the defective allele in his X chromosome, but the chance that the allele would be present is 50 percent.

The human fetus floats in liquid within the *amnion,* a sac formed from embryonic membrane. During its development, the fetus regularly sheds cells into the *amniotic fluid* which fills the amnion. Using a method known as **amniocentesis,** a physician withdraws a sample of the amniotic fluid by hypodermic syringe through the mother's abdomen (Fig. 14.32). The cells in amniotic fluid are mostly of the fibroblast type, and

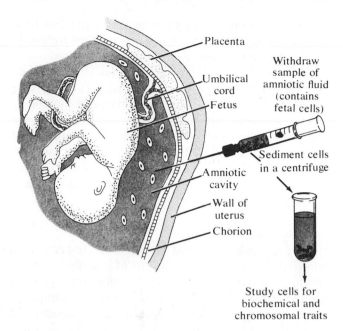

Figure 14.32
Amniocentesis. A sample of amniotic fluid is withdrawn and processed for cytological and biochemical analysis of fetal traits.

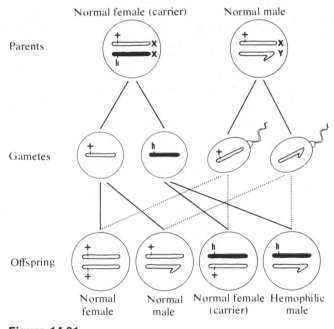

Figure 14.31
Sex-linked recessive afflictions such as hemophilia are expressed more frequently in males than in females because males have only one X chromosome and therefore have no masking allele, as XX females do.

they can be cultured in a flask or dish. These cells are then examined to determine the fetal karyotype or to measure biochemical activities that may indicate the presence of an inherited disease determined by a defective gene, even if the karyotype is perfectly normal.

In one dramatic case involving a sex-linked genetic defect, a family had lost a son who had **agammaglobulinia,** a condition in which no active immunity system is present to fight infection. After amniocentesis, they discovered that their next child would be a boy, and this child had a 50 percent chance of being born with the affliction. A medical team then prepared a complicated germ-free environment in which the infant could be kept after birth. When tests showed that the baby had the disease, he was kept in the germ-free quarters. At this writing he is about four years old and has been kept in good health by isolation from infection. Doctors hope that some medical treatment can be discovered for his condition very soon.

Other cases in which tests of fetal materials indicate the existence of a hereditary disease are regularly reported. The parents can then be guided by *genetic counseling,* and they can make their decisions on the course of action they will take once they understand the situation. Improved health care delivery now includes information obtained from *prenatal* diagnosis, as well as the more conventional postnatal examination and treatment.

Somatic-Cell Hybridization

Understanding human genetics is important in improving treatment and analysis of inherited diseases and in determining how genes function in the cell and the individual. The usual genetic study technique of breeding analysis cannot be applied to humans, and compiling pedigrees to determine gene location and inheritance patterns is time-consuming.

In 1960 a whole new approach became feasible when G. Barski showed that somatic cells from different mouse cell cultures could fuse to form hybrid cells. This study was the beginning of tremendous advances in cell culture, hybridization techniques, chromosome identification by staining, and the mapping of human genes on the 24 kinds of chromosomes in the complement. More than 50 genes have been located on 18 chromosomes by somatic-cell genetic methods, and these discoveries have added considerably to the mapping information obtained from pedigree studies. Sex-linked inheritance patterns make it easy to assign genes to the X chromosome, but few autosomal gene locations were assigned before somatic-cell hybridization became available.

HYBRID CELL LINES. Human somatic cells are usually hybridized with mouse cell-lines or those from other rodents since the chromosome complements are strikingly different. Rodent cell-lines are readily available, and these mammalian species produce many similar but not identical protein products of their respective genes. When different somatic cell-lines are properly treated and mixed together in a dish, somatic-cell hybrids form spontaneously after a few hours. The frequency of fusion can be increased if cells are first exposed to inactivated Sendai viruses. This treatment causes formation of intercellular bridges between adjacent cells. To encourage growth of the hybrid cells and prevent parental cells from swamping the growing hybrid-cell colonies, two factors are important:

1. A *selective* medium in which hybrids grow and parental cells do not.
2. Parental cell-lines that either do not proliferate normally in culture or have enzyme deficiencies which prevent their growth in the selective medium.

The usual selective medium contains **aminopterin,** which blocks production of the folic acid needed to synthesize DNA nucleotides by the conventional biochemical pathway. Nucleotides can be synthesized through an alternative pathway even in the presence of aminopterin, if precursors **hypoxanthine** and **thymidine** are provided in the growth medium (Fig. 14.33). The alternative biosynthetic pathway can be used only if enzymes are present to utilize these precursors: the enzymes are **hypoxanthine guanine**

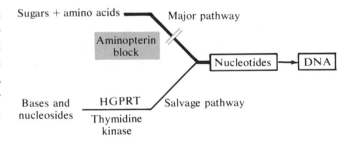

Figure 14.33
Formation of DNA occurs through a salvage pathway if the major pathway is blocked by the drug aminopterin. This process occurs only if the enzymes HGPRT and thymidine kinase are synthesized in the cells.

phosphoribosyltransferase (HGPRT) and **thymidine kinase (TK).**

If human fibroblasts or lymphocytes containing both enzymes are fused with mouse cells deficient in one or both enzymes, the hybrid-cell colonies will develop, while the parental cells do not grow. The human parental cells normally do not proliferate or do so very slowly in culture, and the mouse parental cells lack the enzymes to grow in the HAT (hypoxanthine, aminopterin, thymidine) selective medium. Single hybrid cells grow and each produces a visible colony that includes the original fusion cell and all its mitotic descendants (Fig. 14.34).

Two important features make these hybrid cells uniquely useful for study:

1. Mouse chromosomes are retained, but the human chromsomes are lost preferentially and at random so that only one or a small number remain in the hybrid nucleus.
2. Both mouse and human genes are expressed at the same time in the hybrid cells, each producing

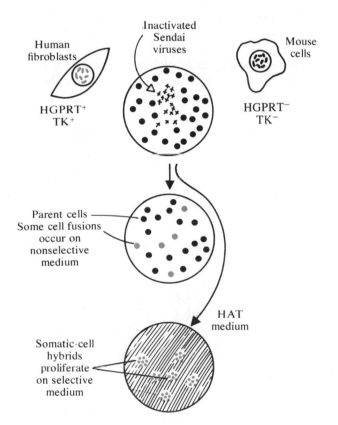

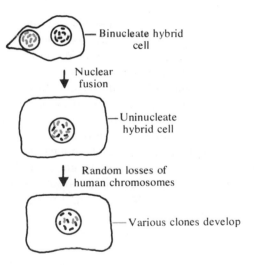

Figure 14.34
Flow scheme showing the method for isolating clones of somatic-cell hybrids on selective HAT medium. The events involving individual cells are shown at the right and clonal events on the medium in the culture dishes are shown at the left.

unique protein translation products, or functionally similar proteins, that can be distinguished by their amino acid composition.

The chromosomes and their gene products are distinguishable in human-rodent hybrid cells. The 40 mouse chromosomes are different in size, shape, and centromere location from the 46 human chromosomes. The particular human chromosomes retained in the hybrid cell can be identified precisely by general morphology and banding pattern after Giemsa or fluorescence staining (see Fig. 14.22). Human-mouse cell hybrids usually have between 41 and 55 chromosomes instead of the 86 chromosomes (40 mouse + 46 human) that must have been present in the beginning. From 1 to 15 of the 46 human chromosomes are retained, therefore, along with all 40 of the mouse chromosomes. Different hybrid-cell lines can be collected and studied cytogenetically since different chromosomes are lost on a random basis.

GENE MAPPING. The procedures for mapping human genes involve determining that two or more genes are present in the same chromosome (linked) and then finding the particular chromosome in which they are located. In general, hybrid-cell lines are examined to see which genes are lost together or expressed together, which indicates they are linked. After this it is a matter of comparing the pattern of expression of these genes in cell lines that have a particular chromosome and those lacking the chromosome.

Because of technical difficulties it has not been possible to collect 24 different hybrid-cell lines in which a different single human chromosome is present in each. Tests are usually conducted using 8–10 different lines. Each line has some unique subset of chromosomes and by comparing gene expression in all of these, the one chromosome common to all cases of expression of the gene is isolated as the chromosome within which this gene is located. The gene will be expressed in every case in which its chromosome is present, and it will not be expressed in every case

in which that certain chromosome is absent from the nucleus.

This method places genes on particular chromosomes but does not indicate their location in a definite region of the chromosome. By using cell lines which have deletions or translocations of chromosome arms or parts of arms, a more precise location can usually be made (Fig. 14.35). For example, the first human gene to be mapped by somatic-cell hybridization was the *TK* gene responsible for production of thymidine kinase. This gene is located in chromosome 17. After a number of mouse-human cell hybrids had been examined, a hybrid with the long arm of human chromosome 17 translocated to a mouse chromosome was found. No other part of chromosome 17 was present. The mouse parent had been deficient in TK, but the hybrid synthesized the *human* form of the enzyme. Since TK was synthesized only when the long arm

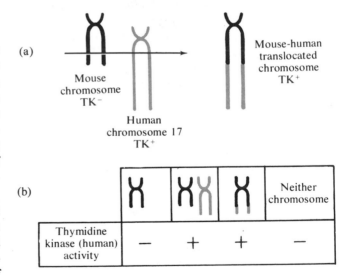

Figure 14.35
Diagram illustrating the method that allowed the localization of the thymidine kinase gene (TK) to human chromosome 17: (a) origin of the mouse-human translocated chromosome; and (b) the observed chromosome found in cells with and without human thymidine kinase activity. See text for details.

of chromosome 17 was present, the gene was assigned to that part of the chromosome.

Most somatic-cell genetic study has concentrated on mapping genes that are expressed all the time, in all or most cell types. The techniques can be extended to studies of genes that are expressed only in response to specific signals and to those expressed only in specialized cell types. In this way, it becomes possible to analyze gene regulation and to analyze patterns related to cell development and differentiation. Such methods also open possibilities for detecting inherited defects that characterize almost any cell type in the organism.

At present the only kinds of defects that can be detected prenatally are those involving gross chromosomal alterations or biochemical malfunctions that happen to be expressed in fibroblasts, the only kind of cell recovered from amniotic fluid in any reasonable amount. If genes are turned off in fibroblasts, their functioning cannot be assayed. If techniques can be developed to turn genes on and off in fibroblasts, then almost any inherited defect could be analyzed from fetal fibroblast cultures. For example, blood proteins are not produced by fibroblasts. If genes which determine blood proteins could be turned on, blood proteins would be produced and could then be analyzed for composition and function. Such possibilities hold considerable promise not only for clinical human genetics, but also for understanding gene modulation that fine-tunes the activity of cells in the functioning human being.

Similarities between Human and Great Ape Chromosomes

With the new staining procedures that distinguish chromosomes by particular banding patterns, more precise comparisons can be made between genomes of related species. The closest living relatives to the human species are the great apes: chimpanzee, gorilla, and orangutan. According to various lines of evidence, it seems that the ape and human families began to diverge from a common ancestor about 20 million years ago. Earlier divergent evolution had produced monkeys and then gibbons, perhaps 40 million years ago. Primates (the order of mammals to which all these groups belong) first appear in fossil deposits which are about 60 million years old.

There are 48 chromosomes in the somatic cells of all the apes, compared with 46 in human cells. Detailed comparisons of the chromosome complements revealed that most human and chimpanzee chromosomes are almost identical in banding patterns (Fig. 14.36). Orangutan chromosomes are very similar to human chromosomes, and since gorilla and chimpanzee chromosomes are almost identical, all the apes have chromosomes remarkably like each other and like humans. One consistent difference between human and ape chromosome complements is the lack of chromosome 2 in apes and the presence of two smaller acrocentric chromosome pairs, which have no counterpart in the human karyotype. When bands of human chromosome 2 are compared with the ape unique acrocentrics, it appears that human chromosome 2 arose by *centric fusion* from the two acrocentrics. This hypothesis could be a simple explanation for the difference in chromosome number, since centric fusion could have occurred in the human lineage, but not in ape evolution, to produce human chromosome 2.

In addition to centric fusion, the other major type of structural rearrangement seems to be *pericentric inversion* (includes the centromere). The D (Nos. 13, 14, 15), F (Nos. 19, 20), and G (Nos. 21, 22) chromosomes are essentially homologous in banding between human and chimpanzee complements, as are autosomes 6, 7, 8, 10, and 11 from the C group, and No. 16 from the E group. The other six autosomes (Nos. 4, 5, 9, 12, 17, and 18) seem to have undergone pericentric inversions. Some other minor rearrangements and small deletions may also have taken place, according to banding interpretations. Based on these comparisons, J. de Grouchy and C. Turleau have attempted to construct the presumptive ancestral karyotype from

HOMO ⟷ PAN

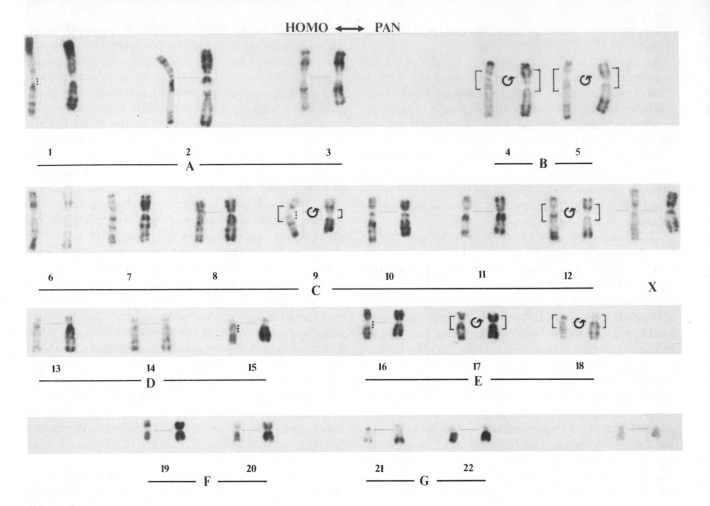

Figure 14.36

Comparison of human (*Homo*) and chimpanzee (*Pan*) karyotypes according to chromosome morphology and R-band pattern. The human chromosome is to the left in each pair. Human chromosome 2 is shown alongside two acrocentric chromosomes (in head-to-head alignment) of the chimpanzee. Pericentric inversions are indicated by curved arrows for chromosomes 4, 5, 9, 12, 17, and 18. Newly acquired material in human chromosomes 9, 15, and 16 are indicated by dots to the right of the chromosome region. (Courtesy of C. Turleau and J. de Grouchy, from Turleau, C., and J. de Grouchy, 1973, *Humangenetik* **20**:151–157, Fig. 1.)

which modern human and ape complements have arisen.

Even more information about the X chromosome is available because about 150 genes on this chromosome are known, as well as its morphology and banding after staining. There has been relatively little genetic or structural change in the X chromosome of various mammalian groups in general, even though mammals first arose over 200 million years ago. The X chromosome in kangaroos, among the most ancient mammalian types, is much like the X in cats, dogs, horses, cattle, rodents, gibbons, great apes, and humans. Gene mutations have obviously occurred, but the overall rate of evolutionary change in X chromosome structure and content is amazingly low. The X chromosome appears to have been retained almost intact in gibbons, great apes, and humans, while gibbon autosomes have undergone profound changes during evolution. Little can be said now about the Y chromosome because its gene content is unknown and its length varies even within different human populations.

On the whole, it seems that more than 98 percent of the band patterns in human and great ape chromosome complements are similar or identical. These patterns are rather crude indicators of homology, of course, and there may be many subtle differences that are not visible with these staining methods. Comparisons of gene content and differences in amino acids of the protein translations are much more reliable indicators of relationships. But the few proteins that have been fully described for amino acid composition and sequence show that ape and human proteins are extremely similar. The greater the evolutionary separation between species, the greater the difference in the amino acids of the same functional proteins.

For example, cytochrome *c* has 104 amino acids in known type and sequence. There is no difference between human and chimpanzee cytochrome *c*. Rhesus monkey has 1 amino acid different from human and chimpanzee. There are 13 amino acid differences between human and horse cytochrome *c*, however. The number and kinds of amino acid differences correspond very closely to relationship and evolutionary rank of the species that have been studied. More codons have mutated in the longer time separating distant relatives, than in the much shorter time since close relatives diverged from a common ancestor in evolution. Similar patterns of change have been found for a few other proteins whose amino acid content is known, such as hemoglobin, one of the histones, and a blood clotting factor.

Studies of human genes and human chromosomes, together with comparative analyses among different groups of animals, continue to reinforce the history of human evolution as seen by more classical methods of geology, comparative anatomy, and embryology. We may someday know our exact lineage, as we now know that our closest living relatives are members of the great ape family.

SUGGESTED READING

Books, Monographs, and Symposia

Crew, F. A. E. 1965. *Sex Determination.* New York: Dover Press.

Ford, E. H. R. 1973. *Human Chromosomes.* New York: Academic Press.

Hamkalo, B. A., and Papaconstantinou, J., eds. 1973. *Molecular Cytogenetics.* New York: Plenum.

McKusick, V. A., and Claiborne, R., eds. 1973. *Medical Genetics.* New York: HP Publications.

Swanson, C. P., Merz, T., and Young, W. J. 1967. *Cytogenetics.* Englewood Cliffs, N.J.: Prentice-Hall.

Articles and Reviews

Barr, M. L., and Bertram, E. G. 1949. A morphological distinction between neurones of the male and female, and the behaviour of the nucleolar satellite during accelerated nucleoprotein synthesis. *Nature* 163:676–677.

Brown, D. D. 1973. The isolation of genes. *Scientific American* 229(2):20–29.

Caspersson, T., Lomakka, G., and Zech, L. 1971. The 24 fluorescence patterns of the human metaphase chromosomes—distinguishing characters and variability. *Hereditas* 67:89–102.

Drets, M. E., and Shaw, M. W. 1971. Specific banding patterns of human chromosomes. *Proceedings of the National Academy of Sciences* 68:2073–2077.

de Grouchy, J. 1974. L'évolution des chromosomes. *La Recherche* 5:325–336.

Ephrussi, B., and Weiss, M. C. 1969. Hybrid somatic cells. *Scientific American* 220(4):26–35.

Friedmann, T. 1971. Prenatal diagnosis of genetic disease. *Scientific American* 225(5):34–42.

German, J. 1970. Studying human chromosomes today. *American Scientist* 58:182–201.

Lejeune, J., Turpin, R., and Gauthier, M. 1959. Le mongolisme, premier exemple d'aberration autosomique humaine. *Annales Génétique* 1:41-49.

Lerner, R. A., and Dixon, F. J. 1973. The human lymphocyte as an experimental animal. *Scientific American* 228(6):82–91.

Ruddle, F. H., and Kucherlapati, R. S. 1974. Hybrid cells and human genes. *Scientific American* 231(1):36–44.

Tijo, J. H., and Levan, A. 1956. The chromosome number of man. *Hereditas* 42:1–6.

Glossary

acetyl coenzyme A a high-energy intermediate in energy-transferring reactions in metabolism; activated acetate linked to coenzyme A

acrocentric centromere situated close to one end of a chromosome; a chromosome whose centromere is located very near one end of the chromosome

actin a major protein of muscle; the principal protein of thin filaments of striated muscle and many nonmuscle cells, involved in cell movements

active site the region on an enzyme at which a substrate binds and reacts with the enzyme to form a temporary enzyme-substrate complex

active transport the energy-requiring assisted passage of molecules across a membrane, in the direction of higher concentration of the molecules

adenosine triphosphatase ATPase; the enzyme that hydrolyzes ATP to form ADP and inorganic phosphate and catalyzes the reverse reaction of ATP formation

adenosine triphosphate ATP; a nucleoside triphosphate which is a high-energy intermediate in energy-transferring metabolism and one of the precursors for ribonucleic acid synthesis

allele one of the alternative forms of a gene

amino acid the basic building block of proteins; a carboxylic acid with one or more NH_2 groups

aminoacyl-tRNA synthetase one of a group of enzymes

which catalyzes the activation of an amino acid to the aminoacyl form and the joining of the aminoacyl group to specific transfer RNA carriers

amphipathic molecules having spatially separated hydrophilic and hydrophobic regions

anaphase stage of mitosis or meiosis in which the chromosomes move to opposite poles of the cell

anaplerotic reactions that replenish depleted intermediates for other metabolic pathways

aneuploidy having one or more chromosomes in excess of or less than the usual number

apoenzyme the protein component of a holoenzyme which also contains a nonprotein coenzyme or prosthetic group

aster the region at the poles of dividing cells, including microtubules surrounding a clear zone within which a pair of centrioles is located

ATP *see* **adenosine triphosphate**

ATPase *see* **adenosine triphosphatase**

autoradiography a method for localizing radioactive atoms in microscope preparations of biological materials by exposing of a photographic film emulsion to radioactive atoms incorporated in the biological specimen

autosome any chromosome of the complement that is not a sex chromosome

bacteriophage any virus which requires a bacterial host for its replication

Barr body any condensed or inactivated X chromosome in the interphase nucleus; also called sex chromatin

basal body *see* **centriole**

bioenergetics thermodynamics applied to living systems

bivalent a synapsed pair of homologous chromosomes seen in Meiosis I

breakage and reunion a mechanism of crossing over between linked genes giving rise to recombinations of the genes

C-banding a method for staining chromosomes differentially showing locations of heterochromatin in the stained banded regions of the chromosome

C_3 **cycle** part of the dark reactions of photosynthesis; CO_2 is reduced to carbohydrate via a 3-carbon intermediate, 3-phosphoglyceric acid; also known as the Calvin cycle

C_4 **cycle** an accessory CO_2-reducing pathway in photosynthesis occurring in plants lacking photorespiration; also known as the Hatch-Slack pathway

Calvin cycle *see* C_3 **cycle**

carotenoid an accessory photosynthetic pigment

carrier a transport protein within the membrane that binds temporarily with a molecule to be transported across the membrane

catalyst any agent that modulates the rate of a chemical reaction without altering the equilibrium point of that reaction (*see also* **enzyme**)

cell concept a generalization of the cell theory, stating that the cell is the ultimate structural unit of the organism

cell cycle the sequence of events in dividing cells in which an interphase consisting of G_1, S, and G_2 periods separates one mitosis from the mitosis of a successive cell cycle

cell division formation of two daughter cells from a parent cell by enclosure of the two nuclei in separate cell compartments (*see also* **cell plate formation, furrowing**)

cell plate formation a cell division process typical of higher plants; new wall materials are laid down first in the cell center and then continue to be laid down in an outward direction to the cell periphery

cell wall rigid or semirigid structure enclosing the living protoplast of plant, algal, fungal, and prokaryotic cells

centric fusion the mechanism of formation of one large metacentric chromosome from two smaller acrocentrics after breakage and subsequent fusion

centriole a microtubule-containing organelle located at the spindle poles in dividing cells or embedded at the cell periphery and forming the basal portion of a cilium or flagellum

centromere the region of the chromosome to which the spindle fibers attach and which is required for chromosome movement to the poles at anaphase

chiasma site of exchange between two chromatids of a bivalent, visible in diplotene, diakinesis, and metaphase I of meiosis as a crossover figure

chlorophyll the principal photosynthetic light-capturing pigment, located in chloroplast thylakoids or in prokaryotic photosynthetic folded membranes

chloroplast the chlorophyll-containing photosynthetic organelle in eukaryotes

cholesterol a major lipid constituent of animal cell plasma membrane

chromatid one half of a replicated chromosome, joined to the other chromatid at the centromere region

chromatin the deoxyribonucleoprotein material of the chromosomes

chromatin fiber the continuous deoxyribonucleoprotein molecule of the chromosome

chromomere a beadlike or knobby region best seen in early meiotic chromosomes

chromosome the gene-containing structure in the nucleus

cilium (pl. **cilia**) a whiplike locomotor organelle produced by a centriole

cisterna (pl. **cisternae**) a flattened membranous sac filled with fluid contents

cleavage *see* **furrowing**

codon a triplet of nucleotides that specifies an amino acid in proteins

coenzyme a small organic molecule associated with the protein portion of a holoenzyme, which is weakly bound at the active site of the enzyme and is required for enzyme activity (*see also* **prosthetic group**)

coenzyme A a small organic molecule that participates in energy-transfer reactions, usually as a carrier of activated metabolites (e.g., acetate)

co-linearity the spatial correlation between codons in DNA and amino acids in the polypeptide translated from the DNA blueprint

complementary base pairing specific hydrogen bond interactions between a particular purine and a particular pyrimidine component in nucleic acids; for example, guanine and cytosine or adenine and thymine or uracil

constitutive constant or unchanging; for example, a constitutive enzyme which is synthesized at a constant rate and not subject to regulation, or constitutive heterochromatin which is permanently condensed nuclear chromatin

co-repressor a metabolite that combines with repressor protein and blocks transcription of messenger RNA specifying a repressible enzyme synthesis

coupling factor F_1 factor; the headpiece of the mitochondrial inner membrane subunit which has ATPase activity and is a catalyst in mitochondrial ATP synthesis during oxidative phosphorylation coupled with electron transport

covalent bond interaction between atoms with shared electron shells

cristae infoldings of the mitochondrial inner membrane and the site of enzymes of coupled oxidative phosphorylation and electron transport

crossing over exchange of homologous chromosome segments leading to recombination of linked genes

cyclic adenosine monophosphate cyclic AMP; a nucleotide which is active in regulating transcription of messenger RNA

cycloheximide an organic molecule that inhibits protein synthesis on cytoplasmic ribosomes of eukaryotic cells

cytochrome oxidase cytochrome a-a_3; the terminal enzyme of aerobic respiration which transfers electrons to reduced oxygen

cytochromes electron-transport enzymes containing heme or related prosthetic group components which undergo valency changes of the iron atom

cytogenetics the study of biological systems using the combined methods of cytology and genetics

cytokinesis *see* **cell division**

cytology the study of the cell and its parts using microscopy

cytoplasm the protoplasmic contents of the cell, exclusive of the nucleus

cytosol the unstructured portion of the cytoplasm in which the organelles are bathed; the cytoplasmic matrix

dalton unit of molecular weight approximately equal to the weight of a hydrogen atom

deletion loss of part of a chromosome or DNA molecule from the genome

denaturation weakening and disruption of secondary or tertiary structure, or both, of proteins and nucleic acids leading to loss of function (*see also* **melting**)

deoxyribonucleic acid DNA; the genetic material

diakinesis last of the stages of prophase in Meiosis I

dictyosome a stack of cisternae that forms part of the Golgi apparatus

diploid a cell or individual or species having two sets of homologous chromosomes in the nucleus

diplotene a stage of prophase in Meiosis I when synapsed chromosomes "open out" and remain associated only at the chiasmata

D-loop synthesis the mode of DNA replication in mitochondria in which staggered copying of the parental strands produces a displacement (D) loop

Down's syndrome a clinical condition in humans usually associated with an extra chromosome 21 in the nucleus; trisomy-21, formerly called "mongolism"

duplication an extra copy of one or more genes in the chromosome complement

dyad a replicated chromosome consisting of two chromatids

dynein protein component, which has ATPase activity, of the arms of subfiber-A in microtubule doublets of the cilium or flagellum

effector a modulator or regulatory metabolite that activates or inhibits an enzyme by binding at an allosteric site on the enzyme, rather than at the active site where the substrate binds

electron transport chain a group of electron carriers, such as cytochromes, which transfer electrons from donor to acceptor with an accompanying release of energy at each transfer step along the chain

electrophoresis a method of separating macromolecules or particles according to their charge, size, and shape as they migrate through a gel or other medium in an electrical field

endocytosis intake of solutes or particles by enclosure in a portion of plasma membrane bringing these materials into the cell

endomembrane system concept that states there is a physical continuity among all the membranes of the eukaryotic cell, each membrane being a differentiated region of a single membrane system in the whole cell

endoplasmic reticulum *ER;* sheet(s) of folded membrane distributed within the cytoplasm of the eukaryotic cell that function as sites of protein synthesis and transport

endopolyploidy replication of chromosomes without later separation into different nuclei, leading to multiple sets of chromosomes in a common nucleus

endproduct repression a control mechanism in which the synthesis of an enzyme required for a metabolic pathway is inhibited by the presence of the final product of that metabolic pathway, thereby stopping further pathway reactions

enzymes the unique protein catalysts of biological systems

equilibrium density gradient centrifugation a method used to separate macromolecules and cell components on the basis of differences which cause them to come to rest at equilibrium in a region of the gradient that has a density of solute corresponding to their own buoyant density in the solute; the gradation of densities is produced by centrifugation at very high speeds

ER see **endoplasmic reticulum**

euchromatin noncondensed, active chromosomes or chromosome regions of the interphase nucleus

eukaryotes organisms with a well-defined nucleus enclosed in a nuclear membrane and usually having one or more other membranous subcellular compartments; any cellular organism that is not prokaryotic

exocytosis a mode of transport of substances out of the cell by enclosure in a portion of the plasma membrane and subsequent expulsion to the outside

F₁ coupling factor *see* **coupling factor**

facilitated diffusion assisted passage (transport) of molecules across the membrane toward their lower concentration along a gradient

FAD, FADH₂ *see* **flavin adenine dinucleotide**

fatty acids long hydrocarbon chain components of many lipids that are lacking double bonded carbon atoms (saturated) or have one or more such double bonds (unsaturated)

feedback inhibition a control mechanism that regulates enzyme activity through inhibition of an enzyme sequence by the end product in the sequence (usually by inhibiting the first enzyme in a sequence), thereby stopping the sequence and further production of the end product

fermentation oxidation of carbohydrate in oxygen-independent pathways (*see also* **glycolysis**)

First Law of thermodynamics energy can be neither created nor destroyed; statement of the principle of the conservation of energy

flagellum (pl. **flagella**) whiplike locomotor organelle produced by a centriole; ultrastructurally identical to a cilium but usually much longer than a cilium

flavin adenine dinucleotide FAD; an electron carrier molecule that acts in energy-transfer reactions as a coenzyme portion of holoenzyme; the reduced form of the redox couple is FADH₂

fluid mosaic membrane model of cell membranes which postulates the distribution of proteins in and on a phospholipid bilayer that has a consistency of a light oil and therefore permits movements of particles within the membrane

formylmethionyl-tRNA fmet-tRNA; the first amino acid brought to the ribosome at the initiation of polypeptide chain synthesis

free diffusion the unassisted passage of molecules from a region of their higher concentration along a concentration gradient toward a region of lower concentration

free energy the usable energy in biological systems, which is released as a result of chemical reactions and becomes available to do work in the cell

freeze-fracture method procedure for preparing materials for electron microscopy by rapid freezing and sectioning to induce fracture formation; the exposed fracture faces are treated by physical methods before being observed and photographed in the electron microscope; the fracture faces may. or may not be further etched before observations are made

furrowing a cell division mechanism typical of animal groups which involves a pinching in, or cleavage, to form two daughter cells from the parent cell

G-banding a method for staining chromosomes with Giemsa stain which reveals patterns of deeply stained bands, separated by lightly stained regions

gamete a reproductive cell that can develop only after uniting with another reproductive cell to produce the new individual of the next sexual generation; eggs and sperm are gametes

gap junction a region of differentiation involving portions of plasma membranes of adjacent cells, containing a spacc between the two membranes that permits cell-to-cell communication

gene amplification differential replication of some genes, producing many copies of these genes, at the same time that other genes in the chromosome complement do not replicate

gluconeogenesis synthesis of carbohydrates from noncarbohydrate precursors such as fats or proteins

glyceride *see* **neutral fats**

glycolysis the oxidation of sugar to lactic acid in fermentation reactions, typical of animal cell metabolism, that are independent of oxygen; also known as the Embden-Meyerhof pathway

glycoprotein a conjugated protein containing one or more sugar residues; a component of the plasma membrane and of chloroplast membranes in particular

glyoxylate cycle an anaplerotic pathway replenishing intermediary metabolites, which involves three enzymes also found in Krebs cycle reactions and two unique enzymes, malate synthase and isocitrate lyase

glyoxysome *see* **microbody**

Golgi apparatus a region of smooth endoplasmic reticulum that functions in processing and packaging proteins synthesized for export, such as zymogens

haploids cell or individual having one copy of each chromosome present

Hatch-Slack pathway *see* C_4 **cycle**

heme an iron-containing porphyrin derivative which serves as a prosthetic group in hemoglobins and in enzymes such as catalase and cytochromes

heterochromatin chromatin or chromosomes in the interphase nucleus which is either condensed at all times (constitutive) or, in some cells, at some times (facultative)

histone a major protein component of the chromosome, having a high content of the basic amino acids arginine and lysine

homologous having the same or similar gene content

hybridization, molecular formation of a double-stranded structure, DNA-DNA, DNA-RNA, or RNA-RNA, by hydrogen-bonding of complementary single-stranded molecules or parts of molecules; hybridization-competition and hybridization-saturation are the major assay methods for complementarity

hydrogen bond a weak chemical interaction that occurs when a covalently-linked hydrogen atom interacts with a nearby electronegative atom

hydrogenosome *see* **microbody**

hydrolysis the process by which one molecule is processed into two smaller molecules by the addition of water, usually resulting in the release of energy

hydrophilic molecules or parts of molecules that readily associate with water; usually containing polar groups that form hydrogen bonds in water

hydrophobic molecules or parts of molecules that do not readily associate with water; usually nonpolar, poorly soluble or insoluble in water

hyperchromic shift an increase in DNA absorption at 260 nm by disruption of hydrogen bonds to give single-stranded structures; indicates melting of duplex DNA

idiogram a diagrammatic representation of a chromosome complement of a cell or individual based on a karyotype of the entire chromosome complement present

independent assortment a Mendelian principle of inheritance for genes on different chromosomes which leads to random combinations of parental alleles in their progeny

inducible enzyme type that is synthesized only in the presence of its substrate (inducer), for example, β-galactosidase

inner membrane subunit IMS; particulate components of the inner surface of the mitochondrial inner membrane consisting of a headpiece and a stalk (*see also* **coupling factor**)

interphase the state of the eukaryotic nucleus when it is not engaged in mitosis or meiosis; consists of G_1, S, and G_2 periods in cycling cells

inversion structural rearrangement of part of a chromosome so that genes within that part end up in inverse order; paracentric inversions exclude the centromere, while pericentric inversions include the centromere in the inverted region

isomers alternative molecular forms of a chemical compound

isotopes alternative nuclear forms of an atom, all having the same atomic number (proton number) but different atomic weight (neutron number varies); a radioactive isotope is an unstable form that emits radiation as stable isotopes are formed during its decay, while a heavy isotope is a stable form that has one or more neutrons in excess of the number in the common isotope form of the atom; ^{14}C and ^3H are radioactive isotopes, ^{15}N is a heavy isotope

isozymes alternative molecular forms of an enzyme

junction differentiations of plasma membranes of adjacent cells involved in cell-to-cell communication phenomena; major types of communication junctions are gap junctions, septate junctions, and tight junctions; desmosomes are adhesion junctions between contiguous cells

karyotype a distribution of photographed chromosomes from a cell or individual showing chromosomes in pairs and in order of decreasing size

Kleinfelter's syndrome a clinical condition in the human male having one or more extra X chromosomes

Krebs cycle most common pathway for oxidative metabolism of pyruvic acid which is an endproduct of glucose fermentation; part of aerobic respiration, also known as the citric acid cycle or the tricarboxylic acid cycle

lampbrush chromosomes giant bivalents with extensively looped-out regions of chromatin fiber, especially prominent in amphibian oocyte nuclei during the diplotene stage of Meiosis I prophase

leptotene the first of the prophase stages in the first of the two divisions of meiosis, before chromosome synapsis begins

ligase enzyme that joins together the parts of single strands of DNA between the 5' end of one strand and the 3' end of another

lipid a heterogeneous class of organic compounds that are poorly soluble or insoluble in water but soluble in nonaqueous solvents such as ether

lysosome a membrane-bounded cytoplasmic organelle in eukaryotic cells that contains a variety of hydrolytic enzymes capable of digesting all types of organic compounds found in biological systems

matrix the essentially unstructured substance of a cell or organelle consisting of a suspension of molecules and particles in a watery medium

meiocyte any cell capable of undergoing meiosis; oocytes, spermatocytes, and sporocytes are meiocytes

meiosis the reduction division of the nucleus in sexual organisms which produces daughter nuclei having half the number of chromosomes as the original meiocyte nucleus; consists of two divisions in succession

melting the separation of the two strands of duplex DNA to form single strands upon disruption of hydrogen bonds between the duplex strands (*see also* **denaturation**)

meromyosin, heavy the portion of the myosin molecule with ATPase activity and Ca^{2+}-binding properties produced by trypsin digestion of myosin and used in assays to identify actin filaments with which it binds in "decorated" displays

mesosome an extensively infolded portion of the prokaryotic plasma membrane which has been implicated in respiratory and cell division functions

messenger RNA mRNA; the complementary copy of DNA which is made during transcription and which codes for protein during translation

metacentric median location of the centromere in the chromosome; a chromosome with a median centromere and therefore with two equal-length arms

metaphase the stage of mitosis or meiosis when chromosomes are aligned along the equatorial plane of the spindle

microbody a membrane-bounded eukaryotic cytoplasmic organelle with varied enzyme content and functions; usually contains catalase and may contain the unique enzymes of the glyoxylate cycle; glyoxysomes, hydrogenosomes, and peroxisomes are the major types of microbody in the cell

microsome a membrane fragment of the endoplasmic reticulum produced during centrifugation having ribosomes attached to the outer surface of the vesicle

microtubule an unbranched hollow cylindrical assembly of protofilaments which is involved in cell movement phenomena; spindle fibers, ciliary subfibers, and centriole subfibers are microtubules

microvilli fingerlike projections of plasma membranes of animal epithelial cells, particularly of the gut

mitochondrion the double-membrane cytoplasmic organelle in eukaryotic cells characterized by inner membrane infolds called cristae; center of aerobic respiration metabolism in which sugars processed partway during fermentation are oxidized completely to CO_2 and H_2O, with free energy stored in ATP

mitosis the division of the nucleus that produces two daughter nuclei exactly like the original parental nucleus; somatic nuclear division

mole grams molecular weight of a substance

monomer the basic unit of a larger functional molecule or particle or cell structure

mRNA *see* **messenger RNA**

multienzyme system a group of enzymes active in sequential steps of a metabolic pathway and in physical proximity to one another

myofibril multinucleated cell unit of the muscle fiber of striated muscle containing bundles of myofilaments

myofilament individual thick (myosin) and thin (actin) filaments of the myofibril

myosin the muscle protein making up the thick filaments of striated muscle and of a few other cell systems; has ATPase activity (*see also* **meromyosin**)

NAD⁺, NADP⁺ *see* **nicotinamide adenine dinucleotide**

neutral fats glycerides; fatty acid esters of the alcohol glycerol; a major storage form of fats

nicotinamide adenine dinucleotide NAD^+; an electron carrier molecule that acts in energy-transfer reactions as a coenzyme portion of a holoenzyme; the reduced form of the redox couple is NADH; nicotinamide adenine dinucleotide phosphate, or $NADP^+$, acts in a similar manner in oxidation-reductions that take place in biosynthetic pathways; its reduced form is NADPH

nonchromosomal inheritance pattern, gene, or mutation not associated with chromosomes of the nucleus and not adhering to Mendelian inheritance rules

nondisjunction faulty separation of homologous chromosomes or of sister chromatids during nuclear division producing cells or individuals having an aneuploid number of chromosomes

NOR *see* **nucleolar-organizing region**

nuclear envelope the double membrane surrounding the eukaryotic nucleus

nucleic acid polymer of nucleotides in an unbranched chain; DNA and RNA

nucleoid the region of segregated DNA in prokaryotic cells which is not separated from the cytoplasm by a membrane

nucleolar-organizing chromosome a chromosome in the complement containing various genes, including the genes for ribosomal RNA, with the capacity to generate a nucleolus at the site of ribosomal RNA genes

nucleolar-organizing region NOR; the specific part of the nucleolar-organizing chromosome where ribosomal RNA genes are situated and where the nucleolus is produced

nucleolus a discrete structure in the nucleus that is associated with ribosomal RNA synthesis

nucleoplasm the unstructured matrix portion of the nucleus in which the chromosomes and nucleoli are bathed

nucleoside molecule containing a nitrogenous base linked to a pentose sugar

nucleotide a nucleoside phosphate, any of the monomeric units that make up DNA and RNA

nucleus the major membrane-bounded compartment of the eukaryotic cell housing the chromosomes and nucleoli

operator a specific nucleotide sequence in the operon which binds repressor and thereby exerts control over transcription of its adjacent structural gene(s)

operon a cluster of associated genes and recognition sites that participate in regulating and specifying amino acid polymerization into polypeptides; includes regulatory gene, promotor site, operator site, and structural gene(s)

organelle a structural differentiation of the cell containing particular enzymes and performing particular functions for the whole cell or individual

oxidant an oxidizing agent, which loses electrons, or hydrogens, to a reducing agent or reductant

oxidation reactions involving loss of electrons or hydrogens

oxidative phosphorylation synthesis of ATP during coupled electron transport in aerobic respiration (*see also* **coupling factor, inner membrane subunit**)

pachytene the stage of prophase I of meiosis when synapsis of homologous chromosomes is completed

peptide bond the universal link between amino acids in proteins, formed when the amino group of one monomer joins with the carboxyl group of the adjacent amino acid in a dehydration reaction

permease a type of carrier protein situated within a membrane and involved in transport of specific substrate molecules across that membrane

peroxisome *see* **microbody**

pH measure of hydrogen ion concentration in aqueous solutions, ranging from $[H^+] = 1 \, M$ at pH 0 to $[H^+] = 0$ at pH 14; $\log_{10}[H^+]$

phage *see* **bacteriophage**

phosphate group $-O-\overset{\displaystyle O^-}{\underset{\displaystyle O^-}{P}}{=}O$; $-PO_4^{2-}$

phosphoryl group $-\overset{\displaystyle O^-}{\underset{\displaystyle O^-}{P}}{=}O$; involved in energy transfer reactions

photophosphorylation process of formation of ATP from ADP and inorganic phosphate in the light reactions of photosynthesis; occurs by a cyclic or noncyclic pathway involving an electron transport system

photorespiration uptake of oxygen and release of carbon dioxide by photosynthetic cells or individuals in the light

photosynthesis manufacture of sugar from CO_2 and H_2O in the light and in the presence of chlorophyll in eukaryotic chloroplasts or within prokaryotic photosynthetic membranes

photosystem I PS I; a photochemical reaction system in photosynthesis which produces $NADPH^+$ but does not evolve O_2

photosystem II PS II; a photochemical reaction system in photosynthesis which splits water, producing O_2 and a weak reductant; coupled to photosystem I

phycobilin accessory photosynthetic pigment present in red and blue-green algae

plasmalemma plasma membrane of the cell; the membrane that surrounds the living protoplast

plasmodesmata cytoplasmic channels between plant cells

plastid eukaryotic organelle of one or more types but most commonly referring to the chloroplast

polymer an association of monomer units in a large molecule

polymerase enzyme catalyzing the synthesis of DNA or RNA from nucleoside triphosphate precursors

polypeptide a polymer of amino acids in a long unbranched chain

polyploid cell or individual having one or more whole complements of chromosomes in excess of the usual number for a species or species group

polysome polyribosome; an aggregation of ribosomes, connected by a strand of messenger RNA, which is active in protein synthesis

polytene chromosome a multistranded endoreplicated chromosome

pore an opening in a membrane or other structure; usually referring to the nuclear pore complex of the nuclear envelope

primary structure the sequence of amino acids in a polypeptide chain

procentriole an immature centriole

prokaryotes organisms of the bacteria and blue-green algae groups lacking a nucleus separated from the cytoplasm by a membrane; any cellular organism that is not a eukaryote

prometaphase the stage of nuclear division when congression of the chromosomes occurs just before their alignment on the equatorial plane of the spindle

promotor a specific nucleotide sequence in the operon to which RNA polymerase binds, which is therefore a regulatory component of the operon

prophase the first stage of mitosis or meiosis, after DNA replication and before chromosomes align on the equatorial plane of the spindle

proplastid an immature plastid

prosthetic group a relatively small molecule that remains very tightly bound to the active site of an enzyme and which is required for the enzyme interaction with substrate (*see also* **coenzyme**)

protist any eukaryotic organism not classified as belonging to the fungi, plants, or animals; usually a unicellular organism, such as euglenoids and protozoa

protoplasm the living material of the cell

protoplast the living structure of the cell, made of protoplasm, contained within but including the plasma membrane

PS I, PS II *see* **photosystem I, photosystem II**

puff a region of expanded chromosome undergoing active transcription, usually observed in giant polytene chromosomes

pumps, ion systems that underwrite active transport of molecules across a membrane by expelling one substance out of the cell and thereby helping to drive many kinds of molecules into the cell along an energy gradient

purine parent compound of the nitrogen-containing bases adenine and guanine

pyrimidine parent compound of the nitrogen-containing bases cytosine, thymine, and uracil

Q-banding method of staining chromosomes using a fluorescent dye so that stained banded regions are differentiated from unstained areas between bands

quantum the energy of a photon, its amount being inversely proportional to the wavelength of emitted radiation

quaternary structure specific assemblages of different polypeptide chains which, when combined in the protein, do not have the same structural or chemical properties as in the individual chains

R-banding band pattern formation after staining chromosomes which is reversed from the usual pattern of G-bands produced by Giemsa staining

rDNA *see* **ribosomal DNA**

reannealing renaturation; specifically, the restoration of duplex DNA regions through complementary base pairing of single stranded DNA molecules

redox couple compounds that occur in both the oxidized and reduced forms and which are participants in oxidation-reductions, such as NAD^+-NADH

redox potential *see* **standard electrode potential**

reducing equivalent a general term for electrons or hydrogens, or both, that are transferred in oxidation-reduction reactions

reductant a reducing agent, which accepts electrons or hydrogens in oxidation-reduction reactions

reduction reactions involving gain of electrons or hydrogens

regulation the modulation of metabolism or gene action through control mechanisms

repetitious DNA repeated sequences of nucleotides that may occur in hundreds, thousands, or even millions of reiterated copies in a chromosome complement

replication fork a site within a replicating duplex DNA molecule at which synthesis of complementary strands is proceeding at that moment

repressible enzyme type that is synthesized in the absence of its substrate which represses synthesis of the enzyme

repressor a protein product of the regulator gene of the operon which binds to the operator site thus regulating transcription of structural genes

respiration the principal energy-yielding reactions of aerobic cells, involving the transfer of electrons from organic fuel molecules to O_2

ribonucleic acid RNA; nucleic acid polymers that function in transcription and translation of coded DNA

ribosomal DNA rDNA; the genes at the nucleolar-organizing region that code for ribosomal RNA

ribosomal RNA rRNA; ribonucleic acids that are part of the ribosome structure and which function in protein synthesis

ribosome a complex structure composed of RNA and protein, which is the site of protein synthesis in the cytoplasm, mitochondria, and chloroplasts

RNA *see* **ribonucleic acid**

rough *ER* that portion of the endoplasmic reticulum with attached ribosomes

rRNA see **ribosomal RNA**

S period the interval during the cell cycle when DNA replication occurs

S value *see* **sedimentation coefficient**

sarcolemma the plasma membrane of the muscle fiber

sarcomere the contractile unit of muscle fiber, extending from one Z-line to the adjacent Z-line

sarcoplasm the cytoplasm of the muscle fiber

sarcoplasmic reticulum the endoplasmic reticulum of the muscle fiber

secondary constriction any pinched-in site along a chromosome other than the primary constriction at the centromere region

secondary structure the local structure of the polypeptide chain, over a distance of only a few amino acid residues; may exist either as an irregular chain or as a regular and repeating shape such as the α-helix

Second Law of thermodynamics entropy never decreases; the principle that all physical and chemical change proceeds in such a direction that the entropy of the universe increases to the maximum possible, at which point equilibrium exists and there is maximum chaos

sedimentation coefficient S; a quantitative measure of the rate of sedimentation of a given substance in a centrifugal field, expressed in Svedberg units

self assembly the organization of structure in the absence of a template or parent structure

semiconservative replication the usual mode of duplex DNA synthesis resulting in daughter duplex molecules which contain one parental strand and one newly formed strand

septate junction *see* **junction**

sex chromatin *see* **Barr body**

sex chromosome any chromosome involved in sex determination; the X and Y chromosomes are sex chromosomes

sliding filament mechanism a proposal applied to explanation of cellular movement phenomena, particularly to muscle contraction and to ciliary bending

smooth *ER* that portion of the endoplasmic reticulum which lacks ribosomes

spindle an aggregate of microtubules seen during nuclear division which functions in the alignment and movement of chromosomes at metaphase and anaphase

spindle fiber a microtubule in dividing cells which extends from one pole to an attachment in the centromere region of a chromosome or which extends from pole to pole in the mitotic or meiotic cell

standard electrode potential E_0; the oxidation-reduction potential of a substance relative to a hydrogen electrode; expressed in volts

standard free-energy change ΔG^0; a thermodynamic constant representing the difference between the standard free energy of the reactants and the standard free energy of the products of a reaction; energy-releasing reactions have a negative ΔG^0 while energy-requiring reactions have a positive ΔG^0 value

stroma the unstructured matrix of the chloroplast which bathes the grana and stroma thylakoids

submetacentric centromere located a short distance from the middle of the chromosome length; a chromosome whose centromere is not median in location leading to one chromosome arm being somewhat longer than the other

synapsis the specific pairing of homologous chromosomes, typically during zygotene of prophase I in meiosis

synaptinemal complex a complex structural component situated between a pair of synapsed homologous chromosomes during pachytene of Meiosis I

T$_m$ midpoint melting temperature; temperature at the midpoint of transition of a preparation of duplex DNA molecules to single strands during melting

T system invaginations of the sarcolemma in muscle fibers of striated muscle, producing a system of transverse tubular infoldings that maximize signal reception in the entire muscle fiber

telocentric centromere located at one end of a chromosome; a chromosome with its centromere in a terminal location and having, therefore, only one chromosome arm

telophase the stage of nuclear division when nuclear reorganization occurs

tertiary structure the three-dimensional folding of the polypeptide chain into a complex structural form brought about by interactions among side-chains of amino acids at some distance from one another in the primary structure

thermodynamics the branch of physical science that deals with exchanges of energy in collections of matter (*see also* **bioenergetics**)

thylakoid a closed membrane sac that may be disc-shaped in grana or may be greatly elongated in a chloroplast or in the prokaryotic cell cytoplasm; contains the systems of the light-requiring reactions of photosynthesis

tight junction *see* **junction**

transcription a process by which the base sequence of DNA is copied into a single-stranded RNA complementary molecule

transfer RNA the RNA molecule that carries an amino acid to a specific codon in messenger RNA during protein synthesis at the ribosome

translation the process by which the sequence of amino acids is assembled into a polypeptide at the ribosome, under the direction of the coded base sequence copied from DNA into messenger RNA

translocation a structural rearrangement involving nonhomologous chromosomes or chromosome segments

transport assisted passage across a membrane (*see also* **active transport, facilitated diffusion**)

trisomic cell or individual having three copies of one chromosome instead of the usual two copies present in the diploid nucleus

trisomy condition of a cell or individual having three copies of a particular chromosome, as in trisomy-21 or Down's syndrome; an aneuploid condition

tritium ^{3}H; the radioactive isotope of hydrogen

tubulin the major protein component of microtubules

Turner's syndrome clinical condition in the human female having only one X chromosome (*45, X*)

unineme concept of chromosome structure as having only one chromatin fiber in the unreplicated chromosome or in each chromatid of a replicated chromosome

unit membrane any membrane showing a tripartite, or dark-light-dark, pattern of electron density in the electron microscope; a model of membrane structure proposing that a phospholipid bilayer is coated on its outer and inner surfaces by proteins in the extended β-configuration (*see also* **fluid mosaic membrane**)

unwinding protein structural protein that binds to single-stranded regions of duplex DNA during replication and recombination

vacuole a region in the cytoplasm surrounded by a membrane and filled with molecules and particles in a watery medium, particularly frequent in plant cells

vesicle a small, spherical membranous element filled with protein in a watery medium

zygote product of the fusion of two gametes; the cell from which the new individual develops in each sexual generation

zygotene stage during prophase of Meiosis I when homologous chromosomes undergo synapsis

zymogen a digestive enzyme precursor lacking catalytic activity in this form, for example, trypsinogen (converted to active trypsin)

Index

Numbers in boldface indicate pictorial information.

SOME SYMBOLS AND ABBREVIATIONS USED IN THE TEXT

acetyl CoA — acetyl coenzyme A

AMP, ADP, ATP — adenosine 5'-mono-, di-, and triphosphate

ATPase — adenosine triphosphatase

Å — angstrom

ρ — buoyant density, g/cm³

^{14}C — carbon-14

CsCl — cesium chloride

CoA — coenzyme A

cyt. — cytochrome

DNA — deoxyribonucleic acid

e^- — electron

ER — endoplasmic reticulum

FAD, $FADH_2$ — flavin adenine dinucleotide, oxidized and reduced forms

fmet — N-formylmethionine

ΔG — free-energy change

g — gram

H^+ — hydrogen ion

pH — hydrogen-ion concentration

OH^- — hydroxyl ion

P_i — inorganic phosphate, $H_3PO_4^{2-}$

kcal — kilocalorie

mRNA — messenger RNA

μm — micrometer

M — molar concentration

nm — nanometer

NAD^+, NADH — nicotinamide adenine dinucleotide, oxidized and reduced forms

$NADP^+$, NADPH — nicotinamide adenine dinucleotide phosphate, oxidized and reduced forms

^{15}N — nitrogen-15, "heavy" nitrogen

NOR — nucleolar-organizing region

λ — phage lambda *or* wavelength of radiation

PS I, PS II — photosystems I and II

rDNA — ribosomal DNA

RNA — ribonucleic acid

rRNA — ribosomal RNA

rough *ER* — rough endoplasmic reticulum

S — sedimentation coefficient, in Svedberg units

smooth *ER* — smooth endoplasmic reticulum

E_0 — standard electrode potential

ΔE_0 — standard electrode potential change, pH 0 or unspecified

$\Delta E_0'$ — standard electrode potential change, pH 7

ΔG^0 — standard free-energy change, pH 0 or unspecified

$\Delta G^{0'}$ — standard free-energy change, pH 7

tRNA — transfer RNA

^3H — tritium